NEUROSTEROID EFFECTS IN THE CENTRAL NERVOUS SYSTEM

THE ROLE OF THE $GABA_A$ RECEPTOR

METHODS & NEW FRONTIERS IN NEUROSCIENCE

Series Editors
Sidney A. Simon, Ph.D.
Miguel A.L. Nicolelis, M.D., Ph.D.

Published Titles

Apoptosis in Neurobiology
Yusuf A. Hannun, M.D., Professor of Biomedical Research and Chairman/Department of Biochemistry and Molecular Biology, Medical University of South Carolina
Rose-Mary Boustany, M.D., tenured Associate Professor of Pediatrics and Neurobiology, Duke University Medical Center

Methods for Neural Ensemble Recordings
Miguel A.L. Nicolelis, M.D., Ph.D., Professor of Neurobiology and Biomedical Engineering, Duke University Medical Center

Methods of Behavioral Analysis in Neuroscience
Jerry J. Buccafusco, Ph.D., Alzheimer's Research Center, Professor of Pharmacology and Toxicology, Professor of Psychiatry and Health Behavior, Medical College of Georgia

Neural Prostheses for Restoration of Sensory and Motor Function
John K. Chapin, Ph.D., Professor of Physiology and Pharmacology, State University of New York Health Science Center
Karen A. Moxon, Ph.D., Assistant Professor/School of Biomedical Engineering, Science, and Health Systems, Drexel University

Computational Neuroscience: Realistic Modeling for Experimentalists
Eric DeSchutter, M.D., Ph.D., Professor/Department of Medicine, University of Antwerp

Methods in Pain Research
Lawrence Kruger, Ph.D., Professor of Neurobiology (Emeritus), UCLA School of Medicine and Brain Research Institute

Motor Neurobiology of the Spinal Cord
Timothy C. Cope, Ph.D., Professor of Physiology, Emory University School of Medicine

Nicotinic Receptors in the Nervous System
Edward D. Levin, Ph.D., Associate Professor/Department of Psychiatry and Pharmacology and Molecular Cancer Biology and Department of Psychiatry and Behavioral Sciences, Duke University School of Medicine

Methods in Genomic Neuroscience
Helmin R. Chin, Ph.D., Genetics Research Branch, NIMH, NIH
Steven O. Moldin, Ph.D, Genetics Research Branch, NIMH, NIH

Methods in Chemosensory Research
Sidney A. Simon, Ph.D., Professor of Neurobiology, Biomedical Engineering, and Anesthesiology, Duke University
Miguel A.L. Nicolelis, M.D., Ph.D., Professor of Neurobiology and Biomedical Engineering, Duke University

The Somatosensory System: Deciphering the Brain's Own Body Image
Randall J. Nelson, Ph.D., Professor of Anatomy and Neurobiology, University of Tennessee Health Sciences Center

The Superior Colliculus: New Approaches for Studying Sensorimotor Integration
William C. Hall, Ph.D., Department of Neuroscience, Duke University
Adonis Moschovakis, Ph.D., Institute of Applied and Computational Mathematics, Crete

New Concepts in Cerebral Ischemia
Rick C. S. Lin, Ph.D., Professor of Anatomy, University of Mississippi Medical Center

DNA Arrays: Technologies and Experimental Strategies
Elena Grigorenko, Ph.D., Technology Development Group, Millennium Pharmaceuticals

Methods for Alcohol-Related Neuroscience Research
Yuan Liu, Ph.D., National Institute of Neurological Disorders and Stroke, National Institutes of Health
David M. Lovinger, Ph.D., Laboratory of Integrative Neuroscience, NIAAA

In Vivo Optical Imaging of Brain Function
Ron Frostig, Ph.D., Associate Professor/Department of Psychobiology, University of California, Irvine

Primate Audition: Behavior and Neurobiology
Asif A. Ghazanfar, Ph.D., Primate Cognitive Neuroscience Lab, Harvard University

Methods in Drug Abuse Research: Cellular and Circuit Level Analyses
Dr. Barry D. Waterhouse, Ph.D., MCP-Hahnemann University

Functional and Neural Mechanisms of Interval Timing
Warren H. Meck, Ph.D., Professor of Psychology, Duke University

Biomedical Imaging in Experimental Neuroscience
Nick Van Bruggen, Ph.D., Department of Neuroscience Genentech, Inc., South San Francisco
Timothy P.L. Roberts, Ph.D., Associate Professor, University of Toronto

The Primate Visual System
John H. Kaas, Department of Psychology, Vanderbilt University
Christine Collins, Department of Psychology, Vanderbilt University

Neurosteroid Effects in the Central Nervous System
Sheryl S. Smith, Ph.D., Department of Physiology, SUNY Health Science Center

NEUROSTEROID EFFECTS IN THE CENTRAL NERVOUS SYSTEM

THE ROLE OF THE $GABA_A$ RECEPTOR

Sheryl S. Smith, Ph.D.,

Associate Professor,
Department of Physiology and Pharmacology,
SUNY Downstate Medical Center,
Brooklyn, New York

CRC PRESS

Boca Raton London New York Washington, D.C.

Library of Congress Cataloging-in-Publication Data

Neurosteroid effects in the central nervous system : the role of the GABA-A receptor / edited by Sheryl S. Smith
p. ; cm.—(Methods & new frontiers in neuroscience)
Includes bibliographical references and index.
ISBN 0-8493-2392-4 (alk. paper)
1. Neurohormones—Physiological effect. 2. Steroids—Physiological effect. 3. GABA—Receptors. 4. Neurotransmitter receptors. I. Smith, Sheryl S. II. Methods & new frontiers in neuroscience series.
[DNLM: 1. Receptors, GABA-A—physiology. 2. Central Nervous System—physiology. QU 60 N4947 2003]
QP356.4.N4965 2003
612.8′042—dc22 2003058472

Direct all inquiries to CRC Press LLC, 2000 N.W. Corporate Blvd., Boca Raton, Florida 33431.

International Standard Book Number 0-8493-2392-4
Library of Congress Card Number 2003058472
Printed in the United States of America 1 2 3 4 5 6 7 8 9 0
Printed on acid-free paper

Series Preface

Our goal in creating the **Methods & New Frontiers in Neuroscience** series is to present the insights of experts on emerging experimental techniques and theoretical concepts that are, or will be, at the vanguard of neuroscience. Books in the series cover topics ranging from methods to investigate apoptosis, to modern techniques for neural ensemble recordings in behaving animals. The series also covers new and exciting multidisciplinary areas of brain research, such as computational neuroscience and neuroengineering, and describes breakthroughs in classical fields like behavioral neuroscience. We want these books to be the books every neuroscientist will use in order to get acquainted with new methodologies in brain research. These books can be given to graduate students and postdoctoral fellows when they are looking for guidance to start a new line of research.

Each book is edited by an expert and consists of chapters written by the leaders in a particular field. The books are richly illustrated and contain comprehensive bibliographies. Chapters provide substantial background material relevant to the particular subject. Hence, they are not only "methods books," but they also contain detailed "tricks of the trade" and information as to where these methods can be safely applied. In addition, they include information about where to buy equipment and about Web sites helpful in solving both practical and theoretical problems.

We hope that as the volumes become available, the effort put in by us, by the publisher, by the book editors, and by individual authors will contribute to the further development of brain research. The extent to which we achieve this goal will be determined by the utility of these books.

Sidney A. Simon, Ph.D.
Miguel A.L. Nicolelis, M.D., Ph.D.
Series Editors

Editor

Sheryl S. Smith, Ph.D., is an associate professor of physiology and pharmacology at the State University of New York Medical Center in Brooklyn (Downstate).

Dr. Smith received her B.S. in biology at the College of William and Mary in Williamsburg, Virginia, where she was a member of Phi Beta Kappa Honorary Society. She later graduated from University of Texas Southwestern Medical Center at Dallas with a Ph.D. in physiology, and was a postdoctoral fellow in the Department of Cell Biology at the same institution. She then joined the faculty of the Department of Neurobiology and Anatomy at the Medical College of Pennsylvania–Hahnemann University in Philadelphia, where she became an associate professor in 1994. In 2000, she acquired her present position at SUNY Downstate.

She is a member of the Neuroscience Society and serves as a grant reviewer on an NIH study section. She is currently a manuscript reviewer for numerous journals, including the *Journal of Neuroscience*, the *Journal of Physiology*, and *Neuropharmacology*. She has recently received special recognition with an award for innovation from the State of New York Research Foundation. She has published over 50 papers in peer-reviewed journals and 12 reviews/chapters, and has presented her work at 30 national and international meetings. Her work has been funded by NIH, NIAAA, and the USAF, as well as by pharmaceutical companies such as Astra-Zeneca, Merck, and Lundbeck. Her current research interest is the role of chronic neurosteroid exposure in altering inhibition in the brain with emphasis on plasticity of the $GABA_A$ receptor. She also teaches endocrinology and medical neuroscience at Downstate.

Contributors

Yvette Akwa
INSERM, U488
Le Kremlin-Bicetre, France

Etienne-Emile Baulieu
INSERM, U488
Le Kremlin-Bicetre, France

Matt T. Bianchi
Neuroscience Graduate Program
University of Michigan
Ann Arbor, Michigan

Giovanni Biggio
Department of Experimental Biology
Section of Neuroscience and Center of Excellence for Neurobiology of Drug Dependence
University of Cagliari
and
CNR, Institute of Neuroscience
Cagliari, Italy

Daniel Bitran
Department of Psychology
College of Holy Cross
Worcester, Massachusetts

Elisabetta Cagetti
Department of Molecular and Medical Pharmacology
UCLA School of Medicine
Los Angeles, California

Ann S. Clark
Department of Psychological and Brain Sciences
Dartmouth College
Hanover, New Hampshire

Douglas F. Covey
Department of Molecular Biology and Pharmacology
Washington University School of Medicine
St. Louis, Missouri

Laura Dazzi
Department of Experimental Biology
Section of Neuroscience and Center of Excellence for Neurobiology of Drug Dependence
University of Cagliari
Cagliari, Italy

Lawrence N. Eisenman
Department of Neurology
Washington University School of Medicine
St. Louis, Missouri

David H. Farb
Laboratory of Molecular Neurobiology
Department of Pharmacology and Experimental Therapeutics
Boston University School of Medicine
Boston, Massachusetts

Deborah A. Finn
Portland Alcohol Research Center
Department of Veterans Affairs Medical Center and Department of Behaviorial Neuroscience
Oregon Health and Science University
and
VA Medical Center, Research Service
Portland, Oregon

Cheryl A. Frye
Department of Psychology
SUNY at Albany
Albany, New York

Terrell T. Gibbs
Laboratory of Molecular Neurobiology
Department of Pharmacology
and Experimental Therapeutics
Boston University School of Medicine
Boston, Massachusetts

Yejun He
Department of Psychiatry
Washington University School
of Medicine
St. Louis, Missouri

Leslie P. Henderson
Departments of Physiology and of
Biochemistry
Dartmouth Medical School
Hanover, New Hampshire

Brian L. Jones
Department of Physiology
Dartmouth Medical School
Hanover, New Hampshire

Jaideep Kapur
Department of Neurology
University of Virginia Health Sciences
Center
Charlottesville, Virginia

Rahul Khisti
Bowles Center for Alcohol Studies
and Departments of Psychiatry
and Pharmacology
University of North Carolina at Chapel
Hill
Chapel Hill, North Carolina

George F. Koob
Department of Neuropharmacology
The Scripps Research Institute
La Jolla, California

Robert L. Macdonald
Departments of Neurology, Molecular
Physiology and Biophysics
and Pharmacology
Vanderbilt University
Nashville, Tennessee

Douglas B. Matthews
Bowles Center for Alcohol Studies
and Departments of Psychiatry
and Pharmacology
University of North Carolina at Chapel
Hill
Chapel Hill, North Carolina
and
Department of Psychology
University of Memphis
Memphis, Tennessee

Margaret M. McCarthy
Department of Physiology and Program
in Neuroscience
University of Maryland, Baltimore
Baltimore, Maryland

Janelle R. McDaniel
Department of Psychology
University of Memphis
Memphis, Tennessee

Synthia H. Mellon
Department of Obstetrics
Gynecology and Reproductive Sciences
The Center for Reproductive Sciences
and
The Metabolic Research Unit
University of California
San Francisco, California

Steven Mennerick
Department of Psychiatry
and Department of Anatomy
and Neurobiology
Washington University School
of Medicine
St. Louis, Missouri

Zakaria Mtchedlishvili
Department of Neurology
University of Virginia Health Sciences
Center
Charlottesville, Virginia

A. Leslie Morrow
Bowles Center for Alcohol Studies
and Departments of Psychiatry
and Pharmacology
University of North Carolina at Chapel
Hill
Chapel Hill, North Carolina

Joseph L. Nunez
Department of Physiology
University of Maryland, Baltimore
Baltimore, Maryland

Richard W. Olsen
Department of Molecular and Medical
Pharmacology
UCLA School of Medicine
Los Angeles, California

Tara S. Perrot-Sinal
Department of Psychology
Dalhousie University
Halifax, Nova Scotia, Canada

Sandra M. Petralia
Department of Psychology
University of Albany, SUNY
Albany, New York

Robert H. Purdy
Department of Neuropharmacology
The Scripps Research Institute
La Jolla, California

Mariangela Serra
Department of Experimental Biology
Section of Neuroscience and Center
of Excellence for Neurobiology
of Drug Dependence
University of Cagliari
and
CNR, Institute of Neuroscience
Cagliari, Italy

Ethan M. Skowronski-Lutz
Department of Psychology
College of Holy Cross
Worcester, Massachusetts

Inger Sundström-Poromaa
Department of Clinical Science
Obstetrics and Gynecology
Umeå University Hospital
Umeå, Sweden

Sayaka Tokunaga
Department of Psychology
University of Memphis
Memphis, Tennessee

Stefano Vicini
Department of Physiology
and Biophysics
Georgetown University School
of Medicine
Washington, D.C.

Martin Wallner
Department of Molecular and Medical
Pharmacology
UCLA School of Medicine
Los Angeles, California

Paul Yang
Department of Physiology
Dartmouth Medical School
Hanover, New Hampshire

Charles F. Zorumski
Department of Psychiatry
and Department of Anatomy
and Neurobiology
Washington University School
of Medicine
St. Louis, Missouri

Contents

Introduction

Neuroactive steroids are defined as steroids that produce rapid actions on neuronal substrates within the central nervous system (CNS). In contrast to their function as substances that act via intracellular receptors,[1] they include classes of steroids that act on ligand-gated or voltage-gated channels to enhance or depress neuronal activity, as well as those with sustained effects on developmental or regenerative processes. In many cases, these steroid hormones are produced by endocrine sources across naturally occurring hormonal rhythms, i.e., the estrous or circadian cycles, or in response to stress or olfactory cues. Steroids are lipophilic molecules that are bound to carrier proteins that can pass the blood–brain barrier where they gain access to neuronal and glial populations.

Neurosteroids are also metabolized in the brain from precursor "parent" compounds originating from endocrine sources and also *de novo* within the CNS[2] from cholesterol via side chain cleavage enzyme. The effect of these diverse steroids on neurotransmitter systems may open the possibility for diverse actions of steroids on novel behavioral end points related to mood, seizure susceptibility, cognition, and sensori-motor control.

GABA-MODULATORY STEROIDS

3α,5α-THP: Historical Overview of Physiological Effect

One of the first neuroactive steroids to be identified is the well-characterized 3α,5α-THP (3α-OH-5α-pregnan-20-one, 3α-OH-DHP or allopregnanolone), a steroid metabolite of progesterone that has no direct action on conventional progesterone receptor binding sites. This steroid was first demonstrated to be a positive modulator of the $GABA_A$ receptor by studies that described effects on receptor binding[3] and steroid-induced enhancement of GABA-gated current recorded from cultured hippocampal neurons using whole-cell patch clamp procedures,[4,5] where effects on synaptic current were identified. The physiological relevance of these effects was suggested by *in vivo* studies demonstrating GABA-modulatory effects of systemic progesterone[7] dependent upon conversion to 3α,5α-THP.[8] Subsequent studies investigated steroid binding characteristics,[9] delineated subunit requirements for their modulatory effects in recombinant $GABA_A$ receptors,[10] and characterized single-channel gating properties modulated by neuroactive steroids.[11] The GABA-modulatory actions of these endogenous steroids were foreshadowed by early reports of the GABA-modulatory effects of steroid anesthetics, such as alphaxalone,[12] and by Hans Selye, who first suggested that steroids such as progesterone had anesthetic effects when administered at high concentrations.[13] These anesthetic effects of progesterone were subsequently shown to be due to its GABA-modulatory metabolite 3α,5α-THP,[14] while other behavioral effects of this steroid include anxiolytic[15] and anticonvulsant[16] properties. As such, potentiating actions of this neuroactive steroid on $GABA_A$ receptor function may provide the underlying mechanism for disorders associated with anxiety and seizure activity associated with fluctuations in GABA-modulatory steroids, such as premenstrual syndrome and catamenial epilepsy.

The $GABA_A$ Receptor

The postsynaptic $GABA_A$ receptor mediates fast (1 to 100 msec) synaptic inhibition in most areas of the adult CNS and is well characterized both structurally and pharmacologically. These receptors are pentameric structures comprised of alpha-helical peptides with four membrane-spanning domains (Color Figure 1). The single-channel gating properties as well as ligand binding properties of this receptor determine the macroscopic time course for GABA-gated current and thus determine the extent of inhibitory processes. Identified $GABA_A$ receptor subunits include α (1 to 6), β (1 to 3), γ (1 to 4), δ, ε, θ, ρ (1 to 3), and π. The most

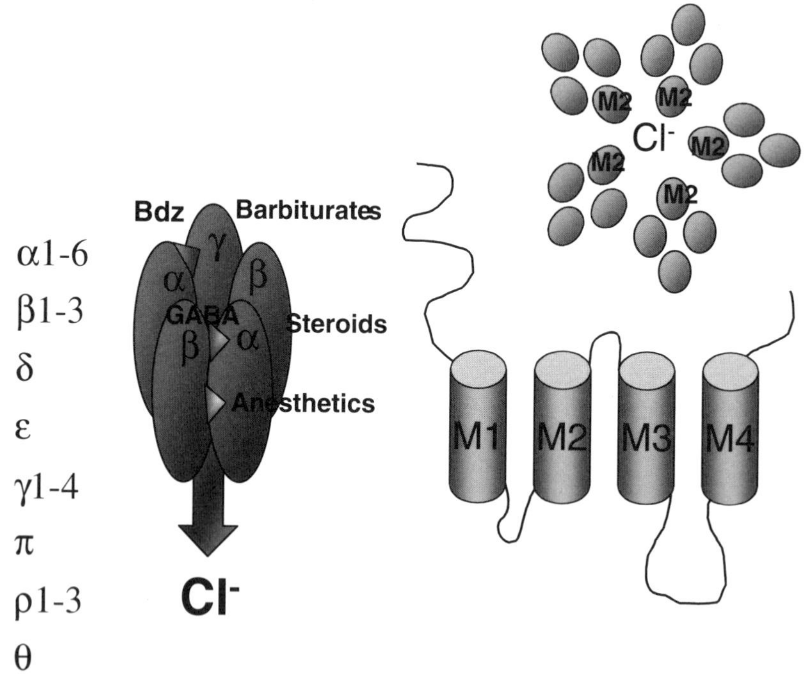

Diagram of the $GABA_A$ receptor (GABAR): A pentameric structure comprised of combinations of various α and β subunits which also contain δ, γ, ε, π, or θ. Atypical GABAR can also be formed of ρ subunits. Each subunit in turn is comprised of 4 trans-membrane spanning α-helices, where M2 lines a central chloride ion channel. Binding of two GABA molecules at the interface between α and β gates open the channel to permit chloride flow. Many modulators exist, including neurosteroids such as 3α,5α-THP, which enhance chloride current, thereby increasing inhibition. The particular combination of subunits for each receptor can influence both the biophysical and pharmacological properties of the receptor. (See color insert following page 74.)

common stoichiometry is 2α, 2β, and 1γ,[17] with $\alpha_1\beta_2\gamma_2$ the most common isoform expressed in the CNS,[18] although numerous other possibilities exist. These receptors represent the primary target for depressant, sedative drugs such as benzodiazepines, barbiturates, anesthetics, and, in some cases, alcohol. The relative subunit composition of a given receptor can produce subtle or marked changes in intrinsic GABA receptor function, including changes in conductance level and response to modulators such as neuroactive steroids.

This compendium will cover an array of topics focused on the actions of steroids that act as modulators of the $GABA_A$ receptor, such as 3α,5α-THP, THDOC, and anabolic steroids, which in some cases can also impact upon other transmitter receptor systems, such as NMDA and dopamine. In addition to describing neurosteroid synthesis, other chapters will describe actions of steroids: at recombinant $GABA_A$ receptor isoforms, on single-channel properties, on macroscopic receptor deactivation and desensitization kinetics, and on synaptic and extrasynaptic receptor populations of receptors. The effects of chronic steroid exposure and withdrawal will be described, as will neurosteroid effects on developmental processes. Other chapters will explore states that alter neurosteroid synthesis, including stress and social isolation, ethanol ingestion, and activation of peripheral benzodiazepine receptors. Finally, a number of chapters will discuss the implications of these GABA-modulatory neurosteroid effects on behavioral outcomes, including anxiety, reproductive behavior, epilepsy, and premenstrual syndrome.

Sheryl S. Smith

REFERENCES

1. Agmo, A. and Pfaff, D.W., Research on the neurobiology of sexual behavior at the millennium, *Behav. Brain Res.*, 105, 1, 2000.
2. Baulieu, E.-E. and Schumacher, M., Neurosteroids, with special reference to the effect of progesterone on myelination in peripheral nerves, *Mult. Scler.*, 3, 105, 1997.
3. Gee, K.W., Chang, W.-C., Brinton, R.E., and McEwen, B.S., GABA-dependent modulation of the Cl^- ionophore by steroids in rat brain, *Eur. J. Pharmacol.*, 136, 419, 1987.
4. Majewska, M.D., Harrison, N.L., Schwartz, R.D., Barker, J.L., and Paul, S.M., Steroid hormone metabolites are barbiturate-like modulators of the GABA receptor, *Science*, 232, 1004, 1986.
5. Callachan, H., Cottrell, G.A., Hather, N.Y., Lambert, J.J., Nooney, J.M., and Peters, J.A., Modulation of the GABAA receptor by progesterone metabolites, *Proc. R. Soc. Lond. B Biol. Sci.*, 231, 359, 1987.
6. Harrison, N.L., Vicini, S., and Barker, J.L., A steroid anesthetic prolongs inhibitory postsynaptic currents in cultured rat hippocampal neurons, *J. Neurosci.*, 7, 604, 1987.
7. Smith, S.S., Waterhouse, B.D., and Woodward, D.J., Progesterone alters GABA and glutamate responsiveness: a possible mechanism for its anxiolytic action, *Brain Res.*, 400, 353, 1987.
8. Smith, S.S., Female sex steroids: from receptors to networks to performance: actions on the sensorimotor system, *Prog. Neurobiol.*, 44, 55, 1994.

9. Turner, D.M., Ransom, R.W., Yang, J.S., and Olsen, R.W., Steroid anesthetics and naturally occurring analogs modulate the gamma-aminobutyric acid receptor complex at a site distinct from barbiturates, *J. Pharmacol. Exp. Ther.*, 248, 960, 1989.
10. Puia, G., Santi, M.R., Vicini, S., Pritchett, D.B., Paul, S.M., Seeburg, P.H., and Costa, E., Neurosteroids act on recombinant human GABA-A receptors, *Neuron*, 4, 759, 1990.
11. Twyman, R.E. and Macdonald, R.L., Neurosteroid regulation of $GABA_A$ receptor-single channel kinetic properties of mouse spinal cord neurons in culture, *J. Physiol.*, 456, 215, 1992.
12. Harrison, N.L. and Simmonds, M.A., Modulation of the GABA receptor complex by a steroid anaesthetic, *Brain Res.*, 323, 287, 1984.
13. Selye, H., Correlations between the chemical structure and pharmacological actions of the steroids, *Endocrinology*, 30, 437, 1942.
14. Norberg, L., Wahlstrom, G., and Backstrom, T., The anaesthetic potency of 3 alpha-hydroxy-5 alpha-pregnan-20-one and 3 alpha-hydroxy-5 beta-pregnan-20-one determined with an intravenous EEG-threshold method in male rats, *Pharmacol. Toxicol.*, 61, 42, 1987.
15. Bitran, D., Hilvers, R.J., and Kellog, C.K., Anxiolytic effects of 3α-hydroxy-5α[β]-pregnan-20-one: endogenous metabolites of progesterone that are active at the $GABA_A$ receptor function, *Brain Res.*, 561, 157, 1993.
16. Belelli, D., Bolger, M.B., and Gee, K.W., Anticonvulsant profile of the progesterone metabolite 5 alpha-pregnan-3 alpha-ol-20-one, *Eur. J. Pharmacol.*, 166, 325, 1989.
17. Chang, Y., Wang, R., Barot, S., and Weiss, D.S., Stoichiometry of a recombinant $GABA_A$ receptor, *J. Neurosci.*, 16, 534, 1990.
18. Hevers, W. and Luddens, H., The diversity of $GABA_A$ receptors. Pharmacological and electrophysiological properties of $GABA_A$ channel subtypes, *Md. Neurobiol.*, 18, 35, 1998.

Commonly Used Steroid Abbreviations in the Book

Alphaxalone* 3α-OH-5α-pregnan-11,20-dione. A synthethetic steroid anesthetic

DHEAS Dehydroepiandrosterone sulfate. An androgen metabolite

5α-DHP (DHP) Dihydroprogesterone (allopregnanedione). A metabolite of progesterone

DHT Dihydrotestosterone. Converted from testosterone (androgen)

E_2 (E) 17β-estradiol (ovarian hormone)

20HP 20α-hydroxyprogesterone. A metabolite of progesterone

17α-MeT 17α-methyltestosterone. An androgen metabolite

P Progesterone (ovarian/adrenal hormone, neurosteroid)

3α,5β-PC* (3α,5β)-20-oxo-pregnane-3-carboxylic acid. A synthetic steroid analog

5α-pregnane-3α,20α-diol* Allopregnanediol

PS Pregnenolone sulfate. The sulfated precursor to progesterone

THDOC* 3α,21-dihydroxy-5α-pregnan-20-one (allotetrahydrodeoxycorticosterone). A metabolite of the mineralocorticoid deoxycorticosterone

3α,5α-THDOC* (3α,5α)-3,21-dihydroxypregnan-20-one (alloTHDOC)

3α,5β-THDOC* (3α,5β)-3,21-dihydoxypregnan-20-one

3α,5α-THP* 3α-OH-5α-pregnan-20-one (allopregnanolone). A metabolite of progesterone (found in rodents and humans)

3α,5α- PROG* Same as above

3α,5β-THP* 3α-OH-5β-pregnan-20-one (pregnanolone). A metabolite of progesterone (found in humans)

3β,5α-THP 3β-OH-5α-pregnan-20-one (epiallopregnanolone). A metabolite of progesterone that is inactive as a $GABA_A$ receptor modulator

* Positive GABA modulator at physiological concentrations.

1 Synthesis, Enzyme Localization, and Regulation of Neurosteroids

Synthia H. Mellon

CONTENTS

0-8493-2392-4/04/$0.00+$1.50

ABSTRACT

The brain, like the gonad, adrenal, and placenta, is a steroidogenic organ. The steroids synthesized by the brain and nervous system — *neurosteroids* — have a wide variety of functions. In general, they mediate their actions not through classic steroid hormone nuclear receptors, but through ion-gated neurotransmitter receptors. This chapter summarizes the biochemistry of the enzymes involved in the biosynthesis of neurosteroids, their localization during development and in the adult, and the regulation of their activity.

1.1 NEUROSTEROIDS

Experiments describing novel functions for certain steroidal compounds at nonclassical $GABA_A$ and NMDA receptors suggested new types of regulation of these neurotransmitter receptors by novel ligands. The demonstration that these novel ligands could be synthesized *de novo* in the brain immediately opened new areas of research. This story of endogenously synthesized neuroactive neurosteroids unfolded with the initial observations by Corpechot and colleagues[1,2] in the 1980s that steroids such as pregnenolone, dehydroepiandrosterone (DHEA), and their sulfate and lipoidal esters were present in higher concentrations in tissue from the nervous system (brain and peripheral nerve) than in the plasma. These steroids remained in the nervous system long after gonadectomy or adrenalectomy, suggesting that these compounds may be synthesized in the brain, rather than being sequestered in the brain following synthesis in steroidogenic tissues. Such steroids were named neurosteroids to refer to their unusual origin and to differentiate them from steroids derived from more classical steroidogenic organs such as gonads, adrenals, and placentae. To test whether steroids were actually made in the brain, several laboratories, including ours, determined directly if enzymes known to be involved in classical steroidogenesis, that is, steroidogenesis in adrenals, gonads, and placentae, were expressed in the nervous system, and therefore, could be responsible for neurosteroid synthesis. Such studies have established that the enzymes found in classic steroidogenic tissues are indeed found in the nervous system. This chapter will describe the enzymatic steps involved in neurosteroid synthesis, the location of the enzymes, and the potential regulation of neurosteroid synthesis.

1.2 ENZYMES INVOLVED IN NEUROSTEROIDOGENESIS: BIOCHEMISTRY AND MOLECULAR BIOLOGY

1.2.1 CYTOCHROME P450

Enzymes involved in neurosteroidogenesis can be classified in two main groups: the cytochrome P450 group and the non-P450 group. Neurosteroidogenic enzymes, P450s and non-P450, are in general either mitochondrial or microsomal, although the exact localization of some enzymes is still controversial.

Cytochrome P450 is a generic term for a large number of oxidative enzymes, all of which have about 500 amino acids and contain a single heme group; they are termed P450 (Pigment 450) because they absorb light at 450 nm in their reduced states. Most cytochrome P450 enzymes are found in the endoplasmic reticulum of the liver, where they metabolize endogenous and exogenous toxins, drugs, xenobiotics, and environmental pollutants (reviewed in Miller[3]). Although there are a huge number of substrates, there are fewer than 200 types of cytochrome P450.[3] Thus, most P450 enzymes can metabolize many substrates and catalyze a broad array of oxidations. Steroidogenic P450s are unusual in that they have limited and specific steroidal substrates. All P450s have similar mechanisms of action. They reduce atmospheric oxygen with electrons from NADPH, and this reduction requires the action of specific co-factors: adrenodoxin reductase and adrenodoxin for mitochondrial P450s, and P450 reductase and b5 for microsomal P450s.

Most, if not all, of the enzymes present in the adrenal, gonad, and placenta have been found in the brain by measuring their enzymatic activity, their mRNA transcript level, or their protein expression. The biosynthetic pathway of neurosteroids is presented in Figure 1.1 and Figure 1.2. The synthesis of specific steroid hormones in the adrenal, gonad, placenta, and brain is dependent on the tissue-, cell-, and developmentally specific expression of these various enzymes. For example, P450c11β and P450c11AS are expressed in the adrenal and not in the gonads or placenta, resulting in glucocorticoid and mineralocorticoid production, while expression of P450c17 in the testis results in androgen production, and P450aro expression in the gonads results in estrogen production.

Many of the steroid hydroxylases have multiple enzymatic activities, as demonstrated by the bioconversions shown in Figure 1.1 and Figure 1.2 (e.g., P450c17, P450c11AS), while some bioconversions are mediated by more than one member of a particular gene family (e.g., 3β HSD, 17β HSD, 3α HSD). Demonstration of a single protein mediating certain reactions has been rigorously demonstrated through purification of the proteins and cloning of the cDNAs encoding these proteins. This is true for P450c17, where cloning and expression of the single P450c17 cDNA demonstrated that this protein had both 17α-hydroxylase activity *and* 17,20 lyase activity, demonstrating that these activities indeed reside within single proteins. On the other hand, similar cDNA cloning has demonstrated the existence of multiple different genes encoding other steroidogenic enzymes, e.g., 17β HSDs and human 3α HSDs.

1.2.2 Hydroxysteroid Dehydrogenases

The hydroxysteroid dehydrogenases (HSDs) have molecular masses of about 35 to 45 kDa, do not have heme groups, and require NAD^+ or $NADP^+$ as co-factors.[4] Whereas most steroidogenic reactions catalyzed by P450 enzymes are due to the action of a single form of P450, each of the reactions catalyzed by hydroxysteroid dehydrogenases can be catalyzed by two or more isozymes, which are very different. Members of this family include the 3α- and 3β-hydroxysteroid dehydrogenases, the two 11β-hydroxysteroid dehydrogenases, and a series of 17β-hydroxysteroid dehydrogenases. The 5α reductases are unrelated to this family.

1.2.3 Neurosteroidogenic Enzymes: P450s

1.2.3.1 P450scc:Cholesterol → Pregnenolone

The first, rate-limiting, and hormonally regulated step in the synthesis of all steroid hormones is the conversion of cholesterol to pregnenolone (Figure 1.1). This reaction is catalyzed by the mitochondrial enzyme cholesterol side chain cleavage, P450scc, in three successive chemical reactions: 20α-hydroxylation, 22-hydroxylation, and scission of the c20–c22 carbon bond cholesterol. These three reactions occur on a single active site that is in contact with the hydrophobic bilayer membrane. The products of this reaction are pregnenolone and isocaproic acid. Homozygous deletion

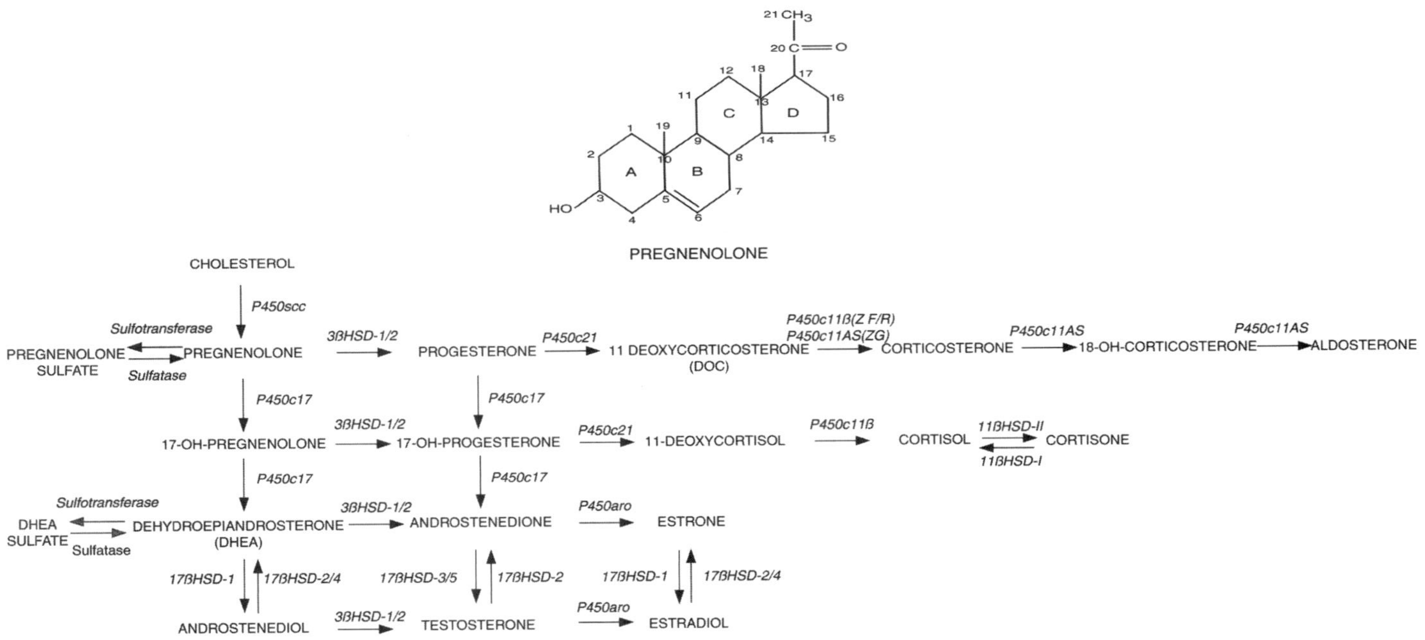

FIGURE 1.1 The biosynthetic pathway of neurosteroids. Top: The cyclopentanoperhydrophenanthrene structure of the steroid pregnenolone. The different rings of the steroid are indicated by letters A to D, and the numbers of the carbon atoms are indicated. Bottom: Pathway for the biosynthesis of neurosteroids in the nervous system. The names for each enzyme are shown by each reaction. P450scc, mitochondrial cholesterol side chain cleavage enzyme, mediates 20α-hydroxylation, 22-hydroxylation, and scission of the c20–c22 bond. 3β HSD mediates both 3β-hydroxysteroid dehydrogenase and Δ5-Δ4 isomerase activities. P450c11β, mitochondrial 11-hydroxylase, mediates 11-hydroxylation. P450c11AS, mitochondrial aldosterone synthase, mediates c11,18-hydroxylation and 18-oxidation. Z F/R and Z G refer to adrenal zona fasciculata/reticularis and zona glomerulosa, respectively, that express the particular P450c11 gene. 17β HSD (also called 17-ketosteroid reductase), mediates c17β-reduction or c17-oxidation. P450aro mediates the conversion of testosterone to estradiol or of androstenedione to estrone. Interconversion of free steroids to their sulfated derivatives, mediated by sulfotransferase and sulfatase, are also shown.

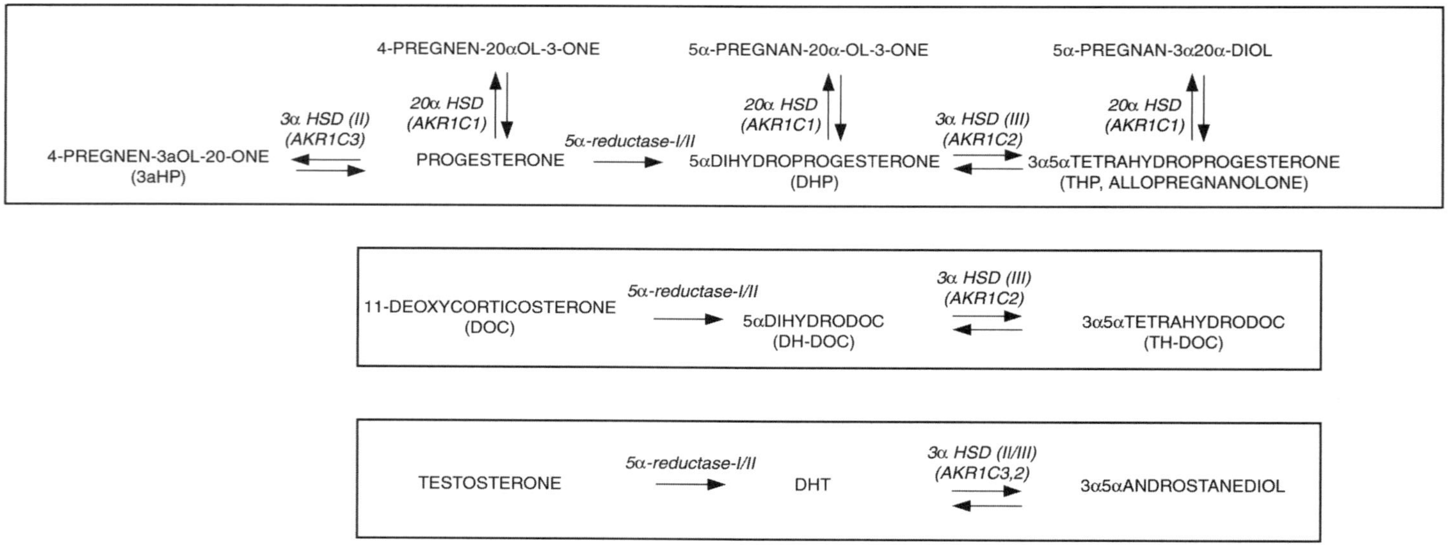

FIGURE 1.2 Formation of neuroactive steroids. The metabolism of progesterone, 11-deoxycorticosterone, and testosterone by two additional enzymes, 5α-reductase and 3α HSD, is shown. The products, allopregnanolone, TH-DOC, and androstanediol, are agonists at the $GABA_A$ receptor and are inactive at intracellular steroid hormone receptors. In addition, progesterone can undergo 3α-reduction by 3αHSD to form the potent neurosteroid 3α-hydroxy-4-pregnen-20-one (3αHP). The reduction of c20 by 20α HSD (shown) or by other 3α HSD isozymes (see Table 1.2) results in the formation of much less potent steroids, and hence c20-reduction is inactivating.

of the gene for P450scc in the rabbit eliminates all steroidogenesis,[5,6] confirming that all steroidogenesis is initiated by this single enzyme, including the brain.[7,8] Homozygous inactivation of P450scc is probably incompatible with human fetal development, and haploinsufficiency of P450scc due to heterozygous mutation causes a late-onset form of congenital lipoid adrenal hyperplasia.[9]

P450scc is one of the slowest enzymes known, with a V_{max} of 1 mol cholesterol/mol enzyme/sec. The slowest part of this reaction may be the entry of cholesterol into the mitochondria and its binding to the active site of P450scc (see Section 1.2.3.3).

The human and rat genomes contain a single gene encoding P450scc,[10–13] which is about 20 kb long, contains nine exons, and in humans is located on chromosome 15. This gene encodes an mRNA of about 2.0 kb, which encodes a 521-amino acid protein. This protein is proteolytically cleaved, removing a 39-amino acid leader peptide that directs the protein to the mitochondria.

1.2.3.2 Transport of Electrons to P450scc: Adrenodoxin Reductase and Adrenodoxin

P450scc functions as the terminal oxidase in a mitochondrial electron transport system. Electrons from NADPH are first accepted by a flavoprotein, adrenodoxin reductase, which is located in the mitochondrial matrix and is loosely associated with the inner membrane.[14–16] Adrenodoxin reductase transfers the electrons to an iron/sulfur protein, adrenodoxin, located in the mitochondrial matrix. Adrenodoxin first forms a 1:1 complex with adrenodoxin reductase, dissociates after oxidation, and then reforms an analogous 1:1 complex with P450scc (or to the other mitochondrial P450s, P450c11β, P450c11AS, and P450c11B3), thus functioning as an indiscriminate diffusible electron shuttle.[3,17] Adrenodoxin reductase and adrenodoxin serve as generic electron transport proteins for all mitochondrial P450s (for example, the vitamin D 1α-, 24-, and 25-hydroxylases) and not just steroidogenic P450s. Hence, these proteins are also often called ferredoxin oxidoreductase and ferredoxin. In humans, there is one gene encoding adrenodoxin reductase, found on chromosome 17, and multiple functional adrenodoxin genes on chromosome 11, encoding identical mRNAs and proteins, and two nonfunctional adrenodoxin pseudogenes on chromosome 20. Adrenodoxin, but not adrenodoxin reductase, is transcriptionally regulated by tropic hormones, acting though cAMP.

1.2.3.3 StAR

A novel protein, steroidogenic acute regulatory protein (StAR) was identified as important for movement of cholesterol into the mitochondria in response to trophic hormone stimulation.[18,19] In adrenals and gonads, StAR expression and function are critical for steroidogenesis. StAR is a 30-kDa phosphoprotein that is synthesized as a 37-kDa precursor, containing an N-terminal mitochondrial targeting sequence. The central role of StAR was proven by showing that it promoted steroidogenesis in non-steroidogenic COS-1 cells co-transfected with StAR and the cholesterol side chain cleavage enzyme system.[20] Individuals who are homozygous for mutations that inactivate StAR have a marked impairment in adrenal and gonadal steroidogenesis, but not in placental steroidogenesis,[21,22] suggesting that there is both StAR-dependent and StAR-independent steroidogenesis. Thus StAR is the acute trigger that is required for the rapid flux of cholesterol from the outer to the inner mitochondrial membrane that is needed for the acute response of aldosterone to angiotensin II, of cortisol to adrenocorticotropic hormone (ACTH), and of sex steroids to an LH pulse. A mitochondrial surface protein, termed the peripheral benzodiazepine receptor (PBR), may also play a permissive role in cholesterol flux into mitochondria.[23] Patients with congenital lipoid adrenal hyperplasia do not appear to have neurological defects, suggesting that StAR is not necessary for neurosteroidogenesis, as it is unnecessary for placental steroidogenesis.[24] Nevertheless, StAR has been recently identified in the rodent brain and has been co-localized with P450scc, suggesting that it indeed may play a role in regulating neurosteroidogenesis.[25,26]

1.2.3.4 3β HSD:

Pregnenolone → Progesterone
17-OH-Pregnenolone → 17-OH-Progesterone
DHEA → Androstenedione

Pregnenolone produced from cholesterol can undergo one of two conversions: it may be 17α-hydroxylated to 17α-hydroxypregnenolone by P450c17 (see below), or converted to progesterone by the enzyme 3β HSD (Figure 1.1). 3β HSD has two distinct enzymatic activities: 3β-dehydrogenation and isomerization of the double bond from C5,6 in the B ring (Δ5 steroids) to C4,5 in the A ring (Δ4 steroids).[27–29] This enzyme is encoded by multiple distinct genes, located on human chromosome 1, that are expressed in a tissue-specific manner. There are at least two forms of human 3β HSD and at least four forms of rodent 3β HSD. The human 3β HSD type I gene is expressed in the placenta, skin, mammary gland, and other tissues, including the brain, while a distinct human type II 3β HSD gene is expressed in adrenals and gonads. In rats, it is unknown whether the type I 3β HSD isoform alone or additional isoforms are expressed in the brain. The enzymes can be classified into two groups: those that function as dehydrogenase/isomerases and those that function as 3-ketosteroid reductases.

1.2.3.5 P450c17:

Pregnenolone → 17-OH-Pregnenolone → DHEA
Progesterone → 17-OH-Progesterone → Androstenedione

Pregnenolone and progesterone may undergo 17α-hydroxylation to 17α-hydroxypregnenolone and 17α-hydroxyprogesterone. These steroids may then undergo scission of the c17,20 bond to form DHEA and androstenedione (Figure 1.1). All four reactions are mediated by a single P450c17 enzyme. P450c17 is bound to the smooth endoplasmic reticulum and accepts electrons from P450 reductase. Since P450c17 has both 17α-hydroxylase and 17,20 lyase activities, it catalyzes a key branch point in steroidogenesis (Figure 1.1). In the human adrenal, the regional expression of P450c17 directs steroidogenesis to the production of mineralocorticoids, glucocorticoids, or sex steroids. In the human zona glomerulosa, P450c17 is not expressed and pregnenolone is metabolized to mineralocorticoids; in the human zona fasciculata, P450c17 is expressed but the majority of the activity is 17α-hydroxylase activity, and hence pregnenolone is directed to glucocorticoids; and in the zona reticularis, P450c17 has both 17α-hydroxylase and c17,20 lyase activities, and hence pregnenolone is metabolized into sex steroids. Several factors are important in determining whether a steroid will undergo 17,20 bond scission after 17α-hydroxylation, such as the presence of the co-factors and the potential competition for substrate between P450c17 and 3β HSD. In the human adrenal, 3β HSD mRNA and activity are low in the zona reticularis and high in the zona fasciculata.[30–32] In addition, there appears to be a gradient of b5 expression in the human adrenal, with the highest concentration in the zona reticularis, indicating that P450c17 expressed in this zone would have greater lyase activity (see Section 1.2.3.8).[33] In addition, phosphorylation of P450c17 will differentially affect the two activities of P450c17: 17α-hydroxylase activity is unaffected by phosphorylation of P450c17, while c17,20 lyase activity requires phosphorylation of P450c17.[34,35] The 17α-hydroxylase reaction occurs more readily than the 17,20 lyase reaction. P450c17 prefers Δ5 substrates, especially for 17,20 bond scission, accounting for the large concentrations of DHEA in the human adrenal.

The single human gene encoding P450c17 is located on chromosome 10 and contains eight exons.[36]

1.2.3.6 P450c21: Progesterone → 11-Deoxycorticosterone 17-OH-Progesterone → 11-Deoxycortisol

Both progesterone and 17-hydroxyprogesterone can be hydroxylated at c21 to yield 11-deoxycorticosterone and 11-deoxycortisol by P450c21 (Figure 1.1). This enzyme has been of great clinical interest because mutations in this enzyme result in congenital adrenal hyperplasia. P450c21 is found in the smooth endoplasmic reticulum. There are two P450c21 genes that lie in the middle of the HLA locus on human chromosome 6. Only one gene is functional in both human beings and mice, but both genes are expressed in cows.

21-Hydroxylase activity has been shown in a large number of extra-adrenal tissues, especially in the fetus and in pregnant women, resulting in the conversion of progesterone to 11-deoxycorticosterone.[37,38] Tissues with this activity include kidney, testis, ovary, skin, urinary bladder, pancreas, thymus, spleen, aorta, and brain. Analysis of RNA from various human fetal tissues using RNase protection assays demonstrated that P450c21 mRNA is not found in these tissues,[39] but more recent studies using reverse transcriptase-polymerase chain reaction (RT-PCR) of human brain tissue have identified P450c21 mRNA in various regions of the human brain.[40,41] The presence of an additional enzyme with 21-hydroxylase activity has been demonstrated by persistence of this activity even in those humans who lack a functional P450c21 gene.[42] Another microsomal P450, P450 2D4, has been shown to have 21-hydroxylase activity and is also found in the brain.[43]

1.2.3.7 P450 Reductase

Both P450c17 and P450c21 receive electrons from a mitochondrial membrane-bound flavoprotein, P450 reductase, which is distinct from adrenodoxin reductase. P450 reductase receives electrons from NADPH and transfers them one at a time to the microsomal P450. The second electron can also be provided by cytochrome b5 (see below). The amount of microsomal P450 reductase is less than that of both P450c21 and P450c17, resulting in competition for this protein. Therefore, factors that influence the association of a specific P450 with the reductase will likely influence the pathway that will be followed, e.g., c17- vs. c21-hydroxylation of progesterone or c17,20 lyase vs. c21-hydroxylation. Electron abundance also influences the activity of P450c17; increasing electron abundance favors both 17α-hydroxylase *and* 17,20 lyase activities, while limiting electron abundance favors only 17α-hydroxylase activity. Therefore, the ratio of P450 reductase to P450c17 seems to be critical in determining the pathway of steroidogenesis.[21,44,45]

1.2.3.8 Cytochrome b5

Cytochrome b5 is a small heme-containing protein that supplies electrons for many cytochrome P450-catalyzed reactions in the liver, and for the reduction of methemoglobin in erythrocytes. Cytochrome b5 is found both in a soluble form, necessary for its function in blood cells, and as a microsomal electron donor, playing the role of co-factor for microsomal P450s. Cytochrome b5 has been shown, *in vitro*, to control the 17,20 lyase activity of P450c17.[46–48] Cytochrome b5 specifically augments the 17,20 lyase activity of P450c17 *in vitro*, and this augmentation is produced by an allosteric modulation of P450c17, not through its function as an electron donor.[48] Those results suggest that expression of b5 may be a mechanism by which c17,20 lyase activity, and hence DHEA production, may be regulated in specific regions of the brain. Cytochrome b5 expression in the human adrenal may be zone specific, as b5 expression appears to be greater in the zona reticularis than in the zona fasciculata.[33] In studies in rhesus macaque, cytochrome b5 expression is restricted to the cells of the zona reticularis, which are also deficient in 3β HSD.[49] Its co-expression with P450c17 in the embryonic macaque adrenal also coincides with DHEA production during development.[50] This differential expression of b5 in primates may account for zone-specific synthesis of glucocorticoids in the zona fasciculata vs. c19 steroids in the zona reticularis.

1.2.3.9 P450c11: 11-Deoxycortisocterone → Corticosterone
Corticosterone → 18-OH-Corticosterone → Aldosterone

The final steps in the synthesis of glucocorticoids and mineralocorticoids are mediated by two distinct adrenocortical enzymes, P450c11β and P450c11AS, encoded by two different genes located on human chromosome 8. The conversion of 11-deoxycorticosterone and 11-deoxycortisol to corticosterone and cortisol is mediated by the mitochondrial 11β-hydroxylase, P450c11β (Figure 1.1). This enzyme is found specifically in the zona fasciculata/reticularis, not in the zona glomerulosa, and is regulated by ACTH. P450c11AS, also referred to as aldosterone synthase, is found exclusively in the zona glomerulosa and has three distinct activities: 11β-hydroxylase, 18-hydroxylase, and 18-oxidase. It therefore converts 11-deoxycorticosterone to aldosterone (Figure 1.1). This enzyme is mainly regulated by the renin/angiotensin system. A third P450c11 gene, called P450c11B3, has been isolated from rat adrenals. P450c11B3 mRNA is only expressed during the early neonatal period and is not expressed in the fetal or adult adrenal. Like P450c11β, P450c11B3 is expressed in the zona fasciculata/reticularis and is regulated by ACTH.[51] However, P450c11B3 has enzymatic activity intermediate between P450c11β and P450c11AS. It has both 11β- and 18-hydroxylase activities, but has no 18-oxidase activity, and therefore can synthesize both corticosterone and 18-OH DOC (from DOC) and 18-OH corticosterone (from corticosterone). A gene corresponding to P450c11B3 has not been found in human beings[52] and has not been reported in other species.

1.2.3.10 P450aro:Testosterone → Estradiol

The aromatization of c18 estrogenic steroids from c19 androgenic steroids is mediated by the enzyme aromatase, P450aro, found in the endoplasmic reticulum (Figure 1.1). P450aro converts androgens to estrogens by two hydroxylations at the c19 methyl and a third hydroxylation at c2. These three hydroxylations result in the loss of c19 and aromatization of the A ring of the steroid. These reactions, all occurring on a single active site of P450aro, utilize three pairs of electrons, donated by three molecules of NADPH and P450 reductase.[53,54]

The gene for P450aro has been cloned, is over 75 kb, and is located on human chromosome 15.[55–57] This gene encodes two mRNAs that differ in the length of their 3' untranslated regions.[55]

1.2.4 Neurosteroidogenic Enzymes: Non-P450s

1.2.4.1 17β HSD:

DHEA ↔ Androstenediol
Androstenedione ↔ Testosterone
Estrone ↔ Estradiol

In the adrenal, DHEA is converted to Δ5 androstenediol and Δ4 androstenedione is converted to testosterone by 17β HSD, also known as 17-ketosteroid reductase (17 KSR). In the ovary, estrone is converted to estradiol by similar mechanisms. These reactions are reversible, and although the reverse reactions are mediated by the same enzyme, they are given the name 17β HSD (Figure 1.1). Like the multiple 3β HSD enzymes, the three 17 KSR/17β HSD reactions are catalyzed by more than one enzyme (reviewed in Labrie et al.).[58] So far, seven types of 17β HSD have been cloned and given the names types I through VII. Type I mainly catalyzes the reductive conversion of estrone to estradiol, but also catalyzes the conversion of DHEA to androstenediol,[59–61] while type IV mainly catalyzes the oxidative conversion of estradiol to estrone.[62–64] Type II catalyzes the oxidative conversion of testosterone to androstenedione, androstenediol to DHEA, and estradiol to estrone. Types III and V are androgenic, as they mainly catalyze the conversion of androstenedione to testosterone (Figure 1.1).

17β HSD is an NADPH-dependent, non-P450 enzyme that is bound to the endoplasmic reticulum. It is widely found in both steroidogenic and non-steroidogenic tissues (reviewed in Labrie et al.).[58] The three-dimensional structure of human 17β HSD-I has been determined by x-ray crystallography.[65,66] No genetic deficiency syndrome for 17β HSD-I or -II has been described. 17β HSD-III is the enzyme that is disordered in the classic syndrome of male pseudohermaphroditism, which is often termed 17-ketosteroid reductase deficiency.[67] An enzyme termed 17β HSD-IV was initially identified as an NAD^+-dependent oxidase with activities similar to 17β HSD-II, but this peroxisomal protein is primarily an enoyl-CoA hydratase and 3-hydroxyacyl-CoA dehydrogenase.[68] 17β HSD-V, originally cloned as a 3α-hydroxy-steroid dehydrogenase, catalyzes the reduction of Δ4 androstenedione to testosterone,[69] but its precise role is unclear, and no deficiency state has been described.

1.2.4.2 11β HSD: Cortisol ↔ Cortisone

The conversion of cortisol to cortisone is mediated by 11β HSD, and the reverse reaction is mediated by an 11-oxidoreductase activity (Figure 1.1). While enzymologic studies suggest that these reactions are mediated by two different proteins, cloning and expression of 11β HSD cDNA showed that one protein has both activities (reviewed in Bujalska et al.[70]). Two isoforms of 11β HSD have been described. The human 11β HSD-I cDNA contains an open reading frame of 876 bases, which predicts a protein of 292 amino acids. The sequence is 77% identical at the amino acid level to rat 11 HSD cDNA.[71] 11β HSD-I is a low-affinity NADH(H)-dependent enzyme[72,73] and is found mainly in human liver, decidua, lung, gonad, pituitary, and cerebellum.[74,75] By contrast, 11β HSD-II is an NAD-dependent enzyme and is localized to the placenta and to mineralocorticoid target tissues (kidney, colon, and salivary gland).[73,75–78] 11β HSD-I can catalyze both the oxidation of cortisol to cortisone, using $NADP^+$ as its co-factor (K_m = 1.6 μM), and the reduction of cortisone to cortisol, using NADPH as its co-factor (K_m = 0.14 μM); the reaction catalyzed depends on which co-factor is available, but the enzyme can only function with high (micromolar) concentrations of steroid.[79] 11β HSD-II catalyzes only the oxidation of cortisol to cortisone using NADH, and can function with low (nanomolar) concentrations of steroid (K_m = 10 to 100 nM).[80] Since 11β HSD converts cortisol to cortisone, it is thought to protect mineralocorticoid receptors, which can be bound by both glucocorticoids and mineralocorticoids, from occupation by glucocorticoids.

Human 11β HSD-I is found on human chromosome 1, consists of six exons, and is at least 9 kb long.[71] Human 11β HSD-II is found on chromosome 16q22 and consists of five exons spread over 6 kb.[81] The coding sequences of human 11β HSD-1 and -II are only 21% identical.

1.2.4.3 5α-Reductase:

Progesterone → 5α-Dihydroprogesterone
11-Deoxycorticosterone → 5α-11-Deoxycorticosterone
Testosterone → DHT

Testosterone is converted to the more potent androgen dihydrotestosterone (DHT) by the enzyme 5α-reductase (Figure 1.1). In the brain, progesterone is converted to 5α DHP, and 11-deoxycorticosterone is likewise reduced at the 5α position. This membrane-bound non-P450 is found mainly in peripheral target tissues such as genital skin and hair follicles.

Cloning and expression studies have demonstrated the existence of two 5α-reductase genes.[82,83] The gene for type 1 is found on the short arm of human chromosome 5, is about 35 kb, and encodes a 29-kDa protein, but does not encode the protein that is required for sexual differentiation.[84] The gene for type 2 is found on the short arm of human chromosome 2 and has the same intron/exon structure as the gene for the type 1 enzyme.[85] 5α-Reductase types 1 and 2 share 50% identity in their amino acid sequences,[86] yet have different kinetic parameters. The type 1 enzyme has a broad pH range between pH 6.0 and 8.5 and a K_m of 1 to 5 μ*M*, while the type 2 enzyme has an acidic optimal pH of 5.5 and a K_m of 0.1 to 1.0 μ*M*. In addition, the human type 1 enzyme is relatively insensitive to the inhibitor finasteride (K_i = 325 m*M* for type 1 and K_i = 12 n*M* for type 2).[86] For studies using rodents, finasteride is 100-fold more potent against the rodent type 1 enzyme than the human type 1 enzyme.[86] 5α-Reductase also has a preference for progesterone as substrate. The apparent K_m for progesterone is between 100 and 300 n*M*, while the apparent K_m for testosterone is 10-fold higher, between 1 and 4 μ*M*.[87]

Mutations in the gene for 5α-reductase type 2 cause classic 5α-reductase deficiency.[88] The human type 1 and 2 genes are differentially regulated during development.[89] The type 1 isozyme is not detectable in the fetus, is only transiently expressed in the newborn skin and scalp, and is permanently expressed in the skin from the time of puberty. The type 2 isozyme is transiently expressed in the skin and scalp of newborns and is the predominant form in fetal genital skin, male accessory sex glands, the normal prostate, prostatic hyperplasia, and adenocarcinoma tissues. In rats, the type 1 mRNA is expressed in basal epithelial cells, while the type 2 mRNA is expressed in the stromal cells of regenerating ventral prostate. In the rodent brain, both type 1 and 2 mRNAs are expressed in a developmentally regulated fashion[90,91] (see below).

1.2.4.4 3α HSD/3α-Hydroxysteroid Oxidoreductase:

5α-Dihydroprogesterone ↔ 3α5α-Tetrahydroprogesterone (Allopregnanolone)
5α DOC ↔ 3α5α-TetrahydroDOC
DHT ↔ Androstanediol

cDNA cloning experiments revealed that most HSDs belong to one of two families: the short-chain dehydrogenase/reductase family (also known as short-chain alcohol dehydrogenases) and the aldo-keto reductase (AKR) family (reviewed in Penning et al.[92]). Mammalian 3α HSDs are members of the AKR1 family of the AKR superfamily. A systematic nomenclature for the AKR superfamily originally was adopted in 1996 and was updated in September 2000.[93] This nomenclature is recognized by the Human Genome Project (HUGO), and genomic information, including chromosomal localization, gene boundaries, human ESTs and SNPs, etc., can also be found on the website: http://www.med.upenn.edu/akr.

The reactions catalyzed by 3α HSD are stereospecific and involve the interconversion of a carbonyl with a hydroxyl group. In the prostate, 3α HSD is involved in *inactivating* DHT by conversion to the weak androgen 3α-androstanediol, while in the nervous system, 3α HSD is involved in *activating* 5α reduced steroids, such as 5α DHP to the potent neurosteroid allopregnanolone. Thus, this enzyme is a key regulator of both steroid hormone receptor and ion-gated receptor occupancy and action.

Most biochemical studies have used the rat liver enzyme for purification, biochemical analysis, generation of antibodies, cDNA cloning, and enzyme structure determination (reviewed in Penning et al.[92]). A single cDNA species from rat liver was cloned, called AKR1C9,[94–97] had an open reading frame of 966 nucleotides, and predicted a protein of 322 amino acids. There are four human 3α HSDs: Type I (EC 1.1.1.213) — AKR1C4: chlordecone reductase (EC 1.1.1.125) and dihydrodiol dehydrogenase 4 (EC 1.3.1.20); Type II (EC 1.1.1.213) — AKR1C3: type V 17β HSD (EC 1.1.1.62) and dihydrodiol dehydrogenase X (EC 1.3.1.20) Type III (EC 1.1.1.213) — AKR1C2: bile acid binding protein and dihydrodiol dehydrogenase 2 (EC1.3.1.20) and 20α (3α) HSD (EC1.1.1.149) — AKR1C1: dihydrodiol dehydrogenase 1 (EC1.3.1.20). These enzymes share >83% amino acid sequence identity and have broad substrate specificities for 3α, 17β, and 20α-hydroxysterols.[98] Since both activating (3α HSD) and inactivating (20α HSD) activities are associated with these enzymes, they play roles in the synthesis of neuroactive steroids, as well as in the conversion of these neuroactive steroids and their precursors to less neuroactive 3α20α-dihydroxy- or 20α-hydroxypregnane via their 20α HSD activity. The rat liver 3α HSD cDNA has high sequence identity (>70%) with the clones for human liver type I 3α HSD (DD4 or chlordecone reductase), human liver DD1 (which is both a 3α and 20α HSD), and human liver DD2 (human bile–acid binding protein).[99,100] Human liver type II 3α HSD,[101] human type III prostatic 3α HSD,[102] and a human type II expressed in the brain (called type II_{brain})[103] cDNAs have also been cloned. Isoforms, for example, in type II isolated from prostate and the type II isolated from the brain, may reflect polymorphisms in the coding sequences of the different cDNAs, rather than being encoded by different genes. Differences in the 3′ untranslated region (UTR) would indicate that the cDNAs were encoded by different genes. The substrate specificities and enzyme activities associated with these different isozymes are listed in Table 1.1[104,105] and Table 1.2.[106] Human type I and type II 3α HSDs differ in their K_m values for 5α DHT, with the type I enzyme having a lower K_m. The human type III and type II also differ in their enzyme activities and substrate specificities. The type II uses 5α-reduced androgens as substrates, and not 5α-reduced progesterone (5α DHP).[104,107] The K_m of the type III for dihydroprogesterone is ~0.6 μ*M*[104,105] and for 5α DHT is ~2 μ*M*, indicating that this enzyme has greater affinity for 5α-dihydroprogesterone. The K_m for the type II (5α DHT as substrate) is also about 2 μ*M*. In addition to having substrate-selective 3α HSD activity,

TABLE 1.1
Kinetic Constants and Substrate Specificities for Human 3α (20α) HSD Isozymes[104,105]

Neurosteroid Substrate (Isozyme)	K_m (μ*M*)	K_{cat} (min^{-1})	K_{cat}/K_m ($min^{-1}\mu M^{-1}$)
3α5α THP			
Human type 2 3α HSD (Oxi)	nd[a]	n.a.	nd
Human type 3 3α HSD (Oxi)	28	0.28	0.01
Human 20α HSD (Red)	2.0	4.4	2.2
5α DHP			
Human type 2 3α HSD (Red)	nd	(0.06)	nd
Human type 3 3α HSD (Red)	0.6	0.48	0.8
Human 20α HSD (Red)	1.1	2.6	2.4
3α5α TH-DOC			
Human type 2 3α HSD (Oxi)	nd	n.a.	nd
Human type 3 3α HSD (Oxi)	6.8	0.13	0.02
Human 20α HSD (Oxi)	2.6	0.22	0.08
5α DH-DOC			
Human type 2 3α HSD (Red)	nd	(0.06)	nd
Human type 3 3α HSD (Red)	0.8	0.49	0.61
Human 20α HSD (Red)	1.2	0.56	0.42

Note: Red = reductive reaction; Oxi = oxidative reaction.

[a] Kinetic constants were not determined (nd) because of no activity (n.a.) or low activity.

TABLE 1.2
Catalytic Efficiencies for Rat and Human 3α (20α) HSDs[106]

	Activity		
Enzyme	3α HSD 5α DHT Reduction (k_{cat}/K_m $mM^{-1}min^{-1}$)	17βHSD Androsterone Reduction (k_{cat}/K_m $mM^{-1}min^{-1}$)	20αHSD Progesterone Reduction (k_{cat}/K_m $mM^{-1}min^{-1}$)
Rat 3α HSD AKR1C9	1000	None	None
Human type 1 3α HSD AKR1C4	231	0.1	0.1
Human type 2 3α HSD AKR1C3	6	42	0.1
Human type 3 3α HSD AKR1C2	9	0.1	0.1
Human 20αHSD AKR1C1	8	9	109

the type II and type III 3α HSD have 20α HSD and 17β HSD activities as well, whereas the rat 3α HSD appears to be a pure 3α HSD, with no 17β HSD or 20α HSD activity at all. All 3α HSD cDNAs share high sequence identity (over 80% at both amino acid and nucleic acid levels). The genes encoding the human type I and II enzymes span approximately 20 and 16 kb, respectively, and both contain nine exons of the same size and intron/exon boundaries,[101] and are found on human chromosome 10p15–10p14. The genes encoding the human type III and IV enzymes are also thought to be localized to the same chromosomal region.

1.2.4.5 Sulfotransferase (HST)

Sulfation of free 3β-hydroxysteroids is a major enzymatic reaction of metabolism, excretion, and homeostasis of steroids and bile acids (reviewed in Hobkirk[108]). Sulfation and sulfohydrolation activities have been reported in lung, kidney, adrenal, and testis, and have been described as crucial during development. 16-Hydroxy DHEA sulfate (DHEAS) originating from the liver, DHEAS originating from the adrenal, and DHEA originating from the placenta serve as precursors for the production of estrone, estradiol, and estriol in the developing human placenta[109–111] (Figure 1.1). Sulfotransferases are a family of cytosolic enzymes that conjugate steroid and phenolic substrates with inorganic sulfate derived from an active donor, PAPS.[112,113] The hydroxysteroid sulfotransferase specifically uses Δ5 steroid substrates that are hydroxylated at C3, 5, 17, or 21.[113] The resulting steroid sulfate esters are hydrophilic. However, sulfation has a role greater than facilitating secretion, since it can also change the pharmacological activity of steroids, such as changing the way in which pregnenolone binds to the $GABA_A$ receptor (reviewed in Majewska[114,115]). HST is mainly found in adrenal in humans and in liver in other mammals.[116]

1.2.4.6 Sulfatase (STS)

The steroid sulfohydrolase is a sterol-sulfate sulfohydrolase, also known as steroid sulfatase or steroid-3-sulfatase. It specifically hydrolyzes sulfate groups in the 3β position of Δ5, steroids such as pregnenolone, DHEA, and androstenediol (Figure 1.1). STS is an important enzyme in steroid metabolism since its activity increases the pool of precursors that can be metabolized by other steroidogenic enzymes to produce biologically active sex steroids.[113,117] The human STS gene has been cloned and mapped to chromosome Xp22.3, proximal to the pseudoautosomal region (PAR), and the genetic aspect of this enzyme has been widely documented.[118] Deficiency of STS activity results in severe ichthyosis caused by accumulation of steroid sulfates in the stratum corneum of the skin.[119,120] The rat and mouse STS have recently been cloned and are fairly dissimilar. The mouse STS cDNA is 75% identical to the rat STS cDNA and only 63% identical to the human STS cDNA.[121,122]

1.2.4.7 Other Neurosteroidogenic Enzymes

Other enzymatic activities that modify steroid hormones have been identified in the brain, but the activities of the resulting neurosteroid products have not yet been characterized. Those enzymes are the 7α-hydroxylase, a novel 7α-hydroxylase expressed primarily in the brain,[123,124] 24-hydroxylase,[125] and 26α-hydroxylase.[126]

1.2.4.7.1 24-Hydroxylase

The turnover of cholesterol in the brain is thought to occur via conversion of excess cholesterol into 24S-hydroxycholesterol. This compound, an oxysterol, is readily secreted from the central nervous system (CNS) into the plasma. Recent cloning of human and mouse cDNA sequences has shown that both proteins are localized to the endoplasmic reticulum, share 95% identity, and represent a new cytochrome P450 subfamily (CYP46).[125] Transfection studies have revealed that the proteins encoded by the cDNAs convert cholesterol into 24S-hydroxycholesterol and, to a lesser extent, 25-hydroxycholesterol. The cholesterol 24-hydroxylase gene contains 15 exons and is located on human chromosome 14q32.1.

Analysis of protein and RNA has demonstrated that cholesterol 24-hydroxylase is expressed predominantly in the brain. In the brain, 24-hydroxylase is expressed in neurons in several regions. The concentrations of 24S-hydroxycholesterol in serum are low in newborn mice, reach a peak between postnatal days 12 and 15, and thereafter decline to baseline levels, whereas the 24-hydroxylase protein is first detected in the brain of mice at birth and continues to accumulate with age.

1.2.4.7.2 26α-Hydroxylase

Another novel member of the cytochrome P450 gene family has recently been cloned from retinoic acid-induced neural differentiation of embryonic stem cells[126] from zebra fish[127] and from a human teratocarcinoma cell line.[128] This cDNA, called CYP26, encodes a 1.9-kb mRNA transcript that is regulated during embryonic stem cell neural differentiation and is also expressed in liver. CYP26 cDNA is 1701 nucleotides (nt) long, containing an open reading frame of 1491 nt that encodes a 56-kDa protein of 497 amino acids. Consistent with its induction by retinoic acid, CYP26 is expressed in the early mouse embryo from at least E8.5. However, there is only low, if any, expression in the body or head of E12.5 embryos. In the adult mouse, CYP26 mRNA is expressed mainly in the liver and brain, with very low amounts in other tissues. By RT-PCR analysis, CYP26 mRNA was detected in a variety of human brain regions, including olfactory bulb, temporal cortex, and hippocampus. This protein appears to be involved in the metabolism of retinoic acid[127]; other substrates have not been reported.

1.2.4.7.3 7α-Hydroxylase

7α-Hydroxylase is a microsomal P450 that hydroxylates steroids at the 7α position. While its activity was originally described in liver protein extract where 7α-hydroxylated bile acids are produced for detoxification, 7α-hydroxylated derivatives of pregnenolone and DHEA appear to modulate glucocorticoid action, functioning as antiglucocorticoids and modulating immune function.[129] The stereospecific 7α-hydroxylation of steroids is mediated by at least two different enzymes, initially identified by competition experiments.[130] In addition, there appears to be at least two different classes of 7α-hydroxylases. 7α-Hydroxylation of cholesterol and Δ4-3-oxosteroids is mediated by one specific P450.[131–134] This enzyme(s) is necessary for conversion of both cholesterol and 27-hydroxycholesterol into bile acids, but does not mediate 7α-hydroxylation of pregnenolone and DHEA. Recently, another brain-enriched P450 7α-hydroxylase, called CYP7B, was identified and its cDNA cloned.[123,124] This latter cDNA encodes a protein that catalyzes the 7α-hydroxylation of pregnenolone and DHEA in the brain, and also catalyzes 7α-hydroxylation of oxysterols. In addition to expression in the brain, CYP7B is found in rat liver and kidney, but is barely detectable in testis, ovary, and adrenal. Expression in the liver is dimorphic, with expression less in female rats than in male rats. CYP7b is widely expressed in the mouse brain and is regionally expressed in the hippocampus[124]; the highest concentrations are found in the corpus callosum. Among the P450s, CYP7b is most similar (39% at the amino acid sequence) to cholesterol 7α-hydroxylase (CYP7) and contains a postulated steroidogenic domain present in other steroid-metabolizing CYPs, but clearly represents a novel type of CYP. The formation of 7α-hydroxylated metabolites of pregnenolone and DHEA is low in prepubertal rats and increases fivefold in adults.[135] It is unknown if brain expression of CYP7b parallels this enzymatic profile.

1.3 ENZYMES INVOLVED IN NEUROSTEROIDOGENESIS: DISTRIBUTION IN THE BRAIN

The presence of functional steroidogenic enzymes in the brain has been established using enzymatic activity measurements, mRNA expression (using RT-PCR, ribonuclease protection assays, or *in situ* hybridization), and protein expression (using Western blotting or immunocytochemistry). These studies by numerous laboratories over the years used mainly the adult rat brains as the source of tissue, but now include analysis of neurosteroidogenesis in the adult lungfish, frog, bird and guinea pig. There are increasing numbers of papers identifying sites of expression of neurosteroidogenic enzymes in human brain samples, which have been recently reviewed.[136] Some of those studies used available tissue samples from patients with neurologic disorders, such as epilepsy, and hence those studies need to be compared with studies using normal brain tissue. There are fewer reports documenting steroidogenic enzyme expression during embryogenesis.[137–144] However, such studies are crucial, since neurosteroids have diverse functions during development as well as in the adult.

1.3.1 Expression of P450scc in the Adult Rat Brain

Initial support for the hypothesis that the brain is a steroidogenic organ derives from experiments demonstrating conversion of radioactive cholesterol to pregnenolone (reviewed in Baulieu and Robel[145]). While these experiments were suggestive that the brain had steroidogenic capacity, the demonstration that CNS tissue and neuronal cells contained P450scc mRNA and protein was hampered by the extremely low amount of this mRNA. Using RT-PCR and Southern blotting, we showed that a P450scc mRNA was regionally expressed in extremely low abundance.[7] In the adult rat, it was found most abundantly in the cortex, and to a lesser extent in the amygdala, hippocampus, and midbrain of both male and female rats. Purification of mixed primary glial cultures showed that type 1 astrocytes synthesized P450scc mRNA.[7] Western blotting and immunocytochemistry showed that P450scc protein was almost as abundant in neonatal cultures of forebrain astrocytes as in mouse adrenocortical Y-1 cells, while P450scc mRNA was orders of magnitude less abundant, suggesting that the protein was stable in the brain.[138]

P450scc has also been identified in Purkinje neurons in the cerebellum.[25,146–148] During early neonatal life (P0 to P7), prior to Purkinje cell differentiation, P450scc-immunoreactive cells were scattered throughout the cerebellum; at P3, P450scc immunoreactivity could be detected in Purkinje cells. Later, P450scc was also found within the molecular layer concurrently with maturation of Purkinje dendrites. Furthermore, synthesis of pregnenolone from cholesterol, i.e., P450scc activity, was also demonstrated in the cerebellum after P7, a time when P450scc protein was detected in maturing Purkinje neurons.[146]

1.3.2 Expression of Adrenodoxin and Adrenodoxin Reductase in the Brain

Adrenodoxin mRNA was found in virtually all rodent tissues examined,[149] including the brain. This is not surprising, as this protein functions with all mitochondrial P450s, not just the steroidogenic P450s. Similarly, as expected, adrenodoxin protein has also been found in brain homogenates.[150] The exact site of expression of these proteins has not been systematically examined either in adult brain or during development. The expression of adrenodoxin reductase has not been documented yet in the CNS or PNS.

1.3.3 Expression of StAR in the Brain

StAR, a 30-kDa protein involved in the acute transport of cholesterol to the inner mitochondrial membrane, is stimulated by tropic hormones in adrenal and gonads.[19] Its lack of expression in the placenta further suggested that StAR was expressed in steroidogenic tissues having an acute response, but not in tissues (placenta, perhaps brain also) in which there is not an acute response.[24] In addition, patients lacking StAR activity had normal placental steroidogenesis and had no apparent neurological defects, suggesting that StAR may not be crucial for neurosteroidogenesis.[24,151] Similar phenotypes and lack of neurological impairment were seen in a mouse StAR knockout line.[152] As early studies failed to detect StAR mRNA in the brain,[151] its role in neurosteroidogenesis was questioned. However, more recent studies *have* demonstrated StAR transcripts in rat brains, at levels two to three orders of magnitude less than in adrenals.[25,26] This is similar to the relative abundance of P450scc in the brain vs. the adrenal. These studies demonstrated StAR's localization in the rat brain and co-localization with P450scc and 3β HSD, using *in situ* hybridization. StAR transcripts were abundant in the cerebral cortex, hippocampus, dentate gyrus, olfactory bulb, cerebellar granule cell layer, and Purkinje cells. StAR has been found in cultured Schwann cells.[153] P450scc and 3β HSD co-localized with StAR in the hippocampus, dentate gyrus, cerebellar granule layer, and Purkinje cells. It is unclear if StAR plays the same role in the brain as it does in the adrenal and gonad, if it is constitutively expressed in the brain, or if its function is acutely regulated by some as yet undefined factor. In culture Schwann cells, unlike in adrenals and gonads, cAMP inhibited the expression of StAR mRNA.[153] It is unclear if this type of negative regulation by cAMP is universal in the brain, or if it is unique to Schwann cells.

1.3.4 P450c17 Expression in the Nervous System

Initial studies failed to detect expression of P450c17 mRNA in any region of the adult rat brain, using highly sensitive RT-PCR analyses.[154] Those studies also indicated that the steroidogenic enzyme proteins may be more easily detected than their mRNAs, since using an antibody directed against human P450c17, we determined the pattern of expression of P450c17 in developing rodent embryos.[155] P450c17 expression is restricted in neurons in specific regions of the developing brain. We first observed P450c17 in E10.5 cells migrating from the neural crest and condensing in DRGs. Immunopositive cells were also observed in the neural tube in the position of the lateral motor column. Neuronal cell bodies immunopositive for P450c17 were restricted to the hindbrain mesopontine system (from E14.5), the thalamus, and the neocortical subplate (from E16.5). Fibers extended in the areas of projections of these nuclei. One particular region in which P450c17 was expressed in the embryo was the neocortical subplate. This region of the developing CNS has been reported to be the region that receives thalamic projections, produces signals for cortical projections, and may produce signals for efferent thalamic projections from the cortex.[156,157] We also found fibers coming from the trigeminal ganglion and the retina.

In all the CNS regions P450c17 expression was transient and P450c17 immunostaining gradually disappeared in the first week of life. In the neonatal mouse and rat (days 0 to 7), P450c17 was mainly detected in the brainstem and in the cerebellum, internal capsule, olfactory tract, hippocampus, stria terminalis, and thalamus. Few immunopositive cell bodies could be detected in any region of the adult rat brain, although fiber tracts could still readily be observed. In the PNS, P450c17 persisted in the adult, both in cell bodies and fibers. Thus, unlike P450scc, our data suggest that P450c17 is expressed in the CNS only during development, but persists in the PNS throughout life.

Recent studies with human brain samples have suggested that P450c17 mRNA could be detected by RT-PCR.[41] In that study, P450c17 mRNA was detected in all regions evaluated: caudate nucleus, cerebellum, corpus callosum, hippocampus, spinal cord, and thalamus. The human tissues that were used in that study were obtained commercially and were pooled tissue samples from 10- to 78-year-old normal males and females. Studies in postnatal rat brains gave somewhat different results.[158] Using RT-PCR in that study as well, the authors found that P450c17 mRNA was found mainly in the mesencephalon, consistent with results demonstrating P450c17 protein in various nuclei in the rodent embryonic hindbrain.[155] P450c17 mRNA could be detected in 3-, 10-, and 61-day-old rats. Other regions of the brain — the cerebrum, diencephalon, and cerebellum — contained much less, if any, P450c17 mRNA at the three ages studied. There were no comparisons between abundance of P450c17 mRNA in the embryonic vs. adult rat brain regions. It is unknown from the studies of mRNA analysis if P450c17 protein was expressed in the same regions.

1.3.5 Expression and Activity of P450 Reductase and b5 in the Brain

P450 reductase activity, protein,[159] and mRNA[159,160] have been reported in the brain. Cytochrome P450 reductase immunoreactivity was detected mainly in neurons, but also in some glial populations, and it appears to be expressed widely in the rat CNS.[159] Although no studies specifically addressed the co-localization of P450 reductase and P450c17 in the same cells, the regional distribution of these proteins suggests that P450c17 may have functional activities in the cortical thalamic and pontine regions, where we described its expression. The expression of b5 has not been documented yet in the CNS or PNS.

1.3.6 3β HSD Expression in the Adult Nervous System

Demonstration of the conversion of pregnenolone to progesterone in cultured rat oligodendrocytes,[161,162] astrocytes,[163] cultured rodent glia and neurons,[164] and discrete regions of the rat and monkey brain[165,166] suggested that 3β HSD is expressed in those regions. Expression of 3β HSD protein and mRNA has been studied in brains from rats[167–170] and in considerable detail in frogs[171] and lungfish,[172] demonstrating that this enzyme is expressed in the brains from both mammalian and nonmammalian vertebrates. There are conflicting reports as to whether 3β HSD type I alone[167] or types I, II, and IV[173] are the isoforms expressed in the adult rat brain. While 3β HSD activity has been reported in rodent embryonic glia[174] and in amphibian neurons,[175,176] 3β HSD mRNA and protein has been identified in neurons.[167–169] 3β HSD mRNA was found in the brain in much lower amounts than in the ovary, adrenal, and liver. Results from different laboratories are conflicting and relate to the specificity of the probe/antibody to a particular subtype of the enzyme and to the sensitivity of the technique used.[167–169] An extensive study using immunocytochemistry in the frog showed the restricted expression of 3β HSD in neurons,[171] in regions consistent with the pattern of expression of 3β HSD type I mRNA in the rat,[167] except in the cerebellum and in the cortex, where only 3β HSD mRNA was reported.[167,169] However, reports of 3β HSD activity did not always correlate with expression of 3β HSD protein. The lack of significant activity in the rat cerebellum and cortex[165,166] correlates with the lack of 3β HSD protein in these regions in the frog,[171] but not with the presence of 3β HSD mRNA in the rat.[167] Transient 3β HSD activity has been reported in cerebellum slices obtained from neonates.[169] High levels of 3β HSD activity have been reported in the rat amygdala and hippocampus[165]; 3β HSD mRNA was detected in the rat hippocampus, but 3β HSD protein was not detected in this region in the frog.[171]

3β HSD is also expressed in Schwann cells.[177] However, its expression is profoundly regulated by contact with sensory neurons. Experiments with cryolesioned sciatic nerves and with transected nerves support the notion that Schwann cells need to be in contact with neurons (axons) for full expression of neurosteroidogenic enzymes.[177] In the absence of this contact, 3β HSD expression is not induced. Studies of oligodenydrocyte precursor cells showed that only oligodendrocyte pre-progenitors and progenitors, but not fully differentiated oligodendrocytes, expressed 3β HSD mRNA and were able to synthesize progesterone from pregnenolone.[178]

1.3.7 P450c11: P450c11β and P450c11AS Expression in the Adult Nervous System

Analysis of RNA from different regions of the adult rat brain indicated that the abundance of P450c11β mRNA was greater than that of P450scc mRNA, since it could be detected readily by RNase protection assays.[7] We showed that P450c11β mRNA was found in virtually all regions analyzed (cortex, hippocampus, hypothalamus, amygdala, cerebellum, midbrain), with the greatest amount in the cortex. Others have shown that P450c11AS mRNA and protein are indeed expressed in adult rat[179] and human brains.[41] In the rat brain, P450c11β and P450c11AS proteins were detected in the cerebellum, especially in Purkinje cells, as well as in the hippocampus. In the human brain, P450c11β and P450c11AS mRNAs were not detected when whole-brain samples were used, but were detected in the caudate nucleus, amygdala, corpus callosum, spinal cord, and thalamus. Neither was detected in the cerebellum and hippocampus.[41] Moreover, aldosterone synthase activity has also been demonstrated in the brain.[180] Thus, the adult rat brain has the ability to synthesize aldosterone. There have been no reports documenting sites of expression of either P450c11β or P450c11AS during development.

1.3.8 Expression of 17β HSD in the Brain

Demonstration of 17β HSD activity in the brain of rats and monkeys[181,182] indicated that 17β HSD protein and mRNA would also be present. Recent analysis of both rat and frog brains has demonstrated that 17β HSD type I is expressed in the brain.[183–185] In adult rat brains, 17β HSD-I protein was detected in nonneuronal cells (GFAP-positive cells).[183] In the frog brain, 17β HSD-I was also found in glia.[184] Regional distribution of 17β HSD protein and mRNA in the rodent and frog is consistent. In human beings, 17β HSD activity was detected in the cortex, and greater activity was found in the subcortical white matter.[186] Analysis of 17β HSD mRNAs in human brains demonstrated that 17β HSD-I, -III, and -IV were expressed in the temporal lobe of children and adults, whereas 17β HSD-II and the pseudogene of 17β HSD-I were not.[186] In adults, 17β HSD-III and -IV mRNA concentrations were significantly higher in the subcortical white matter than in the cortex, similar to what was found for enzyme activity, while 17β HSD-I concentrations were similar in both tissues. In another study in human beings, 17β HSD-I, -III, and -IV mRNAs, but not 17β HSD-II mRNA, were detected in the hippocampus.[187] Similar amounts of these mRNAs were detected in hippocampi from men and women.

1.3.9 Expression of P450aro in the Brain

Several lines of evidence demonstrate that testosterone is aromatized to estradiol in the brain of many species, from frogs and songbirds to human beings (reviewed in Lephart[188] and Hutchison et al.[189]). Aromatase activity in the rat brain was shown to be limited to discrete regions.[190] Aromatase activity was first detected on embryonic day 15, increased up to embryonic day 19, and then declined to adult levels.[191–193]

While three distinct P450aro mRNA species of 2.7, 2.2, and 1.7 kb exist in many tissues, only the 2.7-kb P450aro mRNA was found in the brain.[193] In human beings, the major transcript in the hypothalamic preoptic area and amygdala contained exon I-f, referred to as the brain-specific exon I,[194] while a minor transcript contained exon I-b.[194,195] In the rat cortex, an additional novel P450aro transcript was identified that contained exons IV through X, but initiated transcription within intron III and did not contain exons I through III.[196] Thus the rat brain contains P450aro transcripts that initiate at three different sites. It is not clear if the three different P450aro transcripts correspond to different sites of transcription initiation or if they are variants that lack the heme binding domain.[197]

The regulation of P450aro mRNA expression during rodent development was studied by *in situ* hybridization.[140] In neurons, P450aro mRNA appeared to parallel aromatase activity, but P450aro mRNA was also found in regions not previously associated with activity. The regional distribution of P450aro mRNA was restricted to the preoptic/hypothalamic area on E16 and was more widely distributed by E18 to E20, but was still absent from the cortex, midbrain, and hindbrain structures. From postnatal day 2 to adulthood, P450aro mRNA abundance decreased in the preoptic area, but remained constant in other areas. In the adult, P450aro mRNA was still present with a similar pattern of expression. Female rats had the same distribution of P450aro mRNA, but the number of cells expressing P450aro mRNA in each region was less than that in male rats.[198]

The regions of the brain that contained P450aro mRNA but did not contain activity or were not examined include the mediodorsal thalamus, subfornical organ and cingulate cortex, and hippocampus. P450aro immunoreactivity has been demonstrated in the same areas in which P450aro mRNA has been detected,[199] indicating that the lack of aromatase activity may be due to the sensitivity of the assay.

1.3.10 Expression of 11β HSD in the Brain

Both 11β HSD-I[200,201] and 11β HSD-II[77,202,203] activities and mRNA have been detected in adult rat brains. During development, 11β HSD-II mRNA was widely expressed in the CNS throughout the neuroepithelium, thalamus, and spinal cord up to E12.5. It declined thereafter.[77] Expression in the thalamus and cerebellum persisted postnatally. 11β HSD-I mRNA is absent from the CNS until the third trimester, and activity is low or undetectable.[204] The expression of the two forms of 11β HSD must also be correlated with expression of the glucocorticoid and mineralocorticoid receptors in the brain, since 11β HSD-I is a glucocorticoid *activating* enzyme, while 11β HSD-II is a glucocorticoid *inactivating* enzyme. Glucocorticoid receptor mRNA is highly expressed throughout the brain from midgestation, but mineralocorticoid receptor mRNA expression is absent until the last few days of gestation. High 11β HSD-II at midgestation may protect the developing brain from activation of glucocorticoid receptors, while late in gestation, repression of 11β HSD-II gene expression and increased expression of 11β HSD-I may allow increasing activation of both the glucocorticoid and mineralocorticoid receptors, thereby permitting glucocorticoid-dependent neuronal and glial maturational events to occur late in gestation.

11β HSD-I is highly expressed in the hippocampus. Studies with knockout mice demonstrate that 11β HSD-I is the enzyme that results in the activation of glucocorticoids (i.e., in mice, it converts the inactive 11-dehydrocorticosterone to corticosterone).[205] Analysis of hippocampal function in 11β HSD-I knockout mice suggests that 11β HSD-I attenuates the deleterious effects of chronic glucocorticoid excess upon cognitive function.[206] Aged wild-type mice developed elevated plasma corticosterone levels that correlated with learning deficits in the water maze; this glucocorticoid-associated learning deficit was ameliorated in aged 11β HSD-I knockout mice, despite having elevated plasma corticosterone levels throughout life. These data suggest that the lack of hippocampal 11β HSD-I results in decreased corticosterone concentrations, which result in increases in learning and memory in aged animals. The data further suggest that local brain concentrations of steroids do not necessarily reflect plasma concentrations, and that these locally synthesized steroids have profound effects on learning and memory.

1.3.11 Expression of 5α-Reductase in the Brain

The rat CNS is capable of converting progesterone into 5α′-reduced metabolites (reviewed in Celotti et al.[207]), indicating that this enzyme is present in the CNS. 5α-′Reductase activity, protein, and mRNA have been detected in neurons, astrocytes, and glia,[141,207–210] and the predominant isoform is type I.[211] The occurrence of 5α-reductase activity has also been reported in human temporal neocortex and subcortical white matter,[212] hippocampi from patients with chronic temporal lobe epilepsy,[213] and the brain of teleost fish,[214,215] and the distribution of 5α-reductase-like immunoreactivity has been recently described in the brain of the lungfish *Protopterus annectens*.[172] Type I mRNA is more abundant during development than in the adult, but its overall level of expression during development remains relatively constant.[91,216] Expression of 5α-reductase type II mRNA in the brain is, however, developmentally regulated.[90,91] RT-PCR analysis of embryonic and neonatal rat brain mRNAs demonstrated that whereas type I mRNA was constitutively expressed in the rat brain throughout late gestation, the neonatal period, and adulthood, type II mRNA was transiently expressed only at the end of gestation (E18) through the early postnatal period, dramatically falling by postnatal day 14. This pattern of expression of type II mRNA correlated with testosterone synthesis in the fetal testis and may also suggest that 5α-reductase type II may be involved in regulating brain sex differentiation at a critical period. In the adult rat, immunocytochemical staining detected 5α-reductase type I protein throughout the brain, with no differences between male and female rats.

The pattern of expression of 5α-reductase during embryogenesis was greatly regulated during development:[141,217] First observed at E12 in the spinal cord and neuroepithelial walls of the brain ventricular system, its distribution widened to highly proliferative zones of the CNS from E14 to E16. At E16.5, 5α-reductase mRNA and protein were detected in the cortex, thalamus, cerebellum, medulla, spinal cord, and peripheral ganglia. 5α-Reductase mRNA was found in lesser amounts in the trigeminal ganglia and inferior ganglia of the nodose-petrosal nerves (IX and X) than in the spiral ganglia.[141] However, 5α-reductase type I protein expression was found at similar levels in all peripheral ganglia (trigeminal, DRGs, nodose-petrosal, or spiral ganglia). In late embryogenesis (E18 to E20), 5α-reductase type I mRNA expression and protein were similar in distribution, but decreased in all areas except in the trigeminal ganglia, the ventricular zones of striatum and amygdala, and the thalamus. In the neonate, expression of 5α-reductase was higher in the PNS (trigeminal ganglia, DRGs) than in the CNS, and by 2 weeks of age, 5α-reductase mRNA expression changed completely and was evenly distributed throughout the brain. At this time, white matter structures (e.g., optic chiasma, lateral olfactory tract, internal capsule, corpus callosum) were highly positive for 5α-reductase mRNA, but other regions (hippocampus, striatal neuroepithelium) were also positive.[141] Our results demonstrate that 5α-reductase is expressed in the same location as its mRNA.[217] We and others[141] further established a transient expression of 5α-reductase in the proliferative zones of the developing CNS, suggesting that 5α-reduced steroid may have a role in neuronal proliferation.

Like P450scc, 5α-reductase is expressed in several cell types, including neurons in the hindbrain, cerebellum, spinal cord, and PNS and in a majority of glial cells in the other regions of the CNS. 5α-Reductase and P450scc are co-localized in the same cells in the developing neocortex. In other locations such as the neuroepithelium, spinal cord, and peripheral ganglia, where both P450scc and 5α-reductase type I were expressed, P450scc and 5α-reductase were not always found in the same cells. In the adult DRGs, 5α-reductase was also found in cells expressing P450c17 but not P450scc. The co-localization of P450c17 and 5α-reductase in the regions of the developing brain where they both are expressed remains to be tested. However, the expression of both enzymes in those regions suggests that steroids other than 5α DHP or allopregnanolone may be synthesized in regions such as the hindbrain, the hippocampus, and the spinal cord. Further detection of 17β HSD and P450aro in those same regions could argue for local androgen and estrogen production during embryogenesis in sexually dimorphic regions. In the lungfish brain, 5α-reductase- and 3β HSD-immunoreactive neurons are located in the same brain regions, including the lateral pallium, thalamic nuclei, periventricular preoptic nucleus, dorsal and ventral hypothalamic nuclei, mesencephalic tectum, periaqueductal gray, and nucleus cerebelli.[172]

Brain-expressed 5α-reductase type I may be crucial for regulation of estrogen synthesis during pregnancy. Recent studies created a transgenic mouse that had the 5α-reductase type I gene ablated.[216] About half of homozygous mutant mice died in midgestation between E10.75 and E11, due to estrogen excess. Administration of exogenous estrogen to normal wild-type mice also caused fetal death, and antagonists of estrogen action could prevent fetal demise in the homozygous knockout animals, demonstrating that estrogen, rather than a lack of 5α-reduced steroids, was causative of fetal death. Thus, 5α-reductase type I plays a major role in the fetus in inactivating testosterone, thereby reducing available substrate for estrogen synthesis. The results from the 5α-reductase type I knockout mouse also indicate that it is the inactivation of testosterone mediated by 5α-reductase *type I*, and not by 5α-reductase *type II*, that is crucial for fetal development and survival. 5α-Reductase type I mRNA is induced in the brain with pregnancy. Its increased expression in the brain may regulate estrogen levels in normal wild-type animals, perhaps through feedback mechanisms involving the hypothalamic pituitary gonadal axis.

1.3.12 Expression of 3α HSD in the Brain

Human 3α HSD type II, type II_{brain}, and type III mRNAs have been found in the brain, while human 3α HSD-I mRNA was found only in the liver.[101,103] In humans, both type II_{brain} and type III are found in various regions, but are found in different amounts. Most regions of the human brain express 3α HSD type II_{brain} and type III, assessed by Northern blots of polyA^+ mRNA, including the cerebellum, cortex, medulla, spinal cord, putamen, amygdala, caudate nucleus, corpus callosum, hippocampus, substantia nigra, thalamus, subthalamic nucleus, and occipital, frontal, and temporal lobes. Others have also found 3α HSD-II mRNA in the hippocampus and temporal lobe of patients with temporal lobe epilepsy.[213,218] Extensive biochemical studies demonstrated 3α HSD activity in various regions of the rodent brain, and its regulation by estrogens and lactation (reviewed in Karavolas and Hodges[219]). 3α HSD activity was found mainly in type I astrocytes, but was also detected in oligodendrocytes.[209] Western blots of proteins extracted from different rodent brain regions demonstrated that the olfactory bulb contained the greatest abundance of 3α HSD protein.[101]

1.3.13 Expression of HST in the Brain

The presence of abundant quantities of steroid ester sulfates in the brain of gonadectomized and adrenalectomized animals has been noticed since the early 1980s, when the concept of neurosteroid was proposed.[1,2] We and others have observed that several neurosteroids and their respective sulfate esters have different pharmacologic properties (reviewed in Majewska[114,115]) and different biological effects in the developing brain,[220] suggesting that both steroid sulfotransferase and steroid sulfohydrolase activities may play crucial roles in the modulation of the physiological effects of neurosteroids. However, very little data are available on the expression and enzymatic activity of steroid sulfotransferases in human or rodent fetal or adult brains. Recently, the distribution of HST-like immunoreactivity has been described in the brain of the frog *Rana ridibunda* using an antiserum against rat liver HST.[221] Two populations of HST-immunoreactive cell bodies were found in the diencephalon, namely, in the anterior preoptic area and in the dorsal part of the magnocellular preoptic nucleus. Several bundles of HST-positive nerve fibers were visualized in the telencephalon and diencephalon. The perikarya containing HST-like immunoreactivity were not labeled with antisera against glial fibrillary acidic protein or galactocerebrosides I and II, two selective markers of astrocytes and oligodendrocytes, respectively. In addition, the immunoreactive processes exhibited the varicose aspect of beaded nerve fibers. These observations clearly indicate that, in the frog brain, the HST-immunoreactive material is expressed exclusively in neurons.[221]

Despite the presence of a higher concentration of steroid ester sulfate than of free steroids in the brain, and despite the ubiquitous presence of PAPS, a co-substrate of the steroid sulfotransferase, there is no conclusive evidence for expression of HST in the brain of mammals. Several studies failed to show HST activity in the human fetal brain.[109] Immunocytochemistry did not reveal any expression of HST in fetal or adult human brains or in adult rat brains.[110,222] However, using human fetal brain slices, HST activity was detected.[223] Subsequently, others have characterized brain and liver HST activities from immature rats.[224] These data demonstrated that the brain HST activity (kinetics, pH, and specificity for substrate) was not comparable to the liver HST activity, suggesting that brain and liver sulfotransferases may be different isozymes. These authors found that the brain HST activity was relatively high in late embryogenesis and remained high until puberty, and then decreased in the adult, consistent with previous failures to detect HST activity in the adult rat brain. This ontogeny of HST activity in the brain is the opposite of the ontogeny in the liver, where HST activity is absent during embryogenesis and rises after birth.[224] Recently, HST activity has been detected in the brain of the frog *Rana ridibunda*, indicating that biosynthesis of sulfated neurosteroids in the CNS occurred early during evolution.[221]

1.3.14 Activity and Expression of STS in the Brain

Previous studies have demonstrated steroid sulfatase activity in the brain as well as in a variety of tissues other than its major site of expression in the placenta.[109] We demonstrated the sites of expression of STS mRNA using a mouse STS riboprobe.[139] The sites of STS expression provide more precise indications of the target tissues in which active, nonsulfated steroid hormones resulting from STS activity may act during embryogenesis. In addition to the known roles of STS in the placenta and in the skin, our results emphasized other possible roles for STS in the function of other organs during embryogenesis. In the CNS and PNS, the expression of m STS mRNA may be related to the expression of steroidogenic enzymes involved in neurosteroid synthesis. We previously showed the expression of P450c17 in restricted areas of the developing nervous system,[137] particularly in the neocortex, and demonstrated a role for both DHEA and DHEAS in neocortical neuron differentiation. While DHEA promoted axonal growth, DHEAS promoted dendritic growth and cell clustering.[220] The expression of STS in the neocortex, hippocampus, and thalamus may regulate the ratio of DHEA:DHEAS. Furthermore, as the thalamus has been shown to influence the survival of cortical subplate,[225] where we detected P450c17-expressing neurons, the ratio of these two steroids may play crucial roles in the controlling axonal vs. dendritic growth of cortico-thalamic fibers during neocortex organization. The afferent and efferent connections of the cerebral neocortex develop simultaneously toward the end of embryogenesis.

1.4 REGULATION OF NEUROSTEROIDOGENIC ENZYMES

1.4.1 Regulation of 3α HSD

Changes in neurosteroid concentrations in the brain and in plasma have been associated with the menstrual cycle in women.[226–228] In addition, changes in allopregnanolone concentrations, but not in progesterone concentrations, have been suggested to play a role in premenstrual dysphoric disorder (PDD).[229–236] One successful treatment for women with PDD uses either continuous or intermittent selective serotonin reuptake inhibitor (SSRI) therapy (reviewed in Dimmock et al.[237]). The mechanism by which SSRIs are effective in treating the symptoms of PDD is unknown, but has been thought to be due solely to modulation of the serotonergic system. However, the cyclic nature of the disease — and the fact that intermittent dosing of SSRIs suggested to others that perhaps ovarian steroids or neurosteroids are metabolites of the ovarian steroid progesterone — may be involved in the etiology of PDD. This hypothesis was tested in laboratory animals.[238] In that study, the authors found that fluoxetine could increase the abundance of allopregnanolone in rat brains without substantially affecting the brain concentrations of either progesterone or 5α-dihydroprogesterone. Furthermore, this effect was rapid, occurring within 15 min. These data suggested that fluoxetine may affect 3α HSD activity, resulting in increased conversion of 5α-dihydroprogesterone to allopregnanolone.

We tested that hypothesis using cloned cDNAs encoding rat 3α HSD and human type III 3α HSD, the enzyme that can convert 5α-dihydroprogesterone to allopregnanolone.[107] Our data indicated that three SSRIs, Prozac, Paxil, and Zoloft, all increased allopregnanolone production by increasing the efficiency of conversion of 5α-dihydroprogesterone to allopregnanolone. As 3α HSD catalyzes a reversible reaction, we also tested the hypothesis that these SSRIs increased allopregnanolone concentrations by inhibiting the oxidation of allopregnanolone to 5α-dihydroprogesterone. Our data suggested that Prozac, but not Paxil or Zoloft, increased the oxidation of allopregnanolone as well, but that the overall effect (reductive and oxidative reactions) was to increase allopregnanolone.

Unlike human 3α HSD type III, the type II enzyme does not convert 5α-dihydroprogesterone to allopregnanolone. However, the type II enzyme has considerable 20α HSD activity and converts progesterone to 20α-hydroprogesterone. Using progesterone as a substrate for 3α HSD type II, we found that Prozac, but not Paxil, decreased the 20α HSD activity of 3α HSD type II, and hence inhibits the conversion of progesterone to 20α-hydroprogesterone.

Thus, these three SSRIs tested increase allopregnanolone production through increased efficiency of conversion of 5α-dihydroprogesterone to allopregnanolone. Prozac may also have additional effects through inhibition of the competing 20α-reduction pathway.

Recent studies by another group have used the pure SSRI in similar types of enzymatic assays.[239] Those authors found no effects of fluoxetine, paroxetine, or sertraline on human 3α HSD activity, using various concentrations of the SSRIs. Differences between the two sets of results could be due to the fact that Trauger et

TABLE 1.3
Inhibitors of Human 3α (20α) HSD Isozymes[104,105]

	IC_{50} μ*M*		
Drug	**Human Type 2 3α HSD AKR1C3**	**Human Type 3 3α HSD AKR1C2**	**Human 20α HSD AKR1C1**
Diazepam	84	2.4	5.6
Medazepam	116	19	4.7
Estazolam	47	8.8	6.6
Flunitrazepam	58	13	15
Nitrazepam	69	59	21
Cloxazolam	2.5	(10% inhibition with 10 μ*M*)	(38% inhibition with 10 μ*M*)
Indomethacin	4.1	75	130
Tetrabromo-phenolphthalein	2.3	2.4	0.33
Benzobromarone	5.5	0.36	0.048

al.[239] used the pure active ingredient of the SSRI medicine, while we extracted the medicine with organic solvents and used the material extracted by those means. This may suggest either that another component of the different SSRI medicines may affect 3α HSD activity or that there were other differences in the assays.

1.4.2 Inhibition of 3α HSD by Benzodiazepines

Recent studies[104,105] have examined the effects of benzodiazepines on the activities of the various 3α HSDs (ARK1C1-4) (Table 1.3). In general, several benzodiazepines inhibit the 3α HSD isoforms to varying extents and have IC_{50} values in the micromolar range. Type IV (20α HSD) and type III differ by seven amino acids,[102,240] and several benzodiazepines inhibited these enzymes similarly. Two exceptions were medazepam and nitrazepam, in which the IC_{50} for type IV was three- to fourfold lower than that for type III. The data suggest that these benzodiazepines would be beneficial for inhibiting 20α HSD activity, and hence inhibiting the conversion of potent neuroactive steroids to less potent neurosteroids. Inhibition of type I (liver) and type II by benzodiazepines was different than the inhibition of the type III and IV enzymes; in general, the benzodiazepines had higher IC_{50}s for the type I and II enzymes, and type II was specifically potently inhibited by cloxazolam. Since the type II enzyme uses DHT, and not 5α DHP or 5α DH-DOC, as substrate, these data would suggest that cloxazolam would inhibit the synthesis of the neurosteroid androstanediol, but not the synthesis of allopregnanolone or 3α5α TH-DOC.

Inhibitions of 3α HSD isoforms by the benzodiazepines listed in Table 1.3 were noncompetitive with respect to the substrate. However, some of the inhibitions were uncompetitive with respect to the coenzyme. Such patterns of inhibition suggest that the benzodiazepines can bind to the enzyme–coenzyme complex and to the enzyme–coenzyme–substrate complex. In addition, some benzodiazepines may interact with the enzyme in such a fashion that they induce a structural change in the enzyme, and thereby inhibit the enzyme's activity.

1.4.3 Regulation of 3βHSD by GABA

The roles of classical neurotransmitters and neuropeptides in the regulation of 3β HSD activity have been extensively investigated in the frog brain (reviewed in Mensah-Nyagan et al.[241]). Immunohistochemical labeling of hypothalamic slices with antisera against various subunits of the $GABA_A$ receptor revealed that most 3β HSD-containing neurons (assessed by immunocytochemistry using an antiserum against human placental type I 3β HSD) also exhibited $GABA_A$ receptor α_3- and β_2/β_3-like immunoreactivities,[242] indicating that GABA may regulate 3β HSD activity in the frog central nervous system. Frog hypothalamic explants incubated with increased concentrations of GABA induced a dose-dependent inhibition of the conversion of [^{3}H]pregnenolone into radioactive metabolites, including progesterone and 17-hydroxyprogesterone. This effect of GABA on neurosteroid biosynthesis was mimicked by the $GABA_A$ receptor agonist muscimol and inhibited by the $GABA_A$ receptor antagonists bicuculline and SR95531. By contrast, the selective $GABA_B$ receptor agonist baclofen did not affect the formation of steroids in hypothalamic explants,[242] demonstrating that GABA, acting through $GABA_A$ receptors, inhibits the activity of 3β HSD in the frog brain. These data suggest the existence of an ultrashort regulatory loop by which neurosteroids may regulate their own biosynthesis through modulation of $GABA_A$ receptor activity at the level of 3β HSD neurons.

1.4.4 Regulation of Allopregnanolone Production by Ethanol

There is both behavior and neurochemical data that suggest that alcohol exerts pharmacological effects that are similar to those that modulate $GABA_A$ receptors.[243,244] This would suggest that alcohol directly stimulates $GABA_A$ receptors or that alcohol somehow enhances effects at those receptors. Studies by various laboratories concerning alcohol's effects at $GABA_A$ receptors have yielded conflicting answers. Hence, an approach others have taken was to determine if alcohol could enhance the concentration of $GABA_A$ receptor modulators, such as neurosteroids[245] (see Chapter 10). VanDoren et al.[245] found that systemic alcohol administration to rats elevated both cortical and plasma concentrations of allopregnanolone in a time- and dose-dependent fashion. Use of the 5α-reductase inhibitor finasteride prevented the increase in allopregnanolone concentrations and also prevented some of the behavioral effects of alcohol. Thus, increased synthesis of allopregnanolone was most likely responsible for many of the behavioral effects of alcohol. However, it is not clear where in the neurosteroidogenic cascade alcohol exerts its effects. Use of specific enzyme inhibitors suggested that the effect was downstream of 5α-reductase since finasteride reduced alcohol-stimulated increases in cortical concentrations of allopregnanolone, whereas trilostane, an inhibitor of 3β HSD and hence an inhibitor of progesterone production, had no effect. Thus, the effect of alcohol would appear to be either at the level of 5α-reductase or at 3α HSD. The rapid time course of the effect suggests that alcohol directly affects neurosteroidogenic enzyme activity in some fashion, rather than through increases in gene transcription. The mechanism by which enzyme activity could be affected could be through increased protein translation, posttranslational modification of proteins, or some allosteric mechanism. At this time, it is still unclear which enzyme activity(s) is affected by alcohol.

REFERENCES

1. Corpechot, C., Robel, P., Axelson, M., Sjovall, J., and Baulieu, E.E., Characterization and measurement of dehydroepiandrosterone sulfate in rat brain, *Proc. Natl. Acad. Sci. U.S.A.*, 78, 4704–4707, 1981.
2. Corpechot, C., Synguelakis, M., Talha, S., Axelson, M., Sjovall, J., Vihko, R., Baulieu, E.E., and Robel, P., Pregnenolone and its sulfate ester in the rat brain, *Brain Res.*, 270, 119–125, 1983.
3. Miller, W.L., Molecular biology of steroid hormone synthesis, *Endocr. Rev.*, 9, 295–318, 1988.
4. Penning, T.M., Molecular endocrinology of hydroxysteroid dehydrogenases, *Endocr. Rev.*, 18, 281–305, 1997.
5. Pang, S., Yang, X., Wang, M., Tissot, R., Nino, M., Manaligod, J., Bullock, L.P., and Mason, J.I., Inherited congenital adrenal hyperplasia in the rabbit: absent cholesterol side-chain cleavage cytochrome P450 gene expression, *Endocrinology*, 131, 181–186, 1992.
6. Yang, X., Iwamoto, K., Wang, M., Artwohl, J., Mason, J.I., and Pang, S., Inherited congenital adrenal hyperplasia in the rabbit is caused by a deletion in the gene encoding cytochrome P450 cholesterol side-chain cleavage enzyme, *Endocrinology*, 132, 1977–1982, 1993.
7. Mellon, S.H. and Deschepper, C.F., Neurosteroid biosynthesis: genes for adrenal steroidogenic enzymes are expressed in the brain, *Brain Res.*, 629, 283–292, 1993.
8. Mellon, S.H., Neurosteroids: biochemistry, modes of action, and clinical relevance, *J. Clin. Endocrinol. Metab.*, 78, 1003–1008, 1994.
9. Tajima, T., Fujieda, K., Kouda, N., Nakae, J., and Miller, W.L., Heterozygous mutation in the cholesterol side chain cleavage enzyme (p450scc) gene in a patient with 46,XY sex reversal and adrenal insufficiency, *J. Clin. Endocrinol. Metab.*, 86, 3820–3825, 2001.
10. Matteson, K.J., Chung, B.C., Urdea, M.S., and Miller, W.L., Study of cholesterol side-chain cleavage (20,22 desmolase) deficiency causing congenital lipoid adrenal hyperplasia using bovine-sequence P450scc oligodeoxyribonucleotide probes, *Endocrinology*, 118, 1296–1305, 1986.
11. Morohashi, K., Sogawa, K., Omura, T., and Fujii-Kuriyama, Y., Gene structure of human cytochrome P-450(SCC), cholesterol desmolase, *J. Biochem. (Tokyo)*, 101, 879–887, 1987.
12. Oonk, R.B., Parker, K.L., Gibson, J.L., and Richards, J.S., Rat cholesterol side-chain cleavage cytochrome P-450 (P-450scc) gene: structure and regulation by cAMP in vitro, *J. Biol. Chem.*, 265, 22392–22401, 1990.

13. Chung, B.C., Matteson, K.J., Voutilainen, R., Mohandas, T.K., and Miller, W.L., Human cholesterol side-chain cleavage enzyme, P450scc: cDNA cloning, assignment of the gene to chromosome 15, and expression in the placenta, *Proc. Natl. Acad. Sci. U.S.A.*, 83, 8962–8966, 1986.
14. Omura, T., Sanders, S., Estabrook, R.W., Cooper, D.Y., and Rosenthal, O., Isolation from adrenal cortex of a non-heme iron protein and a flavoprotein functional as a reduced triphosphopyridine mucleotide-cytochrome P-450 reductase, *Arch. Biochem. Biophys.*, 117, 660, 1966.
15. Nakamura, Y., Otsuka, H., and Tamaoki, B., Requirement of a new flavoprotein and a non-heme iron-containing protein in the steroid 11β- and 18-hydroxylase system, *Biochim. Biophys. Acta*, 122, 34–42, 1966.
16. Kimura, T. and Suzuki, K., Components of the electron transport system in adrenal steroid hydroxylase, *J. Biol. Chem.*, 242, 485–491, 1967.
17. Coghlan, V.M. and Vickery, L.E., Expression of human ferredoxin and assembly of the [2Fe-2S] center in Escherichia coli, *Proc. Natl. Acad. Sci. U.S.A.*, 86, 835–839, 1989.
18. Stocco, D.M. and Clark, B.J., Regulation of the acute production of steroids in steroidogenic cells, *Endocr. Rev.*, 17, 221–244, 1996.
19. Clark, B.J., Wells, J., King, S.R., and Stocco, D.M., The purification, cloning, and expression of a novel luteinizing hormone-induced mitochondrial protein in MA-10 mouse Leydig tumor cells: characterization of the steroidogenic acute regulatory protein (StAR), *J. Biol. Chem.*, 269, 28314–28322, 1994.
20. Lin, D., Sugawara, T., Strauss, J.F.I., Clark, B.J., Stocco, D.M., Saenger, P., Rogol, A., and Miller, W.L., Role of steroidogenic acute regulatory protein in adrenal and gonadal steroidogenesis, *Science*, 267, 1828–1831, 1995.
21. Lin, D., Black, S.M., Nagahama, Y., and Miller, W.L., Steroid 17 alpha-hydroxylase and 17,20-lyase activities of P450c17: contributions of serine106 and P450 reductase, *Endocrinology*, 132, 2498–2506, 1993.
22. Bose, H.S., Sugawara, T., Strauss, J.F.R., and Miller, W.L., The pathophysiology and genetics of congenital lipoid adrenal hyperplasia. International Congenital Lipoid Adrenal Hyperplasia Consortium, *N. Engl. J. Med.*, 335, 1870–1878, 1996.
23. Papadopoulos, V., Peripheral-type benzodiazepine/diazepam binding inhibitor receptor: biological role in steroidogenic cell function, *Endocr. Rev.*, 14, 222–240, 1993.
24. Saenger, P., Klonari, Z., Black, S.M., Compagnone, N., Mellon, S.H., Fleischer, A., Abrams, C.A., Shackelton, C.H., and Miller, W.L., Prenatal diagnosis of congenital lipoid adrenal hyperplasia, *J. Clin. Endocrinol. Metab.*, 80, 200–205, 1995.
25. Furukawa, A., Miyatake, A., Ohnishi, T., and Ichikawa, Y., Steroidogenic acute regulatory protein (StAR) transcripts constitutively expressed in the adult rat central nervous system: colocalization of StAR, cytochrome P-450SCC (CYP XIA1), and 3beta-hydroxysteroid dehydrogenase in the rat brain, *J. Neurochem.*, 71, 2231–2238, 1998.
26. King, S.R., Manna, P.R., Ishii, T., Syapin, P.J., Ginsberg, S.D., Wilson, K., Walsh, L.P., Parker, K.L., Stocco, D.M., Smith, R.G., and Lamb, D.J., An essential component in steroid synthesis, the steroidogenic acute regulatory protein, is expressed in discrete regions of the brain, *J. Neurosci.*, 22, 10613–10620, 2002.
27. Luu-The, V., Lachance, Y., Labrie, C., Leblanc, G., Thomas, J.L., Strickler, R.C., and Labrie, F., Full length cDNA structure and deduced amino acid sequence of human 3 beta-hydroxy-5-ene steroid dehydrogenase, *Mol. Endocrinol.*, 3, 1310–1312, 1989.

28. Thomas, J.L., Myers, R.P., and Strickler, R.C., Human placental 3 beta-hydroxy-5-ene-steroid dehydrogenase and steroid 5— —4-ene-isomerase: purification from mitochondria and kinetic profiles, biophysical characterization of the purified mitochondrial and microsomal enzymes, *J. Steroid Biochem.*, 33, 209–217, 1989.
29. Lorence, M.C., Murry, B.A., Trant, J.M., and Mason, J.I., Human 3 beta-hydroxysteroid dehydrogenase/delta 5— —4isomerase from placenta: expression in nonsteroidogenic cells of a protein that catalyzes the dehydrogenation/isomerization of C21 and C19 steroids, *Endocrinology*, 126, 2493–2498, 1990.
30. Voutilainen, R., Ilvesmaki, V., and Miettinen, P.J., Low expression of 3 beta-hydroxy-5-ene steroid dehydrogenase gene in human fetal adrenals in vivo; adrenocorticotropin and protein kinase C-dependent regulation in adrenocortical cultures, *J. Clin. Endocrinol. Metab.*, 72, 761–767, 1991.
31. Endoh, A., Kristiansen, S.B., Casson, P.R., Buster, J.E., and Hornsby, P.J., The zona reticularis is the site of biosynthesis of dehydroepiandrosterone and dehydroepiandrosterone sulfate in the adult human adrenal cortex resulting from its low expression of 3 beta-hydroxysteroid dehydrogenase, *J. Clin. Endocrinol. Metab.*, 81, 3558–3565, 1996.
32. Coulter, C.L., Goldsmith, P.C., Mesiano, S., Voytek, C.C., Martin, M.C., Mason, J.I., and Jaffe, R.B., Functional maturation of the primate fetal adrenal in vivo. II. Ontogeny of corticosteroid synthesis is dependent upon specific zonal expression of 3 beta-hydroxysteroid dehydrogenase/isomerase, *Endocrinology*, 137, 4953–4959, 1996.
33. Yanase, T., Sasano, H., Yubisui, T., Sakai, Y., Takayanagi, R., and Nawata, H., Immunohistochemical study of cytochrome b5 in human adrenal gland and in adrenocortical adenomas from patients with Cushing's syndrome, *Endocr. J.*, 45, 89–95, 1998.
34. Zhang, L.H., Rodriguez, H., Ohno, S., and Miller, W.L., Serine phosphorylation of human P450c17 increases 17,20-lyase activity: implications for adrenarche and the polycystic ovary syndrome, *Proc. Natl. Acad. Sci. U.S.A.*, 92, 10619–10623, 1995.
35. Pandey, A.V., Mellon, S.H., and Miller, W.L., Protein phosphatase 2A and phosphoprotein SET regulate androgen production by P450c17, *J. Biol. Chem.*, 278, 2837–2844, 2003.
36. Matteson, K.J., Picado-Leonard, J., Chung, B.C., Mohandas, T.K., and Miller, W.L., Assignment of the gene for adrenal P450c17 (steroid 17 alpha-hydroxylase/17,20 lyase) to human chromosome 10, *J. Clin. Endocrinol. Metab.*, 63, 789–791, 1986.
37. Casey, M.L. and MacDonald, P.C., Extraadrenal formation of a mineralocorticosteroid: deoxycorticosterone and deoxycorticosterone sulfate biosynthesis and metabolism, *Endocr. Rev.*, 3, 396–403, 1982.
38. Casey, M.L., Winkel, C.A., and MacDonald, P.C., Conversion of progesterone to deoxycorticosterone in the human fetus: steroid 21-hydroxylase activity in fetal tissues, *J. Steroid Biochem.*, 18, 449–452, 1983.
39. Mellon, S.H. and Miller, W.L., Extraadrenal steroid 21-hydroxylation is not mediated by P450c21, *J. Clin. Invest.*, 84, 1497–1502, 1989.
40. Beyenburg, S., Watzka, M., Clusmann, H., Blumcke, I., Bidlingmaier, F., Elger, C.E., and Stoffel-Wagner, B., Messenger RNA of steroid 21-hydroxylase (CYP21) is expressed in the human hippocampus, *Neurosci. Lett.*, 308, 111–114, 2001.
41. Yu, L., Romero, D.G., Gomez-Sanchez, C.E., and Gomez-Sanchez, E.P., Steroidogenic enzyme gene expression in the human brain, *Mol. Cell. Endocrinol.*, 190, 9–17, 2002.
42. Zhou, Z., Agarwal, V.R., Dixit, N., White, P., and Speiser, P.W., Steroid 21-hydroxylase expression and activity in human lymphocytes, *Mol. Cell. Endocrinol.*, 127, 11–18, 1997.

43. Hiroi, T., Kishimoto, W., Chow, T., Imaoka, S., Igarashi, T., and Funae, Y., Progesterone oxidation by cytochrome P450 2D isoforms in the brain, *Endocrinology*, 142, 3901–3908, 2001.
44. Yanagibashi, K. and Hall, P.F., Role of electron transport in the regulation of the lyase activity of C21 side-chain cleavage P-450 from porcine adrenal and testicular microsomes, *J. Biol. Chem.*, 261, 8429–8433, 1986.
45. Yamano, S., Aoyama, T., McBride, O.W., Hardwick, J.P., Gelboin, H.V., and Gonzalez, F.J., Human NADPH-P450 oxidoreductase: complementary DNA cloning, sequence and vaccinia virus-mediated expression and localization of the CYPOR gene to chromosome 7, *Mol. Pharmacol.*, 36, 83–88, 1989.
46. Onoda, M. and Hall, P.F., Cytochrome b5 stimulates purified testicular microsomal cytochrome P-450 (C21 side-chain cleavage), *Biochem. Biophys. Res. Commun.*, 108, 454–460, 1982.
47. Kominami, S., Ogawa, N., Morimune, R., De-Ying, H., and Takemori, S., The role of cytochrome b5 in adrenal microsomal steroidogenesis, *J. Steroid Biochem. Mol. Biol.*, 42, 57–64, 1992.
48. Auchus, R.J., Lee, T.C., and Miller, W.L., Cytochrome b5 augments the 17,20-lyase activity of human P450c17 without direct electron transfer, *J. Biol. Chem.*, 273, 3158–3165, 1998.
49. Mapes, S., Corbin, C.J., Tarantal, A., and Conley, A., The primate adrenal zona reticularis is defined by expression of cytochrome b5, 17alpha-hydroxylase/17,20-lyase cytochrome P450 (P450c17) and NADPH-cytochrome P450 reductase (reductase) but not 3beta-hydroxysteroid dehydrogenase/delta5-4 isomerase (3beta-HSD), *J. Clin. Endocrinol. Metab.*, 84, 3382–3385, 1999.
50. Mapes, S., Tarantal, A.F., Parker, C.R., Moran, F.M., Bahr, J.M., Pyter, L., and Conley, A.J., Adrenocortical cytochrome b5 expression during fetal development of the rhesus macaque, *Endocrinology*, 143, 1451–1458, 2002.
51. Mellon, S.H., Bair, S.R., and Monis, H., P450c11B3 mRNA, transcribed from a third P450c11 gene, is expressed in a tissue-specific, developmentally, and hormonally regulated fashion in the rodent adrenal and encodes a protein with both 11-hydroxylase and 18-hydroxylase activities, *J. Biol. Chem.*, 270, 1643–1649, 1995.
52. Zhang, G. and Miller, W.L., The human genome contains only two CYP11B (P450c11) genes, *J. Clin. Endocrinol. Metab.*, 81, 3254–3256, 1996.
53. Thompson, E.A. and Siiteri, P.K., Studies on the aromatization of C-19 androgens, *Ann. N.Y. Acad. Sci.*, 212, 378–391, 1973.
54. Thompson, E.A., Jr. and Siiteri, P.K., The involvement of human placental microsomal cytochrome P-450 in aromatization, *J. Biol. Chem.*, 249, 5373–5378, 1974.
55. Means, G.D., Mahendroo, M.S., Corbin, C.J., Mathis, J.M., Powell, F.E., Mendelson, C.R., and Simpson, E.R., Structural analysis of the gene encoding human aromatase cytochrome P-450, the enzyme responsible for estrogen biosynthesis, *J. Biol. Chem.*, 264, 19385–19391, 1989.
56. Chen, S.A., Besman, M.J., Sparkes, R.S., Zollman, S., Klisak, I., Mohandas, T., Hall, P.F., and Shively, J.E., Human aromatase: cDNA cloning, Southern blot analysis, and assignment of the gene to chromosome 15, *DNA*, 7, 27–38, 1988.
57. Mahendroo, M.S., Means, G.D., Mendelson, C.R., and Simpson, E.R., Tissue-specific expression of human P-450AROM: the promoter responsible for expression in adipose tissue is different from that utilized in placenta, *J. Biol. Chem.*, 266, 11276–11281, 1991.

58. Labrie, F., Luu-The, V., Lin, S.X., Labrie, C., Simard, J., Breton, R., and Belanger, A., The key role of 17 beta-hydroxysteroid dehydrogenases in sex steroid biology, *Steroids*, 62, 148–158, 1997.
59. Luu-The, V., Labrie, C., Zhao, H.F., Couet, J., Lachance, Y., Simard, J., Leblanc, G., Cote, J., Berube, D., Gagne, R., et al., Characterization of cDNAs for human estradiol 17 beta-dehydrogenase and assignment of the gene to chromosome 17: evidence of two mRNA species with distinct 5'-termini in human placenta, *Mol. Endocrinol.*, 3, 1301–1309, 1989.
60. Peltoketo, H., Isomaa, V., Maentausta, O., and Vihko, R., Complete amino acid sequence of human placental 17 beta-hydroxysteroid dehydrogenase deduced from cDNA, *FEBS Lett.*, 239, 73–77, 1988.
61. Tremblay, Y., Ringler, G.E., Morel, Y., Mohandas, T.K., Labrie, F., Strauss, J.F., III, and Miller, W.L., Regulation of the gene for estrogenic 17-ketosteroid reductase lying on chromosome 17cen——q25, *J. Biol. Chem.*, 264, 20458–20462, 1989.
62. Carstensen, J.F., Tesdorpf, J.G., Kaufmann, M., Markus, M.M., Husen, B., Leenders, F., Jakob, F., de Launoit, Y., and Adamski, J., Characterization of 17 beta-hydroxysteroid dehydrogenase IV, *J. Endocrinol.*, 150 (Suppl.), S3–S12, 1996.
63. Adamski, J., Carstensen, J., Husen, B., Kaufmann, M., de Launoit, Y., Leenders, F., Markus, M., and Jungblut, P.W., New 17 beta-hydroxysteroid dehydrogenases: molecular and cell biology of the type IV porcine and human enzymes, *Ann. N.Y. Acad. Sci.*, 784, 124–136, 1996.
64. Leenders, F., Adamski, J., Husen, B., Thole, H.H., and Jungblut, P.W., Molecular cloning and amino acid sequence of the porcine 17 beta-estradiol dehydrogenase, *Eur. J. Biochem.*, 222, 221–227, 1994.
65. Ghosh, D., Pletnev, V.Z., Zhu, D.W., Wawrzak, Z., Duax, W.L., Pangborn, W., Labrie, F., and Lin, S.X., Structure of human estrogenic 17 beta-hydroxysteroid dehydrogenase at 2.20 A resolution, *Structure*, 3, 503–513, 1995.
66. Sawicki, M.W., Erman, M., Puranen, T., Vihko, P., and Ghosh, D., Structure of the ternary complex of human 17beta-hydroxysteroid dehydrogenase type 1 with 3-hydroxyestra-1,3,5,7-tetraen-17-one (equilin) and NADP+, *Proc. Natl. Acad. Sci. U.S.A.*, 96, 840–845, 1999.
67. Geissler, W.M., Davis, D.L., Wu, L., Bradshaw, K.D., Patel, S., Mendonca, B.B., Elliston, K.O., Wilson, J.D., Russell, D.W., and Andersson, S., Male pseudohermaphroditism caused by mutations of testicular 17 beta-hydroxysteroid dehydrogenase 3, *Nat. Genet.*, 7, 34–39, 1994 (see Comments).
68. Leenders, F., Tesdorpf, J.G., Markus, M., Engel, T., Seedorf, U., and Adamski, J., Porcine 80-kDa protein reveals intrinsic 17 beta-hydroxysteroid dehydrogenase, fatty acyl-CoA-hydratase/dehydrogenase, and sterol transfer activities, *J. Biol. Chem.*, 271, 5438–5442, 1996.
69. Dufort, I., Rheault, P., Huang, X.F., Soucy, P., and Luu-The, V., Characteristics of a highly labile human type 5 17beta-hydroxysteroid dehydrogenase, *Endocrinology*, 140, 568–574, 1999.
70. Bujalska, I., Shimojo, M., Howie, A., and Stewart, P.M., Human 11 beta-hydroxysteroid dehydrogenase: studies on the stably transfected isoforms and localization of the type 2 isozyme within renal tissue, *Steroids*, 62, 77–82, 1997.
71. Tannin, G.M., Agarwal, A.K., Monder, C., New, M.I., and White, P.C., The human gene for 11 beta-hydroxysteroid dehydrogenase: structure, tissue distribution, and chromosomal localization, *J. Biol. Chem.*, 266, 16653–16658, 1991.

72. Moore, C.C., Mellon, S.H., Murai, J., Siiteri, P.K., and Miller, W.L., Structure and function of the hepatic form of 11 beta-hydroxysteroid dehydrogenase in the squirrel monkey, an animal model of glucocorticoid resistance, *Endocrinology*, 133, 368–375, 1993.
73. Stewart, P.M., Murry, B.A., and Mason, J.I., Human kidney 11 beta-hydroxysteroid dehydrogenase is a high affinity nicotinamide adenine dinucleotide-dependent enzyme and differs from the cloned type I isoform, *J. Clin. Endocrinol. Metab.*, 79, 480–484, 1994.
74. Tannin, G.M., Agarwal, A.K., Monder, C., New, M.I., and White, P.C., The human gene for 11 beta-hydroxysteroid dehydrogenase: structure, tissue distribution, and chromosomal localization, *J. Biol. Chem.*, 266, 16653–16658, 1991.
75. Whorwood, C.B., Mason, J.I., Ricketts, M.L., Howie, A.J., and Stewart, P.M., Detection of human 11 beta-hydroxysteroid dehydrogenase isoforms using reverse-transcriptase-polymerase chain reaction and localization of the type 2 isoform to renal collecting ducts, *Mol. Cell. Endocrinol.*, 110, R7–R12, 1995.
76. Brown, R.W., Chapman, K.E., Edwards, C.R., and Seckl, J.R., Human placental 11 beta-hydroxysteroid dehydrogenase: evidence for and partial purification of a distinct NAD-dependent isoform, *Endocrinology*, 132, 2614–2621, 1993.
77. Brown, R.W., Diaz, R., Robson, A.C., Kotelevtsev, Y.V., Mullins, J.J., Kaufman, M.H., and Seckl, J.R., The ontogeny of 11 beta-hydroxysteroid dehydrogenase type 2 and mineralocorticoid receptor gene expression reveal intricate control of glucocorticoid action in development, *Endocrinology*, 137, 794–797, 1996.
78. Albiston, A.L., Obeyesekere, V.R., Smith, R.E., and Krozowski, Z.S., Cloning and tissue distribution of the human 11 beta-hydroxysteroid dehydrogenase type 2 enzyme, *Mol. Cell. Endocrinol.*, 105, R11–R17, 1994.
79. Moore, C.C.D., Mellon, S.H., Murai, J., Siiteri, P.K., and Miller, W.L., Structure and function of the hepatic form of 11 beta-hydroxysteroid dehydrogenase in the squirrel monkey, an animal model of glucocorticoid resistance, *Endocrinology*, 133, 368–375, 1993.
80. Rusvai, E. and Naray-Fejes-Toth, A., A new isoform of 11 beta-hydroxysteroid dehydrogenase in aldosterone target cells, *J. Biol. Chem.*, 268, 10717–10720, 1993.
81. Agarwal, A.K., Rogerson, F.M., Mune, T., and White, P.C., Gene structure and chromosomal localization of the human HSD11K gene encoding the kidney (type 2) isozyme of 11 beta-hydroxysteroid dehydrogenase, *Genomics*, 29, 195–199, 1995.
82. Andersson, S. and Russell, D.W., Structural and biochemical properties of cloned and expressed human and rat steroid 5 alpha-reductases, *Proc. Natl. Acad. Sci. U.S.A.*, 87, 3640–3644, 1990.
83. Andersson, S., Berman, D.M., Jenkins, E.P., and Russell, D.W., Deletion of steroid 5 alpha-reductase 2 gene in male pseudohermaphroditism, *Nature*, 354, 159–161, 1991.
84. Jenkins, E.P., Andersson, S., Imperato-McGinley, J., Wilson, J.D., and Russell, D.W., Genetic and pharmacological evidence for more than one human steroid 5 alpha-reductase, *J. Clin. Invest.*, 89, 293–300, 1992.
85. Thigpen, A.E., Davis, D.L., Gautier, T., Imperato-McGinley, J., and Russell, D.W., Brief report: the molecular basis of steroid 5 alpha-reductase deficiency in a large Dominican kindred, *N. Engl. J. Med.*, 327, 1216–1219, 1992.
86. Russell, D.W. and Wilson, J.D., Steroid 5 alpha-reductase: two genes/two enzymes, *Annu. Rev. Biochem.*, 63, 25–61, 1994.

87. Lephart, E., A review of brain 5α–reductase: cellular, enzymatic, and molecular perspectives and implications for biological function, *Mol. Cell. Neurosci.*, 4, 473–484, 1993.
88. Thigpen, A.E., Davis, D.L., Milatovich, A., Mendonca, B.B., Imperato-McGinley, J., Griffin, J.E., Francke, U., Wilson, J.D., and Russell, D.W., Molecular genetics of steroid 5 alpha-reductase 2 deficiency, *J. Clin. Invest.*, 90, 799–809, 1992.
89. Thigpen, A.E., Silver, R.I., Guileyardo, J.M., Casey, M.L., McConnell, J.D., and Russell, D.W., Tissue distribution and ontogeny of steroid 5 alpha-reductase isozyme expression, *J. Clin. Invest.*, 92, 903–910, 1993.
90. Poletti, A., Coscarella, A., Negri-Cesi, P., Colciago, A., Celotti, F., and Martini, L., 5 Alpha-reductase isozymes in the central nervous system, *Steroids*, 63, 246–251, 1998.
91. Poletti, A., Negri-Cesi, P., Rabuffetti, M., Colciago, A., Celotti, F., and Martini, L., Transient expression of the 5alpha-reductase type 2 isozyme in the rat brain in late fetal and early postnatal life, *Endocrinology*, 139, 2171–2178, 1998.
92. Penning, T.M., Bennett, M.J., Smith-Hoog, S., Schlegel, B.P., Jez, J.M., and Lewis, M., Structure and function of 3 alpha-hydroxysteroid dehydrogenase, *Steroids*, 62, 101–111, 1997.
93. Hyndman, D., Bauman, D.R., Heredia, V.V., and Penning, T.M., The aldo-keto reductase superfamily homepage, *Chem. Biol. Interact.*, 143/144, 621–631, 2003.
94. Pawlowski, J., Huizinga, M., and Penning, T.M., Isolation and partial characterization of a full-length cDNA clone for 3 alpha-hydroxysteroid dehydrogenase: a potential target enzyme for nonsteroidal anti-inflammatory drugs, *Agents Actions*, 34, 289–293, 1991.
95. Cheng, K.C., White, P.C., and Qin, K.N., Molecular cloning and expression of rat liver 3 alpha-hydroxysteroid dehydrogenase, *Mol. Endocrinol.*, 5, 823–828, 1991.
96. Stolz, A., Rahimi-Kiani, M., Ameis, D., Chan, E., Ronk, M., and Shively, J.E., Molecular structure of rat hepatic 3 alpha-hydroxysteroid dehydrogenase: a member of the oxidoreductase gene family, *J. Biol. Chem.*, 266, 15253–15257, 1991.
97. Usui, E., Okuda, K., Kato, Y., and Noshiro, M., Rat hepatic 3 alpha-hydroxysteroid dehydrogenase: expression of cDNA and physiological function in bile acid biosynthetic pathway, *J. Biochem. (Tokyo)*, 115, 230–237, 1994.
98. Penning, T.M., Burczynski, M.E., Jez, J.M., Hung, C.F., Lin, H.K., Ma, H., Moore, M., Palackal, N., and Ratnam, K., Human 3alpha-hydroxysteroid dehydrogenase isoforms (AKR1C1-AKR1C4) of the aldo-keto reductase superfamily: functional plasticity and tissue distribution reveals roles in the inactivation and formation of male and female sex hormones, *Biochem. J.*, 351 (Pt. 1), 67–77, 2000.
99. Deyashiki, Y., Ogasawara, A., Nakayama, T., Nakanishi, M., Miyabe, Y., Sato, K., and Hara, A., Molecular cloning of two human liver 3 alpha-hydroxysteroid/dihydrodiol dehydrogenase isoenzymes that are identical with chlordecone reductase and bile-acid binder, *Biochem. J.*, 299 (Pt. 2), 545–552, 1994.
100. Hara, A., Matsuura, K., Tamada, Y., Sato, K., Miyabe, Y., Deyashiki, Y., and Ishida, N., Relationship of human liver dihydrodiol dehydrogenases to hepatic bile-acid-binding protein and an oxidoreductase of human colon cells, *Biochem. J.*, 313 (Pt. 2), 373–376, 1996.
101. Khanna, M., Qin, K.N., Wang, R.W., and Cheng, K.C., Substrate specificity, gene structure, and tissue-specific distribution of multiple human 3 alpha-hydroxysteroid dehydrogenases, *J. Biol. Chem.*, 270, 20162–20168, 1995.

102. Dufort, I., Soucy, P., Labrie, F., and Luu-The, V., Molecular cloning of human type 3 3 alpha-hydroxysteroid dehydrogenase that differs from 20 alpha-hydroxysteroid dehydrogenase by seven amino acids, *Biochem. Biophys. Res. Commun.*, 228, 474–479, 1996.
103. Griffin, L.D. and Mellon, S.H., Selective serotonin reuptake inhibitors directly alter activity of neurosteroidogenic enzymes, *Proc. Natl. Acad. Sci. U.S.A.*, 96, 13512–13517, 1999.
104. Usami, N., Yamamoto, T., Shintani, S., Ishikura, S., Higaki, Y., Katagiri, Y., and Hara, A., Substrate specificity of human 3(20)alpha-hydroxysteroid dehydrogenase for neurosteroids and its inhibition by benzodiazepines, *Biol. Pharm. Bull.*, 25, 441–445, 2002.
105. Higaki, Y., Usami, N., Shintani, S., Ishikura, S., El-Kabbani, O., and Hara, A., Selective and potent inhibitors of human 20alpha-hydroxysteroid dehydrogenase (AKR1C1) that metabolizes neurosteroids derived from progesterone, *Chem. Biol. Interact.*, 143/144, 503–513, 2003.
106. Jin, Y., Cooper, W.C., and Penning, T.M., Examination of the differences in structure-function of human and rat 3alpha-hydroxysteroid dehydrogenase, *Chem. Biol. Interact.*, 143/144, 383–392, 2003.
107. Griffin, L.D. and Mellon, S.H., Selective serotonin reuptake inhibitors directly alter activity of neurosteroidogenic enzymes, *Proc. Natl. Acad. Sci. U.S.A.*, 96, 13512–13517, 1999.
108. Hobkirk, R., Steroid sulfotransferases and steroid sulfate sulfatases: characteristics and biological roles, *Can. J. Biochem. Cell Biol.*, 63, 1127–1144, 1985.
109. Iwamori, M., Moser, H.W., and Kishimoto, Y., Steroid sulfatase in brain: comparison of sulfohydrolase activities for various steroid sulfates in normal and pathological brains, including the various forms of metachromatic leukodystrophy, *J. Neurochem.*, 27, 1389–1395, 1976.
110. Parker, C.R., Jr., Falany, C.N., Stockard, C.R., Stankovic, A.K., and Grizzle, W.E., Immunohistochemical localization of dehydroepiandrosterone sulfotransferase in human fetal tissues, *J. Clin. Endocrinol. Metab.*, 78, 234–236, 1994.
111. Barker, E.V., Hume, R., Hallas, A., and Coughtrie, W.H., *Dehydroepiandrosterone Sulfotransferase in the Developing Human Fetus: Quantitative Biochemical and Immunological Characterization of the Hepatic, Renal, and Adrenal Enzymes*, 1994.
112. Mulder, G., *The Sulfatation of Drugs and Other Compounds*, CRC Press, Boca Raton, FL, 1981.
113. De Meio, R., Sulfate activation and transfer, in *Metabolism of Sulfur Compounds*, Greenberg, D.M., Ed., Academic Press, New York, 1975, pp. 287–359.
114. Majewska, M., Neurosteroids: GABAA-agonistic and GABAA-antagonistic modulators of the GABAA receptor, in *Neurosteroids and Brain Function*, Costa, E. and Paul, S.M., Eds., Thieme, New York, 1991, pp. 109–117.
115. Majewska, M.D., Neurosteroids: endogenous bimodal modulators of the GABAA receptor: mechanism of action and physiological significance, *Prog. Neurobiol.*, 38, 379–395, 1992.
116. Ogura, K., Kajita, J., Narihata, H., Watabe, T., Ozawa, S., Nagata, K., Yamazoe, Y., and Kato, R., Cloning and sequence analysis of a rat liver cDNA encoding hydroxysteroid sulfotransferase, *Biochem. Biophys. Res. Commun.*, 165, 168–174, 1989.
117. Shapiro, L., Steroid Sulfatase, paper presented at *International Symposium on DHEA Transformation into Androgens and Estrogens in Target Tissues: Intracrinology*, Québec City, 1995.

118. Ballabio, A. and Shapiro, L.J., Steroid sulfatase deficiency and X-linked ichtyosis, in *The Metabolic and Molecular Bases of Inherited Disease*, Scriver, C.R., Beaudet, A.L., Sly, W.S., and Valle, D., Eds., McGraw-Hill, New York, 1995, pp. 2999–3022.
119. Mohandas, T., Geller, R.L., Yen, P.H., Rosendorff, J., Bernstein, R., Yoshida, A., and Shapiro, L.J., Cytogenetic and molecular studies on a recombinant human X chromosome: implications for the spreading of X chromosome inactivation, *Proc. Natl. Acad. Sci. U.S.A.*, 84, 4954–4958, 1987.
120. Shapiro, L., Steroid sulfatase deficiency and x-linked ichthyosis, in *The Metabolic Basis of Inherited Diseases*, Stanbury, J.W., Fredrickson, D.S., Goldstein, J.L., Brown, M.S., Eds., McGraw Hill, New York, 1982, pp. 1027–1034.
121. Li, X.M., Salido, E.C., Gong, Y., Kitada, K., Serikawa, T., Yen, P.H., and Shapiro, L.J., Cloning of the rat steroid sulfatase gene (Sts), a non-pseudoautosomal X-linked gene that undergoes X inactivation, *Mamm. Genome*, 7, 420–424, 1996.
122. Salido, E.C., Li, X.M., Yen, P.H., Martin, N., Mohandas, T.K., and Shapiro, L.J., Cloning and expression of the mouse pseudoautosomal steroid sulphatase gene (Sts), *Nat. Genet.*, 13, 83–86, 1996.
123. Rose, K.A., Stapleton, G., Dott, K., Kieny, M.P., Best, R., Schwarz, M., Russell, D.W., Bjorkhem, I., Seckl, J., and Lathe, R., Cyp7b, a novel brain cytochrome P450, catalyzes the synthesis of neurosteroids 7alpha-hydroxy dehydroepiandrosterone and 7alpha-hydroxy pregnenolone, *Proc. Natl. Acad. Sci. U.S.A.*, 94, 4925–4930, 1997.
124. Stapleton, G., Steel, M., Richardson, M., Mason, J.O., Rose, K.A., Morris, R.G., and Lathe, R., A novel cytochrome P450 expressed primarily in brain, *J. Biol. Chem.*, 270, 29739–29745, 1995.
125. Lund, E.G., Guileyardo, J.M., and Russell, D.W., cDNA cloning of cholesterol 24-hydroxylase, a mediator of cholesterol homeostasis in the brain, *Proc. Natl. Acad. Sci. U.S.A.*, 96, 7238–7243, 1999.
126. Ray, W.J., Bain, G., Yao, M., and Gottlieb, D.I., CYP26, a novel mammalian cytochrome P450, is induced by retinoic acid and defines a new family, *J. Biol. Chem.*, 272, 18702–18708, 1997.
127. White, J.A., Guo, Y.D., Baetz, K., Beckett-Jones, B., Bonasoro, J., Hsu, K.E., Dilworth, F.J., Jones, G., and Petkovich, M., Identification of the retinoic acid-inducible all-trans-retinoic acid 4-hydroxylase, *J. Biol. Chem.*, 271, 29922–29927, 1996.
128. White, J.A., Beckett-Jones, B., Guo, Y.D., Dilworth, F.J., Bonasoro, J., Jones, G., and Petkovich, M., cDNA cloning of human retinoic acid-metabolizing enzyme (hP450RAI) identifies a novel family of cytochromes P450, *J. Biol. Chem.*, 272, 18538–18541, 1997.
129. Morfin, R. and Courchay, G., Pregnenolone and dehydroepiandrosterone as precursors of native 7-hydroxylated metabolites which increase the immune response in mice, *J. Steroid Biochem. Mol. Biol.*, 50, 91–100, 1994.
130. Doostzadeh, J., Cotillon, A.C., and Morfin, R., Hydroxylation of pregnenolone at the 7 alpha- and 7 beta-positions by mouse liver microsomes: effects of cytochrome p450 inhibitors and structure-specific inhibition by steroid hormones, *Steroids*, 63, 383–392, 1998.
131. Tzung, K.W., Ishimura-Oka, K., Kihara, S., Oka, K., and Chan, L., Structure of the mouse cholesterol 7 alpha-hydroxylase gene, *Genomics*, 21, 244–247, 1994.
132. Wood, A.W., Ryan, D.E., Thomas, P.E., and Levin, W., Regio- and stereoselective metabolism of two C19 steroids by five highly purified and reconstituted rat hepatic cytochrome P-450 isozymes, *J. Biol. Chem.*, 258, 8839–8847, 1983.

133. Swinney, D.C., Ryan, D.E., Thomas, P.E., and Levin, W., Regioselective progesterone hydroxylation catalyzed by eleven rat hepatic cytochrome P-450 isozymes, *Biochemistry*, 26, 7073–7083, 1987.
134. Arlotto, M.P., Greenway, D.J., and Parkinson, A., Purification of two isozymes of rat liver microsomal cytochrome P450 with testosterone 7 alpha-hydroxylase activity, *Arch. Biochem. Biophys.*, 270, 441–457, 1989.
135. Akwa, Y., Morfin, R.F., Robel, P., and Baulieu, E.E., Neurosteroid metabolism: 7 alpha-hydroxylation of dehydroepiandrosterone and pregnenolone by rat brain microsomes, *Biochem. J.*, 288 (Pt. 3), 959–964, 1992.
136. Stoffel-Wagner, B., Neurosteroid metabolism in the human brain, *Eur. J. Endocrinol.*, 145, 669–679, 2001.
137. Compagnone, N.A., Bulfone, A., Rubenstein, J.L., and Mellon, S.H., Steroidogenic enzyme P450c17 is expressed in the embryonic central nervous system, *Endocrinology*, 136, 5212–5223, 1995.
138. Compagnone, N.A., Bulfone, A., Rubenstein, J.L., and Mellon, S.H., Expression of the steroidogenic enzyme P450scc in the central and peripheral nervous systems during rodent embryogenesis, *Endocrinology*, 136, 2689–2696, 1995.
139. Compagnone, N.A., Salido, E., Shapiro, L.J., and Mellon, S.H., Expression of steroid sulfatase during embryogenesis, *Endocrinology*, 138, 4768–4773, 1997.
140. Lauber, M.E. and Lichtensteiger, W., Pre- and postnatal ontogeny of aromatase cytochrome P450 messenger ribonucleic acid expression in the male rat brain studied by *in situ* hybridization, *Endocrinology*, 135, 1661–1668, 1994.
141. Lauber, M.E. and Lichtensteiger, W., Ontogeny of 5 alpha-reductase (type 1) messenger ribonucleic acid expression in rat brain: early presence in germinal zones, *Endocrinology*, 137, 2718–2730, 1996.
142. Seckl, J.R. and Walker, B.R., Minireview: 11beta-hydroxysteroid dehydrogenase type 1: a tissue-specific amplifier of glucocorticoid action, *Endocrinology*, 142, 1371–1376, 2001.
143. Holmes, M.C., Kotelevtsev, Y., Mullins, J.J., and Seckl, J.R., Phenotypic analysis of mice bearing targeted deletions of 11beta-hydroxysteroid dehydrogenases 1 and 2 genes, *Mol. Cell. Endocrinol.*, 171, 15–20, 2001.
144. Seckl, J.R., 11beta-Hydroxysteroid dehydrogenase in the brain: a novel regulator of glucocorticoid action? *Front Neuroendocrinol.*, 18, 49–99, 1997.
145. Baulieu, E.E. and Robel, P., Neurosteroids: a new brain function? *J. Steroid Biochem. Mol. Biol.*, 37, 395–403, 1990.
146. Ukena, K., Usui, M., Kohchi, C., and Tsutsui, K., Cytochrome P450 side-chain cleavage enzyme in the cerebellar Purkinje neuron and its neonatal change in rats, *Endocrinology*, 139, 137–147, 1998.
147. Tsutsui, K., Ukena, K., Usui, M., Sakamoto, H., and Takase, M., Novel brain function: biosynthesis and actions of neurosteroids in neurons, *Neurosci. Res.*, 36, 261–273, 2000.
148. Sakamoto, H., Ukena, K., and Tsutsui, K., Effects of progesterone synthesized de novo in the developing Purkinje cell on its dendritic growth and synaptogenesis, *J. Neurosci.*, 21, 6221–6232, 2001.
149. Mellon, S.H., Kushner, J.A., and Vaisse, C., Expression and regulation of adrenodoxin and P450scc mRNA in rodent tissues, *DNA Cell Biol.*, 10, 339–347, 1991.
150. Oftebro, H., Stormer, F.C., and Pedersen, J.L., The presence of an adrenodoxin-like ferredoxin and cytochrome P-450 in brain mitochondria, *J. Biol. Chem.*, 254, 4331–4334, 1979.

151. Lin, D., Sugawara, T., Strauss, J.F., III, Clark, B.J., Stocco, D.M., Saenger, P., Rogol, A., and Miller, W.L., Role of steroidogenic acute regulatory protein in adrenal and gonadal steroidogenesis, *Science*, 267, 1828–1831, 1995.
152. Caron, K.M., Soo, S.C., Wetsel, W.C., Stocco, D.M., Clark, B.J., and Parker, K.L., Targeted disruption of the mouse gene encoding steroidogenic acute regulatory protein provides insights into congenital lipoid adrenal hyperplasia, *Proc. Natl. Acad. Sci. U.S.A.*, 94, 11540–11545, 1997.
153. Benmessahel, Y., Guennoun, R., Cadepond, F., Baulieu, E.E., Schumacher, M., and Groyer, G., Expression of steroidogenic acute regulatory protein in cultured Schwann cells and its regulation by cAMP, *Ann. N.Y. Acad. Sci.*, 973, 83–87, 2002.
154. Mellon, S.H. and Deschepper, C.F., Neurosteroid biosynthesis: genes for adrenal steroidogenic enzymes are expressed in the brain, *Brain Res.*, 629, 283–292, 1993.
155. Compagnone, N.A., Bulfone, A., Rubenstein, J.L.R., and Mellon, S.H., Steroidogenic enzyme P450c17 is expressed in the embryonic central nervous system, *Endocrinology*, 136, 5212–5223, 1995.
156. Rubenstein, J.L., Anderson, S., Shi, L., Miyashita-Lin, E., Bulfone, A., and Hevner, R., Genetic control of cortical regionalization and connectivity, *Cereb. Cortex*, 9, 524–532, 1999.
157. Miyashita-Lin, E.M., Hevner, R., Wassarman, K.M., Martinez, S., and Rubenstein, J.L., Early neocortical regionalization in the absence of thalamic innervation, *Science*, 285, 906–909, 1999.
158. Kohchi, C., Ukena, K., and Tsutsui, K., Age- and region-specific expressions of the messenger RNAs encoding for steroidogenic enzymes p450scc, P450c17 and 3beta-HSD in the postnatal rat brain, *Brain Res.*, 801, 233–238, 1998.
159. Norris, P.J., Hardwick, J.P., and Emson, P.C., Localization of NADPH cytochrome P450 oxidoreductase in rat brain by immunohistochemistry and *in situ* hybridization and a comparison with the distribution of neuronal NADPH-diaphorase staining, *Neuroscience*, 61, 331–350, 1994.
160. Simmons, D.L. and Kasper, C.B., Quantitation of mRNAs specific for the mixed-function oxidase system in rat liver and extrahepatic tissues during development, *Arch. Biochem. Biophys.*, 271, 10–20, 1989.
161. Jung-Testas, I., Hu, Z.Y., Baulieu, E.E., and Robel, P., Neurosteroids: biosynthesis of pregnenolone and progesterone in primary cultures of rat glial cells, *Endocrinology*, 125, 2083–2091, 1989.
162. Hu, Z.Y., Bourreau, E., Jung-Testas, I., Robel, P., and Baulieu, E.E., Neurosteroids: oligodendrocyte mitochondria convert cholesterol to pregnenolone, *Proc. Natl. Acad. Sci. U.S.A.*, 84, 8215–8219, 1987.
163. Kabbadj, K., el-Etr, M., Baulieu, E.E., and Robel, P., Pregnenolone metabolism in rodent embryonic neurons and astrocytes, *Glia*, 7, 170–175, 1993.
164. Bauer, H.C. and Bauer, H., Micromethod for the determination of 3-beta-HSD activity in cultured cells, *J. Steroid Biochem.*, 33, 643–646, 1989.
165. Weidenfeld, J., Siegel, R.A., and Chowers, I., In vitro conversion of pregnenolone to progesterone by discrete brain areas of the male rat, *J. Steroid Biochem.*, 13, 961–963, 1980.
166. Robel, P., Bourreau, E., Corpechot, C., Dang, D.C., Halberg, F., Clarke, C., Haug, M., Schlegel, M.L., Synguelakis, M., Vourch, C., et al., Neuro-steroids: 3 beta-hydroxy-delta 5-derivatives in rat and monkey brain, *J. Steroid Biochem.*, 27, 649–655, 1987.

167. Guennoun, R., Fiddes, R.J., Gouezou, M., Lombes, M., and Baulieu, E.E., A key enzyme in the biosynthesis of neurosteroids, 3 beta-hydroxysteroid dehydrogenase/delta 5-delta 4-isomerase (3 beta-HSD), is expressed in rat brain, *Brain Res. Mol. Brain Res.*, 30, 287–300, 1995.
168. Dupont, E., Labrie, F., Luu-The, V., and Pelletier, G., Immunocytochemical localization of 3 beta-hydroxysteroid dehydrogenase/delta 5-delta 4-isomerase in human ovary, *J. Clin. Endocrinol. Metab.*, 74, 994–998, 1992.
169. Ukena, K., Kohchi, C., and Tsutsui, K., Expression and activity of 3beta-hydroxysteroid dehydrogenase/delta5-delta4-isomerase in the rat Purkinje neuron during neonatal life, *Endocrinology*, 140, 805–813, 1999.
170. Coirini, H., Gouezou, M., Liere, P., Delespierre, B., Pianos, A., Eychenne, B., Schumacher, M., and Guennoun, R., 3 Beta-hydroxysteroid dehydrogenase expression in rat spinal cord, *Neuroscience*, 113, 883–891, 2002.
171. Mensah-Nyagan, A.G., Feuilloley, M., Dupont, E., Do-Rego, J.L., Leboulenger, F., Pelletier, G., and Vaudry, H., Immunocytochemical localization and biological activity of 3 beta-hydroxysteroid dehydrogenase in the central nervous system of the frog, *J. Neurosci.*, 14, 7306–7318, 1994.
172. Mathieu, M., Mensah-Nyagan, A.G., Vallarino, M., Do-Rego, J.L., Beaujean, D., Luu-The, V., Pelletier, G., and Vaudry, H., Immunohistochemical localization of 3β-hydroxyseroid dehydrogenase and 5α reductase in the brain of the African lungfish *Protopterus amnectens*, *J. Comp. Neurol.*, 2000, submitted.
173. Sanne, J.L. and Krueger, K.E., Expression of cytochrome P450 side-chain cleavage enzyme and 3 beta-hydroxysteroid dehydrogenase in the rat central nervous system: a study by polymerase chain reaction and *in situ* hybridization, *J. Neurochem.*, 65, 528–536, 1995.
174. Kabbadj, K., el-Etr, M., Baulieu, E.E., and Robel, P., Pregnenolone metabolism in rodent embryonic neurons and astrocytes, *Glia*, 7, 170–175, 1993.
175. Mensah-Nyagan, A.G., Feuilloley, M., Dupont, E., Do-Rego, J.L., Leboulenger, F., Pelletier, G., and Vaudry, H., Immunocytochemical localization and biological activity of 3 beta-hydroxysteroid dehydrogenase in the central nervous system of the frog, *J. Neurosci.*, 14, 7306–7318, 1994.
176. Mensah-Nyagan, A.G., Beaujean, D., Luu-The, V., Pelletier, G., and Vaudry, H., Anatomical and biochemical evidence for the synthesis of unconjugated and sulfated neurosteroids in amphibians, *Brain Res. Brain Res. Rev.*, 37, 13–24, 2001.
177. Robert, F., Guennoun, R., Desarnaud, F., Do-Thi, A., Benmessahel, Y., Baulieu, E.E., and Schumacher, M., Synthesis of progesterone in Schwann cells: regulation by sensory neurons, *Eur. J. Neurosci.*, 13, 916–924, 2001.
178. Gago, N., Akwa, Y., Sananes, N., Guennoun, R., Baulieu, E.E., El-Etr, M., and Schumacher, M., Progesterone and the oligodendroglial lineage: stage-dependent biosynthesis and metabolism, *Glia*, 36, 295–308, 2001.
179. MacKenzie, S.M., Clark, C.J., Fraser, R., Gomez-Sanchez, C.E., Connell, J.M., and Davies, E., Expression of 11beta-hydroxylase and aldosterone synthase genes in the rat brain, *J. Mol. Endocrinol.*, 24, 321–328, 2000.
180. Gomez-Sanchez, C.E., Zhou, M.Y., Cozza, E.N., Morita, H., Foecking, M.F., and Gomez-Sanchez, E.P., Aldosterone biosynthesis in the rat brain, *Endocrinology*, 138, 3369–3373, 1997.
181. Reddy, V.V., Estrogen metabolism in neural tissues of rabbits: 17 beta-hydroxysteroid oxidoreductase activity, *Steroids*, 34, 207–215, 1979.

182. Resko, J.A., Stadelman, H.L., and Norman, R.L., 17 beta-hydroxysteroid dehydrogenase activity in the pituitary gland and neural tissue of Rhesus monkeys, *J. Steroid Biochem.*, 11, 1429–1434, 1979.
183. Pelletier, G., Luu-The, V., and Labrie, F., Immunocytochemical localization of type I 17 beta-hydroxysteroid dehydrogenase in the rat brain, *Brain Res.*, 704, 233–239, 1995.
184. Mensah-Nyagan, A.M., Feuilloley, M., Do-Rego, J.L., Marcual, A., Lange, C., Tonon, M.C., Pelletier, G., and Vaudry, H., Localization of 17beta-hydroxysteroid dehydrogenase and characterization of testosterone in the brain of the male frog, *Proc. Natl. Acad. Sci. U.S.A.*, 93, 1423–1428, 1996.
185. Mensah-Nyagan, A.G., Do-Rego, J.L., Feuilloley, M., Marcual, A., Lange, C., Pelletier, G., and Vaudry, H., In vivo and *in vitro* evidence for the biosynthesis of testosterone in the telencephalon of the female frog, *J. Neurochem.*, 67, 413–422, 1996.
186. Steckelbroeck, S., Stoffel-Wagner, B., Reichelt, R., Schramm, J., Bidlingmaier, F., Siekmann, L., and Klingmuller, D., Characterization of 17beta-hydroxysteroid dehydrogenase activity in brain tissue: testosterone formation in the human temporal lobe, *J. Neuroendocrinol.*, 11, 457–464, 1999.
187. Beyenburg, S., Watzka, M., Blumcke, I., Schramm, J., Bidlingmaier, F., Elger, C.E., and Stoffel-Wagner, B., Expression of mRNAs encoding for 17beta-hydroxisteroid dehydrogenase isozymes 1, 2, 3 and 4 in epileptic human hippocampus, *Epilepsy Res.*, 41, 83–91, 2000.
188. Lephart, E.D., A review of brain aromatase cytochrome P450, *Brain Res. Brain Res. Rev.*, 22, 1–26, 1996.
189. Hutchison, J.B., Wozniak, A., Beyer, C., and Hutchison, R.E., Regulation of sex-specific formation of oestrogen in brain development: endogenous inhibitors of aromatase, *J. Steroid Biochem. Mol. Biol.*, 56, 201–207, 1996.
190. Roselli, C.E., Horton, L.E., and Resko, J.A., Distribution and regulation of aromatase activity in the rat hypothalamus and limbic system, *Endocrinology*, 117, 2471–2477, 1985.
191. George, F.W. and Ojeda, S.R., Changes in aromatase activity in the rat brain during embryonic, neonatal, and infantile development, *Endocrinology*, 111, 522–529, 1982.
192. MacLusky, N.J., Philip, A., Hurlburt, C., and Naftolin, F., Estrogen formation in the developing rat brain: sex differences in aromatase activity during early post-natal life, *Psychoneuroendocrinology*, 10, 355–361, 1985.
193. Lephart, E.D., Simpson, E.R., McPhaul, M.J., Kilgore, M.W., Wilson, J.D., and Ojeda, S.R., Brain aromatase cytochrome P-450 messenger RNA levels and enzyme activity during prenatal and perinatal development in the rat, *Brain Res. Mol. Brain Res.*, 16, 187–192, 1992.
194. Honda, S., Harada, N., and Takagi, Y., Novel exon 1 of the aromatase gene specific for aromatase transcripts in human brain, *Biochem. Biophys. Res. Commun.*, 198, 1153–1160, 1994.
195. Toda, K., Simpson, E.R., Mendelson, C.R., Shizuta, Y., and Kilgore, M.W., Expression of the gene encoding aromatase cytochrome P450 (CYP19) in fetal tissues, *Mol. Endocrinol.*, 8, 210–217, 1994.
196. Kato, J., Yamada-Mouri, N., and Hirata, S., Structure of aromatase mRNA in the rat brain, *J. Steroid Biochem. Mol. Biol.*, 61, 381–385, 1997.
197. Lephart, E.D., Molecular aspects of brain aromatase cytochrome P450, *J. Steroid Biochem. Mol. Biol.*, 61, 375–380, 1997.

198. Wagner, C.K. and Morrell, J.I., Distribution and steroid hormone regulation of aromatase mRNA expression in the forebrain of adult male and female rats: a cellular-level analysis using *in situ* hybridization, *J. Comp. Neurol.*, 370, 71–84, 1996.
199. Sanghera, M.K., Simpson, E.R., McPhaul, M.J., Kozlowski, G., Conley, A.J., and Lephart, E.D., Immunocytochemical distribution of aromatase cytochrome P450 in the rat brain using peptide-generated polyclonal antibodies, *Endocrinology*, 129, 2834–2844, 1991.
200. Moisan, M.P., Seckl, J.R., and Edwards, C.R., 11 Beta-hydroxysteroid dehydrogenase bioactivity and messenger RNA expression in rat forebrain: localization in hypothalamus, hippocampus, and cortex, *Endocrinology*, 127, 1450–1455, 1990.
201. Lakshmi, V., Sakai, R.R., McEwen, B.S., and Monder, C., Regional distribution of 11 beta-hydroxysteroid dehydrogenase in rat brain, *Endocrinology*, 128, 1741–1748, 1991.
202. Roland, B.L., Li, K.X., and Funder, J.W., Hybridization histochemical localization of 11 beta-hydroxysteroid dehydrogenase type 2 in rat brain, *Endocrinology*, 136, 4697–4700, 1995.
203. Zhou, M.Y., Gomez-Sanchez, E.P., Cox, D.L., Cosby, D., and Gomez-Sanchez, C.E., Cloning, expression, and tissue distribution of the rat nicotinamide adenine dinucleotide-dependent 11 beta-hydroxysteroid dehydrogenase, *Endocrinology*, 136, 3729–3734, 1995.
204. Diaz, R., Brown, R.W., and Seckl, J.R., Distinct ontogeny of glucocorticoid and mineralocorticoid receptor and 11beta-hydroxysteroid dehydrogenase types I and II mRNAs in the fetal rat brain suggest a complex control of glucocorticoid actions, *J. Neurosci.*, 18, 2570–2580, 1998.
205. Rajan, V., Edwards, C.R., and Seckl, J.R., 11 beta-Hydroxysteroid dehydrogenase in cultured hippocampal cells reactivates inert 11-dehydrocorticosterone, potentiating neurotoxicity, *J. Neurosci.*, 16, 65–70, 1996.
206. Yau, J.L., Noble, J., Kenyon, C.J., Hibberd, C., Kotelevtsev, Y., Mullins, J.J., and Seckl, J.R., Lack of tissue glucocorticoid reactivation in 11beta-hydroxysteroid dehydrogenase type 1 knockout mice ameliorates age-related learning impairments, *Proc. Natl. Acad. Sci. U.S.A.*, 98, 4716–4721, 2001.
207. Celotti, F., Melcangi, R.C., and Martini, L., The 5 alpha-reductase in the brain: molecular aspects and relation to brain function, *Front Neuroendocrinol.*, 13, 163–215, 1992.
208. Melcangi, R.C., Celotti, F., Castano, P., and Martini, L., Differential localization of the 5 alpha-reductase and the 3 alpha-hydroxysteroid dehydrogenase in neuronal and glial cultures, *Endocrinology*, 132, 1252–1259, 1993.
209. Melcangi, R.C., Celotti, F., and Martini, L., Progesterone 5-alpha-reduction in neuronal and in different types of glial cell cultures: type 1 and 2 astrocytes and oligodendrocytes, *Brain Res.*, 639, 202–206, 1994.
210. Pelletier, G., Luu-The, V., and Labrie, F., Immunocytochemical localization of 5 alpha-reductase in rat brain, *Mol. Cell. Neurosci.*, 5, 394–399, 1994.
211. Normington, K. and Russell, D.W., Tissue distribution and kinetic characteristics of rat steroid 5 alpha-reductase isozymes: evidence for distinct physiological functions, *J. Biol. Chem.*, 267, 19548–19554, 1992.
212. Stoffel-Wagner, B., Watzka, M., Steckelbroeck, S., Wickert, L., Schramm, J., Romalo, G., Klingmuller, D., and Schweikert, H.U., Expression of 5alpha-reductase in the human temporal lobe of children and adults, *J. Clin. Endocrinol. Metab.*, 83, 3636–3642, 1998.

213. Stoffel-Wagner, B., Beyenburg, S., Watzka, M., Blumcke, I., Bauer, J., Schramm, J., Bidlingmaier, F., and Elger, C.E., Expression of 5alpha-reductase and 3alpha-hydroxisteroid oxidoreductase in the hippocampus of patients with chronic temporal lobe epilepsy, *Epilepsia*, 41, 140–147, 2000.
214. Pasmanik, M. and Callard, G.V., Aromatase and 5 alpha-reductase in the teleost brain, spinal cord, and pituitary gland, *Gen. Comp. Endocrinol.*, 60, 244–251, 1985.
215. Pasmanik, M. and Callard, G.V., Changes in brain aromatase and 5 alpha-reductase activities correlate significantly with seasonal reproductive cycles in goldfish (Carassius auratus), *Endocrinology*, 122, 1349–1356, 1988.
216. Mahendroo, M.S., Cala, K.M., Landrum, D.P., and Russell, D.W., Fetal death in mice lacking 5alpha-reductase type 1 caused by estrogen excess, *Mol. Endocrinol.*, 11, 917–927, 1997.
217. Compagnone, N.A. and Mellon, S.H., Neurosteroids: biosynthesis and function of these novel neuromodulators, *Front Neuroendocrin.*, 21, 1–58, 2000.
218. Steckelbroeck, S., Watzka, M., Reichelt, R., Hans, V.H., Stoffel-Wagner, B., Heidrich, D.D., Schramm, J., Bidlingmaier, F., and Klingmuller, D., Characterization of the 5alpha-reductase-3alpha-hydroxysteroid dehydrogenase complex in the human brain, *J. Clin. Endocrinol. Metab.*, 86, 1324–1331, 2001.
219. Karavolas, H.J. and Hodges, D.R., Neuroendocrine metabolism of progesterone and related progestins, *Ciba Found. Symp.*, 153, 22–44, 1990.
220. Compagnone, N.A. and Mellon, S.H., Dehydroepiandrosterone: a potential signalling molecule for neocortical organization during development, *Proc. Natl. Acad. Sci. U.S.A.*, 95, 4678–4683, 1998 (see Comments).
221. Beaujean, D., Mensah-Nyagan, A.G., Do-Rego, J.L., Luu-The, V., Pelletier, G., and Vaudry, H., Immunocytochemical localization and biological activity of hydroxysteroid sulfotransferase in the frog brain, *J. Neurochem.*, 72, 848–857, 1999.
222. Sharp, S., Barker, E.V., Coughtrie, M.W., Lowenstein, P.R., and Hume, R., Immunochemical characterisation of a dehydroepiandrosterone sulfotransferase in rats and humans, *Eur. J. Biochem.*, 211, 539–548, 1993.
223. Knapstein, P., David, A., Wu, C.H., Archer, D.F., FLickinger, G.L., and Touchstone, J.C., Metabolism of free and sulfoconjugated DHEA in brain tissue in vivo and in vitro, *Steroids*, 11, 885–896, 1968.
224. Rajkowski, K.M., Robel, P., and Baulieu, E.E., Hydroxysteroid sulfotransferase activity in the rat brain and liver as a function of age and sex, *Steroids*, 62, 427–436, 1997.
225. Price, D.J. and Lotto, R.B., Influences of the thalamus on the survival of subplate and cortical plate cells in cultured embryonic mouse brain, *J. Neurosci.*, 16, 3247–3255, 1996.
226. Bixo, M., Andersson, A., Winblad, B., Purdy, R.H., and Backstrom, T., Progesterone, 5alpha-Pregnane-3,20-dione and 3alpha-hydroxy-5alpha-pregnane-20-one in specific regions of the human female brain in different endocrine states, *Brain Res.*, 764, 173–178, 1997.
227. Korneyev, A., Guidotti, A., and Costa, E., Regional and interspecies differences in brain progesterone metabolism, *J. Neurochem.*, 61, 2041–2047, 1993.
228. Milewich, L., Gomez-Sanchez, C., Crowley, G., Porter, J.C., Madden, J.D., and MacDonald, P.C., Progesterone and 5alpha-pregnane-3,20-dione in peripheral blood of normal young women: daily measurements throughout the menstrual cycle, *J. Clin. Endocrinol. Metab.*, 45, 617–622, 1977.
229. Britton, K.T. and Koob, G.F., Neuropharmacology: premenstrual steroids? *Nature*, 392, 869–870, 1998.

230. Fry, J. and Honour, J., Understanding premenstrual syndrome, *Lancet*, 351, 1511, 1998 (author reply, *Lancet*, 351, 1512, 1998).
231. Lubbert, H., Understanding premenstrual syndrome, *Lancet*, 351, 1511, 1998 (author reply, *Lancet*, 351, 1512, 1998).
232. Rapkin, A.J., Morgan, M., Goldman, L., Brann, D.W., Simone, D., and Mahesh, V.B., Progesterone metabolite allopregnanolone in women with premenstrual syndrome, *Obstet. Gynecol.*, 90, 709–714, 1997.
233. Schmidt, P.J., Purdy, R.H., Moore, P.H., Jr., Paul, S.M., and Rubinow, D.R., Circulating levels of anxiolytic steroids in the luteal phase in women with premenstrual syndrome and in control subjects, *J. Clin. Endocrinol. Metab.*, 79, 1256–1260, 1994.
234. Monteleone, P., Luisi, S., Tonetti, A., Bernardi, F., Genazzani, A.D., Luisi, M., Petraglia, F., and Genazzani, A.R., Allopregnanolone concentrations and premenstrual syndrome, *Eur. J. Endocrinol.*, 142, 269–273, 2000.
235. Wang, M., Seippel, L., Purdy, R.H., and Backstrom, T., Relationship between symptom severity and steroid variation in women with premenstrual syndrome: study on serum pregnenolone, pregnenolone sulfate, 5 alpha-pregnane-3,20-dione and 3 alpha-hydroxy-5 alpha-pregnan-20-one, *J. Clin. Endocrinol. Metab.*, 81, 1076–1082, 1996.
236. Griffin, L.D., Conrad, S.C., and Mellon, S.H., Current perspectives on the role of neurosteroids in PMS and depression, *Int. Rev. Neurobiol.*, 46, 479–492, 2001.
237. Dimmock, P.W., Wyatt, K.M., Jones, P.W., and O'Brien, P.M., Efficacy of selective serotonin-reuptake inhibitors in premenstrual syndrome: a systematic review, *Lancet*, 356, 1131–1136, 2000.
238. Uzunov, D.P., Cooper, T.B., Costa, E., and Guidotti, A., Fluoxetine-elicited changes in brain neurosteroid content measured by negative ion mass fragmentography, *Proc. Natl. Acad. Sci. U.S.A.*, 93, 12599–12604, 1996.
239. Trauger, J.W., Jiang, A., Stearns, B.A., and LoGrasso, P.V., Kinetics of allopregnanolone formation catalyzed by human 3 alpha-hydroxysteroid dehydrogenase type III (AKR1C2), *Biochemistry*, 41, 13451–13459, 2002.
240. Hara, A., Matsuura, K., Tamada, Y., Sato, K., Miyabe, Y., Deyashiki, Y., and Ishida, N., Relationship of human liver dihydrodiol dehydrogenases to hepatic bile-acid-binding protein and an oxidoreductase of human colon cells, *Biochem. J.*, 313 (Pt. 2), 373–376, 1996.
241. Mensah-Nyagan, A.G., Do-Rego, J.L., Beaujean, D., Luu-The, V., Pelletier, G., and Vaudry, H., Neurosteroids: expression of steroidogenic enzymes and regulation of steroid biosynthesis in the central nervous system, *Pharmacol. Rev.*, 51, 63–81, 1999.
242. Do-Rego, J.L., Mensah-Nyagan, G.A., Beaujean, D., Vaudry, D., Sieghart, W., Luu-The, V., Pelletier, G., and Vaudry, H., gamma-Aminobutyric acid, acting through gamma-aminobutyric acid type A receptors, inhibits the biosynthesis of neurosteroids in the frog hypothalamus, *Proc. Natl. Acad. Sci. U.S.A.*, 97, 13925–13930, 2000.
243. Suzdak, P.D., Schwartz, R.D., Skolnick, P., and Paul, S.M., Ethanol stimulates gamma-aminobutyric acid receptor-mediated chloride transport in rat brain synaptoneurosomes, *Proc. Natl. Acad. Sci. U.S.A.*, 83, 4071–4075, 1986.
244. Grobin, A.C., Matthews, D.B., Devaud, L.L., and Morrow, A.L., The role of GABA(A) receptors in the acute and chronic effects of ethanol, *Psychopharmacology (Berl.)*, 139, 2–19, 1998.
245. VanDoren, M.J., Matthews, D.B., Janis, G.C., Grobin, A.C., Devaud, L.L., and Morrow, A.L., Neuroactive steroid 3alpha-hydroxy-5alpha-pregnan-20-one modulates electrophysiological and behavioral actions of ethanol, *J. Neurosci.*, 20, 1982–1989, 2000.

2 Subunit Specificity of Steroid Modulation of $GABA_A$ Receptors

Richard W. Olsen, Elisabetta Cagetti, and Martin Wallner

CONTENTS

0-8493-2392-4/04/$0.00+$1.50

2.1 INTRODUCTION

2.1.1 $GABA_A$ Receptors

The γ-aminobutyric acid (GABA) receptor(s) that mediates fast inhibitory synaptic transmission in the nervous system of the entire animal kingdom is a ligand-gated chloride channel receptor that prevents action potential generation by short-circuiting the depolarization produced by excitatory neurotransmission.[1] This receptor is called $GABA_A$ (GABAR) to distinguish it from a G-protein-coupled receptor ($GABA_B$) that mediates a slower form of inhibition in some cells. Mammalian central nervous system (CNS) GABAR are actually a family of heteropentamers constituted from a family of 19 different but homologous subunit genes α1–6, β1–3, γ1–3, δ, ε, θ, π , and ρ1–3.[2,3] Out of literally thousands of possible combinations, some 12 to 20 subtypes or isoforms appear to exist in significant abundance, apparently due to strict, but poorly understood, rules governing subunit assembly. The common naturally occurring isoforms are described in other chapters. These isoforms differ in their regional expression throughout the nervous system, age of expression, and even subcellular localization.[4] They are regulated differentially at the gene expression level but also at the protein level, with respect to cell trafficking, localization to synapses or extrasynaptic membranes, dendrites vs. cell body vs. axon, apposition to different types of GABAergic innervating cells, and turnover. Many of these aspects are regulated by protein phosphorylation.[5]

GABAR are defined by the agonist muscimol, the competitive antagonist bicuculline, and noncompetitive antagonist picrotoxinin. They are also modulated by a variety of important CNS drugs, namely, benzodiazepines, barbiturates and other intravenous anesthetics like propofol and etomidate, volatile general anesthetics, and neuroactive steroids.[2,6–9] The neuroactive steroids include endogenous hormone metabolites that may biologically interface with the CNS. In particular, metabolites of progesterone, cortisone, and testosterone are able to produce rapid and nongenomic effects on the CNS. Many of these effects involve enhancement of GABAR function. This activity has been exploited by the development many years ago of the clinical anesthetic steroid alphaxalone, a synthetic analog of the GABA-active steroid hormone metabolites.[10,11] Alphaxalone is the first compound with general anesthetic activity whose mechanism of action was unambiguously established as enhancing GABAR-mediated inhibition.[12,13]

2.1.2 Neuroactive Steroids and GABAR

Neuroactive steroids enhance GABAR function[12,14–16] by binding to specific steroid receptor sites on the receptor protein and allosterically affecting GABA binding or channel gating. The exact same steroids that are active in modulating GABAR function have been shown to allosterically modulate binding to all the ligand sites on the GABAR protein, namely, the GABA site, the benzodiazepine site, and the picrotoxin/convulsant site labeled by [^{35}S]TBPS (e.g., Olsen et al., 1991[7]).

The structure–activity relationships for a series of steroids to affect GABAR are described in other chapters and show that the steroid backbone structure requires a 3α-hydroxyl group on ring A and an electronegative group like the ketone at position 17 on ring D.[13,17,18] The steroid hormones lack the structural requirements to interact with GABAR, and the GABAR-active steroids in general lack hormone activity (although *in vivo* some of the compounds may be interconverted metabolically), and it is important to note that steroid hormones may have effects, even rapid ones, on brain function and GABAR mediated by the traditional genomic hormone mechanisms.[19,20]

Steroids able to modulate GABAR function were shown, as had been previously demonstrated for other drugs like barbiturates,[7] to allosterically enhance the binding of benzodiazepines like [^{3}H]diazepam or flunitrazepam, to allosterically enhance the binding of GABA agonists like [^{3}H]GABA or muscimol, and to allosterically inhibit the binding of channel-blocking convulsants like [^{35}S]TBPS.[21–25]

2.2 HETEROGENEITY IN STEROID–GABAR INTERACTIONS IN ANIMALS, CELLS, SECTIONS/SLICES, OR HOMOGENATES

2.2.1 Radioligand Binding Studies

2.2.1.1 Observations

A regional heterogeneity, e.g., spinal cord vs. cerebellum vs. forebrain,[22,26] was observed for neuroactive steroid modulation of GABAR binding, and this was even more evident using tissue section binding and the autoradiography technique.[27–29] In particular, the concentration dependence curve varied considerably for a given steroid, with respect to the maximum effect and the midpoint (EC_{50} or IC_{50}). Some steroids, like other positive modulators of GABAR,[7] showed biphasic effects on [^{35}S]TBPS binding, with enhancement at low concentrations and inhibition at higher concentrations, and the shape of the curve varied with brain region.

For example, we determined the concentration (10 μ*M*) of the steroid anesthetic alphaxalone to inhibit [^{35}S]TBPS binding to a total rat brain homogenate by 50%, and included that concentration in a binding assay with a rat brain sagittal slice analyzed by autoradiography:[27] the convulsant binding in some regions, such as the cerebellum, colliculus, and dentate gyrus, was inhibited even more than 50%, while other regions, such as the parietal and occipital cortex, were little inhibited and possibly enhanced (Figure 2.1). This was followed up with a major study on heterogeneity of binding characteristics of GABAR, using multiple radioligands and multiple modulators. Both allopreganolone and alphaxalone were shown to have differential maximal enhancement on four polypeptide subunits (α, 51 kDA > 53-kDa band; β, 55 kDa > 58-kDa band) identified on SDS-PAGE following photoaffinity labeling of purified total rat brain GABAR with [^{3}H]muscimol or [^{3}H]flunitrazepam.[28] The same study showed that the extent of enhancement by alphaxalone of [^{3}H]flunitrazepam binding varied from zero (e.g., cerebellar granule cell layer) to ca. 2-fold (pyriform cortex, lateral septum, globus pallidus); the enhancement of

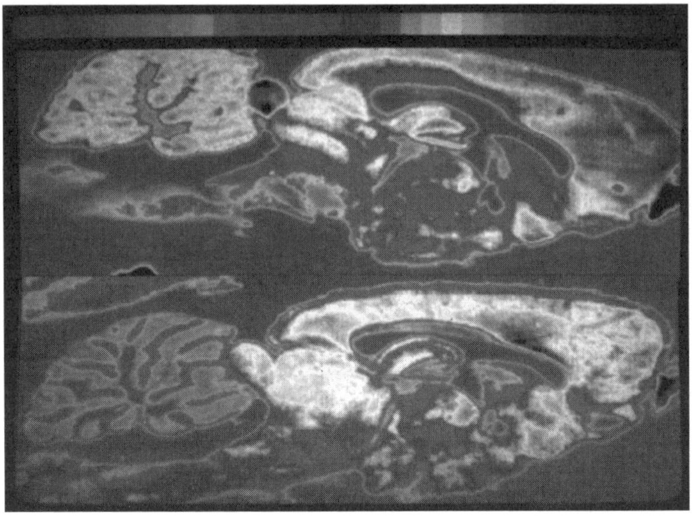

FIGURE 2.1 Brain regional modulation by neuroactive steroid alphaxalone of [^{35}S]TBPS binding to GABAR in rat brain frozen and thawed, unfixed thin sagittal sections by autoradiography. Color scale at the top indicates increased absorbance (radioligand receptor binding). Top: Control, no steroid. Bottom: alphaxalone, 10 μ*M*. (Reproduced from Figure 2 in Sapp, D.W. et al., *J. Pharmacol. Exp. Ther.*, 262, 801, 1992. With permission of the publisher (ASPET). (Shown as Color Figure 2.1, following page 74.)

[^{3}H]muscimol binding likewise varied from 4-fold (deep layers of neocortex) to 2.5-fold (superficial cortex, dentate gyrus) to nil (cerebellum).[28] Some differences in this steroid modulation of GABAR were observed with respect to regional variation when comparing rat brain to postmortem human brain, using the same two radioligands and the neuroactive steroids alphaxalone and 3α-5α-tetrahydrocorticosterone (THDOC). The latter steroid (3 to 10 μ*M*) inhibited binding of [^{3}H]muscimol in temporal lobe neocortex and occipital cortex, while enhancing it in the hippocampus and frontal and sensorimotor cortex; alphaxalone enhanced [^{3}H]muscimol binding everywhere examined, but to a varying extent, and did not enhance [^{3}H]flunitrazepam binding; instead, it had either no effect or inhibition, depending on the region.[29]

2.2.1.2 Possible Interpretations

Part of the heterogeneity in the binding modulation curves was apparently due to the assay conditions, such as method of washing of membrane homogenates or preincubation conditions with tissue sections; these were interpreted to reflect primarily differences in effectiveness of removing endogenous ligands in the tissue, including GABA and neurosteroids themselves. Thus, thoroughly washed, osmotically shocked, frozen and thawed membranes or osmotically shocked sections gave more reproducible baseline and modulated binding results.[30] The effects of unknown concentrations of endogenous GABA in the assay could be minimized by adding a known small amount of GABA to the assays[22,31] (Figure 2.2).

Another interpretation was the possibility of more than one steroid binding site on a given GABAR molecule.[24,32,33] GABAR are heteropentamers with homologous

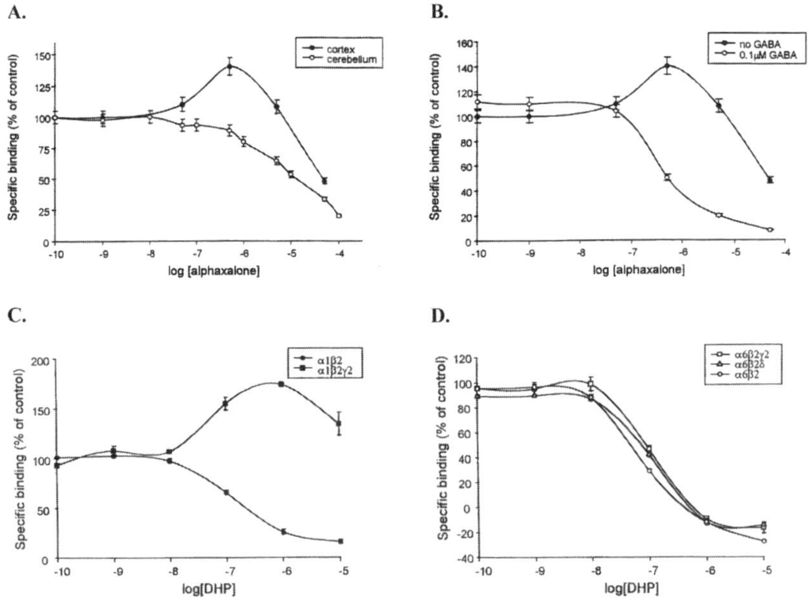

FIGURE 2.2 Concentration dependence for steroid modulation of [^{35}S]TBPS binding to GABAR in membrane homogenates from rat brain regions (A) and effect of exogenous GABA (B) with alphaxalone, and recombinant GABAR of varying subunit composition transiently expressed in Sf9 cells (C and D) with allopregnanolone (DHP, 3α,5α). (A) Compares cortex and cerebellum. (B) Compares cortical membranes to which GABA (0.1 μ*M*) is added or not. (C) Describes α1β2 and α1β2γ2 receptors. (D) Compares α6β2, α6β2γ2, and α6β2δ receptors. (Reproduced from Figure 1A and B and Figure 3C and D in Srinivasan, S. et al., *Neurochem. Res.*, 24, 1363, 1999. With permission of the publisher (Kluwers/Plenum).)

subunits and could possess up to five sites for any ligand. They are believed to contain at least two sites for agonist and show positive cooperativity (Hill number of about 2) for channel currents activated by GABA agonists and some modulators.[2,4]

Importantly, regional heterogeneity in the brain could be due to multiple GABAR isoforms present in the tissue assayed. The presence of different subunit combinations very likely produces pharmacological subtypes[28,34–37] (discussed further below), although other explanations remain possible.

Finally, there might be heterogeneity in GABAR produced by factors other than subunit composition, such as subcellular localization, interaction with associated proteins, or phosphorylation state of the receptors or the associated proteins[13] (see below).

2.2.2 Pharmacology

2.2.2.1 Regional and Age Differences Detected by Electrophysiology

Some evidence has been presented for heterogeneity in steroid sensitivity of GABAR synapses or GABA-activated currents in neurons in different CNS regions. The concentrations of allopregnanolone and THDOC that enhance GABAR synapses are lower (high n*M*) in hippocampal CA1 neurons and cerebellar Purkinje cells than in dentate gyrus granule cells and hypothalamic neurons (low μ*M*).[33] Likewise, steroid sensitivity has been reported to change in a given region during development, i.e., as a function of age. Thus, the neonatal dentate gyrus granule cells are more sensitive to neuroactive steroids than the same cells in older animals.[13,38]

2.2.2.2 Sex Differences

In addition, there are sex differences in behavior and pharmacology that might be due to differential steroid sensitivity.[10,39–48] Sex differences may reflect differences in neurosteroid levels, neurosteroid-sensitive circuitry, or GABAR sensitivity to neurosteroids, possibly mediated by differential GABAR subunit composition.

2.2.2.3 Plasticity

Changes in steroid sensitivity corresponding to plastic changes in the brain are induced by experiences, including chronic GABAergic drug administration, drug withdrawal, and animal models of epilepsy; these are probably mediated by subunit switches.[33,48–52] Some plasticity may be due to changes in neurosteroid levels *in vivo*, with or without changes in GABAR sensitivity, as seen in acute stress[53] or stress induced by social isolation[54] and in pregnancy.[11] Interestingly, chronic administration to rats of progesterone or its GABAR-active metabolite, allopregnanolone, and withdrawal from the steroid lead to behavioral withdrawal signs, tolerance to steroids and benzodiazepines in some assays, and changes in GABAR subunit composition; such changes may be a model of premenstrual dysphoria.[55] The effect of chronic progesterone and withdrawal on GABAR expression and function can be modeled in primary cultured neurons.[56]

2.2.2.4 The GABAR δ Subunit Knockout Mouse

Gene targeting in mouse producing an animal totally lacking the GABAR δ subunit results in drastically reduced sensitivity to neuroactive steroids behaviorally *in vivo*[57] and *in vitro*, measured at the level of cerebellum neuronal GABAR currents,[58] and hippocampal slice pharmacology and *in vitro* receptor binding.[59] This suggests that the δ subunit-containing GABAR play some specific role in the action of neuroactive steroids (see below).

2.2.2.5 Possible Non-GABAR Mechanisms for Differential Sensitivity to Neuroactive Steroids

Since dentate gyrus granule cells are relatively insensitive to endogenous GABAergic steroids, but highly sensitive to synthetic analogs, there may be a selective metabolic destruction of the endogenous steroids in this brain area.[13] Of course, extracellular levels of endogenous GABA or neurosteroids themselves might also have an effect. Assembly, trafficking, and turnover of GABAR may also affect regional sensitivity to neuroactive steroids, as this is likely to be subunit dependent. Many of these processes are regulated by protein phosphorylation–dephosphorylation.[5] The phosphorylation state of the

GABAR system (GABAR protein or some other protein affecting its function), especially involving protein kinase C (PKC), has been reported to regulate sensitivity to modulation by neuroactive steroids.[60] Inhibitors of PKC reduce or block neuroactive steroid modulation in the hippocampus[61] and hypothalamus.[62] Conversely, a mouse knockout for the ε isoform of PKC shows elevated modulation of GABAR by neuroactive steroids and other modulators *in vivo* and *in vitro*.[63] In none of these cases has the protein that is phosphorylated been identified, so the mechanism of modulation is unclear. These effects are sufficiently slow to potentially involve changes in cell biology, possibly involving subunit switches.

2.3 RECOMBINANT RECEPTOR STUDIES

2.3.1 Subunit Composition Provides Partial Explanation for Heterogeneity in Concentration Dependence of Neuroactive Steroid Modulation of GABAR Binding

We showed (Figure 2.2) that the biphasic curve seen for steroid modulation of [^{35}S]TBPS binding in the cortex but not the cerebellum[27] could be due primarily to different subunit compositions dominating the data: the α1β2γ2 combination (biphasic) is more abundant in the cortex (biphasic) than in the cerebellum (monophasic), where binding was largely due to α6-containing isoforms, which are monophasic. Second, omitting the γ subunit from the α1β2γ2 subunit combination (shows enhanced binding by low concentrations of steroid) also resulted in loss of the enhancement part of the curve in the α1β2 receptors.[31] Note that the degree of enhancement or shape of the steroid concentration dependence has been suggested by Srinivasan et al.[31] to have correlation to channel function, but the exact relationship was not established. Actually, the enhancement phase of the concentration dependence curve for neurosteroid modulation of [^{35}S]TBPS binding (α1βγ2 > α1β or α6β, α6βγ, or α6βδ)[31] appears to correlate with *lower efficacy* of enhancement of GABA currents measured in oocytes.[37] Brain regional heterogeneity was observed for other allosteric ligand binding interactions and explained by variable subunit composition.[64]

2.3.2 Subunit Selectivity for Neuroactive Steroid Modulation of Recombinant GABAR Subunits Expressed in Heterologous Cells

2.3.2.1 Subunit Selectivity Involving α

Over the years, all or most of the subunits of GABAR were tested for steroid sensitivity in recombinant expression studies. In general, all combinations of α, β, and γ subunits respond, while low sensitivity was found for receptors containing δ, ε, and ρ. The status on δ has been updated recently and is described in detail below. All the α subunits form receptors with β or with β and γ that are sensitive to steroids, with respect to both enhancement of GABA and direct channel gating. The potencies and efficacies differ by a small degree, perhaps two- to threefold, although not all studies

TABLE 2.1
Comparison of Neuroactive Steroid (Alloprenanolone) Modulation of Various Human GABAR Subunit Combinations

Human Recombinant Receptor Combination	EC_{50} (n*M*)	I_{max} (%)
$\alpha_1\beta_1$	380 ± 10	143 ± 2
$\alpha_1\beta_1\gamma_1$	559 ± 22	62 ± 8
$\alpha_1\beta_1\gamma_{2L}$	89 ± 6	69 ± 4
$\alpha_1\beta_1\gamma_3$	294 ± 36	74 ± 5
$\alpha_1\beta_2\gamma_{2L}$	177 ± 2	75 ± 5
$\alpha_1\beta_3\gamma_{2L}$	195 ± 36	72 ± 4
$\alpha_2\beta_1\gamma_{2L}$	146 ± 11	66 ± 6
$\alpha_3\beta_1\gamma_{2L}$	74 ± 1	67 ± 7
$\alpha_4\beta_1\gamma_{2L}$	317 ± 25	72 ± 6
$\alpha_4\beta_3\gamma_{2L}$	322 ± 27	80 ± 13
$\alpha_5\beta_1\gamma_{2L}$	302 ± 38	81 ± 2
$\alpha_6\beta_1\gamma_{2L}$	220 ± 12	131 ± 6
$\alpha_4\beta_3\delta$	183 ± 69	171 ± 5
$\alpha_1\beta_1\varepsilon$	ND	15 ± 2

Note: The human GABAR subunit combinations were recombinantly expressed in *Xenopus* oocytes and measured by two-electrode voltage clamp recordings. GABA was bath applied at an EC_{10} concentration, and EC_{50} is the concentration of the steroid required for 50% of the maximum enhancement observed for that receptor. I_{max} is the maximum potentiation of GABA by the steroid relative to the maximum of GABA alone. ND = not determined.

Reproduced from Table 1 in Belelli, D. et al., *Neuropharmacology*, 43, 651, 2002. With permission from the publisher (Elsevier).

were done on the same cells or by the same workers. A systematic comparison of the different α subunits was described,[37] and the results are summarized in Table 2.1. In general, the α1, α3, and α6 subunits gave more responses than α2, α4, and α5 (all combined with β1γ2L) to a given concentration of allopregnanolone, tested at an EC_{10} concentration of GABA. The α1β1 combination gave a larger I_{max} (143% of the maximum current response to GABA) than γ-containing GABAR, except the α6β1γ2L. The α1- and α3-containing receptors had a higher affinity for steroid than the other α values, while the α6β1γ2L gave a larger I_{max} than other α-containing receptors, and the α4- and α6-containing GABAR were the only combinations to give increases in current at saturating GABA concentrations. Furthermore, the α4β3γ2L was not exceptional compared to other α types, but the α4β3δ showed the highest potency and efficacy for steroid enhancement of an EC_{10} concentration of GABA.[37]

2.3.2.2 Subunit Selectivity Involving β

The nature of the β subunit had little effect on neurosteroid sensitivity, with under twofold differences between β1, β2, and β3.[37,65–67] This is in contrast to studies showing selectivity of β2 and β3 over β1-containing GABAR (with α and γ) for enhancement by loreclezole, etomidate, and propofol.[68–71] GABAR composed of homomeric β1 subunits was spontaneously open, activated by pentobarbital and propofol but not steroids. It was weakly activated by GABA, and this was potentiated by steroids and the other drugs.[72] In homomeric β3 GABAR, [^{35}S]TBPS binding that is inhibited by neuroactive steroids and other anesthetics is evident, although there is no binding of GABA or benzodiazepines.[73] Homomeric β3 channels were not spontaneously open and were not activated by GABA, but were activated by propofol and pentobarbital, but not steroids.[74] Similarly, β2 and β3, but not β1, homomers were activated by propofol, etomidate, and pentobarbital, but not GABA or steroids.[75] There is some suggestion that the steroid site of action on the GABAR protein may be on the N-terminal side (first 245 amino acids) of the middle of the transmembrane spanning region 2 (TM2) of the β subunit.[76]

2.3.2.3 Subunit Selectivity Involving γ

The presence of the γ subunit has been shown to increase the sensitivity for steroid modulation with respect to apparent affinity but to decrease the maximal effect.[37]

2.3.2.4 Subunit Selectivity Involving δ

The first report of δ-containing GABAR showed that α6β3δ was not stimulated by neuroactive steroids, while α6β3γ2 was.[77] It was thus somewhat surprising to find that mice lacking the δ subunit became markedly less sensitive to neuroactive steroids *in vivo*[57] and *in vitro*.[58,59] Recently, others have found that the α1β3δ[78] and α4β3δ[79] combinations are not only sensitive to steroids, but more sensitive than the corresponding α4β3γ2 or other α subunit-containing GABAR, as shown in Figure 2.3 and Figure 2.4.[37,79,80] This has been confirmed in our lab (Wallner et al., unpublished, 2003) using α4β2δ. It appears that GABAR composed of the α4β2δ combination is found in nature, and this shows a low efficacy for GABA alone (partial agonism), resulting in a corresponding greater enhancement by steroid. These results are also consistent with a physiological role for the α4βδ combination in tonic inhibition involving extrasynaptic receptors that can respond to endogenous extracellular GABA and spillover.[81–83] The receptors involved in tonic inhibition in the DG (α4βδ) are uniquely sensitive to THDOC at 10 n*M*, a physiologically relevant concentration and one that does not affect GABAR composed of other subunit compositions, which generally require ≥100 n*M*.[84] It is interesting that the α4βδ GABAR may also be uniquely sensitive to other modulators such as exogenous anesthetics,[80,85] including ethanol,[86] suggesting a major effect of CNS drugs on this extrasynaptic receptor as target.

2.3.2.5 Subunit Selectivity Involving ε

Receptors containing the ε subunit are poorly responsive to neurosteroids, as well as other anesthetics.[37,87] However, this has been suggested to be due to overexpression of the ε subunit with some vectors, resulting in an altered stoichiometry of subunit assembly, rather than innate insensitivity.[88]

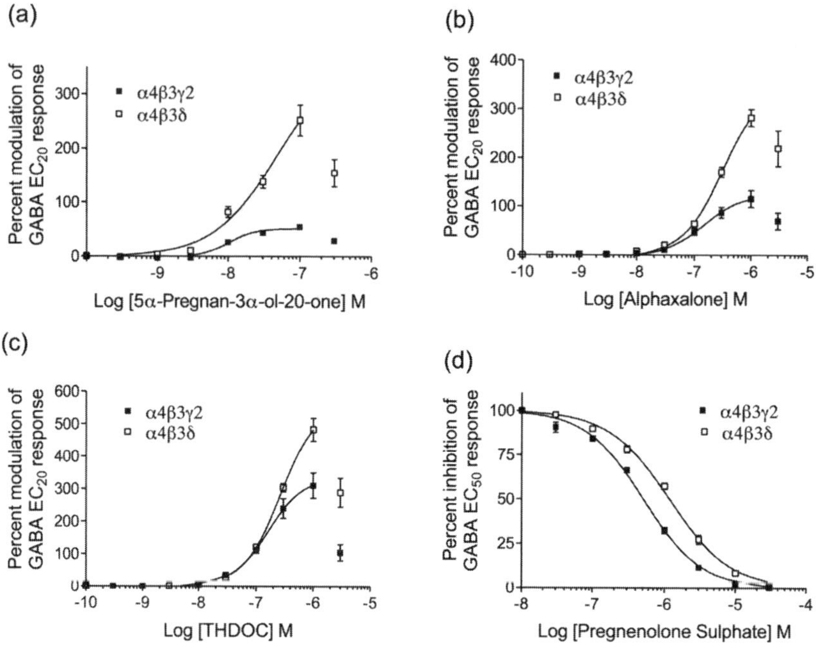

FIGURE 2.3 Effect of the δ subunit on neurosteroid sensitivity. The modulatory effects of various neuroactive steroids on GABA currents were compared in L(-tk) cells stably expressing either α4β2γ2 or α4β3δ GABAR. (Reproduced from Figure 9 in Brown, N. et al., *Br. J. Pharmacol.*, 136, 965, 2002. With permission of the publisher (Nature Publishing Group).

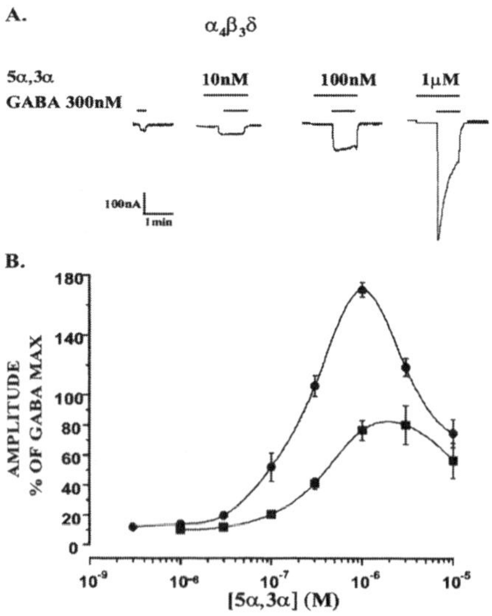

FIGURE 2.4 The influence of the δ subunit on allopregnanlone (5α,3α) modulation of GABA (EC_{10}) in oocytes expressing α4β3δ (●) or α4β3γ2L (■). (Reproduced from Figure 5 in Belelli, D. et al., *Neuropharmacology*, 43, 651, 2002. With permission of the publisher (Elsevier).)

2.3.2.6 Subunit Selectivity Involving ρ

Receptors composed of homomeric ρ subunits are relatively insensitive to neurosteroids (as well as bicuculline, benzodiazepines, and all general anesthetics).[89,90] This property was used in determining residues that might be permissive or required for modulation by alcohol and volatile anesthetics, by comparing chimeras of ρ and non-ρ subunits, and then point mutations. Two important residues were found — S270 and A291 — in the TM2 and TM3 regions of GABAR α subunits. If these residues were mutated into those found in ρ subunits, the sensitivity to anesthetics was lost. The corresponding site(s) for neuroactive steroid modulation is yet to be identified.

REFERENCES

1. Martin, D.L. and Olsen, R.W., *GABA in the Nervous System: The View at Fifty Years*, 1st ed., Lippincott Williams & Wilkins, Philadelphia, 2000.
2. Macdonald, R.L. and Olsen, R.W., $GABA_A$ receptor channels, *Annu. Rev. Neurosci.*, 17, 569, 1994.
3. Olsen, R.W. and Macdonald, R.L., $GABA_A$ receptor complex: structure and function, in *Glutamate and GABA Receptors and Transporters: Structure, Function and Pharmacology*, 1st ed., Krogsgaard-Larsen, P., Egeberg, J., and Schousboe, A., Eds., Taylor & Francis, London, 2002, p. 202.
4. Sieghart, W. and Sperk, G., Subunit composition, distribution and function of $GABA_A$ receptor subtypes, *Curr. Top. Med. Chem.*, 2, 795, 2002.
5. Moss, S.J. and Smart, T.G., Constructing inhibitory synapses, *Nat. Rev. Neurosci.*, 2, 240, 2001.
6. Huidobro-Toro, J.P. et al., Neurochemical actions of anesthetic drugs on the γ-aminobutyric acid receptor-chloride channel complex, *J. Pharmacol. Exp. Ther.*, 242, 963, 1987.
7. Olsen, R.W. et al., Allosteric actions of CNS depressants including anesthetics on subtypes of the inhibitory $GABA_A$ receptor-chloride channel complex, in *Molecular and Cellular Mechanisms of Alcohol and Anesthetics*, Rubin, E., Miller, K.W., and Roth, S.H., Eds., Annals of the New York Academy of Sciences, New York, 1991, p. 145.
8. Haefely, W., Allosteric modulation of the $GABA_A$ receptor channel: a mechanism for interaction with a multitude of central nervous system functions, in *The Challenge of Neuropharmacology*, Mohler, H. and DePrada, M., Eds., Editiones Roche, Basel, 1994, p. 15.
9. Costa, E., From $GABA_A$ receptor diversity emerges a unified vision of GABAergic inhibition, *Annu. Rev. Pharmacol. Toxicol.*, 38, 321, 1998.
10. Paul, S.M. and Purdy, R.H., Neuroactive steroids, *FASEB J.*, 6, 2311, 1992.
11. Concas, A. et al., Role of brain allopregnanolone in the plasticity of γ-aminobutyric acid type A receptor in rat brain during pregnancy and after delivery, *Proc. Natl. Acad. Sci. U.S.A.*, 95, 13284, 1998.
12. Harrison, N.L. and Simmonds, M.A., Modulation of the GABA receptor complex by a steroid anaesthetic, *Brain Res.*, 323, 287, 1984.
13. Lambert, J.J. et al., Modulation of native and recombinant $GABA_A$ receptors by endogenous and synthetic neuroactive steroids, *Brain Res. Brain Res. Rev.*, 37, 68, 2001.

14. Smith, S.S., Waterhouse, B.D., and Woodward, D.J., Locally applied progesterone metabolites alter neuronal responsiveness in the cerebellum, *Brain Res. Bull.*, 18, 739, 1987.
15. Peters, J.A. et al., Modulation of the $GABA_A$ receptor by depressant barbiturates and pregnane steroids, *Br. J. Pharmacol.*, 94, 1257, 1988.
16. Twyman, R.E. and Macdonald, R.L., Neurosteroid regulation of $GABA_A$ receptor single-channel kinetic properties of mouse spinal cord neurons in culture, *J. Physiol.*, 456, 215, 1992.
17. Harrison, N.L. et al., Structure-activity relationships for steroid interaction with the γ-aminobutyric acidA receptor complex, *J. Pharmacol. Exp. Ther.*, 241, 346, 1987.
18. Wittmer, L.L. et al., Enantioselectivity of steroid-induced γ-aminobutyric acidA receptor modulation and anesthesia, *Mol. Pharmacol.*, 50, 1581, 1996.
19. Plassart-Schiess, E. and Baulieu, E.E., Neurosteroids: recent findings, *Brain Res. Brain Res. Rev.*, 37, 133, 2001.
20. Rupprecht, R., Neuroactive steroids: mechanisms of action and neuropsychopharmacological properties, *Psychoneuroendocrinology*, 28, 139, 2003.
21. Majewska, M.D. et al., Steroid hormone metabolites are barbiturate-like modulators of the GABA receptor, *Science*, 232, 1004, 1986.
22. Gee, K.W. et al., Steroid modulation of the chloride ionophore in rat brain: structure-activity requirements, regional dependence and mechanism of action, *J. Pharmacol. Exp. Ther.*, 246, 803, 1988.
23. Turner, D.M. et al., Steroid anesthetics and naturally occurring analogs modulate the γ-aminobutyric acid receptor complex at a site distinct from barbiturates, *J. Pharmacol. Exp. Ther.*, 248, 960, 1989.
24. Morrow, A.L. et al., Characterization of steroid interactions with γ-aminobutyric acid receptor-gated chlorid ion channels: evidence for multiple steroid recognition sites, *Mol. Pharmacol.*, 37, 263, 1990.
25. Hawkinson, J.E. et al., Correlation of neuroactive steroid modulation of [^{35}S]t-butylbicyclophosphorothionate and [^{3}H]flunitrazepam binding and γ-aminobutyric acidA receptor function, *Mol. Pharmacol.*, 46, 977, 1994.
26. Gee, K.W. and Lan, N.C., Gamma-aminobutyric acidA receptor complexes in rat frontal cortex and spinal cord show differential responses to steroid modulation, *Mol. Pharmacol.*, 40, 995, 1991.
27. Sapp, D.W. et al., Regional variation in steroid anesthetic modulation of [^{35}S]TBPS binding to γ-aminobutyric acidA receptors in rat brain, *J. Pharmacol. Exp. Ther.*, 262, 801, 1992.
28. Bureau, M.H. and Olsen, R.W., $GABA_A$ receptor subtypes: ligand binding heterogeneity demonstrated by photoaffinity labeling and autoradiography, *J. Neurochem.*, 61, 1479, 1993.
29. Nguyen, Q. et al., Modulation of $GABA_A$ receptor binding in human brain by neuroactive steroids: species and brain regional differences, *Synapse*, 19, 77, 1995.
30. McCabe, R.T. et al., Osmotic shock: a method to eliminate endogenous γ-aminobutyric acid and account for the influence on benzodiazepine binding affinity in autoradiographic studies, *J. Pharmacol. Exp. Ther.*, 245, 342, 1988.
31. Srinivasan, S. et al., Biphasic modulation of $GABA_A$ receptor binding by steroids suggests functional correlates, *Neurochem. Res.*, 24, 1363, 1999.
32. Prince, R.J. and Simmonds, M.A., Differential antagonism by epipregnanolone of alphaxalone and pregnanolone potentiation of [^{3}H]flunitrazepam binding suggests more than one class of binding site for steroids at $GABA_A$ receptors, *Neuropharmacology*, 32, 59, 1993.

33. Brussaard, A.B. and Herbison, A.E., Long-term plasticity of postsynaptic $GABA_A$-receptor function in the adult brain: insights from the oxytocin neurone, *Trends Neurosci.*, 23, 190, 2000.
34. Lan, N.C. et al., Differential responses of expressed recombinant human γ-aminobutyric acid A receptors to neurosteroids, *J. Neurochem.*, 57, 1818, 1991.
35. Shingai, R., Sutherland, M.L., and Barnard, E.A., Effects of subunit types of the cloned $GABA_A$ receptor on the response to a neurosteroid, *Eur. J. Pharmacol.*, 206, 77, 1991.
36. Puia, G. et al., Does neurosteroid modulatory efficacy depend on $GABA_A$ receptor subunit composition? *Receptors Channels*, 1, 135, 1993.
37. Belelli, D. et al., The influence of subunit composition on the interaction of neurosteroids with $GABA_A$ receptors, *Neuropharmacology*, 43, 651, 2002.
38. Cooper, E.J., Johnston, G.A., and Edwards, F.A., Effects of a naturally occurring neurosteroid on $GABA_A$ IPSCs during development in rat hippocampal or cerebellar slices, *J. Physiol.*, 521 (Pt. 2), 437, 1999.
39. Fink, G. et al., Sex difference in response to alphaxalone anaesthesia may be oestrogen dependent, *Nature*, 298, 270, 1982.
40. Pericic, D., Manev, H., and Lakic, N., Sex differences in the response of rats to drugs affecting GABAergic transmission, *Life Sci.*, 36, 541, 1985.
41. Backstrom, T. et al., Steroids in relation to epilepsy and anaesthesia, *Ciba Found. Symp.*, 153, 225, 1990.
42. Bitran, D., Hilvers, R.J., and Kellogg, C.K., Ovarian endocrine status modulates the anxiolytic potency of diazepam and the efficacy of γ-aminobutyric acid-benzodiazepine receptor-mediated chloride ion transport, *Behav. Neurosci.*, 105, 653, 1991.
43. Johnston, A.L. and File, S.E., Sex differences in animal tests of anxiety, *Physiol. Behav.*, 49, 245, 1991.
44. Kokka, N. et al., Sex differences in sensitivity to pentylenetetrazol but not in $GABA_A$ receptor binding, *Pharmacol. Biochem. Behav.*, 43, 441, 1992.
45. Akinci, M.K. and Johnston, G.A., Sex differences in the effects of acute swim stress on binding to $GABA_A$ receptors in mouse brain, *J. Neurochem.*, 60, 2212, 1993.
46. Finn, D.A. and Gee, K.W., The influence of estrus cycle on neurosteroid potency at the γ-aminobutyric acidA receptor complex, *J. Pharmacol. Exp. Ther.*, 265, 1374, 1993.
47. Maren, S., De Oca, B., and Fanselow, M.S., Sex differences in hippocampal long-term potentiation (LTP) and Pavlovian fear conditioning in rats: positive correlation between LTP and contextual learning, *Brain Res.*, 661, 25, 1994.
48. Reddy, D.S. and Rogawski, M.A., Enhanced anticonvulsant activity of ganaxolone after neurosteroid withdrawal in a rat model of catamenial epilepsy, *J. Pharmacol. Exp. Ther.*, 294, 909, 2000.
49. Devaud, L.L. et al., Sensitization of γ-aminobutyric acidA receptors to neuroactive steroids in rats during ethanol withdrawal, *J. Pharmacol. Exp. Ther.*, 278, 510, 1996.
50. Banerjee, P.K. et al., Absence seizures decrease steroid modulation of t-[^{35}S]butylbicyclophosphorothionate binding in thalamic relay nuclei, *J. Pharmacol. Exp. Ther.*, 287, 766, 1998.
51. Mtchedlishvili, Z., Bertram, E.H., and Kapur, J., Diminished allopregnanolone enhancement of $GABA_A$ receptor currents in a rat model of chronic temporal lobe epilepsy, *J. Physiol.*, 537, 453, 2001.
52. Cagetti, E. et al., Withdrawal from chronic intermittent ethanol treatment changes subunit composition, reduces synaptic function, and decreases behavioral responses to positive allosteric modulators of $GABA_A$ receptors, *Mol. Pharmacol.*, 63, 53, 2003.

53. Barbaccia, M.L. et al., Time-dependent changes in rat brain neuroactive steroid concentrations and $GABA_A$ receptor function after acute stress, *Neuroendocrinology*, 63, 166, 1996.
54. Serra, M. et al., Social isolation-induced decreases in both the abundance of neuroactive steroids and $GABA_A$ receptor function in rat brain, *J. Neurochem.*, 75, 732, 2000.
55. Smith, S.S. et al., $GABA_A$ receptor α4 subunit suppression prevents withdrawal properties of an endogenous steroid, *Nature*, 392, 926, 1998.
56. Follesa, P. et al., Allopregnanolone synthesis in cerebellar granule cells: roles in regulation of $GABA_A$ receptor expression and function during progesterone treatment and withdrawal, *Mol. Pharmacol.*, 57, 1262, 2000.
57. Mihalek, R.M. et al., Attenuated sensitivity to neuroactive steroids in γ-aminobutyrate type A receptor δ subunit knockout mice, *Proc. Natl. Acad. Sci. U.S.A.*, 96, 12905, 1999.
58. Vicini, S., Losi, G., and Homanics, G.E., $GABA_A$ receptor δ subunit deletion prevents neurosteroid modulation of inhibitory synaptic currents in cerebellar neurons, *Neuropharmacology*, 43, 646, 2002.
59. Spigelman, I. et al., Impaired inhibition and neurosteroid modulation of $GABA_A$ receptors in mice lacking the δ subunit, *J. Neurophysiol.*, 90, April 17, 2003.
60. Leidenheimer, N.J. and Chapell, R., Effects of PKC activation and receptor desensitization on neurosteroid modulation of $GABA_A$ receptors, *Brain Res. Mol. Brain Res.*, 52, 173, 1997.
61. Fancsik, A., Linn, D.M., and Tasker, J.G., Neurosteroid modulation of GABA IPSCs is phosphorylation dependent, *J. Neurosci.*, 20, 3067, 2000.
62. Brussaard, A.B. et al., Progesterone-metabolite prevents protein kinase C-dependent modulation of γ-aminobutyric acid type A receptors in oxytocin neurons, *Proc. Natl. Acad. Sci. U.S.A.*, 97, 3625, 2000.
63. Hodge, C.W. et al., Supersensitivity to allosteric $GABA_A$ receptor modulators and alcohol in mice lacking PKCε, *Nat. Neurosci.*, 2, 997, 1999.
64. Korpi, E.R. and Luddens, H., Regional γ-aminobutyric acid sensitivity of t-butylbicyclophosphoro[^{35}S]thionate binding depends on γ-aminobutyric acidA receptor α subunit, *Mol. Pharmacol.*, 44, 87, 1993.
65. Hadingham, K.L. et al., Role of the β subunit in determining the pharmacology of human γ-aminobutyric acid type A receptors, *Mol. Pharmacol.*, 44, 1211, 1993.
66. Sanna, E. et al., Differential subunit dependence of the actions of the general anesthetics alphaxalone and etomidate at γ-aminobutyric acid type A receptors expressed in Xenopus laevis oocytes, *Mol. Pharmacol.*, 51, 484, 1997.
67. Maitra, R. and Reynolds, J.N., Subunit dependent modulation of $GABA_A$ receptor function by neuroactive steroids, *Brain Res.*, 819, 75, 1999.
68. Wingrove, P.B. et al., The modulatory action of loreclezole at the γ-aminobutyric acid type A receptor is determined by a single amino acid in the β2 and β3 subunit, *Proc. Natl. Acad. Sci. U.S.A.*, 91, 4569, 1994.
69. Belelli, D. et al., The interaction of the general anesthetic etomidate with the γ-aminobutyric acid type A receptor is influenced by a single amino acid, *Proc. Natl. Acad. Sci. U.S.A.*, 94, 11031, 1997.
70. Moody, E.J. et al., Distinct loci mediate the direct and indirect actions of the anesthetic etomidate at $GABA_A$ receptors, *J. Neurochem.*, 69, 1310, 1997.
71. Sanna, E. et al., Characterization of the electrophysiological and pharmacological effects of 4-iodo-2,6-diisopropylphenol, a propofol analogue devoid of sedative-anaesthetic properties, *Br. J. Pharmacol.*, 126, 1444, 1999.

72. Sanna, E., Garau, F., and Harris, R.A., Novel properties of homomeric β1 γ-aminobutyric acid type A receptors: actions of the anesthetics propofol and pentobarbital, *Mol. Pharmacol.*, 47, 213, 1995.
73. Slany, A. et al., Rat β3 subunits expressed in human embryonic kidney 293 cells form high affinity [^{35}S]t-butylbicyclophosphorothionate binding sites modulated by several allosteric ligands of γ-aminobutyric acid type A receptors, *Mol. Pharmacol.*, 48, 385, 1995.
74. Davies, P.A., Kirkness, E.F., and Hales, T.G., Modulation by general anaesthetics of rat $GABA_A$ receptors comprised of α1β3 and β3 subunits expressed in human embryonic kidney 293 cells, *Br. J. Pharmacol.*, 120, 899, 1997.
75. Cestari, I.N. et al., The agonistic action of pentobarbital on $GABA_A$ β-subunit homomeric receptors, *Neuroreport*, 7, 943, 1996.
76. Rick, C.E. et al., Neurosteroids act on the $GABA_A$ receptor at sites on the N-terminal side of the middle of TM2, *Neuroreport*, 9, 379, 1998.
77. Zhu, W.J. et al., δ subunit inhibits neurosteroid modulation of $GABA_A$ receptors, *J. Neurosci.*, 16, 6648, 1996.
78. Wohlfarth, K.M., Bianchi, M.T., and Macdonald, R.L., Enhanced neurosteroid potentiation of ternary $GABA_A$ receptors containing the δ subunit, *J. Neurosci.*, 22, 1541, 2002.
79. Adkins, C.E. et al., $\alpha_4\beta_3\delta$ $GABA_A$ receptors characterized by fluorescence resonance energy transfer-derived measurements of membrane potential, *J. Biol. Chem.*, 276, 38934, 2001.
80. Brown, N. et al., Pharmacological characterization of a novel cell line expressing human $\alpha_4\beta_3\delta$ $GABA_A$ receptors, *Br. J. Pharmacol.*, 136, 965, 2002.
81. Jones, A. et al., Ligand-gated ion channel subunit partnerships: $GABA_A$ receptor α6 subunit gene inactivation inhibits δ subunit expression, *J. Neurosci.*, 17, 1350, 1997.
82. Brickley, S.G. et al., Adaptive regulation of neuronal excitability by a voltage-independent potassium conductance, *Nature*, 409, 88, 2001.
83. Nusser, Z. and Mody, I., Selective modulation of tonic and phasic inhibitions in dentate gyrus granule cells, *J. Neurophysiol.*, 87, 2624, 2002.
84. Stell, B.M. and Mody, I., Receptors with different affinities mediate phasic and tonic GABA(A) conductances in hippocampal neurons, *J. Neurosci.*, 22, RC223, 2002.
85. Bai, D. et al., Distinct functional and pharmacological properties of tonic and quantal inhibitory postsynaptic currents mediated by γ-aminobutyric acid(A) receptors in hippocampal neurons, *Mol. Pharmacol.*, 59, 814, 2001.
86. Sundstrom-Poromaa, I. et al., Hormonally regulated $\alpha_4\beta_2\delta$ $GABA_A$ receptors are a target for alcohol, *Nat. Neurosci.*, 5, 721, 2002.
87. Davies, P.A. et al., Insensitivity to anaesthetic agents conferred by a class of $GABA_A$ receptor subunit, *Nature*, 385, 820, 1997.
88. Thompson, S.A. et al., Overexpression of the $GABA_A$ receptor ε subunit results in insensitivity to anaesthetics, *Neuropharmacology*, 43, 662, 2002.
89. Morris, K.D., Moorefield, C.N., and Amin, J., Differential modulation of the γ-aminobutyric acid type C receptor by neuroactive steroids, *Mol. Pharmacol.*, 56, 752, 1999.
90. Mihic, S.J. et al., Sites of alcohol and volatile anaesthetic action on $GABA_A$ and glycine receptors, *Nature*, 389, 385, 1997.

3 THDOC and the $GABA_A$ Receptor

Stefano Vicini

CONTENTS

3.1 HISTORICAL NOTES ON THE EFFECTS OF STEROIDAL DRUGS ON GABAERGIC SYNAPTIC TRANSMISSION: THE PROLONGATION OF DECAY AS MAJOR EFFECTS OF STEROIDS

γ-Aminobutyric acid (GABA), the most abundant inhibitory neurotransmitter of mammalian brain, hyperpolarizes neurons by opening $GABA_A$ receptor's Cl^- channel and is target of many clinically relevant drugs (for review, see Bormann[1] and Korpi et al.[2]). As early as 1984, the steroidal drug alphaxalone was demonstrated to potentiate GABA responses recorded extracellularly from rat brain slices.[3] This finding allowed the additional demonstration that $GABA_A$ receptor function is allosterically modulated by synthetic and endogenous steroids.[4–12] Endogenous steroids, neurosteroids, which may be synthesized in the nervous system in glial cells (for review, see Baulieu and Robel[13] and Mellon et al.[14]), have been demonstrated to have anxiolytic, hypnotic, anesthetic, and anticonvulsant activities (for review, see references 2 and 15–18). Additional studies have shown that neurosteroids may be involved in LTP, memory enhancement, behavioral actions of stress, and neuroprotective effects.[19–23]

Steroids elicit two types of responses in the central nervous system (CNS): an immediate GABAergic action and a delayed response depending on changes in DNA

0-8493-2392-4/04/$0.00+$1.50

transcription mediated by a specific cytoplasmic-nuclear ligand-dependent transcription activator.[24] The GABAergic action of steroids was initially ascribed to a non-specific action on the lipid bilayer of the plasma membrane,[25] but their stereo specificity[26–29] and their failure to act when applied intracellularly[30] supported the existence in the $GABA_A$ receptor of a specific high-affinity steroid binding and regulatory site. However, the kinetic of such a high-efficacy binding site cannot be easily studied with radioligand binding because of the neurosteroid lipophylicity.

Among brain neurosteroids, the mineralocorticoid deoxycorticosterone metabolite allotetrahydrodeoxycorticosterone, 3α,21-dihydroxy-5α-pregnan-20-one (THDOC), is one of the most studied. This and similar steroids at high concentrations can also directly activate Cl^- channels[12,28] similarly to barbiturates but unlike benzodiazepines.[4,9] Early study of the action of THDOC on inhibitory synaptic transmission was performed with voltage-clamp recordings from cultured hippocampal neurons.[7,8] Whole-cell recordings of inhibitory postsynaptic currents (IPSCs) recorded between neurons demonstrated that neurosteroids enhance IPSCs by prolonging the decay time constant with little effects on amplitude or rise time of the synaptic current.[7,8] The longer duration of IPSCs increased the charge transfer and therefore the strength of inhibition.[8,31,32] The increased charge carried by inhibitory synapses very likely underlies sedative and hypnotic effects not too dissimilar from those of barbiturates[4,9] or intravenous and other volatile anesthetic agents.[33–37] Indeed, the prolongation of the decay of $GABA_A$ receptor-mediated IPSCs at central synapses is thought to be a major factor in the production of anesthesia.[38] It is not surprising that the CNS depressant action of THDOC was first discovered many years ago in tests of anesthesia and sedation.[26] In contrast to THDOC, the endogenous progesterone precursor pregnenolone sulfate (PS) negatively modulates the GABA-gated Cl^- currents, although at fairly high concentrations in the micromolar range.[39–41] This negative modulating neurosteroid has a mirror effect on IPSCs, shortening their duration by decreasing the decay time constant. The decrease in charge induced by PS underlies a pro-convulsant action of this endogenous steroid.[39–43]

3.2 THE ISOLATION OF ENDOGENOUS STEROIDS AND MINIMAL STRUCTURAL REQUIREMENT FOR THDOC ACTION

In early work THDOC was shown to be a potent barbiturate-like ligand of the $GABA_A$ receptor–chloride ion channel complex.[4] At nanomolar concentrations the steroid inhibited binding of the convulsant t-butylbicyclophosphorothionate (TBPS) to the GABA receptor complex and increased the binding of the benzodiazepine flunitrazepam. THDOC also stimulated chloride uptake (as measured by uptake of $^{36}Cl^-$) into isolated brain vesicles.[4] The potentiation of GABA-mediated $^{36}Cl^-$ uptake by THDOC was further characterized in rat cerebral cortical synaptoneurosomes.[44] THDOC increased the potency of muscimol to stimulate $^{36}Cl^-$ uptake with biphasic concentration response curves, with Hill coefficients of less than 1. Mathematical modeling of these results suggested that THDOC interacts with multiple binding sites on $GABA_A$ receptors. In addition, THDOC enhanced the ability of the benzodiazepine clonazepam to potentiate muscimol-stimulated $^{36}Cl^-$ uptake. The central benzodiazepine antagonist Ro15-1788 failed to inhibit THDOC-induced potentiation of muscimol-stimulated $^{36}Cl^-$ uptake.[44,45]

Regional differences in rat spinal cord vs. frontal cortex in neuroactive steroid modulation of native $GABA_A$ receptors have been reported with TBPS binding and $^{36}Cl^-$ uptake.[46] At the time of these early studies on neurosteroid modulation of GABA receptors, multiple genes were discovered to encode for the $GABA_A$ receptor subunits, generating the theoretical possibility of an extraordinary structural heterogeneity of these receptors, which is further amplified by alternative splicing (for review, see Barnhard et al.[47] and Whiting et al.[48]). It was therefore appealing to speculate that distinct binding of allosteric modulators on the various $GABA_A$ receptor subunits may allow an even greater level of complexity in the regulation of GABAergic synapses. The role of subunit composition in shaping the intrinsic kinetics of $GABA_A$ receptors,[49–53] together with the finding that the main action of THDOC is the alteration of the kinetics of synaptic responses,[7,8,54] prompts the obvious question of the interactions between all these determinants of GABAergic function.

Early attempts to determine whether specific structural determinants of $GABA_A$ were required to express the allosteric positive modulation of GABA action by neurosteroids[12] demonstrated that although the presence of a heterooligomeric structure that includes γ subunits is required for the expression of the allosteric modulatory action by benzodiazepines,[55] such a structural requirement is not essential for the allosteric modulation of GABA action by neurosteroids.[12] In fact, the progesterone metabolite 3α-OH-DHP was able to modulate GABA-activated Cl^- current either in homomeric (β) or in heterooligomeric ($\alpha\beta$, $\alpha\beta\gamma$) recombinant $GABA_A$ receptors.[12]

In subsequent studies it was investigated whether the steroid modulation varied in potency or in efficacy depending on the type of α or γ subunit included in the $GABA_A$ receptor structure.[56] Again, in contrast with the allosteric modulation of GABA action by benzodiazepines and other anxiolytics,[57] the structure of a given molecular form of the α subunit included in the recombinant $GABA_A$ receptor did not appear to play a major determinant role in optimizing the neurosteroid efficacy or potency as positive modulators, with exceptions noted below. In recombinant $GABA_A$ receptors containing $\alpha1$, $\alpha3$, or $\alpha5$ subunit together with $\beta1\gamma2$ expressed in HEK 293 cells, a neurosteroid at two doses failed to show differences.[56] In a different expression model, i.e., *Xenopus oocytes*, Shingai et al.[58] showed that the efficacy of neurosteroids in modulating GABA responses was greater when $GABA_A$ receptors included the $\alpha1\beta1\gamma2$ subunits than when they expressed the $\alpha3\beta1\gamma2$ subunits. Conversely, using subunit cDNA transfection in HEK 293 cells, Lan et al.[59] demonstrated that the same neurosteroid could increase [^{3}H]-flunitrazepam binding by a greater extent in $\alpha3\beta1\gamma2$ than in $\alpha1\beta1\gamma2$ recombinant receptors. These data were the first indication that although some steroid specificity could be observed with distinct $GABA_A$ receptor subtypes, this specificity was very much dependent upon the heterologous expression system utilized and possibly the different electrophysiological assay.

An intriguing result reported by our laboratory was that the transfection of HEK 293 cells with α1β1 and γ1 subunit cDNAs led to the expression of $GABA_A$ receptors that were modulated more efficaciously by a neurosteroid than when the γ1 subunit was replaced by the γ2 or the γ3 subunits.[57] While this result awaits support from other laboratories, the expression pattern of γ1 subunit in the hypothalamus[60,61] and in glial cells,[62] where neurosteroids are primarily synthesized,[63] suggests that the positive and negative modulation of GABA receptors in these cells may depend on the production of neurosteroids itself. Such a preference for the neurosteroid modulation could be important in those brain pathologies where changes in steroidogenesis are possible.[64]

In recent years a great interest has been raised by the findings of altered sensitivity to neurosteroids and anesthetics by specific $GABA_A$ receptor subunits.[65–67] The presence of the δ subunit diminished the effects of THDOC,[54] and the ε subunit abolished response to anesthetics.[66–68] Incorporation of the ε subunit dramatically reduces[66,69] or has no effect on the GABA-modulatory properties of neurosteroid.[67] Subsequent studies by other groups[70–72] presented evidence for the opposite regulation of recombinant $GABA_A$ receptors expressing the δ subunit by neurosteroids. These results created a discrepancy in the literature on the subtype specificity of neurosteroid action. The underlying reasons for this discrepancy could relate to different expression systems, clones, vectors, and methods of agonist application. Furthermore, reports of neurosteroid modulation of $GABA_A$ receptors being influenced by inhibitors of protein kinases raise the prospect that posttranslational modifications may contribute to this heterogeneity.[73,74] Recent data on transgenic mice lacking the δ subunit may begin to shed some light on the possible reason underlying these discrepancies, as discussed later. In parallel to the definition of subunit requirement, recently the N-terminal side of the middle of TM2 of a subunit has been identified as the recognition site for neurosteroid action.[75]

3.3 THE DISCOVERY OF THE KEY ROLE OF DESENSITIZATION IN GABA CHANNEL DEACTIVATION AND THE ACTION OF THDOC AND OTHER MODULATORS

The activity of ligand-gated ion channels can be described in kinetic terms by defining transition rates between individual conformational states. GABAergic IPSCs are characterized by a biphasic decay.[76,77] Solid evidence demonstrates that the movement among these states and the balance of unbinding, desensitization, and reopening of desensitized receptors underlie the IPSC decay.[78,79] Fast entry and recovery from $GABA_A$ receptor desensitization have been indeed shown to contribute to the IPSC kinetics in hippocampal neurons and cerebellar granule cells.[78,79,51]

With the assumption that drug binding does not introduce new transitions to the kinetic scheme of ligand-gated channel activity, drug action can then be viewed as altering the transition rates between existing states. Pharmacological modulation of the $GABA_A$ receptor by barbiturates has been proposed to alter transition rates between agonist-bound closed states,[16,80] and benzodiazepines increase the agonist binding rate[81] or decrease the agonist unbinding rate and accelerate desensitization.[82] The effects of halothane on GABA responses from HEK 293 cells transfected with α1β2γ2s cDNAs were found to be consistent with a reduction in the microscopic agonist unbinding rate, but not with changes in channel gating steps, such as the channel opening or closing rates, or entry and exit from microscopic desensitization.[38]

As discussed above, neurosteroids potentiate inhibitory synaptic transmission by increasing the duration of the IPSC.[4,8] Neurosteroids affect voltage-gated calcium channels and other presynaptic mechanisms that would influence neurotransmitter release.[83,84] Additional presynaptic factors such as synchronization of transmitter release complicate the interpretation of the mechanism for the effects of allosteric modulators of $GABA_A$ receptor on inhibitory synaptic transmission.[85] Furthermore, the mechanisms of neurosteroid modulation of GABAergic synaptic transmission cannot be easily predicted from studies of neurosteroid effects on steady-state, low-agonist activated currents since synaptic currents are produced by vesicle release that causes transient jumps in high GABA concentrations, inducing desensitizing responses.[78,79]

Neurosteroid effects have been therefore investigated with fast and brief (1 msec) applications of a saturating concentration of GABA (1 m*M*) onto excised nucleated patches[54] to mimic the postsynaptic component of IPSCs.[51,78,86,87] The main effect of THDOC was to prolong the duration of GABA responses just as it prolongs IPSC decay. THDOC also markedly slowed the recovery from desensitization of $GABA_A$ receptors, as estimated with paired pulse GABA application.[54] However, in contrast to neurosteroid-induced potentiation of currents gated by low GABA concentrations,[16,17] THDOC reduced the peak of GABA responses evoked by brief pulses of saturating GABA, and the reduction of peak GABA response was also observed with preperfusion with low concentrations of GABA.[54] This is due to the direct agonist effect of THDOC and other neurosteroids.[4,12] When steroids are preperfused, they induce desensitization of a part of the available receptor population, thereby lowering the peak amplitude.[78,85, 88]

Neurosteroids induce increases in channel open probability when steady-state activation is induced by low agonist concentrations.[11,12,89] This is due to the increased opening frequency for single-channel currents activated at low GABA concentrations in stationary recording conditions, although there is disagreement on neurosteroid effects on burst duration.[12,89] It is possible that distinct effects may relate to distinct molecular structures of the $GABA_A$ receptors investigated in these studies. When the effects of THDOC were studied on GABA-activated channel currents in non-equilibrium conditions, a significant increase was observed in the number of late openings of GABA-gated channels outlasting the removal of the agonist that resulted in a prolonged, slow decay component of the ensemble average.[54] It is likely that the increase in the late opening results from particular combinations of microscopic kinetic parameters of unbinding or entry and exit from desensitized states.

Further demonstration that desensitized states are required for THDOC to modulate GABA responses is derived from the finding that THDOC slowed deactivation only after desensitizing responses to a low-affinity $GABA_A$ receptor agonist — taurine.[90] In fact, THDOC fails to alter deactivation of short-lasting taurine-gated currents.[54] In these responses, the unbinding rate of taurine is very fast and there is little entry into desensitized states, accompanied by an extremely rapid recovery. Therefore, brief pulses (2 msec) of taurine onto outside-out patches evoked responses lacking a second slow exponential decay, as is expected from the reduced desensitization and decreased reopening. A similar mechanism was described for low-affinity agonists L-cysteate and β-alanine in gating N-methyl-D-aspartic acid and $GABA_A$ receptors, respectively.[78,91] When taurine was applied for a long time (2 sec), currents had consistent desensitization, followed by double exponential deactivation kinetics that was prolonged by THDOC.[54] THDOC also increased the fast desensitization rate of prolonged taurine responses but not of GABA responses, probably because fast desensitization of GABA responses is so rapid that THDOC-induced increases of desensitization are difficult to measure.[54]

In comparison to the effect of THDOC, the negative modulating steroid pregnenolone sulfate had distinct effects on currents produced by saturating GABA concentrations.[43] PS speeded deactivation after brief applications of saturating GABA and increased the rate and degree of macroscopic GABA receptor desensitization during prolonged GABA applications. Similarly to THDOC, PS slowed the rate of recovery from desensitization measured with paired GABA applications. It also had only small effects on peak currents and failed to alter deactivation after brief taurine applications.[43] Again, similarly to THDOC, PS augmented macroscopic desensitization and slowed deactivation of currents from prolonged low-affinity agonist applications.[43] These results, together with modeling studies, suggest that PS inhibits GABA-gated chloride currents by enhancing receptor desensitization and stabilizing desensitized states with an increase in the rate of entry into doubly liganded desensitized states.[43]

More insights into the role for desensitization in the mechanisms of regulation of GABA responses derive from studies of fast and brief applications of saturating concentrations of GABA or taurine onto HEK 293 cells transfected with $\alpha 1\beta 2\gamma 2s$ cDNAs in the absence and presence of halothane.[38] In addition to slowing deactivation after a brief pulse of GABA, halothane increased agonist sensitivity and slowed recovery from desensitization but did not alter macroscopic desensitization or maximal activation rate, and only slightly slowed rapid deactivation after taurine application.[38] These results thus appear to support the hypothesis that halothane slows recovery from desensitization. However, slowing the recovery rate from desensitization was found inconsistent by these investigators with the increase in net current and with the increase in the amplitude of low-concentration responses. The pattern of modulation of GABA and taurine responses to halothane was modeled and found to be consistent with slowing dissociation of the agonist from the receptor rather than microscopic desensitization.[38] As discussed by Li and Pearce,[38] given the similarities in actions produced by neurosteroids and halothane, it is possible that neurosteroids also reduce the agonist unbinding rate. Further studies with recombinant $GABA_A$ receptors of defined composition and with parallel kinetic modeling will answer this intriguing proposal.

At present, the results of Zhu and Vicini[54] indicate that the extent of desensitization induced by a brief pulse of GABA directly relates to the prolongation of the slow decay component of GABA-activated currents by neurosteroid. This result suggests that the regulation of desensitization and resensitization could be of crucial importance not only in shaping the IPSCs,[78,79] but also in the allosteric regulation by neurosteroid. THDOC prolongation of decay time of IPSC duration leads to increased inhibition of neuronal firing. However, the result of slowed recovery from desensitization also predicts a reduction of the efficacy of inhibitory synapses during high-frequency activity that may have consequences in the behavioral effects of THDOC.

3.4 EVIDENCE FOR HETEROGENEITY OF THDOC ACTION IN NATIVE GABA RECEPTORS

Different sensitivity to neurosteroid modulation of native $GABA_A$ receptors has been reported since the very beginning of the discovery of the action of these compounds on GABA.[46] A reduced sensitivity of native $GABA_A$ receptor to neurosteroids during the development of granule neurons in culture has been related to the developmental increase of δ subunit expression.[54] A reduced effect of neurosteroids was also reported for $GABA_A$ receptors involved in inhibitory synaptic currents recorded in hippocampal granule neurons in developing rat brain slices,[31] a cell type with abundant expression of δ subunit.[92] These results taken together suggested that the expression of the δ subunit underlined the heterogeneity of THDOC action in native GABA receptors.

It was logical then to investigate the action of the neurosteroid of the δ subunit knockout mice. The deletion of the δ subunit produced overtly normal mice with occasional seizures and electroencephalographic abnormalities. Additionally, a severely reduced sensitivity to the behavioral effects of neurosteroid was observed, including anesthesia, anxiolysis, and pro-absence seizure activity.[93] This result supported studies that indicated that the presence of the δ subunit confers greater sensitivity to the effects of THDOC.[69–72]

Results from studies of the effects of THDOC on spontaneous IPSCs (sIPSCs) from hippocampal and cerebellar granule neurons in mice with a deletion of the δ subunit indicate a reduced neurosteroid sensitivity.[94,95] However, anatomical data show that the δ subunit containing $GABA_A$ receptors in granule neurons is likely extrasynaptic, at least for cerebellar granule neurons.[96] Yet sIPSCs display decreased sensitivity to THDOC in these neurons in δ subunit knockout mice.[95] No difference in neurosteroid modulation was found between stellate cells (which do not express the δ subunit) and granule cells from wild-type mice, and THDOC potentiation was significantly reduced in both cell types from δ knockout mice compared to controls.[95] Taken together, these data suggest that some sort of compensatory response involving the maintenance of the capability of neurosteroids to modulate $GABA_A$ receptors could underlie the observed reduction of THDOC action in δ knockout mice. We proposed that the compensatory effect could be generated by an enhanced action of THDOC[65] on extrasynaptic receptors lacking the δ subunit in the knockout mice and is transferred to synaptic receptors in both granule and stellate neurons.[95] It remains to be seen if this reduction in THDOC sensitivity is similarly transferred to inhibitory synapses throughout the CNS. This could explain the observed reduction of neurosteroid-induced behavioral responses in δ knockout mice.[93] An alternative explanation[97] proposes that developmental changes secondary to the δ knockout are responsible for the altered sensitivity to the neurosteroid, since the total number of $GABA_A$ receptors in the whole brain (even in regions where δ is not expressed) was halved in the knockout.[93]

Recently the modulation of currents resulting from tonic activation of the $GABA_A$ receptor in cerebellar granule neurons deriving from spillover of GABA from inhibitory synapses[98] has been reported.[97] This current was potentiated by THDOC in cells from P6 rats but was unaffected in cells from adult rats by both THDOC and PS.[97] In addition, the early component of the IPSC in granule neurons, which is thought to be generated by synaptic receptors[99] that lack δ subunits, shows high sensitivity to neurosteroids.[97] It is likely that the receptors mediating the tonic and spillover IPSC currents contain α6 and δ subunits that are extrasynaptic.[96]

This supports the data on the presence of the δ subunit decreasing the sensitivity of α6 subunit-containing receptors to neurosteroids.[65] Normal neurosteroid sensitivity of sIPSCs in neurons from δ –/– mice was restored by experimentally stimulating intracellular protein kinase C.[95] These data supported the earlier evidence that neurosteroid sensitivity can be modulated with feedback-regulatory mechanisms.[74] Additional functional heterogeneity of neurosteroid action unrelated to the expression of the δ subunit comes from reported striking regional differences in the modulation of ^{35S}TBPS binding between strata radiatum and strata oriens of the hippocampus, even within the apical vs. basal dendrites of pyramidal neurons.[100]

In conclusion, it is quite likely that neurosteroid activity is under control of both subunit composition and posttranslational mechanisms, and this could in part explain discrepancies in the literature of results with recombinant receptors.

REFERENCES

1. Bormann, J., The 'ABC' of GABA receptors, *Trends Pharmacol. Sci.*, 21, 16, 2000.
2. Korpi, E.R., Grunder, G., and Luddens, H., Drug interactions at $GABA_A$ receptors, *Prog. Neurobiol.*, 67, 113, 2002.
3. Harrison, N.L. and Simmonds, M.A., Modulation of the GABA receptor complex by a steroid anaesthetic, *Brain Res.*, 323, 287, 1984.
4. Majewska, M.D. et al., Steroid hormone metabolites are barbiturate-like modulators of the GABA receptor, *Science*, 232, 1004, 1986.
5. Crawley, J.N. et al., Anxiolytic activity of an endogenous adrenal steroid, *Brain Res.*, 398, 382, 1986.
6. Callachan, H. et al., Modulation of the $GABA_A$ receptor by progesterone metabolites, *Proc. R. Soc. Lond. B Biol. Sci.*, 231, 359, 1987.
7. Barker, J.L. et al., Potentiation of gamma-aminobutyric-acid-activated chloride conductance by a steroid anaesthetic in cultured rat spinal neurons, *J. Physiol. (Lond.)*, 386, 485, 1987.
8. Harrison, N.L., Vicini, S., and Barker J.L., A steroid anesthetic prolongs inhibitory post-synaptic currents in cultured rat hippocampal neurons, *J. Neurosci.*, 7, 604, 1987.
9. Peters, J.A. et al., Modulation of the $GABA_A$ receptor by depressant barbiturates and pregnane steroids, *Br. J. Pharmacol.*, 94, 1257, 1988.

10. Turner, D.M. et al., Steroid anesthetics and naturally occurring analogs modulate the gamma-aminobutyric acid receptor complex at a site distinct from barbiturates, *J. Pharmacol. Exp. Ther.*, 248, 960, 1989.
11. Mistry, D.K. and Cottrell, G.A., Actions of steroids and bemegride on the GABAA receptor of mouse spinal neurones in culture, *Exp. Physiol.*, 75, 199, 1990.
12. Puia, G. et al., Neurosteroids act on recombinant human $GABA_A$ receptors, *Neuron*, 4, 759, 1990.
13. Baulieu, E.E. and Robel, P., Neurosteroids: a new brain function? *J. Steroid Biochem. Mol. Biol.*, 37, 395, 1990.
14. Mellon, S.H., Griffin, L.D., and Compagnone, N.A., Biosynthesis and action of neurosteroids, *Brain Res. Brain Res. Rev.*, 3, 2001.
15. Paul, S.M. and Purdy, R.H., Neuroactive steroids, *FASEB J.*, 6, 2311, 1992.
16. MacDonald, R.L. and Olsen R.W., $GABA_A$ receptor channels, *Annu. Rev. Neurosci.*, 17, 569, 1994.
17. Lambert J.J. et al., Neurosteroids and $GABA_A$ receptor function, *Trends Pharmacol. Sci.*, 16, 295, 1995.
18. Lambert J.J. et al., Modulation of native and recombinant $GABA_A$ receptors by endogenous and synthetic neuroactive steroids, *Brain Res. Brain Res. Rev.*, 37, 68, 2001.
19. Purdy, R.H. et al., Stress-induced elevations of $GABA_A$ receptor-active steroids in the rat brain, *Proc. Natl. Acad. Sci. U.S.A.*, 88, 4553, 1991.
20. Frye, C.A., The neurosteroid 3α,5α-THP has antiseizure and possible neuroprotective effects in an animal model of epilepsy, *Brain Res.*, 696, 113, 1995.
21. Steffensen, S.C., Dehydroepiandrosterone sulfate suppresses hippocampal recurrent inhibition and synchronizes neuronal activity to theta rhythm, *Hippocampus*, 5, 320, 1995.
22. Yoo, A., Harris, J., and Dubrovsky, B., Dose-response study of dehydroepiandrosterone sulfate on dentate gyrus long-term potentiation, *Exp. Neurol.*, 137, 151, 1996.
23. Reddy, D.S. and Rogawski, M.A., Stress-induced deoxycorticosterone-derived neurosteroids modulate $GABA_A$ receptor function and seizure susceptibility, *J. Neurosci.*, 22, 3795, 2002.
24. Yamamoto, K.R., Steroid receptor regulated transcription of specific genes and gene networks, *Annu. Rev. Genet.*, 19, 209, 1985.
25. Lawrence, D.K. and Gill, E.W., Structurally specific effects of some steroid anesthetics on spin-labeled liposomes, *Mol. Pharm.*, 11, 280, 1975.
26. Selye, H., Studies concerning the anaesthetic action of steroid hormones, *J. Pharmacol. Exp. Ther.*, 105, 27, 1941.
27. Selye, H., Correlations between the chemical structure and the pharmacological actions of the steroids, *Endocrinology*, 3, 437, 1942.
28. Harrison, N.L. et al., Structure-activity relationships for steroid interaction with the gamma-aminobutyric $acid_A$ receptor complex, *J. Pharmacol. Exp. Ther.*, 241, 346, 1987.
29. Gee, K.W. et al., Steroid modulation of the chloride ionophore in rat brain: structure-activity requirements, regional dependence and mechanism of action, *J. Pharmacol. Exp. Ther.*, 246, 803, 1988.
30. Lambert, J.J. et al., Steroid modulation of the $GABA_A$ receptor complex: electrophysiological studies, *Ciba Found. Symp.*, 153, 56, 1990 (discussion, 153, 71, 1990).
31. Cooper, E.J., Johnston, G.A., and Edwards, F.A., Effects of a naturally occurring neurosteroid on $GABA_A$ IPSCs during development in rat hippocampal or cerebellar slices, *J. Physiol. (Lond.)*, 521, 437, 1999.

32. Vicini, S. et al., $GABA_A$ receptor α1 subunit deletion prevents developmental changes of inhibitory synaptic currents in cerebellar neurons, *J. Neurosci.*, 21, 3009, 2001.
33. Zimmerman, S.A., Jones, M.V., and Harrison, N.L., Potentiation of $GABA_A$ receptor Cl^- current correlates with *in vivo* anesthetic potency, *J. Pharmacol. Exp. Ther.*, 270, 987, 1994.
34. Gage, P.W. and Robertson, B., Prolongation of inhibitory postsynaptic currents by pentobarbitone, halothane and ketamine in CA1 pyramidal cells in rat hippocampus, *Br. J. Pharmacol.*, 85, 675, 1985.
35. Mody, I., Tanelian, D.L., and MacIver, M.B., Halothane enhances tonic neuronal inhibition by elevating intracellular calcium, *Brain Res.*, 538, 319, 1991.
36. Jones, M.V. and Harrison, N.L., Effects of volatile anesthetics on the kinetics of inhibitory postsynaptic currents in cultured rat hippocampal neurons, *J. Neurophysiol.*, 70, 1339, 1993.
37. Pearce, R.A., Volatile anesthetic enhancement of paired-pulse depression investigated in the rat hippocampus in vitro, *J. Physiol. (Lond.)*, 492, 823, 1996.
38. Li, X. and Pearce, R.A., Effects of halothane on $GABA_A$ receptor kinetics: evidence for slowed agonist unbinding, *J. Neurosci.*, 20, 899, 2000.
39. Majewska, M.D. and Schwartz, R.D., Pregnenolone-sulfate: an endogenous antagonist of the gamma-aminobutyric acid receptor complex in brain? *Brain Res.*, 404, 355, 1987.
40. Majewska, M.D., Mienville, J.M., and Vicini, S., Neurosteroid pregnenolone-sulfate antagonizes electrophysiological responses to GABA in neurons, *Neurosci. Lett.*, 90, 279, 1988.
41. Wu, F.S., Gibbs, T.T., and Farb, D.H., Inverse modulation of γ-aminobutyric acid- and glycine-induced currents by progesterone, *Mol. Pharmacol.*, 37, 597, 1990.
42. Park-Chung, M. et al., Sulfated and unsulfated steroids modulate $GABA_A$ receptor function through distinct sites, *Brain Res.*, 830, 72, 1999.
43. Shen W. et al., Pregnenolone sulfate modulates inhibitory synaptic transmission by enhancing $GABA_A$ receptor desensitization, *J. Neurosci.*, 20, 3571, 2000.
44. Morrow, A.L. et al., Characterization of steroid interactions with gamma-aminobutyric acid receptor-gated chloride ion channels: evidence for multiple steroid recognition sites, *Mol. Pharmacol.*, 37, 263, 1990.
45. Morrow, A.L. et al., Neuroactive steroid modulators of the stress response, *Ann. N.Y. Acad. Sci.*, 771, 257, 1995.
46. Gee, K.W. and Lan, N.C., Gamma-aminobutyric $acid_A$ receptor complexes in rat frontal cortex and spinal cord show differential responses to steroid modulation, *Mol. Pharmacol.*, 40, 995, 1991.
47. Barnard, E.A. et al., International Union of Pharmacology. XV. Subtypes of gamma-aminobutyric $acid_A$ receptors: classification on the basis of subunit structure and receptor function, *Pharmacol. Rev.*, 50, 291, 1988.
48. Whiting, P.J. et al., Molecular and functional diversity of the expanding $GABA_A$ receptor gene family, *Ann. N.Y. Acad. Sci.*, 868, 645, 1999.
49. Verdoorn, T.A., Formation of heteromeric $GABA_A$ receptors containing two different alpha subunits, *Mol. Pharmacol.*, 45, 475, 1994.
50. Gingrich, K.J., Roberts, W.A., and Kass, R.S., Dependence of the $GABA_A$ receptor gating kinetics on the α-subunit isoform: implications for structure-function relations and synaptic transmission, *J. Physiol. (Lond.)*, 489.2, 529, 1995.
51. Tia, S. et al., Distinct deactivation and desensitization kinetics of recombinant $GABA_A$ receptors, *Neuropharmacology*, 35, 1375, 1996.

52. Lavoie, A.M. et al., Activation and deactivation rates of recombinant $GABA_A$ receptor channels are dependent on α subunit isoform, *Biophys. J.*, 73, 2518, 1997.
53. McClellan, A.M and Twyman, R., Receptor system response kinetics reveal functional subtypes of native murine and recombinant human $GABA_A$ receptors, *J. Physiol. (Lond.)*, 515, 711, 1999.
54. Zhu, W.J. and Vicini S., Neurosteroid prolongs $GABA_A$ channel deactivation by altering kinetics of desensitized states, *J. Neurosci.*, 17, 4022, 1997.
55. Pritchett, D.B., Luddens, H., and Seeburg, P.H., Type I and Type II $GABA_A$-benzodiazepine receptors produced in transfected cells, *Science*, 245, 1389, 1989.
56. Puia, G. et al., Does neurosteroid modulatory efficacy depend on GabaA receptor subunit composition? *Receptors Channels*, 1, 135, 1993.
57. Puia, G. et al., Influence of recombinant $GABA_A$ receptor subunit composition on the action of allosteric modulators of GABA-gated Cl^- currents, *Mol. Pharm.*, 36, 691, 1991.
58. Shingai, R., Sutherland, M.L., and Barnard, E., Effects of subunit types of the cloned GABA receptor on the response to a neurosteroid, *Eur. J. Pharm.*, 206, 77, 1991.
59. Lan, N.C. et al., Differential responses of expressed recombinant human γ- aminobutyric $acid_A$ receptors to neurosteroids, *J. Neurochem.*, 57, 1818, 1991.
60. Wisden, W. et al., The distribution of 13 $GABA_A$ receptor subunit mRNAs in the rat brain. I. Telencephalon, diencephalon, mesencephalon, *J. Neurosci.*, 12, 1040, 1992.
61. Nett, S.T. et al., Properties and sex-specific differences of $GABA_A$ receptors in neurons expressing $\gamma 1$ subunit mRNA in the preoptic area of the rat, *J. Neurophysiol.*, 81, 192, 1999.
62. Bovolin, P. et al., Expression patterns of gamma-aminobutyric acid type A receptor subunit mRNAs in primary cultures of granule neurons and astrocytes from neonatal rat cerebella, *Proc. Natl. Acad. Sci. U.S.A.*, 89, 9344, 1992.
63. Jung-Testas, I. et al., Neurosteroids: biosynthesis of pregnenolone and progesterone in primary cultures of rat glial cells, *Endocrinology*, 125, 2083, 1989.
64. Majewska, M.D., Neurosteroids: endogenous bimodal modulators of the $GABA_A$ receptor: mechanism of action and physiological significance, *Prog. Neurobiol.*, 38, 379, 1992.
65. Zhu, W.J. et al., Delta subunit inhibits neurosteroid modulation of $GABA_A$ receptors, *J. Neurosci.*, 16, 6648, 1996.
66. Davies, P.A. et al., Insensitivity to anaesthetic agents conferred by a class of $GABA_A$ receptor subunit, *Nature*, 385, 820, 1997.
67. Whiting, P.J. et al., Neuronally restricted RNA splicing regulates the expression of a novel $GABA_A$ receptor subunit conferring atypical functional properties, *J. Neurosci.*, 17, 5027, 1997.
68. Thompson, S.A. et al., Overexpression of the $GABA_A$ receptor epsilon subunit results in insensitivity to anaesthetics, *Neuropharmacology*, 43, 662, 2002.
69. Belelli, D. et al., The influence of subunit composition on the interaction of neurosteroids with $GABA_A$ receptors, *Neuropharmacology*, 43, 651, 2002.
70. Adkins, C.E. et al., $\alpha l4\beta 3\delta$ $GABA_A$ receptors characterized by fluorescence resonance energy transfer-derived measurements of membrane potential, *J. Biol. Chem.*, 276, 38934, 2001.
71. Brown, N. et al., Pharmacological characterization of a novel cell line expressing human $\alpha 4\beta 3\delta$ $GABA_A$ receptors, *Br. J. Pharmacol.*, 136, 965, 2002.
72. Wohlfarth, K.M., Bianchi, M.T., and Macdonald, R.L., Enhanced neurosteroid potentiation of ternary $GABA_A$ receptors containing the δ subunit, *J. Neurosci.*, 22, 1541, 2002.

73. Leidenheimer, N.J. and Chapell, R., Effects of PKC activation and receptor desensitization on neurosteroid modulation of $GABA_A$ receptors, *Brain Res. Mol. Brain Res.*, 52, 173, 1997.
74. Fancsik, A., Linn, D.M., and Tasker, J.G. Neurosteroid modulation of GABA IPSCs is phosphorylation dependent, *J. Neurosci.*, 20, 3067, 2000.
75. Rick, C.E. et al., Neurosteroids act on the $GABA_A$ receptor at sites on the N-terminal side of the middle of TM2, *Neuroreport*, 9, 379, 1998.
76. Edwards, F.A., Konnerth, A., and Sakmann, B., Quantal analysis of inhibitory synaptic transmission in the dentate gyrus of rat hippocampal slices: a patch clamp study, *J. Physiol. (Lond.)*, 430, 213, 1990.
77. Pearce, R.A., Physiological evidence for two distinct $GABA_A$ responses in rat hippocampus, *Neuron*, 10, 189, 1993.
78. Jones, M.V. and Westbrook, G.L., Desensitization states prolong $GABA_A$ channels response to brief agonist pulses, *Neuron*, 15, 181, 1995.
79. Jones, M.V. and Westbrook, G.L. The impact of receptor desensitization on fast synaptic transmission, *Trends Neurosci.*, 19, 96, 1996.
80. MacDonald, R.L., Rogers, C.J., and Twyman, R.E., Barbiturate regulation of kinetic properties of the $GABA_A$ receptor channel of mouse spinal neurones in culture, *J. Physiol. (Lond.)*, 417, 483, 1989.
81. Rogers, C.J., Twyman, R.E., and Macdonald, R.L., Benzodiazepine and β-carboline regulation of single $GABA_A$ receptor channels in spinal neurones in culture, *J. Physiol. (Lond.)*, 475, 69, 1994.
82. Mellor, J.R. and Randall, A.D., Frequency-dependent actions of benzodiazepines on $GABA_A$ receptors in cultured murine cerebellar granule cells, *J. Physiol. (Lond.)*, 503, 353, 1997.
83. Ffrench-Mullen, J.M., Interaction of 3β,5α-tetrahydrodeoxycorticosterone in rat and guinea-pig neurons: a comparison of Ca^{2+}- and $GABA_A$-Cl^--channel current modulation, *Steroids*, 64, 76, 1999.
84. Poisbeau, P., Feltz, P., and Schlichter, R., Modulation of $GABA_A$ receptor-mediated IPSCs by neuroactive steroids in a rat hypothalamo-hypophyseal coculture model, *J. Physiol. (Lond.)*, 500, 475, 1997.
85. Mody, I. et al., Bridging the cleft at GABA synapses in the brain, *Trends Neurosci.*, 17, 517, 1994.
86. Puia, G., Costa, E., and Vicini, S., Functional diversity of GABA-activated Cl^- currents in Purkinje versus granule neurons in rat cerebellar slices, *Neuron*, 12, 117, 1994.
87. Maconochie, D.J., Zempel, J.M., and Steinbach, J.H., How quickly can $GABA_A$ receptors open? *Neuron*, 12, 61, 1994.
88. Celentano, J.J. and Wong R.K., Multiphasic desensitization of the GABAA receptor in outside-out patches, *Biophys. J.*, 66, 1039, 1994.
89. Twyman, R.E. and Macdonald, R.L., Neurosteroid regulation of $GABA_A$ receptor single-channel kinetic properties of mouse spinal cord neurons in culture, *J. Physiol. (Lond.)*, 456, 215, 1992.
90. Huxtable, R.J., Physiological actions of taurine, *Physiol. Rev.*, 72, 101, 1992.
91. Lester, R.A. and Jahr, C.E., NMDA channel behavior depends on agonist affinity, *J. Neurosci.*, 12, 635, 1992.
92. Laurie, D.J., Seeburg, P.H., and Wisden, W., The distribution of thirteen $GABA_A$ receptor subunit mRNAs in the rat brain. II. Olfactory bulb and cerebellum, *J. Neurosci.*, 12, 1063, 1992.

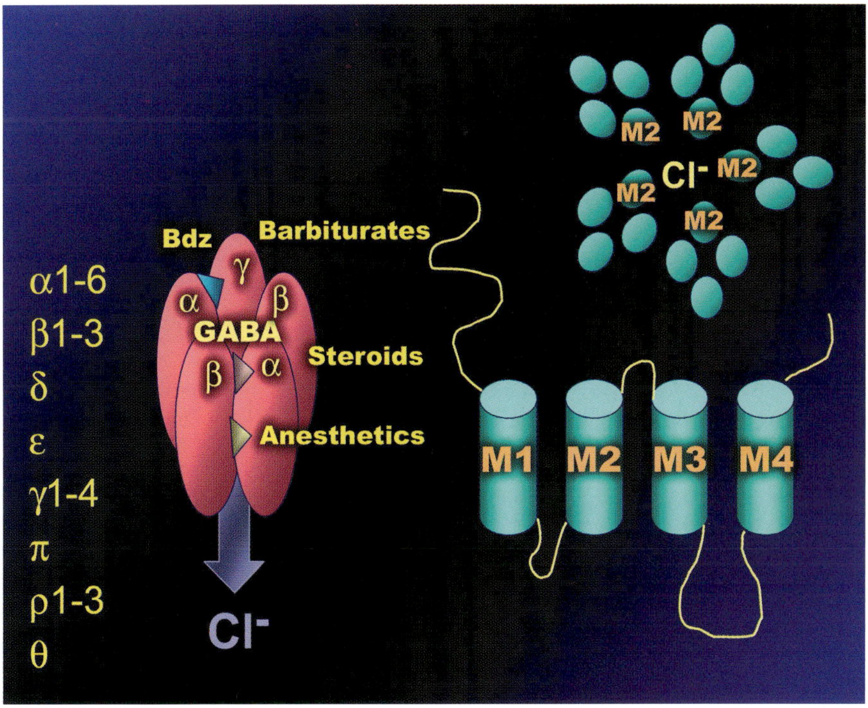

Diagram of the GABA$_A$ receptor (GABAR): A pentameric structure comprised of combinations of various α and β subunits, which also contain δ, γ, ε, π, or θ. Atypical GABAR can also be formed of ρ subunits. Each subunit in turn is comprised of 4 trans-membrane spanning α-helices, where M2 lines a central chloride ion channel. Binding of two GABA molecules at the interface between α and β gates opens the channel to permit chloride ion flow. Many modulators exist, including neurosteroids such as 3α, 5α-THP, which enhance chloride current, thereby increasing inhibition. The particular combination of subunits for each receptor can influence both the biophysical and pharmacological properties of the receptor (Bdz, benzodiazepine).

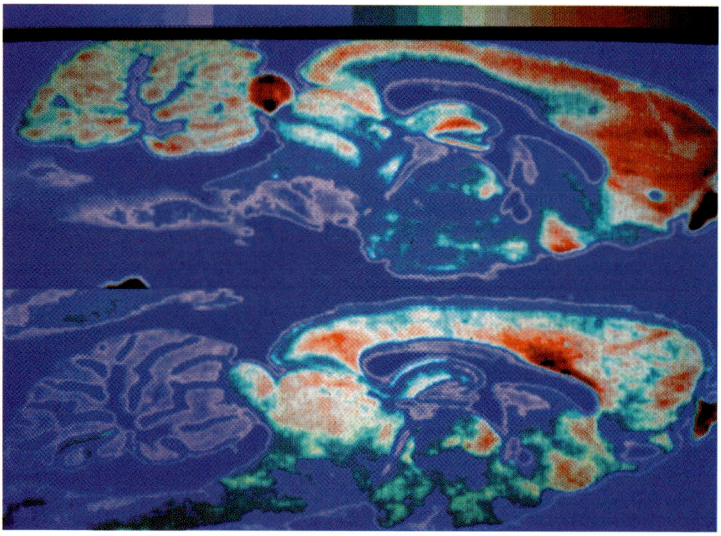

COLOR FIGURE 2.1 Brain regional modulation by neuroactive steroid alphaxalone of [^{35}S]TBPS binding to GABAR in rat brain frozen and thawed, unfixed thin sagittal sections by autoradiography. Color scale at the top indicates increased absorbance (radioligand receptor binding). Top: Control, no steroid. Bottom: alphaxalone, 10 μ*M*. (Reproduced from Figure 2 in Sapp, D.W. et al., *J. Pharmacol. Exp. Ther.*, 262, 801, 1992. With permission of the publisher (ASPET).)

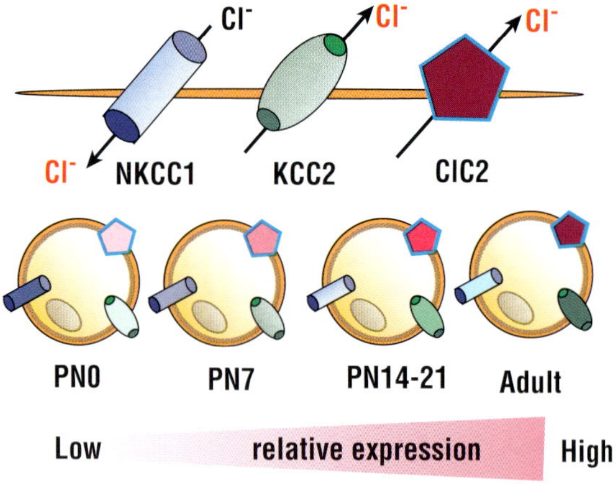

COLOR FIGURE 8.2 Chloride cotransporters. The transmembrane concentration gradient of chloride is the primary determinant of a depolarizing vs. hyperpolarizing effect of $GABA_A$ receptor activation. Developmentally, a depolarizing response gradually gives way to hyperpolarization as neurons mature. This change in membrane excitability following $GABA_A$ receptor activation is dictated by changes in the expression and activity of two electroneutral chlorides, potassium and sodium cotransporters. The cotransporter NKCC1 transports chloride into cells and is expressed at high levels in immature neurons but gradually declines as development progresses. In contrast, KCC2, which transports chloride out of cells, is found at very low levels in immature neurons but increases in expression as neurons mature. The inwardly rectifying chloride channel, ClC2, transports chloride out of cells as well and is expressed in mature adult neurons.

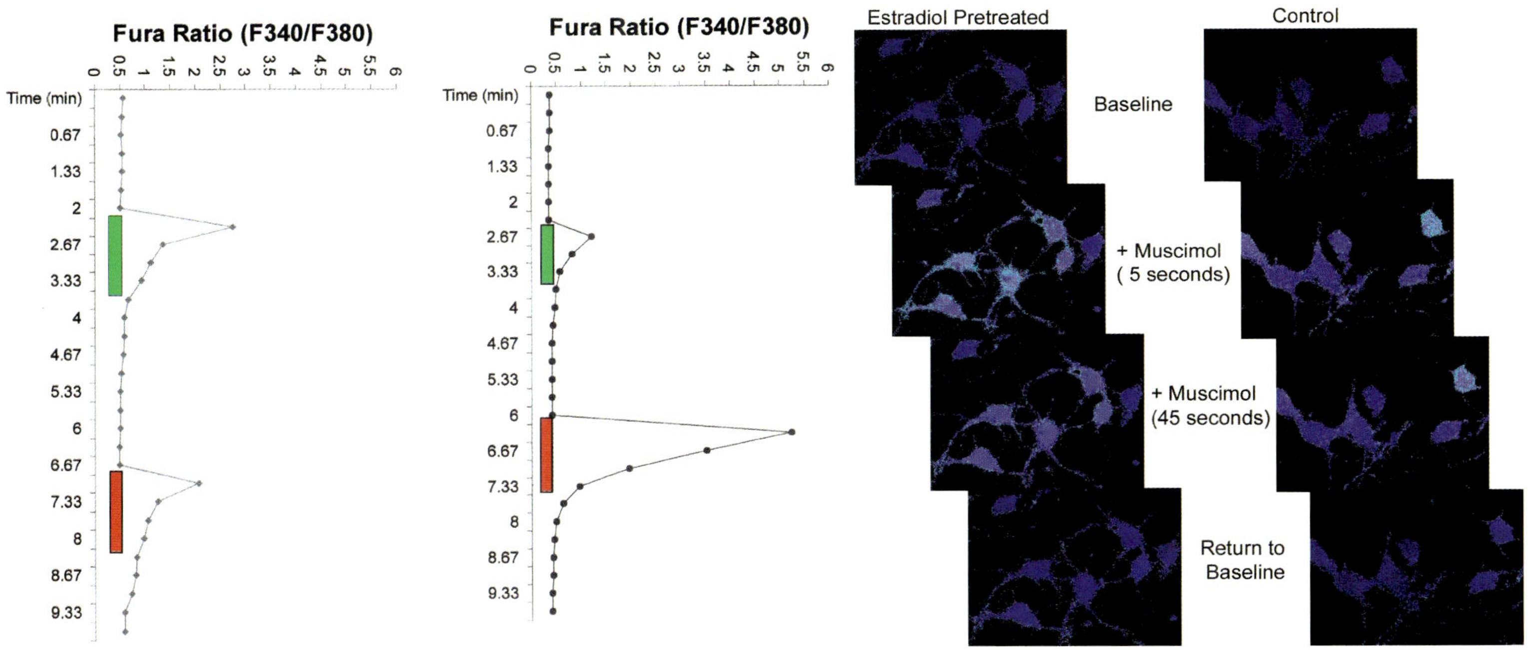

COLOR FIGURE 8.4 Estradiol increases calcium influx induced by depolarizing GABA. Intracellular free calcium was monitored with the use of the calcium-sensitive dye, Fura-AM. Application of 10 μ*M* glutamate induces a moderate influx of calcium (green bar, left panel, top) compared to equimolar muscimol (red bars). In cells that are pretreated with estradiol (left panel, top), the influx of calcium is about twice the magnitude seen in cells that were not pretreated with estradiol (left panel, bottom). The photomicrograph in the right panel illustrates the time course of calcium flux in estradiol pretreated and control cells, with the lighter color indicating greater activation of Fura by free calcium.

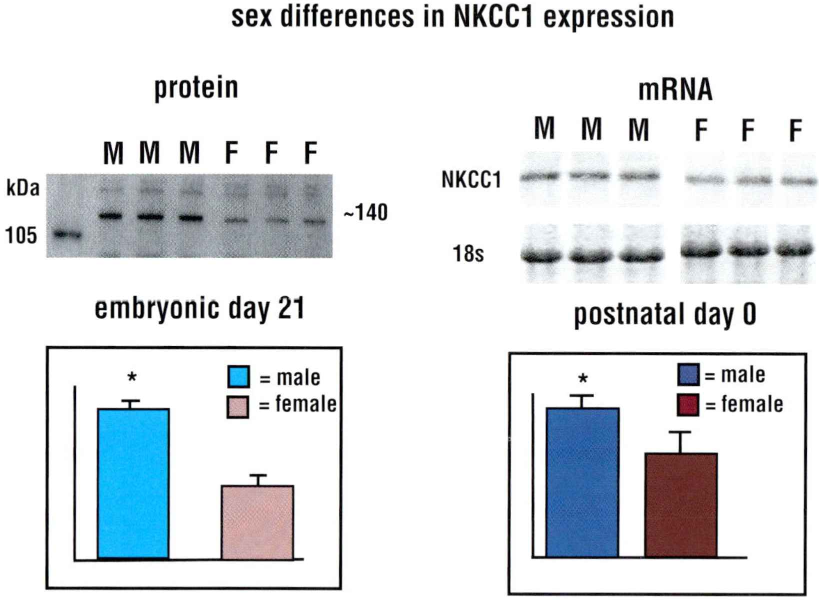

COLOR FIGURE 8.5 NKCC1 levels are higher in the hypothalamus of neonatal males. The cotransporter NKCC1 transports chloride into cells and increases the ability of $GABA_A$ receptor activation to induce a depolarizing excitatory response. Analysis of NKCC1 protein levels by Western blot of gestational day 21 fetuses indicates higher levels in the hypothalamus of males compared to females. Analysis of mRNA levels by Northern blot on the day of birth indicates that the sex difference persists to this developmental time point.

93. Mihalek, R.M. et al., Attenuated sensitivity to neuroactive steroids in γ-aminobutyrate type A receptor δ subunit knockout mice, *Proc. Natl. Acad. Sci. U.S.A.*, 96, 12905, 1999.
94. Spigelman, I. et al., Behavior and physiology of mice lacking the $GABA_A$-receptor δ subunit, *Epilepsia*, 43, 3, 2002.
95. Vicini, S., Losi G., and Homanics, G., $GABA_A$ receptor delta subunit deletion prevents neurosteroid modulation of inhibitory synaptic currents in cerebellar neurons, *Neuropharmacology*, 43, 646, 2002.
96. Nusser, Z., Sieghart, W., and Somogyi P., Segregation of different $GABA_A$ receptors to synaptic and extrasynaptic membranes of cerebellar granule cells, *J. Neurosci.*, 18, 1693, 1998.
97. Hamann, M., Rossi, D.J., and Attwell, D., Tonic and spillover inhibition of granule cells control information flow through cerebellar cortex, *Neuron*, 33, 625, 2002.
98. Brickley, S.G., Cull-Candy, S.G., and Farrant, M., Development of a tonic form of synaptic inhibition in rat cerebellar granule cells resulting from persistent activation of $GABA_A$ receptors, *J. Physiol. (Lond.)*, 497, 753, 1996
99. Rossi, D.J. and Hamann, M., Spillover-mediated transmission at inhibitory synapses promoted by high affinity α6 subunit $GABA_A$ receptors and glomerular geometry, *Neuron*, 20, 783, 1998.
100. Orchinik, M. et al., Heterogeneity of hippocampal $GABA_A$ receptors: regulation by corticosterone, *J. Neurosci.*, 21, 330, 2001.

4 Neurosteroid Modulation of δ Subunit-Containing $GABA_A$ Receptor Channels

Matt T. Bianchi and Robert L. Macdonald

CONTENTS

4.1 INTRODUCTION

In the more than 15 years since the first demonstration of neurosteroid interaction with $GABA_A$ receptor channels (Majewska et al., 1986), significant progress has been made in terms of the diversity of steroid compounds capable of acutely and reversibly modulating receptor function, the isoform dependence of neurosteroid interactions, and the potential roles of endogenous neurosteroids in the physiology and pathophysiology of central nervous system (CNS) function. Together with increased understanding of the regional as well as subcellular expression patterns of various $GABA_A$ receptor isoforms, there is continued opportunity for investigating the complex effects of neurosteroids on GABA-ergic inhibition. Although extensive work has targeted the most (experimentally) salient aspect of inhibition, the inhibitory postsynaptic response, there is increasing evidence for the potential functional significance of so-called tonic or nonsynaptic inhibition. $GABA_A$ receptors containing the δ subunit have been reported to be restricted to extrasynaptic membranes, and also to confer enhanced neurosteroid sensitivity, suggesting that tonic inhibition

0-8493-2392-4/04/$0.00+$1.50

may be subject to dynamic regulation by endogenous neurosteroids. The focus of this chapter is the emerging role of δ subunit-containing $GABA_A$ receptors in neuronal inhibition, with particular emphasis on modulation by neurosteroids.

Neurosteroids are synthesized within the brain and have among their molecular targets ionotropic neurotransmitter receptors, including the $GABA_A$ receptor. Unfortunately, the mechanistic basis for neurosteroid effects on brain function, including anesthetic, hypnotic, anxiolytic, and anticonvulsant effects, remains poorly understood (Baulieu and Robel, 1990; Paul and Purdy, 1992; Lambert et al., 1995). However, extensive electrophysiological investigations have attempted to clarify the importance of various factors, such as subunit specificity and phosphorylation, for neurosteroid regulation of inhibition using both native and recombinant receptor preparations (Lambert et al., 1995, 2001). Neurosteroid modulation of $GABA_A$ receptor function can be via positive or negative allosterism (with sulfated pregnane steroids mediating the latter effect) (Majewska et al., 1986; Gee et al., 1988; Puia et al., 1990; Gee and Lan, 1991; Park-Chung et al., 1999). Among the positive allosteric agents, there are likely to be at least three binding sites on the $GABA_A$ receptor based on functional studies. For example, low (n*M*) concentrations can enhance GABA-evoked currents, while higher concentrations (~1 μ*M*) also can directly activate $GABA_A$ receptor currents in the absence of GABA (Lambert et al., 1995). Even higher concentrations (~10 μ*M*) are thought to involve open-channel block based on single-channel experiments (Twyman and Macdonald, 1992) as well as whole-cell methods (Wohlfarth et al., 2002). Although no specific binding sites within the receptor complex have been identified to date, functional studies suggested that neurosteroid modulation likely occurs via different binding sites from those of GABA, benzodiazepines, and barbiturates (Gee et al., 1988; Turner et al., 1989). Several studies have shown that various combinations of α, β, and γ subunits can alter the sensitivity of $GABA_A$ receptor currents to neurosteroid modulation (reviewed in Lambert et al., 2001; Twyman and Macdonald, 1992). However, recent studies of receptors containing the δ subunit have stimulated new interest in αβδ isoforms as an important target for neurosteroid modulation. The significance of these investigations as well as the potential physiological importance of αβδ receptors and tonic inhibition in the brain will be highlighted in this chapter.

4.2 δ SUBUNIT EXPRESSION AND RECEPTOR ASSEMBLY

$GABA_A$ receptors are pentameric complexes of α, β, and one other type of subunit (for example, γ, δ, or ε) in a likely stoichiometric ratio of 2:2:1 (Chang et al., 1996; Tretter et al., 1997; Seighart et al., 1999). Although both synaptic and nonsynaptic receptors are likely to share this stoichiometry and pentameric structure, there is some evidence that subcellular targeting may be influenced by the presence of a γ (synaptic) or δ (extrasynaptic) subunit (Nusser et al., 1995, 1998). However, further work in this area is required, since there is also evidence for γ subunit-containing receptors in extrasynaptic membranes, and in most brain regions evidence is lacking for detailed localization of δ subunits. Subunit expression patterns have been established based on detection of mRNA (Laurie et al., 1992a, 1992b; Wisden et al., 1992) or protein

(Pirker et al., 2000) in rodent brain slices. The δ subunit is expressed most strongly in cerebellar granule cells, hippocampal dentate gyrus, and thalamus, although protein can be detected in the olfactory bulb, amygdala, cortex, substantia nigra, hypothalamus, and basal forebrain (Pirker et al., 2000). Biochemical studies suggested that δ subunits may preferentially assemble with particular α subunits, but that γ and δ subunits may not coexist *in vivo* (Benke et al., 1991; Araujo et al., 1998). For example, in the cerebellum, where most of the work on assembly and subcellular localization has been performed, the likely assembly patterns included α1α6βδ and α6βδ isoforms (Quirk et al., 1995; Nusser et al., 1998; Jechlinger et al., 1998). Sur et al. (1999) showed evidence that in the thalamus, the α4β3δ isoform was dominant. Recombinant studies have demonstrated assembly of the δ subunit together with α1, α4, and α6 subtypes (at least) in mammalian expression systems (Saxena and Macdonald, 1994, 1996; Adkins et al., 2001; Brown et al., 2002). The molecular rubric dictating these assembly preferences remains to be elucidated and may differ among regions of the brain, or even within compartments of individual cells.

4.3 TONIC AND PHASIC INHIBITION

GABA-ergic inhibition in the brain can be loosely divided into tonic and phasic types. Tonic inhibition is mediated by extrasynaptic receptors that may be exposed to low concentrations of extracellular GABA (such as from spillover of synaptic transmission) over relatively long time frames (seconds to minutes or longer). Phasic inhibition, in contrast, is mediated by receptors targeted to synaptic specializations that respond to brief transients of high-concentration GABA in the synaptic cleft (Clements, 1996). The former case would require a receptor capable of sustained response to activation by GABA, while the latter would involve the capacity to activate rapidly to generate a robust response to a brief stimulus. Several recent studies have begun to explore the properties and potential roles of tonic inhibition in various brain regions. For example, studies in the cerebellum (Brickley et al., 1996, 2001; Rossi and Hamann, 1998; Leao et al., 2000; Hamann et al., 2002; Rossi et al., 2003), hippocampal granule cells (Stell and Mody, 2002), hippocampal dentate gyrus (Nusser and Mody, 2002), CA1 pyramidal neurons (Bai et al., 2001), and cultured hippocampal neurons (Yeung et al., 2003) have begun to characterize the relative significance and distinctive properties of tonic inhibition. However, this is a daunting task given the (theoretical and empirical) likelihood of multiple $GABA_A$ receptor isoforms contributing to this sort of current in a given area, not to mention the region- and cell-type-specific expression of subunits from which to assemble receptors. Bai et al. (2001) suggested that tonic currents may play a significant role as targets of anesthetic and sedative agents. Overstreet and Westbrook (2001) also showed that the anticonvulsant vigabatrin may have a more significant effect on tonic than synaptic inhibition via enhancing extracellular concentrations of GABA.

It has been known for some time that various neurotransmitters, including GABA, circulate in the extracellular space in 0.5 to 1 μM (or higher) concentrations (Lerma et al., 1986). Ambient extracellular neurotransmitters are known to activate slower signaling through metabotropic neurotransmitter receptors such as $GABA_B$ and mGlu receptors (reviewed in Isaacson, 2000). However, ionotropic neurotransmitter receptors are also found in extrasynaptic locations (Nusser et al., 1995), and may therefore mediate slow electrical signaling as well. The concentration of an extracellular neurotransmitter may be affected by spillover from synaptic transmission, by membrane transporters, or by release from glial cells (Attwell et al., 1993; Zoli et al., 1999; Liu et al., 2000; Bach-y-Rita, 2001). The potential importance of tonic forms of inhibition is highlighted by the suggestion that it may contribute far more overall inhibitory drive (that is, charge carried) than synaptic events (Brickley et al., 1996; Hamann et al., 2002). Also, it is estimated that a large proportion of $GABA_A$ receptors on the surface of neurons are not associated with synapses (Nusser et al., 1995, 1998; Scotti and Reuter, 2001). Recent functional studies also suggested that tonic GABA-ergic inhibition plays a critical role in cerebellar processing (Hamann et al., 2002). One of the most impressive demonstrations of the importance of tonic inhibition came from investigating mice lacking the α6 subunit, and therefore also the δ subunit (Jones et al., 1997), resulting in a total absence of GABA-ergic tonic inhibition in the cerebellum. Remarkably, these animals had no seizure phenotype or other neurological sequelae, and this was shown to be due to compensatory up-regulation of an inhibitory potassium conductance (Brickley et al., 2001).

Recombinant studies have revealed distinct properties conferred by the δ subunit that would be consistent with a role in tonic forms of inhibition. Receptors containing the δ subunit desensitize much slower and less extensively than αβγ isoforms and have a lower EC_{50} for GABA (Saxena and Macdonald, 1996; Haas and Macdonald, 1999) — ideal properties for a receptor that senses low GABA concentration for sustained time periods. Additionally, the α6 subunit confers an even higher sensitivity for GABA, and thus α6β3δ receptors exhibit very low GABA EC_{50} values (~200 n*M*). This isoform (or other α6 subtype-containing isoforms) may be the extracellular target for low concentrations of GABA (Rossi and Hamann, 1998) and perhaps other GABA-mimetic amino acids (see below).

Another interesting aspect of extrasynaptic neurotransmission is that the concentrations of amino acids such as β-alanine and taurine may be as high as 10 to 20 μ*M* (Lerma et al., 1986). These compounds are agonists at $GABA_A$ receptors (as well as glycine receptors), although the affinity of β-alanine and the efficacy of taurine appear to be much lower than those of GABA. Recent data indicated that tonic currents in the cerebellum were mediated entirely by $GABA_A$ receptors (Wall and Usowitz, 1997), suggesting that these "promiscuous" ligands may in fact interact with extrasynaptic $GABA_A$ receptors under physiological conditions. In particular, the high affinity of α6 subunit-containing receptors may play an important role in detection of ambient extracellular ligands (Rossi and Hamann, 1998).

Unfortunately, there has been no direct demonstration of δ subunit-containing receptors mediating tonic currents measured in brain slices. Further developments of subunit-specific antagonists may shed light on this area. A recent study by Yeung et al. (2003) suggested that tonic inhibition observed in cultured embryonic hippocampal neurons was insensitive to block by penicillin. Although the effects of penicillin have yet to be reported for αβδ receptor currents, the tonic current observed in that study was also sensitive to benzodiazepines, raising questions as to the subunit composition contributing this current. Also, since neurosteroids clearly enhance synaptic currents (Harrison et al., 1987; Zorumski et al., 1998; Cooper et al., 1999; Fansczik et al., 2000), it may be difficult to dissect the relative importance of neurosteroid modulation of tonic vs. synaptic forms of inhibition for effects such as the anticonvulsant, anxiolytic, or hypnotic actions. The discovery of highly selective neurosteroid agents and isoform-specific antagonists, coupled with additional analysis of knockout animals (see below), may allow more precise experimental manipulations of tonic and phasic inhibition.

4.4 FUNCTIONAL PROPERTIES OF δ SUBUNIT-CONTAINING RECEPTORS

The distinctive functional properties of $GABA_A$ receptors containing the δ subunit were evident in early recombinant expression studies comparing αβδ and αβγ isoforms (Saxena and Macdonald, 1994, 1996; Fisher and Macdonald, 1997), the two major isoforms expressed in the brain (McKernan and Whiting, 1996). In comparison with receptors containing a γ subunit, α1β3δ receptors had smaller current amplitudes, lower GABA EC_{50} values, increased outward rectification, and slower macroscopic desensitization. Subsequent studies defined more rigorously the biophysical properties of α1β3δ receptors. Single-channel recordings indicated that α1β3δ receptors exhibit only brief, isolated openings, even in the presence of saturating GABA concentrations (Fisher and Macdonald, 1997) (Figure 4.1A). Kinetic analysis showed these openings to be well described by two exponential functions, implying that two short-lived open states are available to the fully liganded receptor. This characteristic was in contrast to the long-duration bursts of openings observed with α1β3γ2L receptor single channels (Angelotti and Macdonald, 1993; Fisher and Macdonald, 1997) (Figure 4.1B). Fast desensitization had been proposed to regulate synaptic receptor function and therefore inhibitory postsynaptic current (IPSC) shape (Jones and Westbrook, 1995; Haas and Macdonald, 1999). Experiments using ultrafast perfusion methods demonstrated that α1β3γ2L receptor currents showed fast and pronounced desensitization, consistent with their proposed localization primarily to synaptic locations (Haas and Macdonald, 1999) (Figure 4.1C). In contrast, similar experiments demonstrated the absence of fast desensitization of α1β3δ receptor currents (Haas and Macdonald, 1999) (Figure 4.1D), a novel property that contrasted all native and recombinant heterotrimeric $GABA_A$ receptor isoforms that have been examined to date. The increased sensitivity to GABA and minimal desensitization supported a role for αβδ receptors as extrasynaptic sensors of persistent low GABA concentration that might underlie tonic forms of inhibition.

Pharmacological investigations have shown that αβδ receptors have distinct properties with respect to allosteric modulation, compared to αβγ isoforms. For example, δ subunit-containing receptors display insensitivity to benzodiazepines (Saxena and Macdonald, 1996), consistent with the structural requirement of a γ subunit in combination with α subtypes 1 to 3 or 5 (Sigel and Buhr, 1997). Also, there was increased sensitivity to zinc and lanthanum antagonism (Saxena and Macdonald, 1994, 1996). Perhaps the most interesting pharmacological property conferred by the δ subunit was the enhanced (relative to αβγ isoforms) sensitivity to positive allosteric modulation by a large variety of agents acting at distinct

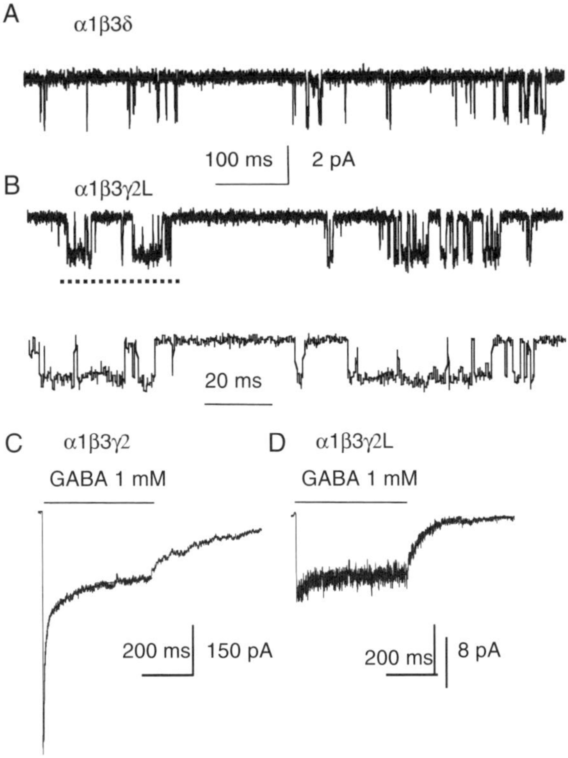

FIGURE 4.1 Single-channel properties of α1β3δ and α1β3γ2L receptors. Typical single-channel openings are shown from excised outside-out membrane patches obtained from L929 fibroblasts transiently expressing α1β3δ (A) and α1β3γ2L (B) $GABA_A$ receptor channels. Patches were voltage-clamped at –75 mV (openings are downward) in the presence of GABA (1 m*M*). A segment of α1β3γ2L openings (dotted line below middle trace of panel B) is expanded (bottom trace of panel B) to demonstrate the "bursting" single-channel activity, in contrast to the brief isolated openings of α1β3δ receptor channels. Concentration jump experiments illustrate the rapid and extensive desensitization of α1β3γ2L receptor currents (C), in comparison to the minimally desensitizing α1β3δ currents (D). Data were obtained from excised outside-out patches containing multiple channels, exposed to GABA (1 m*M*) for 400 msec. (Current traces were adapted from Haas, K.F. and Macdonald, R.L., *J. Physiol.*, 514, 27–45, 1999 (A, B); Bianchi, M.T. et al., *J. Neurosci.*, 21, 1127–1136, 2001 (C, D). With permission.)

binding sites, including neurosteroids (Adkins et al., 2001; Brown et al., 2002; Wohlfarth et al., 2002) (see Figure 4.2A and C), barbiturates (Adkins et al., 2001; Feng et al., 2002), the nonbenzodiazepine anxiolytic tracazolate (Thompson et al., 2002), ethanol (Sundstrom-Poromaa et al., 2002), etomidate (Brown et al., 2002), the anesthetic isofluorane (Lees and Edwards, 1998), and protons (Krishek et al., 1995). Interestingly, there is recent evidence that GABA itself may interact differently with δ subunit-containing receptors. Adkins et al. (2001) found that the synthetic GABA analog THIP showed greater efficacy at α4β3δ receptors, raising the possibility that incorporation of the δ subunit may render GABA a partial

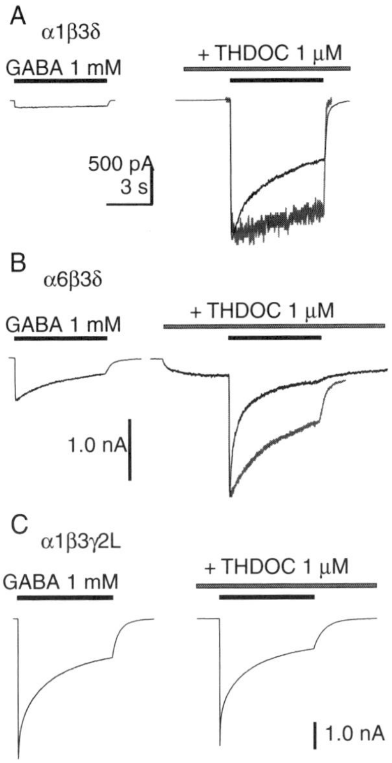

FIGURE 4.2 Isoform-dependent modulation of $GABA_A$ receptor currents by the neurosteroid THDOC. Whole-cell currents evoked by a saturating concentration of GABA (1 m*M*) in the absence (left traces of each panel) and presence (right traces) of THDOC (1 μ*M*) for HEK 293T cells transiently expressing α1β3δ (A), α6β3δ (B), and α1β3γ2L (C) $GABA_A$ receptors. In these experiments, cells were lifted from the recording dish to improve solution exchange times. In each case, THDOC (1 μ*M*) was preapplied for at least 2 sec prior to GABA application. To demonstrate the effects of THDOC on macroscopic desensitization, currents evoked by GABA alone (gray traces) were scaled to the peak of THDOC-modulated currents and overlaid for panels A and B. (Data were taken from Wohlfarth, K.M. et al., *J. Neurosci.*, 22, 1541–1549, 2002; Bianchi, M.T. et al., *Neuropharmacology*, 43, 492–502, 2002. With permission.)

agonist. Finally, just as there is evidence for α subtype dependence of αβγ receptor functional properties, a recent study indicated that the α subtype can further specify the enhancement of peak currents and desensitization by neurosteroids (Figure 4.2B), as well as the rate and extent of desensitization, the rate of deactivation, and the voltage dependence of current amplitude and desensitization (Bianchi et al., 2002). It is becoming increasingly clear that δ subunit-containing receptors have distinct properties, and that co-assembly with different α subtypes available in different brain regions may provide further functional diversity. The emerging pharmacological diversity of αβδ receptors raises the possibility not only of dynamic endogenous regulation of tonic inhibition, but also of the potential for selective therapeutic targeting of tonic inhibitory tone.

4.5 EFFECTS OF NEUROSTEROIDS ON αβδ RECEPTOR CURRENTS

Although an early study suggested that the δ subunit conferred a novel resistance to positive allosteric modulation by neurosteroids (Zhu et al., 1996), a report of decreased neurosteroid sensitivity of mice lacking the δ subunit prompted a reexamination of the issue (Mihalek et al., 1999). Adkins et al. (2001) compared α4β3δ and α4β3γ2 receptors stably expressed in L(-tk) cells (under the control of a dexamethasone-inducible promoter), showing that the neurosteroids alphaxalone and pregnanolone enhanced the δ subunit-containing receptors to a greater extent than the γ subunit-containing receptors. This result was confirmed by Brown et al. (2002) and Bellili et al. (2002), in both cases expressing the δ subunit with α4 and β3 subunits. Enhanced sensitivity to neurosteroids was shown also to include the α1β3δ and α6β3δ isoforms by Wohlfarth et al. (2002). An intriguing feature of these studies was that neurosteroid modulation could enhance αβδ currents to larger amplitudes than were obtained with a saturating concentration of GABA (Figure 4.2A and B). Although this substantial alteration in the efficacy of GABA was rarely observed with native or recombinant αβγ isoforms, the observation of supramaximal modulation was not new to αβδ receptors. Indeed, both the anesthetic isofluorane (Lees and Edwards, 1998) and the nonbenzodiazepine anxiolytic tracazolate (Thompson et al., 2002) exhibited similar modulatory effects on αβδ receptor currents.

Consistent with the general idea of modulating the efficacy of αβδ receptor currents, allosteric enhancement of agonist efficacy has been reported for αβγ isoforms in the context of activation by non-GABA agonists. The efficacy (judged by whole-cell current amplitude) of P4S, a synthetic GABA analog with partial agonist activity (relative to that of GABA), was increased by various modulators, including benzodiazepines, the neurosteroid allopregnanolone, and the anticonvulsant loreclezole (Macksay et al., 1995). In that study, αβγ receptor currents evoked by saturating concentrations of P4S were potentiated by these modulators, similar to the effect of neurosteroids and other modulators on αβδ receptor currents (activated by GABA). Thus there is growing evidence that allosteric modulation of agonist efficacy may be a general property of $GABA_A$ receptors independent of subunit composition. It remains to be seen whether modulation of agonist efficacy can be demonstrated as a means of $GABA_A$ receptor regulation *in vivo*.

These observations raised questions not only about the nature of neurosteroid modulation of αβδ receptors, but also about the nature of GABA as an agonist of δ subunit-containing receptors. Initial characterization of αβδ receptor currents using single-channel, whole-cell, and rapid perfusion techniques (Saxena and Macdonald, 1994; Fisher and Macdonald, 1997; Haas and Macdonald, 1999) suggested that the presence of a δ subunit imposed structural limitations on $GABA_A$ receptor function (low efficacy gating and minimal desensitization) (Bianchi et al., 2001). However, the effect of THDOC modulation of αβδ receptor currents, remarkable for striking effects on current amplitude, desensitization, and single-channel gating, pointed to an alternative interpretation: GABA may be a partial agonist at these receptors. In other words, incorporation of a δ subunit does in fact impose these restrictions in the context of activation by GABA, but they do not represent intrinsic structural limitations since they can be overcome by allosteric modulation.

Wohlfarth et al. (2002) further investigated the activity of the neurosteroid THDOC on αβδ receptor currents using a combination of rapid perfusion techniques for better resolution of whole-cell current properties, and single-channel recording to examine the underlying mechanisms of modulation. THDOC produced a ~10-fold enhancement of maximal (evoked by 1 m*M* GABA) α1β3δ currents by the neurosteroid THDOC. This robust enhancement of efficacy was accompanied by a marked increase in the rate and extent of desensitization (Figure 4.2A). Single-channel experiments demonstrated a shift in gating pattern that contributed to the enhancement of macroscopic currents. Specifically, the two short-lived open states available to the GABA-bound receptors was converted to a bursting pattern characterized by three open states, reminiscent of αβγ receptor single-channel behavior (Figure 4.3A and B; compare to Figure 4.1A and B).

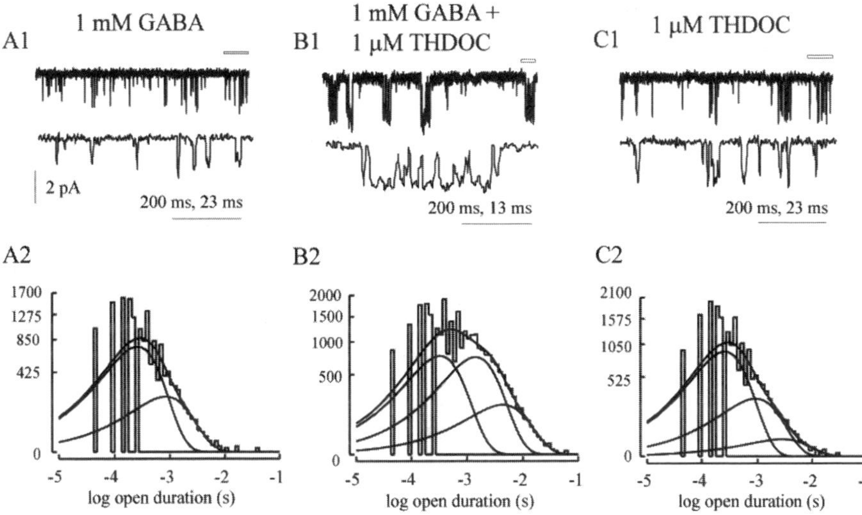

FIGURE 4.3 Modulation of α1β3δ receptor single-channel currents by the neurosteroid THDOC. (A1) Single-channel openings of α1β3δ receptor channels recorded from outside-out patches voltage-clamped at –75 mV in the presence of GABA (1 m*M*). For each panel, a segment of the top current trace is expanded in the trace below. For each condition, open-duration histograms are shown, overlapped with the exponential fitting of the data sets. α1β3δ receptor openings were of short duration and could be described well by two exponential functions with time constants of ~0.3 and ~1.0 msec (A2). (B1) In the presence of GABA (1 m*M*) and THDOC (1 μ*M*), α1β3δ receptor single-channel currents were characterized by longer-duration openings and tended to occur in bursts. THDOC modulation (1 μ*M*) resulted in openings that were described well by three exponential functions, with the third exponential function having a time constant of ~6 msec (B2). (C1) Activation of single α1β3δ receptor channels by THDOC (1 μ*M*) alone. (C2) These openings were also well described by three time constants. (Data were taken from Wohlfarth, K.M. et al., *J. Neurosci.*, 22, 1541–1549, 2002. With permission.)

Although the effects on open frequency could not be evaluated due to the concomitant introduction of desensitization, it was possible that opening rate constants were also altered by THDOC, as has been suggested previously in single-channel studies of native $GABA_A$ receptor channels (see below). Interestingly, direct activation of α1β3δ receptors by THDOC alone was able to evoke some long-duration openings, further evidence for a significant impact of neurosteroids on the functional capacity of αβδ receptors (Figure 4.3C).

Prior studies of the mechanism of neurosteroid enhancement of $GABA_A$ receptors indicated changes in frequency (Puia et al., 1990), using recombinant expression; open duration (Mistry and Cottrell, 1990); or both frequency and duration of openings (Twyman and Macdonald, 1992), using spinal neurons. These data suggested distinct mechanisms from the benzodiazepines, thought to strictly modulate opening frequency through actions on GABA binding steps, and barbiturates, thought to modulate open duration by shifting the steady-state occupancy toward the two longer-duration open states while preserving the relative proportions of these two states (Twyman et al., 1989a, 1989b). The ratio of the two longer-duration open states was found to be GABA concentration independent, while the relative contribution of those states (compared to the short-duration state) increased with GABA concentration (Macdonald et al., 1989). Detailed single-channel analysis (Twyman and Macdonald, 1992) revealed that neurosteroids decreased the likelihood of entry into the briefest open state while increasing entry into the longest open state.

4.6 INSIGHTS FROM δ-SUBUNIT KNOCKOUT MICE

The current understanding of subunit-specific properties of $GABA_A$ receptors is based predominantly on expression of recombinant cloned subunits in artificial expression systems such as mammalian cells, insect cells, and *Xenopus* oocytes. Although many biophysical and pharmacological properties observed in these systems were similar to those observed in neuronal preparations, the ultimate determinant of subunit roles is likely to depend on functional evaluations within a neuronal context. Transgenic and knockout animals have provided unique insights, at the level of the whole organism, into the involvement of various subunits in development, seizure activity, behavior, and responsiveness to drugs, and even unexpected interactions between subunit expression patterns. This section summarizes some of the major contributions.

The first report of mice deficient for the δ subunit demonstrated a marked decrease in the behavioral sensitivity to neurosteroid administration (Mihalek et al., 1999). This observation was surprising since it had been reported that the δ subunit was unique in conferring insensitivity to neurosteroids (Zhu et al., 1996). However, the aforementioned recent studies clearly showed that αβδ receptors were actually more sensitive than αβγ receptors (for α1, α4, and α6 subtypes) (Adkins et al., 2001; Wohlfarth et al., 2002). Further investigation of these mice (or creation of conditional knockouts) may determine whether the neurosteroid insensitivity was directly related to the absence of the δ subunit, or whether some compensatory changes lead indirectly to this phenomenon. Interestingly, the same δ subunit-deficient mice studied in 1999 showed several behavioral defects related to ethanol, including decreased consumption, decreased withdrawal, and reduced anticonvulsant effects (Mihalek et al., 2001). Whether this effect on ethanol sensitivity is related to the attenuated neurosteroid effects is unknown, although there is evidence for a connection between ethanol and neurosteroid modulation of $GABA_A$ receptor function (see below).

Biochemical analysis indicated that deletion of the δ subunit did not alter muscimol binding, but increased Ro-15-4313 binding, due to a compensatory increase in the γ2 subunit in the cerebellum (Tretter et al., 2001). However, no alteration in the expression of the α6 subtype was detected, indicating that the previously described subunit partnership between α6 and δ subunits is unidirectional (Jones et al., 1997). Deletion of the α6 subunit had been shown to decrease expression of the δ subunit, due to increased degradation at the posttranslational level (mRNA levels were normal) (Jones et al., 1997). Interestingly, despite the presence of other subunits (such as α1 and β2/3 subunits) that are known to co-assemble with δ subunits in recombinant systems, the observation of δ subunit degradation indicates no functional association with non-α6 subunits *in vivo*, at least in the cerebellum.

Subsequent studies confirmed that δ knockout mice had increased expression of the γ2 subunit, and also showed that the expression of the α4 subtype decreased (Korpi et al., 2002; Peng et al., 2002). In the forebrain, there was increased association of the γ2 subunit with the α4 subtype compared to wild-type animals, suggesting preferential assembly of α4 and δ subunits in that area under normal conditions.

The potential importance of the δ subunit in epilepsy was suggested by knockout mice exhibiting an increase in susceptibility to pentylenetetrazol (PTZ)-induced seizures (Spigelman et al., 2002). In that study, dentate granule cell IPSCs from mutant mice were found to have shorter durations than wild-type controls. Notably, evoked IPSP duration was reduced to a larger extent than that of spontaneous potentials. Assuming that the difference between decay in these populations was related to the activation of extrasynaptic receptors from spillover that was more likely to occur in the evoked case, the results implied that either δ subunit-containing receptors are localized mostly to extrasynaptic membranes or that δ subunit-containing receptors are preferentially targeted to synapses distant from the soma (where mIPSCs are more likely recorded). They also showed that alphaxalone (10 μ*M*) prolongation of IPSC duration was attenuated in the knockout cells. Whether the increased seizure susceptibility was a consequence of the accelerated synaptic currents or from the attenuated neurosteroid sensitivity (assuming endogenous neurosteroid presence) remains unknown. It is important to note, with all of these knockout animal studies, that the physiological and behavioral phenotypes cannot necessarily be attributed directly to the absence of the δ subunit protein. Clearly, compensatory changes in $GABA_A$ receptor subunit expression, as well as other ion channels (Brickley et al., 2001), are occurring in this context, necessitating cautionary interpretation.

4.7 ROLE OF PHOSPHORYLATION AND CO-MODULATION

Despite the clear evidence for enhanced sensitivity of recombinant αβδ isoforms, data from neuronal preparations remain inconsistent with regard to the role of the δ subunit. For example, Zhu et al. (1996) used single-cell polymerase chain reaction

(PCR) techniques in combination with electrophysiology to show that expression of the δ subunit was inversely related to modulation of $GABA_A$ receptor currents by the neurosteroid THDOC. Also, neurosteroid sensitivity decreased with development over a time when the δ subunit expression was increasing (Cooper et al., 1999). Although it is possible that these differences relate to recombinant expression systems failing to reflect the stoichiometry or subunit composition of native receptors, these data raise the possibility that factors present in a neuronal environment (and absent in expression systems) play a significant role in neurosteroid modulation. One such factor may be phosphorylation, a modification that has been shown to affect neurosteroid modulation of IPSCs (Fanscik, 2000; Vicini et al., 2002) and recombinant $GABA_A$ receptors (Leidenheimer and Chapell, 1997). Another possibility is that other allosteric cofactors may influence the sensitivity of $GABA_A$ receptors to neurosteroids. For example, there is increasing evidence that ethanol may potentiate the modulatory effects of neurosteroids (VanDoren et al., 2000; Morrow et al., 2001; Akk and Steinbach, 2003). Although ethanol is not an endogenous ligand, the results provide further support for the possibility of cofactors adjusting the impact of neurosteroid modulation.

Another finding that supports the influence of a neuronal environment in regulating neurosteroid sensitivity of $GABA_A$ receptors was that dentate granule cells became less sensitive to THDOC enhancement with developmental maturation (Cooper et al., 1999) despite the coincident increase in δ subunit expression (Laurie et al., 1992a). It remains unclear whether steroid modulation is related to receptor assembly issues (such as δ subunits co-assembling with γ subunits) or whether some other factor such as phosphorylation alters the neurosteroid sensitivity. In fact, by combining electrophysiological recordings and single-cell PCR techniques, Zhu et al. (1996) observed an inverse relation between the presence of δ subunit mRNA and neurosteroid potentiation in cultured cerebellar granule cells. Also, there is evidence for a phosphorylation dependence of allopregnanolone modulation of GABA-ergic IPSCs in the hypothalamus (Fancsik et al., 2000). Such posttranslational receptor modifications may, among other cellular processes, contribute to potential differences between neuronal preparations and recombinant systems.

Vicini et al. (2002) showed that IPSCs recorded from cerebellar granule cells from mice lacking the δ subunit were insensitive to modulation by neurosteroids (while IPSC duration was clearly prolonged in control animals). Although this was consistent with the importance of αβδ isoforms for neurosteroid action, it was surprising given the evidence for strict extrasynaptic localization of the δ subunit in the cerebellum (Nusser et al., 1998). Other findings from that study suggest a more complex basis for the action of neurosteroids. For example, neurosteroid actions were also decreased in IPSCs of stellate neurons from the knockout mice, a cell type that does not normally express the δ subunit. Additionally, neurosteroid modulation of IPSCs did not differ between stellate and granule cells taken from wild-type animals despite the difference in δ subunit expression. Interestingly, stimulation of protein kinase C (PKC) pathways in these animals prolonged IPSC decays in both cell types from the knockout animals to durations seen with wild-type neurons. Because the IPSCs measured in wild-type animals were insensitive to modulation by the same PKC pathway, it is possible that expression of the δ subunit is somehow related to PKC activity on synaptic $GABA_A$ receptors. However, the need for caution in generalizing observations from individual anatomical regions is emphasized by other studies, such as Koksma et al. (2003), who showed that PKC activation was inversely related to neurosteroid sensitivity of $GABA_A$ receptor currents from the rat supraoptic nucleus.

Other work in the cerebellum suggested that $GABA_A$ receptors mediating tonic inhibition, probably containing the α6 and δ subunits, were insensitive to neurosteroid modulation (Hamann et al., 2002). Although these data also support a distinct pharmacology of αβδ receptors in a neuronal milieu, it is also possible that the long-duration (30 to 60 sec) applications resulted in pronounced desensitization, reducing the apparent effects on the tonic current. Neurosteroids have been shown to produce nearly complete desensitization in α6β3δ receptors during a 4-sec application of GABA (Bianchi et al., 2002). Additional work, ideally including the unambiguous identification of the $GABA_A$ receptor isoforms responsible for tonic currents in the cerebellum and elsewhere in the brain, is required to clarify these issues.

4.8 CONCLUSIONS

Neurosteroids appear to be involved with many aspects of CNS function. They also appear to interact with multiple receptor types at the cellular level. How the cellular effects translate into CNS modulation is the subject of ongoing study. Although it is clear that $GABA_A$ receptors are an important target for neurosteroid action, many important questions remain unanswered. For example, the binding site(s) on the receptor complex for neurosteroids has yet to be elucidated. Also, specific antagonists of δ-containing isoforms are currently unavailable, precluding pharmacological isolation of αβδ-mediated currents in brain slice or whole-animal experimental preparations. Also, because multiple $GABA_A$ receptor isoforms may underlie tonic inhibition, the possibility of selective pharmacological manipulation of this type of inhibition remains uncertain. Finally, the complex interaction of neurosteroids with $GABA_A$ receptors may not be based on strict structural determinants (such as the clear isoform selectivity that has been demonstrated for the benzodiazepines). Although this possibility may present a significant challenge for investigation of structure–function relationships and the design of novel subunit-specific compounds, it may signal a powerful functional flexibility for modulation of inhibition in the brain under diverse physiological conditions.

REFERENCES

Adkins, C.E., Pillai, G.V., Kerby, J., Bonnert, T.P., Haldon, C., McKernan, R.M., Gonzalez, J.E., Oades, K., Whiting, P.J., and Simpson, P.B. (2001). α4β3δ $GABA_A$ receptors characterized by fluorescence resonance energy transfer-derived measurements of membrane potential. *J. Biol. Chem.*, 276, 38934–38939.

Akk, G. and Steinbach, J.H. (2003). Low doses of ethanol and a neuroactive steroid positively interact to modulate rat GABA(A) receptor function. *J. Physiol.*, 546, 641–646.

Angelotti, T.P. and Macdonald, R.L. (1993). Assembly of $GABA_A$ receptor subunits: $\alpha1\beta1$ and $\alpha1\beta1\gamma2S$ subunits produce unique ion channels with dissimilar single-channel properties. *J. Neurosci.*, 13, 1429–1440.

Araujo, F., Ruano, D., and Vitorica, J. (1998). Absence of association between delta and gamma2 subunits in native GABA(A) receptors from rat brain. *Eur. J. Pharmacol.*, 347, 347–353.

Attwell, D., Barbour, B., and Szatkowski, M. (1993). Nonvesicular release of neurotransmitter. *Neuron*, 11, 401–407.

Bach-y-Rita, P. (2001). Nonsynaptic diffusion neurotransmission in the brain: functional considerations. *Neurochem. Res.*, 26, 871–873.

Bai, D., Zhu, G., Pennefather, P., Jackson, M.F., MacDonald, J.F., and Orser, B.A. (2001). Distinct functional and pharmacological properties of tonic and quantal inhibitory postsynaptic currents mediated by gamma-aminobutyric acid(A) receptors in hippocampal neurons. *Mol. Pharmacol.*, 59, 814–824.

Baulieu, E.-E. and Robel, P. (1990). Neurosteroids: a new brain function? *J. Steroid Biochem. Mol. Biol.*, 37, 395–403.

Benke, D., Meterns, S., Trzeciak, A., Gillesen, D., and Mohler, H. (1991). Identification and immunohistochemical mapping of $GABA_A$ receptor subtypes containing the δ subunit in rat brain, *FEBS Lett.*, 283, 145–149.

Bianchi, M.T., Haas, K.F., and Macdonald, R.L. (2001). Structural determinants of fast desensitization and desensitization-deactivation coupling in $GABA_A$ receptors. *J. Neurosci.*, 21, 1127–1136.

Bianchi, M.T., Haas, K.F., and Macdonald, R.L. (2002). Alpha1 and alpha6 subunits specify distinct desensitization, deactivation and neurosteroid modulation of GABA(A) receptors containing the delta subunit. *Neuropharmacology*, 43, 492–502.

Brickley, S.G., Cull-Candy, S.G., and Farrant, M. (1996). Development of a tonic form of synaptic inhibition in rat cerebellar granule cells resulting from persistent activation of $GABA_A$ receptors. *J. Physiol.*, 497 (Pt. 3), 753–759.

Brickley, S.G., Revilla, V., Cull-Candy, S.G., Wisden, W., and Farrant, M. (2001). Adaptive regulation of neuronal excitability by a voltage-independent potassium conductance. *Nature*, 409, 88–92.

Brown, N., Kerby, J., Bonnert, T.P., Whiting, P.J., and Wafford, K.A. (2002). Pharmacological characterization of a novel cell line expressing human alpha(4)beta(3)delta GABA(A) receptors. *Br. J. Pharmacol.*, 136, 965–974.

Chang, Y., Wang, R., Barot, S., and Weiss, D.S. (1996). Stoichiometry of a recombinant $GABA_A$ receptor. *J. Neurosci.*, 16, 5415–5424.

Clements, J.D. (1996). Transmitter timecourse in the synaptic cleft: its role in central synaptic function. *Trends Neurosci.*, 19, 163–171.

Cooper, E.J., Johnston, G.A.R., and Edwards, F.A. (1999). Effects of naturally occurring neurosteroid on $GABA_A$ IPSCs during development in rat hippocampal or cerebellar slices. *J. Physiol.*, 521, 437–449.

Fancsik, A., Linn, D.M., and Tasker, J.G. (2000). Neurosteroid modulation of GABA IPSCs is phosphorylation dependent. *J. Neurosci.*, 20, 3067–3075.

Feng, H.-J., Bianchi, M.T., and Macdonald, R.L. (2002). Pentobarbital exerts distinct modulation of desensitization and channel gating in gamma and delta subunit-containing GABA-A receptors. *Society for Neuroscience Abstract*, 41.12.

Fisher, J.L. and Macdonald, R.L. (1997). Single channel properties of $GABA_A$ receptors containing $\gamma2$ or δ subtypes expressed with $\alpha1$ and $\beta3$ subtypes in L929 cells. *J. Physiol.*, 505, 283–297.

Gee, K.W. and Lan, N.C. (1991). γ-Aminobutyric $acid_A$ receptor complexes in rat cortex and spinal cord show differential responses to steroid modulation. *Mol. Pharmacol.*, 40, 995–999.

Gee, K.W., Bolger, M.B., Brinton, R.E., Coirini, H., and McEwen, B.S. (1988). Steroid modulation of the chloride ionophore in rat brain: structure-activity requirements, regional dependence and mechanism of action. *J. Pharmacol. Exp. Ther.*, 246, 803–812.

Haas, K.F. and Macdonald, R.L. (1999). $GABA_A$ receptor subunit gamma2 and delta subtypes confer unique kinetic properties on recombinant $GABA_A$ receptor currents in mouse fibroblasts. *J. Physiol.*, 514, 27–45.

Hamann, M., Rossi, D.J., and Attwell, D. (2002). Tonic and spillover inhibition of granule cells control information flow through cerebellar cortex. *Neuron*, 33, 625–633.

Harrison, N.L., Vicini, S., and Barker, J.L. (1987). A steroid anesthetic prolongs inhibitory postsynaptic currents in cultured rat hippocampal neurons. *J. Neurosci.*, 7, 604–609.

Isaacson, J.S. (2000). Spillover in the spotlight. *Curr. Biol.*, 10, R475–R477.

Jechlinger, M., Pelz, R., Tretter, V., Klausberger, T., and Sieghart, W. (1998). Subunit composition and quantitative importance of hetero-oligomeric receptors: $GABA_A$ receptors containing alpha6 subunits. *J. Neurosci.*, 18, 2449–2457.

Jones, A., Korpi, E.R., McKernan, R.M., Pelz, R., Nusser, Z., Makela, R., Mellor, J.R., Pollard, S., Bahn, S., Stephenson, F.A., Randall, A.D., Sieghart, W., Somogyi, P., Smith, A.J., and Wisden, W. (1997). Ligand-gated ion channel subunit partnerships: $GABA_A$ receptor alpha6 subunit gene inactivation inhibits delta subunit expression. *J. Neurosci.*, 17, 1350–1362.

Jones, M.V. and Westbrook, G.L. (1995). Desensitized states prolong $GABA_A$ channel responses to brief agonist pulses. *Neuron*, 15, 181–191.

Koksma, J.J., van Kesteren, R.E., Rosahl, T.W., Zwart, R., Smit, A.B., Luddens, H., and Brussaard, A.B. (2003). Oxytocin regulates neurosteroid modulation of GABA(A) receptors in supraoptic nucleus around parturition. *J. Neurosci.*, 23, 788–797.

Korpi, E.R., Mihalek, R.M., Sinkkonen, S.T., Hauer, B., Hevers, W., Homanics, G.E., Sieghart, W., and Luddens, H. (2002). Altered receptor subtypes in the forebrain of GABA(A) receptor delta subunit-deficient mice: recruitment of gamma 2 subunits. *Neuroscience*, 109, 733–743.

Lambert, J.J., Belelli, D., Hill-Venning, C., and Peters, J.A. (1995). Neurosteroids and $GABA_A$ receptor function. *Trends Pharmacol. Sci.*, 16, 295–303.

Lambert, J.J., Belelli, D., Harney, S.C., Peters, J.A., and Frenguelli, B.G. (2001). Modulation of native and recombinant GABA(A) receptors by endogenous and synthetic neuroactive steroids. *Brain Res. Brain Res. Rev.*, 37, 68–80.

Laurie, D.J., Wisden, W., and Seeburg, P.H. (1992a). The distribution of thirteen $GABA_A$ receptor subunit mRNAs in the rat brain. III. Embryonic and postnatal development. *J. Neurosci.*, 12, 4151–4172.

Laurie, D.J., Seeburg, P.H., and Wisden, W. (1992b). The distribution of 13 $GABA_A$ receptor subunit mRNAs in the rat brain. II. Olfactory bulb and cerebellum. *J. Neurosci.*, 12, 1063–1076.

Leao, R.M., Mellor, J.R., and Randall, A.D. (2000). Tonic benzodiazepine-sensitive GABAergic inhibition in cultured rodent cerebellar granule cells. *Neuropharmacology*, 39, 990–1003.

Lees, G. and Edwards, M.D. (1998). Modulation of recombination human gamma-aminobutyric acidA receptors by isoflurane: influence of the delta subunit. *Anesthesiology*, 88, 206–217.

Leidenheimer, N.J. and Chapell, R. (1997). Effects of PKC activation and receptor desensitization on neurosteroid modulation of GABA(A) receptors. *Brain Res. Mol. Brain Res.*, 52, 173–181.

Lerma, J., Herranz, A.S., Herreras, O., Abraira, V., and Martin del Rio, R. (1986). In vivo determination of extracellular concentration of amino acids in the rat hippocampus: a method based on brain dialysis and computerized analysis. *Brain Res.*, 384, 145–155.

Liu, Q.Y., Schaffner, A.E., Chang, Y.H., Maric, D., and Barker, J.L. (2000). Persistent activation of GABA(A) receptor/Cl(–) channels by astrocyte-derived GABA in cultured embryonic rat hippocampal neurons. *J. Neurophysiol.*, 84, 1392–1403.

Macdonald, R.L., Rogers, C.J., and Twyman, R.E. (1989). Kinetic properties of the $GABA_A$ receptor main conductance state of mouse spinal cord neurones in culture. *J. Physiol.*, 410, 479–499.

Majewska, M.D., Harrison, N.L., Schwartz, R.D., Barker, J.L., and Paul, S.M. (1986). Steroid hormone metabolites are barbiturate-like modulators of the GABA receptor. *Science*, 232, 1004–1007.

McKernan, R.M. and Whiting, P.J. (1996). Which $GABA_A$-receptor subtypes really occur in the brain? *Trends Neurosci.*, 19, 139–143.

Mihalek, R.M., Banerjee, P.K., Korpi, E.R., Quinlan, J.J., Firestone, L.L., Mi, Z.P., Lagenaur, C., Tretter, V., Sieghart, W., Anagnostaras, S.G., Sage, J.R., Fanselow, M.S., Guidotti, A., Spigelman, I., Li, Z., DeLorey, T.M., Olsen, R.W., and Homanics, G.E. (1999). Attenuated sensitivity to neuroactive steroids in gamma-aminobutyrate type A receptor delta subunit knockout mice. *Proc. Natl. Acad. Sci. U.S.A.*, 96, 12905–12910.

Mihalek, R.M., Bowers, B.J., Wehner, J.M., Kralic, J.E., VanDoren, M.J., Morrow, A.L., and Homanics, G.E. (2001). GABA(A)-receptor delta subunit knockout mice have multiple defects in behavioral responses to ethanol. *Alcohol. Clin. Exp. Res.*, 25, 1708–1718.

Mistry, D.K. and Cottrell, G.A. (1990). Actions of steroids and bemegride on the GABAA receptor of mouse spinal neurones in culture. *Exp. Physiology*, 75, 199–209.

Morrow, A.L., VanDoren, M.J., Penland, S.N., Matthews, D.B. (2001). The role of GABAergic neuroactive steroids in ethanol action, tolerance and dependence. *Brain Res. Brain Res. Rev.*, 37, 98–109.

Nusser, Z., Roberts, J.D., Baude, A., Richards, J.G., and Somogyi, P. (1995). Relative densities of synaptic and extrasynaptic $GABA_A$ receptors on cerebellar granule cells as determined by a quantitative immunogold method. *J. Neurosci.*, 15, 2948–2960.

Nusser, Z., Sieghart, W., and Somogyi, P. (1998). Segregation of different $GABA_A$ receptors to synaptic and extrasynaptic membranes of cerebellar granule cells. *J. Neurosci.*, 18, 1693–1703.

Nusser, Z. and Mody, I. (2002). Selective modulation of tonic and phasic inhibitions in dentate gyrus granule cells. *J. Neurophysiol.*, 87, 2624–2628.

Overstreet, L.S. and Westbrook, G.L. (2001). Paradoxical reduction of synaptic inhibition by vigabatrin. *J. Neurophysiol.*, 86, 596–603.

Park-Chung, M., Malayev, A., Purdy, R.H., Gibbs, T.T., and Farb, D.H. (1999). Sulfated and unsulfated steroids modulate γ-aminobutyric $acid_A$ receptor function through distinct sites. *Brain Res.*, 830, 72–87.

Paul, S.M. and Purdy, R.H. (1992). Neuroactive steroids. *FASEB J.*, 6, 2311–2322.

Peng, Z., Hauer, B., Mihalek, R.M., Homanics, G.E., Sieghart, W., Olsen, R.W., and Houser, C.R. (2002). GABA(A) receptor changes in delta subunit-deficient mice: altered expression of alpha4 and gamma2 subunits in the forebrain. *J. Comp. Neurol.*, 446, 179–197.

Pirker, S., Schwarzer, C., Wieselthaler, A., Sieghart, W., and Sperk, G. (2000). GABA(A) receptors: immunocytochemical distribution of 13 subunits in the adult rat brain. *Neuroscience*, 101, 815–850.

Puia, G., Santi, M.R., Vicini, S., Pritchett, D.B., Purdy, R.H., Paul, S.M., Seeburg, P.H., and Costa, E. (1990). Neurosteroids act on recombinant human $GABA_A$ receptor. *Neuron*, 4, 759–765.

Quirk, K., Whiting, P.J., Ragan, C.I., and McKernan, R.M. (1995). Characterisation of delta-subunit containing $GABA_A$ receptors from rat brain. *Eur. J. Pharmacol.*, 290, 175–181.

Rossi, D.J. and Hamann, M. (1998). Spillover-mediated transmission at inhibitory synapses promoted by high affinity alpha6 subunit GABA(A) receptors and glomerular geometry. *Neuron*, 20, 783–795.

Rossi, D., Hamann, M., and Attwell, D. (2003). Multiple modes of GABAergic inhibition of rat cerebellar granule cells. *J Physiol. (Lond.)*, epub ahead of print.

Saxena, N.C. and Macdonald, R.L. (1994). Assembly of $GABA_A$ receptor subunits: role of the delta subunit. *J. Neurosci.*, 14, 7077–7086.

Saxena, N.C. and Macdonald, R.L. (1996). Properties of putative cerebellar γ-aminobutyric $acid_A$ receptor isoforms. *Mol. Pharmacol.*, 49, 567–579.

Scotti, A.L. and Reuter, H. (2001). Synaptic and extrasynaptic gamma-aminobutyric acid type A receptor clusters in rat hippocampal cultures during development. *Proc. Natl. Acad. Sci. U.S.A.*, 98, 3489–3494.

Sieghart, W., Fuchs, K., Tretter, V., Ebert, V., Jechlinger, M., Hoger, H., and Adamiker, D. (1999). Structure and subunit composition of GABA(A) receptors. *Neurochem. Int.*, 34, 379–385.

Sigel, E. and Buhr, A. (1997). The benzodiazepine binding site of $GABA_A$ receptors. *Trends Pharmacol. Sci.*, 18, 425–429.

Spigelman, I., Li, Z., Banerjee, P.K., Mihalek, R.M., Homanics, G.E., and Olsen, R.W. (2002). Behavior and physiology of mice lacking the $GABA_A$-receptor delta subunit. *Epilepsia*, 43 (Suppl. 5), 3–8.

Stell, B.M. and Mody, I. (2002). Receptors with different affinities mediate phasic and tonic GABA(A) conductances in hippocampal neurons. *J. Neurosci.*, 22, RC223.

Sundstrom-Poromaa, I., Smith, D.H., Gong, Q.H., Sabado, T.N., Li, X., Light, A., Wiedmann, M., Williams, K., and Smith, S.S. (2002). Hormonally regulated alpha(4)beta(2)delta GABA(A) receptors are a target for alcohol. *Nat. Neurosci.*, 5, 721–722.

Sur, C., Farrar, S.J., Kerby, J., Whiting, P.J., Atack, J.R., and McKernan, R.M. (1999). Preferential coassembly of alpha4 and delta subunits of the gamma-aminobutyric acidA receptor in rat thalamus. *Mol. Pharmacol.*, 56, 110–115.

Thompson, S.A., Wingrove, P.B., Connelly, L., Whiting, P.J., and Wafford, K.A. (2002). Tracazolate reveals a novel type of allosteric interaction with recombinant gamma-aminobutyric acid(A) receptors. *Mol. Pharmacol.*, 61, 861–869.

Tretter, V., Ehya, N., Fuchs, K., and Sieghart, W. (1997). Stoichiometry and assembly of a recombinant $GABA_A$ receptor subtype. *J. Neurosci.*, 17, 2728–2737.

Tretter, V., Hauer, B., Nusser, Z., Mihalek, R.M., Hoger, H., Homanics, G.E., Somogyi, P., and Sieghart, W. (2001). Targeted disruption of the GABA(A) receptor delta subunit gene leads to an up-regulation of gamma 2 subunit-containing receptors in cerebellar granule cells. *J. Biol. Chem.*, 276, 10532–10538.

Turner, D.M., Ransom, R.W., Yang, J.S.-J., and Olsen, E.W. (1989). Steroid anesthetics and naturally occurring analogs modulate the γ-aminobutyric acid receptor complex at a site distinct from barbiturates. *J. Pharmacol. Exp. Ther.*, 248, 960–966.

Twyman, R.E., Rogers, C.J., and Macdonald, R.L. (1989a). Pentobarbital and picrotoxin have reciprocal actions on single $GABA_A$ receptor channels. *Neurosci. Lett.*, 96, 89–95.

Twyman, R.E., Rogers, C.J., and Macdonald, R.L. (1989b). Differential regulation of gamma-aminobutyric acid receptor channels by diazepam and phenobarbital. *Ann. Neurol.*, 25, 213–220.

Twyman, R.E. and Macdonald, R.L. (1992). Neurosteroid regulation of $GABA_A$ receptor single-channel kinetic properties of mouse spinal cord neurons in culture. *J. Physiol.*, 456, 215–245.

VanDoren, M.J., Matthews, D.B., Janis, G.C., Grobin, A.C., Devaud, L.L., and Morrow, A.L. (2000). Neuroactive steroid 3alpha-hydroxy-5alpha-pregnan-20-one modulates electrophysiological and behavioral actions of ethanol. *J. Neurosci.*, 20, 1982–1989.

Vicini, S., Losi, G., and Homanics, G.E. (2002). GABA(A) receptor delta subunit deletion prevents neurosteroid modulation of inhibitory synaptic currents in cerebellar neurons. *Neuropharmacology*, 43, 646–650.

Wall, M.J. and Usowicz, M.M. (1997). Development of action potential-dependent and independent spontaneous $GABA_A$ receptor-mediated currents in granule cells of postnatal rat cerebellum. *Eur. J. Neurosci.*, 9, 533–548.

Wisden, W., Laurie, D.J., Monyer, H., and Seeburg, P.H. (1992). The distribution of 13 $GABA_A$ receptor subunit mRNAs in the rat brain. I. Telencephalon, diencephalon, mesencephalon. *J. Neurosci.*, 12, 1040–1062.

Wohlfarth, K.M., Bianchi, M.T., and Macdonald, R.L. (2002). Enhanced neurosteroid potentiation of ternary $GABA_A$ receptors containing the delta subunit. *J. Neurosci.*, 22, 1541–1549.

Yeung, J.Y., Canning, K.J., Zhu, G., Pennefather, P., MacDonald, J.F., and Orser, B.A. (2003). Tonically activated $GABA_A$ receptors in hippocampal neurons are high-affinity, low-conductance sensors for extracellular GABA. *Mol. Pharmacol.*, 63, 2–8.

Zhu, W.J., Wang, J.F., Krueger, K.E., and Vicini, S. (1996). Delta subunit inhibits neurosteroid modulation of $GABA_A$ receptors. *J. Neurosci.*, 16, 6648–6656.

Zoli, M., Jansson, A., Sykova, E., Agnati, L.F., and Fuxe, K. (1999). Volume transmission in the CNS and its relevance for neuropsychopharmacology. *Trends Pharmacol. Sci.*, 20, 142–150.

Zorumski, C.F., Mennerick, S., and Covey, D.F. (1998). Enantioselective modulation of GABAergic synaptic transmission by steroids and benz[e]indenes in hippocampal microcultures. *Synapse*, 29, 162–171.

5 Potentiation and Inhibition of GABA$_A$ Receptor Function by Neuroactive Steroids

Lawrence N. Eisenman, Yejun He, Douglas F. Covey, Charles F. Zorumski, and Steven Mennerick

CONTENTS

0-8493-2392-4/04/$0.00+$1.50

5.1 INTRODUCTION

Certain steroids have long been known to rapidly and potently influence nervous system function[1,2] through mechanisms unrelated to intracellular steroid receptor activation and associated transcriptional activity.[3–5] In this chapter, synthetic and endogenous steroids that possess these rapid actions will be termed neuroactive steroids.[6] Our more specific focus will be on steroids with actions at $GABA_A$ receptors, which we will refer to as GABA-active steroids. As addressed in other chapters, there are two main classes of GABA-active steroids. The first class potentiates $GABA_A$ receptor function. A second class inhibits $GABA_A$ receptor function. Representatives of both GABA-potentiating and GABA-inhibiting GABA-active steroids are synthesized endogenously and may tonically regulate receptor function. Additionally, exogenous representatives of both classes may have clinical applications. After a brief review of the two classes of GABA-active steroids, the actions of neuroactive steroids at other potentially relevant molecular targets, and the potential clinical utility of neuroactive steroids, this chapter will focus on recent attempts by our groups and others to understand structure–activity relationships of GABA-potentiating and GABA-inhibiting steroids. We also review our studies of interactions between the two classes and recent studies of the mechanism of GABA-inhibiting steroids.

5.2 ENDOGENOUS GABA-POTENTIATING AND GABA-INHIBITING STEROIDS

In this section, we introduce the two classes of GABA-active steroids and some of the endogenous representatives. Later sections will expand on some of the structural requirements for GABA-active steroids that we introduce here. Specifics of the endogenous synthetic pathways have been reviewed recently.[7,8] Neuroactive steroids are derived from pregnenolone, which is synthesized endogenously from cholesterol by a P450 side chain cleavage enzyme. Processing of pregnenolone by 3β-hydroxysteroid dehydrogenase/isomerase produces progesterone. GABA-active steroids are in turn synthesized from progesterone. One of the better-known GABA-potentiating steroids is (3α,5α)-3-hydroxypregnan-20-one (3α5αP), also called allopregnanolone (although *Chemical Abstracts* uses this trivial name for the 3β-epimer); 3α-hydroxydihydroprogesterone; or 3α,5α-tetrahydroprogesterone (Figure 5.1A). The 5β-epimer (3α5βP) is also found endogenously and potently augments $GABA_A$ responses. These GABA-active steroids are synthesized from progesterone by the actions of 5α- or 5β-reductases and a 3α-hydroxysteroid dehydrogenase. Parallel enzymatic processing of 11-deoxycorticosterone produces the GABA-potentiating steroids (3α,5α)-3,21-dihydroxypregnan-20-one (3α5αTHDOC) and (3α,5β)-3,21-dihydroxypregnan-20-one (3α5βTHDOC). All of these GABA-potentiating steroids are potent and efficacious modulators of $GABA_A$ receptors. Detectable potentiation of $GABA_A$-mediated responses is achieved at 10 to 100 n*M*, with maximum potentiation achieved at ~10 μ*M*.[4] With relatively low GABA concentrations (~EC_{10} concentrations), maximum potentiation of GABA-gated currents can be 10- to 20-fold.[4] To place this in the context of other, perhaps better-known, modulators of $GABA_A$ receptors, benzodiaz-

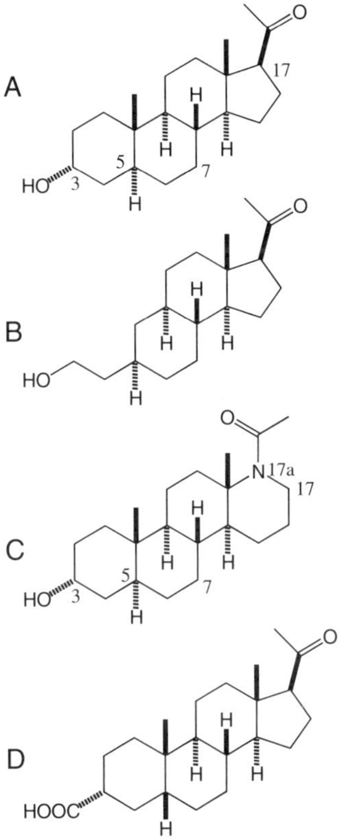

FIGURE 5.1 Representative GABA-active potentiating steroids and steroid analogues. (A) (3α,5α)-3-Hydroxypregnan-20-one (3α5αP). Carbons 3, 5, 7, and 17 are labeled for reference. Dashed bonds denote alpha configuration; bold bonds denote beta stereochemistry. Note that the stereochemistry at C5 is not important for potentiating activity, but the alpha configuration at C3 is essential. See text for details. (B) The benz[*e*]indene analogue of 3α5αP. (C) A 17a-aza-D-homosteroid analogue of 3α5αP. (D) (3α,5β)-20-Oxo-pregnane-3-carboxylic acid (3α5βPC), a steroid analogue with both GABA-potentiating and GABA-inhibiting properties (see also Figure 5.3).

epines are slightly more potent (EC_{50} of 1 to 10 n*M*) but yield a weaker (~2-fold) maximum potentiation.[9] Barbiturates are less potent (EC_{50} of ~10 μ*M*) but yield a maximum potentiation similar to that of GABA-potentiating steroids.[10] Of course, to some degree these are simplifications, as subunit composition of receptors and other factors will influence the potentiation by these modulators.

Most endogenous GABA-inhibiting steroids are characterized by a sulfate moiety (rather than the hydroxyl group of GABA-potentiating steroids) at carbon 3 (C3). Sulfotransferases transfer a sulfate group to the hydroxyl group of pregnenolone or dehydroepiandrosterone to create two of the better-known GABA-inhibiting steroids: pregnenolone sulfate (PS) (Figure 5.2A) and dehydroepiandrosterone sulfate (DHEAS). Active endogenous GABA-inhibiting steroids are also synthesized by a sulfate moiety addition to the hydroxyl group (in either the 3α or 3β configuration)

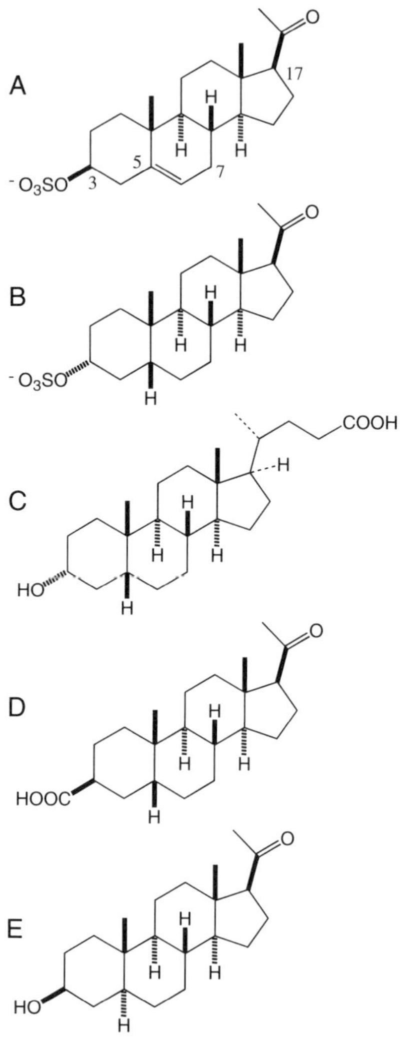

FIGURE 5.2 Representative GABA-active inhibiting steroids. (A) Pregnenolone sulfate (PS). (B) (3α,5β)-3-Hydroxypregnan-20-one sulfate (3α5βPS). Note that the stereochemistry at neither C3 nor C5 is important for inhibition. See text for details. (C) The bile steroid lithocholic acid. (D) (3β,5β)-20-Oxo-pregnane-3-carboxylic acid. This is the 3β-epimer of 3α5βPC and possesses purely GABA-inhibiting activity.[85] (E) (3β,5α)-3-Hydroxypregnan-20-one (3β5αP). We have recently found that 3β-hydroxysteroids are weak inhibitors of $GABA_A$ responses and share mechanistic features with sulfated steroids.[28]

of 5α- or 5β-reduced pregnane steroids (e.g., Figure 5.2B). All of these GABA-inhibiting steroids act in the low micromolar range to depress $GABA_A$ receptor-mediated currents.[11,12] Although the mechanism of these GABA-inhibiting steroids has long been known to be noncompetitive with respect to GABA,[13] the precise mechanism of inhibition is unclear.

5.3 NON-$GABA_A$ RECEPTOR TARGETS OF NEUROACTIVE STEROIDS

As outlined below, neuroactive steroids have been explored as neuro-palliatives in a variety of settings, including use as anesthetics, anticonvulsants, anxiolytics, antidepressants, neuroprotectants, and promnestics. Actions at $GABA_A$ receptors have enjoyed a central role in hypotheses regarding the behavioral effects of neuroactive steroids. However, it is worth bearing in mind that a necessary and sufficient role for $GABA_A$ receptors in complex behavioral states such as anesthesia, although widely accepted, is far from proven. Thus, other molecular targets of neuroactive steroids are reviewed briefly here. We keep the emphasis on those steroids also active at $GABA_A$ receptors, highlighting effects of other neuroactive steroids only when they serve to elucidate a particular point. When relevant, we review the actions of both GABA-potentiating and GABA-inhibiting steroids at these alternate targets. Another extensive treatment of many of these issues has recently been given.[14]

5.3.1 Glycine Receptors

Glycine receptors, like $GABA_A$ receptors, are ligand-gated Cl^- channels and share considerable structural homology with $GABA_A$ receptors. Therefore, it is reasonable to expect that glycine receptors may be important targets for GABA-active steroids. Some GABA-potentiating steroids, including endogenous GABA-potentiating steroids, are nearly completely inert at glycine receptors.[10,15–17] On the other hand, some synthetic GABA-potentiating steroids augment glycine responses generated by recombinant glycine receptors.[14] Concentrations needed to modulate glycine receptor function by the most potent of these steroids are 20-fold higher than those needed to generate equivalent $GABA_A$ receptor potentiation.

The GABA-inhibiting steroid PS reportedly blocks glycine-gated currents in chick spinal cord neurons.[18] The concentration range over which glycine responses are blocked is similar to the range over which GABA responses are blocked. However, PS addition results in a rightward shift in the glycine concentration response curve, with no change in the maximum glycine current generated. This is different from the noncompetitive profile for PS actions at $GABA_A$ receptors, and it seems likely that the mechanism of PS interaction with glycine receptors is dissimilar from the mechanism of $GABA_A$ receptor block.

5.3.2 Ionotropic Glutamate Receptors

Although not a target for GABA-potentiating steroids, both NMDA and non-NMDA classes of ionotropic glutamate receptors are targets for PS and other GABA-inhibiting steroids.[5,11,19–21] Interestingly, PS potentiates NMDA receptor-mediated currents, but other sulfated neuroactive steroids, such as 3α5αPS, inhibit NMDA currents.[19] Substantial progress has been made toward elucidating structural attributes of both the steroid molecule and the NMDA receptor responsible for steroid sensitivity.[19,22–25] This topic is covered extensively in another chapter and therefore will not be reviewed here. Generally, high micromolar concentrations of sulfated steroids are necessary to achieve actions on NMDA receptor-mediated currents.

Non-NMDA receptors are also targets for several sulfated GABA-inhibiting steroids.[11,20] PS and other sulfated steroids depress non-NMDA receptor-mediated responses at high micromolar concentrations. Compared with the NMDA receptor, less is known about the structural requirements of the ligand or receptor for these effects, although apparently both AMPA and kainate receptor function are depressed by PS.[23]

5.3.3 Nicotinic Acetylcholine Receptors

Micromolar concentrations of GABA-potentiating steroids (100- to 1000-fold higher than GABA-potentiating concentrations) inhibit neuronal nicotinic responses.[26,27] In general, the ligand stereochemistry required for nicotinic receptor block is different from that required for $GABA_A$ receptor potentiation. For instance, several 3β-hydroxysteroids are effective nicotinic antagonists,[27] in addition to showing GABA-inhibiting activity.[28] Steroid inhibition of neuronal nicotinic receptors correlates with the ability of the steroids to induce anesthesia in tadpoles, although the concentrations needed for nicotinic effects are much higher than anesthetic concentrations.[27]

Notably, the non-GABA-active estrogen steroids 17β-estradiol and 17α-ethynylestradiol potentiate nicotinic responses generated at neuronal nicotinic α4β2 receptors.[26,29] Chimera strategies exploiting differential sensitivity of human and rat nicotinic receptors locate the binding site for these steroids on the C terminus of the α4 subunit.[26] Mutation of the site did not affect inhibition of the nicotinic responses by steroids like progesterone. The results are noteworthy because they are the clearest example of the identification of a neuroactive steroid binding site on a receptor.

5.3.4 $5\text{-}HT_3$ Serotonin Receptors

The ionotropic serotonin receptor has recently been found to be a target of certain neuroactive steroids. 3α5αP was found to inhibit serotonin responses, along with non-GABA-active steroids 17α- and 17β-estradiol, 17α-ethynylestradiol, testosterone, mestranol, and progesterone.[30,31] On the other hand, PS and cholesterol were inactive at $5\text{-}HT_3$ receptors.

5.3.5 Voltage-Gated Ca^{2+} Channels

Voltage-gated Ca^{2+} channels are depressed by tens of micromolar concentrations of some steroids with GABA-potentiating and GABA-inhibiting activity.[32–35] The effect of one synthetic GABA-potentiating steroid, (3α,5α,17β)-3-hydroxyandrostane-17-carbonitrile, was examined in detail[35] and found to have actions on N-, Q-, and R-, but not L- or P-, type high-voltage-activated (HVA) Ca^{2+} currents.

In dorsal root ganglion neurons, where low-voltage-activated (LVA) Ca^{2+} currents are readily isolated, the same steroid, (3α,5α,17β)-3-hydroxyandrostane-17-carbonitrile, was found to inhibit LVA currents.[36] Another synthetic steroid, (3β,5α,17β)-17-hydroxyestrane-3-carbonitrile, was devoid of GABA activity yet retained good ability to inhibit LVA currents. With this steroid, there was relatively little activity on HVA currents, with the exception of L-type currents. This might suggest that similar steroids could be used to selectively inhibit certain subclasses of voltage-gated Ca^{2+} channels.

5.3.6 σ1 Receptors

The σ_1 receptor is an enigmatic entity that has neuromodulatory effects on excitatory transmission in the central nervous system (CNS). σ_1 receptor ligands modulate cell death, memory function, stress, and depression. Several neuroactive steroids modulate the binding of σ_1 receptor radioligands in various assays. Progesterone was found to be the most effective inhibitor, and GABA-active steroids PS and DHEAS may also interact with the receptor.[37] Modulating endogenous levels of progesterone with inhibitors of synthetic enzymes alters behavioral effects of σ_1 ligands.[38]

There have been recent reports that PS and DHEAS potently (high n*M* concentration range) interact with σ_1 receptors, producing activation of pertussis toxin-sensitive GTP binding proteins.[39] Effects of PS and DHEAS on σ_1 receptors may underlie presynaptic enhancement of synaptic glutamate release,[40,41] although these presynaptic effects of PS and DHEAS at σ_1 receptors occur in the micromolar range (>10 μ*M*). Pregnenolone sulfate effects are enantioselective, with the naturally occurring enantiomer being more active.[40]

5.3.7 Summary of Other Targets

It can be appreciated that neuroactive steroids can hardly be called selective for $GABA_A$ receptors. However, the levels of GABA-potentiating steroids needed to modulate most of these alternate targets are higher than endogenous concentrations and perhaps clinically relevant concentrations, making GABA receptors still the most likely relevant target for potentiating GABA-active steroids. Also, many of the behavioral effects of GABA-potentiating steroids, including anesthesia, sedation, and anxiolysis, are consistent with actions of other known GABA potentiators, such as barbiturates and benzodiazepines. For GABA-inhibiting steroids, concentrations necessary to modulate most targets, including $GABA_A$ receptors, are in the micromolar range, making modulation by endogenous GABA-inhibiting steroids (which are present at tens to a hundred nanomolar) questionable and obscuring the behaviorally relevant targets for GABA-inhibiting steroids.

5.4 CLINICAL RELEVANCE OF GABA-ACTIVE STEROIDS

A role for endogenous GABA-active steroids in the etiology of several neurological and psychiatric disorders has been proposed, and exogenous GABA-active steroids have been explored as therapeutics in various clinical arenas. Many of these clinical topics are covered in detail in other chapters in this volume. In addition, we[42,43] and others[8,44] have recently reviewed this topic elsewhere. Here we briefly review the evidence for clinical relevance and possibilities for clinical applications of both GABA-potentiating and GABA-inhibiting steroids.

5.4.1 GABA-Potentiating Steroids

Anesthesia was among the first described behavioral effects of neuroactive steroids,[1,2] and development of anesthetic steroids remains a topic of interest.[43] Advantages include blood–brain barrier accessibility and potency/efficacy at $GABA_A$ receptors, which as argued above, are likely to be a major target for many anesthetic drugs. Problems with previous attempts to develop viable steroid anesthetics (and that may plague development of neuroactive steroids for other clinical applications) have included poor bioavailability (due to enzymatic activity and poor water solubility), adverse excitatory actions, and adverse reactions to vehicle. There has been progress on all of these aspects, leading to considerable optimism toward the development of steroid anesthetics.[43,44]

GABA-potentiating steroids may have relevance to several psychiatric and neurological disorders. Endogenous GABA-potentiating steroids have recently been proposed to be involved in the etiology of depressive illnesses, including major depression[45] and postpartum depression.[46] Like other GABAergic agents (e.g., barbiturates and benzodiazepines), GABA-potentiating steroids have anxiolytic, sedative-hypnotic, and anticonvulsant effects. In animal models of anxiety, levels of GABA-potentiating steroids increase,[47] presumably reflecting an endogenous anxiolytic mechanism. Alterations in endogenous GABA-potentiating steroids have not been clearly implicated in any seizure disorders, but there is interest in whether catamenial epilepsy, seizures associated with women's menstrual cycles, might be related to decreases in progesterone-derived GABA-potentiating steroids.[48]

Also of clinical relevance are emerging stories of the relationship between ethanol intoxication and GABA-potentiating steroids. Potentiation of $GABA_A$ responses by ethanol is highly dependent on experimental conditions. Recent work suggests that ethanol's effects on $GABA_A$ receptors may be mediated by GABA-potentiating steroids, either by effects on synthesizing enzymes[49] or by direct interaction with ambient steroids.[50] Indirect effects of ethanol on $GABA_A$ responses could potentially help explain the capriciousness of ethanol's actions at these receptors.[49]

5.4.2 GABA-Inhibiting Steroids

Because GABA-inhibiting steroids require supraphysiological concentrations for actions at most targets, including $GABA_A$ receptors (see above), there is limited evidence for a link between endogenous levels of these steroids and the etiology of human disorders. On the other hand, it is possible that high-affinity effects, like some reported σ_1 receptor effects (see above), may lead to the identification of important physiological and pathophysiological roles for these steroids.

There is interest in the use of PS and DHEAS as promnestic agents,[51–54] originally on the basis of PS's ability to potentiate NMDA receptor-mediated currents, which are critical for the induction of many forms of synaptic plasticity. Interestingly, aging results in a decrease in PS levels in rats, correlated with a loss in memory performance. Normalization of PS levels with exogenous steroid ameliorates the memory disturbance.[52]

Other sulfated GABA-inhibiting steroids inhibit NMDA receptor-mediated currents (see above). Because NMDA receptor overstimulation is thought to participate in excitotoxic damage in a number of clinical disorders, including hypoglycemia, stroke, ischemia, and seizures, sulfated steroids could in principle be candidates for development as anti-excitotoxic agents. Unfortunately, the anti-GABAergic properties of these steroids likely may undermine their usefulness. Our structure–activity studies have led us to develop a steroid with properties that may circumvent this problem. This synthetic steroid, (3α,5β)-20-oxo-pregnane-3-carboxylic acid, blocks NMDA receptors, while retaining GABA-potentiating activity. The properties of this neuroactive steroid are discussed in more detail below.

5.4.3 Summary of Clinical Relevance of Neuroactive Steroids

Given their potent and robust actions at $GABA_A$ receptors, it is not surprising that GABA-potentiating neurosteroids have received attention as potentially useful anesthetics, anticonvulsants, anxiolytics, and sedative-hypnotics. In many cases, the success of neurosteroids in these arenas has been limited by practical matters of appropriate drug delivery. As more is learned about structure–activity relationships of these ligands, new strategies may be developed to increase bioavailability and retain the desirable properties of the drugs. It will be exciting to see the development of ideas regarding the role of endogenous GABA-active steroids in clinically relevant states, particularly depression, ethanol intoxication, and perhaps seizure disorders. In the case of GABA-inhibiting steroids, exogenous administration may be appropriate for treatment of cognitive decline in aging, and further refinement of NMDA receptor antagonists without GABA-inhibiting activity may result in useful treatments for excitotoxic neurodegenerative disorders.

5.5 STUDIES OF STRUCTURE–ACTIVITY RELATIONSHIPS

This section reviews our recent strategies for understanding both GABA-potentiating and GABA-inhibiting actions of GABA-active steroids. We and others have pursued several general avenues to help define requirements for GABA activity. First, molecular biological approaches, including subunit selectivity and mutational approaches, are actively being applied to the problem of steroid modulation and are reviewed in other chapters. Although interesting progress has been made, binding sites on the $GABA_A$ receptor remain elusive. Other strategies include photoaffinity labeling of the putative GABA-active site, a strategy that our group is actively pursuing.[55] A final strategy, one that we have employed most heavily and that we review here, is alteration of the ligand structure itself to gain insight into the nature of the binding site(s) relevant to GABA activity.

5.5.1 GABA-Potentiating Steroids

It has long been known that there are two important structural requirements for robust GABA potentiation by steroids. First, GABA-potentiating steroids require a hydrogen bond donor, usually an –OH group, in the α configuration at C3.[56] Second, active potentiators have a hydrogen bond-accepting group on the β face of the steroid at C17. GABA-potentiating steroids are reduced at C5, but the stereochemistry at C5 appears to have little effect on potentiating ability (but see information below on azasteroids). Our groups have focused on refining the structural requirements using a series of tricyclic ligands, azasteroid analogues, and other structural analogues, all synthesized by Covey and colleagues. In addition, a major effort has centered on characterizing enantiomers of GABA-potentiating steroids. Each of these is reviewed in turn.

Tricyclic benz[*e*]indendes are steroid analogues in which the A ring is partially dismantled (Figure 5.1B). This results in an –OH group that is more flexible than when constrained by the rigid ring. We have used these ligands to probe the importance of the ring structure in correctly placing the hydrogen bonding groups. Benz[*e*]indene counterparts of both 3α5αP and 3α5βP are about as effective as their natural steroid counterparts at noncompetitively displacing [^{35}S]TBPS binding from rat brain membranes, at potentiating responses to low concentrations of GABA in electrophysiological assays, and as anesthetics. On the other hand, the benz[*e*]indene analogue of the synthetic GABA-potentiating steroid alphaxalone proved only a weak potentiator of GABA-mediated currents.[57–63] The conclusion from these structure–function studies is that the A ring structure is not necessary for positioning the hydrogen bond donor, although in the context of some otherwise effective steroid structures, loss of the ring results in greatly diminished activity.

In some of the early work on benz[*e*]indenes, phenanthrene analogues, containing flexible side chains in place of the D ring, were tested and found to have very weak GABA-potentiating activity. In an effort to more systematically map the structural requirements associated with the D ring and necessary for strong biological activity, Covey and colleagues[64] recently prepared several 17a-aza-D-homosteroids (Figure 5.1C). One of the eight synthesized azasteroids possessed properties in both [^{35}S]TBPS displacement and electrophysiological assays equivalent to natural steroids.[64] Several other azasteroids possessed reasonable electrophysiological activity but were weak displacers of TBPS binding. The basis for this discrepancy is not clear, as natural GABA-potentiating steroids correlate well in the two assays.

Interestingly, the results of these studies suggest that while the stereochemistry at C5 has minimal effect on GABA-potentiating activity in natural steroids, the stereochemistry at this position can dramatically influence the activity of the synthetic azasteroids. On the basis of molecular modeling and conformational analyses, we suggested that the distance between hydrogen bonding groups may be an important determinant in the biological activity of the ligands. Distances between the hydrogen bonding groups in the 5β-reduced azasteroids may be too short to support optimal interaction with the putative binding site.

In other work to map the conformational preferences of the hydrogen bond acceptor on the D ring, various conformationally constrained hydrogen bond acceptors were substituted for the 20-keto group.[65] A variety of hydrogen bond acceptors retained reasonable activity in the TBPS binding assay, with groups oriented perpendicular to the plane of the D ring retaining best activity. Other approaches toward conformationally constrained analogues may also be possible.

Another general strategy we have pursued is that of synthesizing unnatural enantiomers of natural GABA-potentiating steroids. Although the Meyer–Overton hypothesis is currently out of favor,[66] it is possible that a hydrophobic entity like a neurosteroid might alter GABA receptor function through its interaction with membrane lipids rather than through a specific binding site on the proteinaceous receptor. This mechanism could be consistent with the inability to identify a binding site through molecular biological approaches and with the known importance of membrane composition in the gating properties of ion channels.[67,68] One piece of evidence often cited in favor of a specific binding site is enantioselectivity of ligand effects. Enantiomers of GABA-potentiating steroids have all eight chiral centers reversed and therefore are true mirror images of the natural counterpart. Enantiomeric pairs have identical physico-chemical properties, with the exception of the rotation of polarized light. This is in contrast with diastereomers, which typically have only one chiral center reversed and therefore do not have identical physico-chemical properties. One generally expects that ligands interacting with protein, which is composed entirely of amino acids in the L-configuration, should show enantioselectivity. On the other hand, ligand interactions within the membrane are essentially non-enantioselective, as evidenced by the fact that no enantioselectivity has been consistently observed for the interactions of membrane sterols and optically active membrane lipids.[69–75]

Results with enantiomers to date have shown clear evidence of enantioselectivity using enantiomeric pairs of 3α5α-potentiating steroids. The natural enantiomer was more potent, by as much as 20- to 80-fold.[76,77] However, much weaker enantioselectivity was found for 3α5β pairs.[77] In all cases, enantioselectivity or lack of enantioselectivity extended across multiple assays, including [^{35}S]TBPS displacement, potentiation of GABA-gated currents, direct gating of GABA receptors by steroid, and anesthesia.

The interpretation of these experiments is not entirely clear. The similarity of 5α- and 5β-diastereomers in their ability to potentiate GABA responses has suggested the parsimonious explanation of a single binding site for 5α and 5β steroids. In order to retain this simple idea, the results could be interpreted to imply a single binding site that excludes only the enantiomer of 3α5α steroids. This option is a formal possibility given evidence that some ligand–receptor interactions can occur in a non-enantioselective manner. For instance, nornicotine enantiomers bind nicotinic acetylcholine receptors with equal potency and act with equal efficacy.[78–80] Our studies would then provide new criteria that must be accommodated by molecular models of a single steroid binding site.

5.5.2 GABA-Inhibiting Steroids

We have also been interested in defining structure–activity relationships and mechanisms of GABA-inhibiting steroids. As alluded to previously, there are several reasons for this interest. First, many GABA-inhibiting steroids are also active at NMDA receptors, a pivotal receptor for development, plasticity, and excitotoxicity. The ability to structurally separate GABA activity from NMDA activity may produce a more useful drug. Also, our studies of the mechanism of GABA-inhibiting steroids have led us to understand that GABA inhibition is somehow activation dependent. This property may allow GABA-inhibiting steroids to be used to reverse behavioral states involving overstimulation of $GABA_A$ receptors, such as anesthesia or drug/alcohol intoxication, without affecting normal GABA function. Finally, better understanding the mechanism of GABA-inhibiting steroids may lead us to deeper insight into $GABA_A$ receptor behavior itself.

In general, $GABA_A$ receptor inhibition seems to have fewer structural requirements than $GABA_A$ receptor potentiation. As mentioned previously, an anionic group (usually a sulfate group) at C3 seems important, though not necessary, as reviewed below. GABA-inhibiting steroids are approximately equally active with the charged moiety in the 3α or 3β configuration. Different steroid classes consisting of a pregnene, 5β- or 5α-reduced pregnane, or 5β- or 5α-reduced androstane can all support inhibition.[12,13,81,82] Enantiomeric pairs for DHEAS, PS, and 3α5βPS have been synthesized. In hippocampal neurons little enantioselectivity was found for PS and 3α5βPS, although significant enantioselectivity (~8-fold) was found for DHEAS.[83]

In some of our recent work, we have begun to explore the possible separation of NMDA receptor activity from GABA receptor inhibition. As noted, several sulfated pregnane steroid derivatives block NMDA receptors and therefore may be considered candidates for ameliorating excitotoxicity associated with stroke, seizures, ischemia, and other neurodegenerative insults.[43,84] Unfortunately, because of the structural promiscuity of the GABA-inhibiting actions of steroids sulfated at C3, the anti-excitatory effects of NMDA receptor blockade are likely to be undermined by more potent anti-GABA effects of the same drug. Extending previous work by Weaver and colleagues,[84] who worked with hemisuccinate analogues of sulfated steroids, we synthesized the steroid analogue (3α,5β)-20-oxo-pregnane-3-carboxylic acid (3α5βPC) (Figures 5.1D and 3A).[85] At physiological pH the carboxylic acid group at C3 should be mostly deprotonated ($-COO^-$) and meet the structural requirements for NMDA receptor block and GABA receptor inhibition. On the other hand, the small fraction of un-ionized 3α5βPC contains a hydrogen bond donor at C3 (–COOH) that might be expected to impart GABA potentiation rather than block. Given the generally more potent actions of GABA-potentiating steroids compared with GABA-inhibiting steroids, we reasoned that despite the small fraction of un-ionized 3α5βPC, GABA-potentiating activity might be biologically relevant.

We tested the effects of 3α5βPC at NMDA and GABA receptors.[85] As expected, the drug inhibited NMDA receptor-mediated responses. This inhibition required ~50 μM steroid for detectable effects and was independent of membrane potential. The actions of 3α5βPC at $GABA_A$ receptors were considerably more complex. At physiological pH and at negative membrane potentials, the drug potentiated responses to low GABA concentrations (Figure 5.3B1) and to synaptically released GABA. As the concentration of steroid was increased or as the membrane potential became more depolarized, GABA-inhibiting actions of the steroid became more apparent (Figure 5.3B1). At lower pH, potentiation increased and inhibition was lost (Figure 5.3B2), as expected if the un-ionized form of 3α5βPC was responsible for potentiation and the ionized form was responsible for inhibition. The 3β-diastereomer (Figure 5.2D) was found to have only GABA-inhibiting action, consistent with the necessity of the 3α configuration for potentiation and the lack of stereoselectivity for inhibition.

We used the 3β-diastereomer (Figure 5.2D) to probe inhibition in more detail. Inhibition required >10 μM steroid, somewhat higher than GABA-inhibiting sulfated steroids.[85] We found that inhibition was quite dependent on GABA concentration and membrane potential. These features were somewhat surprising because at the time, there was only very limited[86] or no[87] evidence for dependence of sulfated

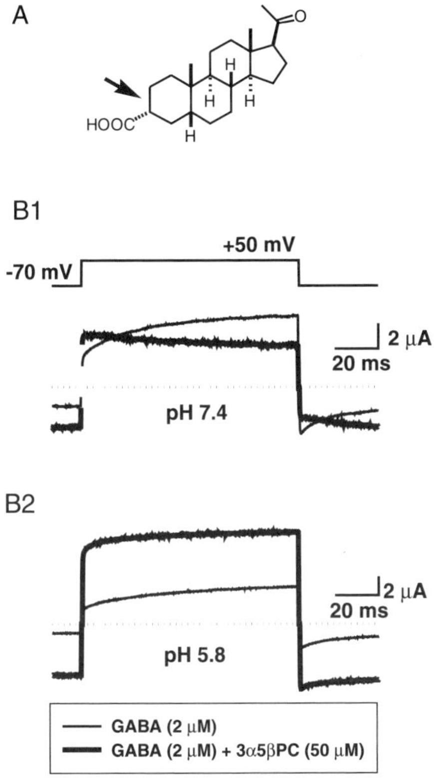

FIGURE 5.3 Demonstration of both the GABA-potentiating and GABA-inhibiting actions of 3α5βPC. (A) Structure of 3α5βPC with emphasis (arrow) on the carboxylic acid group in the α configuration at C3. (B1) The effect of 3α5βPC on GABA responses at –70 and +50 mV at pH 7.4. Membrane currents are responses to 2 μ*M* GABA (applied for 30 sec before the onset of the traces shown) from a *Xenopus* oocyte expressing recombinant α1β2γ2 GABA$_A$ receptor. The thin trace is the response to 2 μ*M* GABA in the absence of 3α5βPC, and the thick trace is the response to GABA plus 50 μ*M* 3α5βPC. Leak conductances in the absence of GABA have been digitally subtracted. The upper trace shows the voltage protocol. The dotted line is the zero current level. Note that 3α5βPC potentiates the inward GABA current observed at –70 mV and potentiates the outward GABA current during the first few milliseconds following a pulse to +50 mV. However, gradually the current at +50 mV relaxes in the presence of 3α5βPC so that there is net inhibition of the outward current by the end of the positive voltage pulse. (B2) Same experiment in the same oocyte, but performed at pH 5.8. Note the greater potentiation and lack of inhibition.

steroid block on channel activation and no evidence for voltage-dependent inhibition.[87] Perhaps the most surprising aspect of the work, and the most dramatic example of the lack of structural specificity required for inhibition, was the observation that the bile steroid lithocholic acid, which contains a carboxylic acid group at C24 rather than C3 (Figure 5.2C), blocks GABA responses similarly to 3β5βPC.[85] Thus, the anionic group can be placed at the opposite end of the molccule with little loss of GABA-inhibiting activity.

Recently, our search for tools to better understand GABA-potentiating steroids intersected rather unexpectedly with our interest in GABA-inhibiting steroids. We were pursuing hints in the literature that 3β-hydroxysteroids (e.g., Figure 5.2E) might act as competitive antagonists with respect to GABA-potentiating steroids.[88–90] We reasoned that if competitive antagonists exist, this would offer an important pharmacological tool to explore endogenous neurosteroid actions and may serve as additional evidence for a specific, receptor-like GABA potentiation site. We undertook an electrophysiological approach to the question and found that 3β-hydroxysteroids at micromolar concentrations did indeed reverse the effects of potentiating steroids.[90] However, the profile appeared to be noncompetitive rather than competitive with respect to potentiating steroid. In fact, inhibition of potentiation was strongest at the highest concentrations of potentiator used. Furthermore, we found that 3β-hydroxysteroids inhibited GABA responses themselves (in the absence of GABA-potentiating steroid), but only when the GABA concentration was elevated to above the GABA EC_{50}. This dependence of inhibition on GABA concentration was reminiscent of the dependence of the 3α5βPC block on GABA concentration. We also examined the sulfated steroid block in more detail[90] and found significant dependence of inhibition on GABA concentration. Also, like 3β-hydroxysteroids, sulfated steroids used at sufficiently low concentrations inhibited the effect of GABA-potentiating steroids without having a detectable effect on the response to GABA alone. The actions of both sulfated steroids and 3β-hydroxysteroids were nearly abolished by a point mutation (V256S) in the α1 subunit of the $GABA_A$ receptor, a mutation that likely inhibits the conformational change promoted by sulfated steroids.[87] Thus, we concluded that 3β-hydroxysteroids share similar, albeit less potent, GABA-inhibiting activity with the better-known sulfated steroid inhibitors. This work once again underscores the lack of structural selectivity for GABA-inhibiting steroids. Charge is apparently not necessary for inhibition (see also Demirgoren et al.[81] and Le Foll et al.[91]).

5.5.3 Summary of Structure–Activity Studies

Our studies of both GABA-potentiating and GABA-inhibiting steroids have focused on altered ligands as probes of the recognition sites associated with $GABA_A$ receptor modulation. Enantiomeric pairs of GABA-potentiating and GABA-inhibiting steroids have been synthesized. In some cases results have shown clear evidence of enantioselective interaction with $GABA_A$ receptors. In other cases, enantioselectivity has been less clear. This information will need to be incorporated in future models of steroid recognitions sites. Synthesis of other novel ligands has helped map the importance of the conformation of the hydrogen bonding groups in positioning GABA-potentiating steroids at their site of action. Generally, GABA-inhibiting steroids show surprising lack of structural specificity, including weak enantioselectivity. Yet a single change in the conformation of a hydroxyl group at C3 can change a GABA-inhibiting steroid into a GABA-potentiating steroid.

5.6 MECHANISM AND SYNAPTIC EFFECTS OF GABA-INHIBITING STEROIDS

The work reviewed above uncovered an interesting property of GABA-inhibiting steroids that has not been previously investigated in detail: the activation dependence of GABA inhibition. Classical activation-dependent blockers act as channel blockers, although traditionally PS and other GABA-inhibiting steroids have been thought not to act as channel blockers.[87,92] When actions of PS are modeled with a popular seven-state kinetic model of the $GABA_A$ receptor, enhancement of desensitization can explain many of PS's effects.[93] Nevertheless, other mechanisms, including channel block, could be possible. We have recently undertaken a characterization of properties of pregnenolone sulfate block, focusing on the nature of activation dependence.[100] In some experiments we compared the behavior of the PS block with that of picrotoxinin, a GABA receptor blocker that has been proposed to share mechanistic properties with PS[94–96] and to possess some features consistent with an open-channel block mechanism.[97]

In initial experiments we sought to address the question of whether PS block is facilitated by a liganded receptor or whether activation (e.g., channel opening or desensitization) is required for inhibition. To address this question, we compared inhibition of GABA responses to inhibition of responses gated by the partial agonist P4S. P4S is a relatively potent agonist, with maximum responses generated by ~1 m*M*. However, at $\alpha1\beta2\gamma2$-containing receptors, P4S gates a maximum response that is <50% of that gated by GABA.[98] Therefore, if agonist–receptor interaction is necessary and sufficient for effective PS inhibition, 1 m*M* P4S and 100 μ*M* GABA, which produce nearly equivalent receptor binding, should be similarly blocked by a given PS concentration. In contrast, we found that P4S responses were weakly blocked, in proportion to the number of open channels, but not in proportion to the number of agonist-bound receptors. Therefore, we conclude that liganded receptors are not sufficient to facilitate PS block. Rather, PS block is more closely associated with open (activated) states of the receptor.

If channel opening facilitates PS block, then kinetics of GABA responses in the combined presence of agonist and inhibitor may reflect the activation dependence. In our studies of PS block we found that when co-applied simultaneously with GABA, PS block developed slowly. At low PS concentrations, the peak response to GABA plus PS was nearly the same as that in the absence of PS, followed by a marked fade in the response and dramatic inhibition of the steady-state phase of the GABA response.[99] This profile of block could be consistent with activation dependence of block, but could be equally consistent with very slow equilibration of PS with receptors. To distinguish these possibilities, we preequilibrated cells in PS alone for 20 to 40 sec, enough time to reach equilibrium, and then applied the combination of GABA with PS. We found that even with preequilibration, responses still showed a peak and decay similar to those elicited without preequilibration in antagonist. Therefore, we conclude that the kinetics of responses in the presence of PS and GABA reflects activation-facilitated block by PS. The presumed channel blocker picrotoxinin also exhibited a kinetic profile in these pre- and co-application experiments similar to that of PS.

In other experiments, we investigated the onset and offset rate constants of PS block during the steady-state phase of GABA responses. We found, contrary to what might be expected from a channel blocker, that offset was slowed at higher GABA concentration, but the onset of PS block did not vary with GABA concentration. Thus, slowed offset of PS inhibition appears to account for the greater steady-state PS inhibition observed at high GABA concentrations. On the other hand, we found that the onset rate constant of block varied with PS concentration, over a range of 2.5 to 20 μ*M*, as expected if binding rate (which should be dependent on blocker concentration) is the rate-limiting step in the inhibition of GABA responses. Similar to PS, the picrotoxinin onset rate increased with picrotoxinin concentration. Nevertheless, PS and picrotoxinin are dissimilar in a number of other respects. These differences, combined with previous single-channel data, suggest that it is unlikely that PS acts through an open-channel block mechanism. In summary, from this work, we conclude that the activation dependence of PS inhibition, while facilitated by states of the receptor beyond simple agonist binding, is likely distinct from channel block. Future work will be aimed at developing a tenable model of the $GABA_A$ receptor and PS inhibition of the receptor.

How do the properties of GABA-inhibiting steroids affect GABA neurotransmission? The activation-dependent properties of block impart some interesting features to the inhibition of inhibitory postsynaptic currents (IPSCs). A primary effect of GABA-inhibiting steroids on IPSCs is truncation of the decay time course.[99] At low steroid concentrations, this is the primary effect, with little depression of the IPSC peak amplitude. At sufficiently low concentrations (e.g., ~2 μ*M* 3β5αPS or 10 μ*M* 3β5βTHDOC), little or no effect on the IPSC decay or peak was detected, yet at this same concentration, IPSCs prolonged by potentiating steroids were significantly accelerated.[28] Like results with exogenous drug applications, these results suggest that GABA-inhibiting steroids may be used under certain conditions to reverse the effects of GABA-potentiating drugs without direct effects on normal GABA synaptic transmission. The results could have implications for treating ethanol or other drug intoxication in which GABA potentiation may have a prominent role.

5.6.1 Summary of Mechanisms and Synaptic Effects of GABA-Inhibiting Steroids

Our work with GABA-inhibiting steroids has revealed some of the properties that underlie the apparent activation dependence of the antagonist actions of PS. The activation dependence requires more than simple binding of agonist to receptor. In addition, the kinetics of steroid association with activated receptors are very slow and independent of GABA concentration. Its properties are dissimilar to those of picrotoxinin, a blocker that has some features typical of channel block. These differences, combined with previous single-channel measurements from others, make it unlikely that PS inhibition is a result of channel block. Rather, it seems more likely that inhibition is a result of allosteric interactions of PS with activated states of the receptor. The features of block may allow GABA-inhibiting steroids to be used, at low concentrations, to reverse exogenous GABA potentiation without altering normal GABA function and tone.

5.7 CONCLUSIONS

Interest in GABA-active steroids has been spurred by a number of factors. The potent actions of these endogenous modulators suggest physiological roles in altering GABA function. Both endogenous and exogenous GABA-active steroids may have relevance to the etiology and treatment, respectively, of CNS disorders. These agents may be used as probes to deepen our understanding of the $GABA_A$ receptor. Although other molecular targets for GABA-active steroids have been identified, the $GABA_A$ receptor remains a prime candidate for many behavioral effects of GABA-potentiating steroids. We have recently found that the actions of sulfated steroids are facilitated by activation of $GABA_A$ receptors. This feature may suggest new applications of these steroids to clinical use.

REFERENCES

1. Seyle, H., Anaesthetic effects of steroid hormones, *Proc. Soc. Exp. Biol. Med.*, 46, 116–121, 1941.
2. Cashin, M.F. and Moravek, V., The physiological actions of cholesterol, *Am. J. Physiol.*, 82, 294–2988, 1927.
3. Majewska, M.D., Harrison, N.L., Schwartz, R.D., Barker, J.L., and Paul, S.M., Steroid hormone metabolites are barbiturate-like modulators of the GABA receptor, *Science*, 232, 1004–1007, 1986.
4. Harrison, N.L. and Simmonds, M.A., Modulation of the GABA receptor complex by a steroid anaesthetic, *Brain Res.*, 323, 287–292, 1984.
5. Irwin, R.P., Maragakis, N.J., Rogawski, M.A., Purdy, R.H., Farb, D.H., and Paul, S.M., Pregnenolone sulfate augments NMDA receptor mediated increases in intracellular Ca^{2+} in cultured rat hippocampal neurons, *Neurosci. Lett.*, 141, 30–34, 1992.
6. Robel, P., Schumacher, M., and Baulieu, E.-E., Neurosteroids: from definition and biochemistry to physiopathological function, in *Neurosteroids: A New Regulatory Function in the Nervous System*, Baulieu, E.-E., Robel, P., and Schumacher, M., Eds., Humana Press, Totowa, NJ, 1999, pp. 1–25.
7. Mellon, S.H., Griffin, L.D., and Compagnone, N.A., Biosynthesis and action of neurosteroids, *Brain Res. Brain Res. Rev.*, 37, 3–12, 2001.
8. Mellon, S.H. and Griffin, L.D., Neurosteroids: biochemistry and clinical significance, *Trends Endocrinol. Metab.*, 13, 35–43, 2002.
9. Wafford, K.A., Whiting, P.J., and Kemp, J.A., Differences in affinity and efficacy of benzodiazepine receptor ligands at recombinant gamma-aminobutyric acid A receptor subtypes, *Mol. Pharmacol.*, 43, 240–244, 1993.
10. Barker, J.L., Harrison, N.L., Lange, G.D., and Owen, D.G., Potentiation of gamma-aminobutyric-acid-activated chloride conductance by a steroid anaesthetic in cultured rat spinal neurones, *J. Physiol.*, 386, 485–501, 1987.
11. Park-Chung, M., Wu, F.S., and Farb, D.H., 3 alpha-Hydroxy-5 beta-pregnan-20-one sulfate: a negative modulator of the NMDA-induced current in cultured neurons, *Mol. Pharmacol.*, 46, 146–150, 1994.
12. Park-Chung, M., Malayev, A., Purdy, R.H., Gibbs, T.T., and Farb, D.H., Sulfated and unsulfated steroids modulate gamma-aminobutyric acid A receptor function through distinct sites, *Brain Res.*, 830, 72–87, 1999.

13. Majewska, M.D., Mienville, J.M., and Vicini, S., Neurosteroid pregnenolone sulfate antagonizes electrophysiological responses to GABA in neurons, *Neurosci. Lett.*, 90, 279–284, 1988.
14. Lambert, J.J., Belelli, D., Harney, S.C., Peters, J.A., and Frenguelli, B.G., Modulation of native and recombinant GABA(A) receptors by endogenous and synthetic neuroactive steroids, *Brain Res. Brain Res. Rev.*, 37, 68–80, 2001.
15. Harrison, N.L., Vicini, S., and Barker, J.L., A steroid anesthetic prolongs inhibitory postsynaptic currents in cultured rat hippocampal neurons, *J. Neurosci.*, 7, 604–609, 1987.
16. Belelli, D., Pistis, M., Peters, J.A., and Lambert, J.J., The interaction of general anaesthetics and neurosteroids with GABA(A) and glycine receptors, *Neurochem. Int.*, 34, 447–452, 1999.
17. Pistis, M., Belelli, D., Peters, J.A., and Lambert, J.J., The interaction of general anaesthetics with recombinant GABAA and glycine receptors expressed in *Xenopus laevis* oocytes: a comparative study, *Br. J. Pharmacol.*, 122, 1707–1719, 1997.
18. Wu, F.S., Chen, S.C., and Tsai, J.J., Competitive inhibition of the glycine-induced current by pregnenolone sulfate in cultured chick spinal cord neurons, *Brain Res.*, 750, 318–320, 1997.
19. Park-Chung, M., Wu, F.S., Purdy, R.H., Malayev, A.A., Gibbs, T.T., and Farb, D.H., Distinct sites for inverse modulation of N-methyl-D-aspartate receptors by sulfated steroids, *Mol. Pharmacol.*, 52, 1113–1123, 1997.
20. Wu, F.S., Gibbs, T.T., and Farb, D.H., Pregnenolone sulfate: a positive allosteric modulator at the N-methyl-D-aspartate receptor, *Mol. Pharmacol.*, 40, 333–336, 1991.
21. Bowlby, M.R., Pregnenolone sulfate potentiation of N-methyl-D-aspartate receptor channels in hippocampal neurons, *Mol. Pharmacol.*, 43, 813–819, 1993.
22. Ceccon, M., Rumbaugh, G., and Vicini, S., Distinct effect of pregnenolone sulfate on NMDA receptor subtypes, *Neuropharmacology*, 40, 491–500, 2001.
23. Yaghoubi, N., Malayev, A., Russek, S.J., Gibbs, T.T., and Farb, D.H., Neurosteroid modulation of recombinant ionotropic glutamate receptors, *Brain Res.*, 803, 153–160, 1998.
24. Weaver, C.E., Land, M.B., Purdy, R.H., Richards, K.G., Gibbs, T.T., and Farb, D.H., Geometry and charge determine pharmacological effects of steroids on N-methyl-D-aspartate receptor-induced Ca(2+) accumulation and cell death, *J. Pharmacol. Exp. Ther.*, 293, 747–754, 2000.
25. Malayev, A., Gibbs, T.T., and Farb, D.H., Inhibition of the NMDA response by pregnenolone sulphate reveals subtype selective modulation of NMDA receptors by sulphated steroids, *Br. J. Pharmacol.*, 135, 901–909, 2002.
26. Paradiso, K., Zhang, J., and Steinbach, J.H., The C terminus of the human nicotinic alpha 4 beta 2 receptor forms a binding site required for potentiation by an estrogenic steroid, *J. Neurosci.*, 21, 6561–6568, 2001.
27. Paradiso, K., Sabey, K., Evers, A.S., Zorumski, C.F., Covey, D.F., and Steinbach, J.H., Steroid inhibition of rat neuronal nicotinic alpha 4 beta 2 receptors expressed in HEK 293 cells, *Mol. Pharmacol.*, 58, 341–351, 2000.
28. Wang, M., He, Y., Eisenman, L.N., Fields, C., Zeng, C.M., Mathews, J., Benz, A., Fu, T., Zorumski, E., Steinbach, J.H., Covey, D.F., Zorumski, C.F., and Mennerick, S., 3-beta-Hydroxypregnane steroids are pregnenolone sulfate-like GABA(A) receptor antagonists, *J. Neurosci.*, 22, 3366–3375, 2002.

29. Curtis, L., Buisson, B., Bertrand, S., and Bertrand, D., Potentiation of human alpha 4 beta 2 neuronal nicotinic acetylcholine receptor by estradiol, *Mol. Pharmacol.*, 61, 127–135, 2002.
30. Wu, F.S., Lai, C.P., and Liu, B.C., Non-competitive inhibition of 5-HT3 receptor-mediated currents by progesterone in rat nodose ganglion neurons, *Neurosci. Lett.*, 278, 37–40, 2000.
31. Wetzel, C.H., Hermann, B., Behl, C., Pestel, E., Rammes, G., Zieglgansberger, W., Holsboer, F., and Rupprecht, R., Functional antagonism of gonadal steroids at the 5-hydroxytryptamine type 3 receptor, *Mol. Endocrinol.*, 12, 1441–1451, 1998.
32. Ffrench-Mullen, J.M. and Spence, K.T., Neurosteroids block Ca^{2+} channel current in freshly isolated hippocampal CA1 neurons, *Eur. J. Pharmacol.*, 202, 269–272, 1991.
33. Ffrench-Mullen, J.M., Danks, P., and Spence, K.T., Neurosteroids modulate calcium currents in hippocampal CA1 neurons via a pertussis toxin-sensitive G-protein-coupled mechanism, *J. Neurosci.*, 14, 1963–1977, 1994.
34. Spence, K.T., Plata-Salaman, C.R., and Ffrench-Mullen, J.M., The neurosteroids pregnenolone and pregnenolone-sulfate but not progesterone, block Ca^{2+} currents in acutely isolated hippocampal CA1 neurons, *Life Sci.*, 49, L235–L239, 1991.
35. Nakashima, Y.M., Todorovic, S.M., Covey, D.F., and Lingle, C.J., The anesthetic steroid (+)-3-alpha-hydroxy-5-alpha-androstane-17-beta-carbonitrile blocks N-, Q-, and R-type, but not L- and P-type, high voltage-activated Ca2+ current in hippocampal and dorsal root ganglion neurons of the rat, *Mol. Pharmacol.*, 54, 559–568, 1998.
36. Todorovic, S.M., Prakriya, M., Nakashima, Y.M., Nilsson, K.R., Han, M.C., Zorumski, C.F., Covey, D.F., and Lingle, C.J., Enantioselective blockade of T-type Ca^{2+} current in adult rat sensory neurons by a steroid that lacks gamma-aminobutyric acid-modulatory activity, *Mol. Pharmacol.*, 54, 918–927, 1998.
37. Su, T.P., London, E.D., and Jaffe, J.H., Steroid binding at sigma receptors suggests a link between endocrine, nervous, and immune systems, *Science*, 240, 219–221, 1988.
38. Maurice, T., Urani, A., Phan, V.L., and Romieu, P., The interaction between neuroactive steroids and the sigma1 receptor function: behavioral consequences and therapeutic opportunities, *Brain Res. Brain Res. Rev.*, 37, 116–132, 2001.
39. Ueda, H., Yoshida, A., Tokuyama, S., Mizuno, K., Maruo, J., Matsuno, K., and Mita, S., Neurosteroids stimulate G protein-coupled sigma receptors in mouse brain synaptic membrane, *Neurosci. Res.*, 41, 33–40, 2001.
40. Meyer, D.A., Carta, M., Partridge, L.D., Covey, D.F., and Valenzuela, C.F., Neurosteroids enhance spontaneous glutamate release in hippocampal neurons: possible role of metabotropic sigma1-like receptors, *J. Biol. Chem.*, 277, 28725–28732, 2002.
41. Partridge, L.D. and Valenzuela, C.F., Neurosteroid-induced enhancement of glutamate transmission in rat hippocampal slices, *Neurosci. Lett.*, 301, 103–106, 2001.
42. Covey, D.F., Evers, A.S., Mennerick, S., Zorumski, C.F., and Purdy, R.H., Recent developments in structure-activity relationships for steroid modulators of GABAA receptors, *Brain Res. Brain Res. Rev.*, 37, 91–97, 2001.
43. Zorumski, C.F., Mennerick, S., Isenberg, K.E., and Covey, D.F., Potential clinical uses of neuroactive steroids, *Curr. Opin. Invest. Drugs*, 1, 360–369, 2000.
44. Gasior, M., Carter, R.B., and Witkin, J.M., Neuroactive steroids: potential therapeutic use in neurological and psychiatric disorders, *Trends Pharmacol. Sci.*, 20, 107–112, 1999.

45. Uzunova, V., Sheline, Y., Davis, J.M., Rasmusson, A., Uzunov, D.P., Costa, E., and Guidotti, A., Increase in the cerebrospinal fluid content of neurosteroids in patients with unipolar major depression who are receiving fluoxetine or fluvoxamine, *Proc. Natl. Acad. Sci. U.S.A.*, 95, 3239–3244, 1998.
46. Griffin, L.D. and Mellon, S.H., Selective serotonin reuptake inhibitors directly alter activity of neurosteroidogenic enzymes, *Proc. Natl. Acad. Sci. U.S.A.*, 96, 13512–13517, 1999.
47. Barbaccia, M.L., Roscetti, G., Trabucchi, M., Purdy, R.H., Mostallino, M.C., Concas, A., and Biggio, G., The effects of inhibitors of GABAergic transmission and stress on brain and plasma allopregnanolone concentrations, *Br. J. Pharmacol.*, 120, 1582–1588, 1997.
48. Reddy, D.S., Kim, H.Y., and Rogawski, M.A., Neurosteroid withdrawal model of perimenstrual catamenial epilepsy, *Epilepsia*, 42, 328–336, 2001.
49. Morrow, A.L., VanDoren, M.J., Penland, S.N., and Matthews, D.B., The role of GABAergic neuroactive steroids in ethanol action, tolerance and dependence, *Brain Res. Brain Res. Rev.*, 37, 98–109, 2001.
50. Criswell, H.E., McCown, T.J., Ming, Z., Mueller, R.A., and Breese, G.R., Interactive role for neurosteroids in ethanol enhancement of gamma-aminobutyric acid-gated currents from dissociated substantia nigra reticulata neurons, *J. Pharmacol. Exp. Ther.*, 291, 1054–1059, 1999.
51. Flood, J.F., Morley, J.E., and Roberts, E., Memory-enhancing effects in male mice of pregnenolone and steroids metabolically derived from it, *Proc. Natl. Acad. Sci. U.S.A.*, 89, 1567–1571, 1992.
52. Vallee, M., Mayo, W., Darnaudery, M., Corpechot, C., Young, J., Koehl, M., Le Moal, M., Baulieu, E.E., Robel, P., and Simon, H., Neurosteroids: deficient cognitive performance in aged rats depends on low pregnenolone sulfate levels in the hippocampus, *Proc. Natl. Acad. Sci. U.S.A.*, 94, 14865–14870, 1997.
53. Vallee, M., Shen, W., Heinrichs, S.C., Zorumski, C.F., Covey, D.F., Koob, G.F., and Purdy, R.H., Steroid structure and pharmacological properties determine the anti-amnesic effects of pregnenolone sulphate in the passive avoidance task in rats, *Eur. J. Neurosci.*, 14, 2003–2010, 2001.
54. Akwa, Y., Ladurelle, N., Covey, D.F., and Baulieu, E.E., The synthetic enantiomer of pregnenolone sulfate is very active on memory in rats and mice, even more so than its physiological neurosteroid counterpart: distinct mechanisms? *Proc. Natl. Acad. Sci. U.S.A.*, 98, 14033–14037, 2001.
55. Darbandi-Tonkabon, R., Hastings, W.R., Zeng, C.M., Akk, G., Manion, B.D., Bracamontes, J.R., Steinbach, J.H., Mennerick, S.J., Covey, D.F., and Evers, A.S., Photoaffinity labeling with a neuroactive steroid analogue: 6-azi-pregnanolone labels VDAC-1 in rat brain, *J. Biol. Chem.*, 278, 13196–13206, 2003.
56. Phillipps, G.H., Structure-activity relationships in steroidal anaesthetics, *J. Steroid Biochem.*, 6, 607–613, 1975.
57. Covey, D.F., Hu, Y., Bouley, M.G., Holland, K.D., Rodgers-Neame, N.T., Isenberg, K.E., and Zorumski, C.F., Modulation of GABAA receptor function by benz[e]indenes and phenanthrenes, *J. Med. Chem.*, 36, 627–630, 1993.
58. Hu, Y., Zorumski, C.F., and Covey, D.F., Neurosteroid analogues: structure-activity studies of benz[e]indene modulators of GABAA receptor function. 1. The effect of 6-methyl substitution on the electrophysiological activity of 7-substituted benz[e]indene-3-carbonitriles, *J. Med. Chem.*, 36, 3956–3967, 1993.

59. Han, M., Hu, Y., Zorumski, C.F., and Covey, D.F., Neurosteroid analogues. 3. The synthesis and electrophysiological evaluation of benz[e]indene congeners of neuroactive steroids having the 5 beta-configuration, *J. Med. Chem.*, 38, 4548–4556, 1995.
60. Rodgers-Neame, N.T., Covey, D.F., Hu, Y., Isenberg, K.E., and Zorumski, C.F., Effects of a benz[e]indene on gamma-aminobutyric acid-gated chloride currents in cultured postnatal rat hippocampal neurons, *Mol. Pharmacol.*, 42, 952–957, 1992.
61. Zorumski, C.F., Wittmer, L.L., Isenberg, K.E., Hu, Y.F., and Covey, D.F., Effects of neurosteroid and benz[e]indene enantiomers on GABA(A) receptors in cultured hippocampal neurons and transfected HEK-293 cells, *Neuropharmacology*, 35, 1161–1168, 1996.
62. Zorumski, C.F., Mennerick, S.J., and Covey, D.F., Enantioselective modulation of GABAergic synaptic transmission by steroids and benz[e]indenes in hippocampal microcultures, *Synapse*, 29, 162–171, 1998.
63. Hu, Y., Zorumski, C.F., and Covey, D.F., Neurosteroid analogues. 2. Total synthesis and electrophysiological evaluation of benz[e]indene analogues of the anesthetic steroid alphaxalone, *J. Org. Chem.*, 60, 3619–3625, 1995.
64. Covey, D.F., Han, M., Kumar, A.S., de La Cruz, M.A., Meadows, E.S., Hu, Y., Tonnies, A., Nathan, D., Coleman, M., Benz, A., Evers, A.S., Zorumski, C.F., and Mennerick, S., Neurosteroid analogues. 8. Structure-activity studies of N-acylated 17a-aza-D-homosteroid analogues of the anesthetic steroids (3-alpha,5-alpha)- and (3alpha,5beta)-3-hydroxypregnan-20-one, *J. Med. Chem.*, 43, 3201–3204, 2000.
65. Anderson, A., Boyd, A.C., Clark, J.K., Fielding, L., Gemmell, D.K., Hamilton, N.M., Maidment, M.S., May, V., McGuire, R., McPhail, P., Sansbury, F.H., Sundaram, H., and Taylor, R., Conformationally constrained anesthetic steroids that modulate GABA(A) receptors, *J. Med. Chem.*, 43, 4118–4125, 2000.
66. Franks, N.P. and Lieb, W.R., Molecular and cellular mechanisms of general anaesthesia, *Nature*, 367, 607–614, 1994.
67. Fong, T.M. and McNamee, M.G., Correlation between acetylcholine receptor function and structural properties of membranes, *Biochemistry*, 25, 830–840, 1986.
68. Rankin, S.E., Addona, G.H., Kloczewiak, M.A., Bugge, B., and Miller, K.W., The cholesterol dependence of activation and fast desensitization of the nicotinic acetylcholine receptor, *Biophys. J.*, 73, 2446–2455, 1997.
69. Agarwal, K., Bali, A., and Gupta, C.M., Effect of phospholipid structure on stability and survival times of liposomes in circulation, *Biochim. Biophys. Acta*, 883, 468–475, 1986.
70. Agarwal, K., Bali, A., and Gupta, C.M., Influence of the phospholipid structure on the stability of liposomes in serum, *Biochim. Biophys. Acta*, 856, 36–40, 1986.
71. Guyer, W. and Bloch, K., Phosphatidylcholine and cholesterol interactions in model membranes, *Chem. Phys. Lipids*, 33, 313–322, 1983.
72. Arnett, E.M. and Gold, J.M., Chiral aggregation phenomena. 4. A search for stereospecific interactions between highly purified enantiomeric and racemic dipalmitoyl phosphatidylcholines and other chiral surfactants in monolayers, vesicles, and gels, *J. Am. Chem. Soc.*, 104, 636–639, 1982.
73. Ghosh, D., Lyman, R.L., and Tinoco, J., Behavior of specific natural lecithins and cholesterol at the air-water interface, *Chem. Phys. Lipids*, 7, 173–184, 1971.
74. Hermetter, A. and Paltauf, F., Interaction of enantiomeric alkylacyl and diacylglycerophosphocholines with cholesterol in bilayer membranes, *Chem. Phys. Lipids*, 31, 283–289, 1982.

75. Mannock, D.A., McIntosh, T.J., Jiang, X., Covey, D.F., and McElhaney, R.N., Effects of natural and enantiomeric cholesterol on the thermotropic phase behavior and structure of egg sphingomyelin bilayer membranes, *Biophys. J.*, 84, 1038–1046, 2003.
76. Wittmer, L.L., Hu, Y., Kalkbrenner, M., Evers, A.S., Zorumski, C.F., and Covey, D.F., Enantioselectivity of steroid-induced gamma-aminobutyric acid A receptor modulation and anesthesia, *Mol. Pharmacol.*, 50, 1581–1586, 1996.
77. Covey, D.F., Nathan, D., Kalkbrenner, M., Nilsson, K.R., Hu, Y., Zorumski, C.F., and Evers, A.S., Enantioselectivity of pregnanolone-induced gamma-aminobutyric acid(A) receptor modulation and anesthesia, *J. Pharmacol. Exp. Ther.*, 293, 1009–1016, 2000.
78. Zhang, X. and Nordberg, A., The competition of (–)–[3H]nicotine binding by the enantiomers of nicotine, nornicotine and anatoxin-a in membranes and solubilized preparations of different brain regions of rat, *Naunyn Schmiedebergs Arch. Pharmacol.*, 348, 28–34, 1993.
79. Badio, B. and Daly, J.W., Epibatidine, a potent analgetic and nicotinic agonist, *Mol. Pharmacol.*, 45, 563–569, 1994.
80. Abreo, M.A., Lin, N.H., Garvey, D.S., Gunn, D.E., Hettinger, A.M., Wasicak, J.T., Pavlik, P.A., Martin, Y.C., Donnelly-Roberts, D.L., Anderson, D.J., Sullivan, J.P., Williams, M., Arneric, S.P., and Holladay, M.W., Novel 3-pyridyl ethers with subnanomolar affinity for central neuronal nicotinic acetylcholine receptors, *J. Med. Chem.*, 39, 817–825, 1996.
81. Demirgoren, S., Majewska, M.D., Spivak, C.E., and London, E.D., Receptor binding and electrophysiological effects of dehydroepiandrosterone sulfate, an antagonist of the GABAA receptor, *Neuroscience*, 45, 127–135, 1991.
82. Majewska, M.D., Demirgoren, S., Spivak, C.E., and London, E.D., The neurosteroid dehydroepiandrosterone sulfate is an allosteric antagonist of the GABAA receptor, *Brain Res.*, 526, 143–146, 1990.
83. Nilsson, K.R., Zorumski, C.F., and Covey, D.F., Neurosteroid analogues. 6. The synthesis and GABAA receptor pharmacology of enantiomers of dehydroepiandrosterone sulfate, pregnenolone sulfate, and (3-alpha,5-beta)-3-hydroxypregnan-20-one sulfate, *J. Med. Chem.*, 41, 2604–2613, 1998.
84. Weaver, C.E., Jr., Marek, P., Park-Chung, M., Tam, S.W., and Farb, D.H., Neuroprotective activity of a new class of steroidal inhibitors of the N-methyl-D-aspartate receptor, *Proc. Natl. Acad. Sci. U.S.A.*, 94, 10450–10454, 1997.
85. Mennerick, S., Zeng, C.M., Benz, A., Shen, W., Izumi, Y., Evers, A.S., Covey, D.F., and Zorumski, C.F., Effects on gamma-aminobutyric acid (GABA)(A) receptors of a neuroactive steroid that negatively modulates glutamate neurotransmission and augments GABA neurotransmission, *Mol. Pharmacol.*, 60, 732–741, 2001.
86. Woodward, R.M., Polenzani, L., and Miledi, R., Effects of steroids on gamma-aminobutyric acid receptors expressed in Xenopus oocytes by poly(A)+ RNA from mammalian brain and retina, *Mol. Pharmacol.*, 41, 89–103, 1992.
87. Akk, G., Bracamontes, J., and Steinbach, J.H., Pregnenolone sulfate block of GABAA receptors: mechanism and involvement of a residue in the M2 region of the α subunit, *J. Physiol.*, 532, 673–684, 2001.
88. Prince, R.J. and Simmonds, M.A., 5 beta-Pregnan-3 beta-ol-20-one, a specific antagonist at the neurosteroid site of the GABAA receptor-complex, *Neurosci. Lett.*, 135, 273–275, 1992.

89. Prince, R.J. and Simmonds, M.A., Differential antagonism by epipregnanolone of alphaxalone and pregnanolone potentiation of [3H]flunitrazepam binding suggests more than one class of binding site for steroids at GABAA receptors, *Neuropharmacology*, 32, 59–63, 1993.
90. Wang, M.D., Backstrom, T., and Landgren, S., The inhibitory effects of allopregnanolone and pregnanolone on the population spike, evoked in the rat hippocampal CA1 stratum pyramidale *in vitro*, can be blocked selectively by epiallopregnanolone, *Acta Physiol. Scand.*, 169, 333–341, 2000.
91. Le Foll, F., Louiset, E., Castel, H., Vaudry, H., and Cazin, L., Electrophysiological effects of various neuroactive steroids on the GABA (A) receptor in pituitary melanotrope cells, *Eur. J. Pharmacol.*, 331, 303–311, 1997.
92. Mienville, J.M. and Vicini, S., Pregnenolone sulfate antagonizes GABAA receptor-mediated currents via a reduction of channel opening frequency, *Brain Res.*, 489, 190–194, 1989.
93. Shen, W., Mennerick, S., Covey, D.F., and Zorumski, C.F., Pregnenolone sulfate modulates inhibitory synaptic transmission by enhancing GABAA receptor desensitization, *J. Neurosci.*, 20, 3571–3579, 2000.
94. Majewska, M.D. and Schwartz, R.D., Pregnenolone-sulfate: an endogenous antagonist of the gamma-aminobutyric acid receptor complex in brain? *Brain Res.*, 404, 355–360, 1987.
95. Majewska, M.D., Demirgoren, S., and London, E.D., Binding of pregnenolone sulfate to rat brain membranes suggests multiple sites of steroid action at the GABAA receptor, *Eur. J. Pharmacol.*, 189, 307–315, 1990.
96. Gee, K.W., Joy, D.S., and Belelli, D., Complex interactions between pregnenolone sulfate and the t-butylbicyclophosphorothionate-labeled chloride ionophore in rat brain, *Brain Res.*, 482, 169–173, 1989.
97. Yoon, K.W., Covey, D.F., and Rothman, S.M., Multiple mechanisms of picrotoxin block of GABA-induced currents in rat hippocampal neurons, *J. Physiol.*, 464, 423–439, 1993.
98. Ebert, B., Wafford, K.A., Whiting, P.J., Krogsgaard-Larsen, P., and Kemp, J.A., Molecular pharmacology of gamma-aminobutyric acid type A receptor agonists and partial agonists in oocytes injected with different alpha, beta, and gamma receptor subunit combinations, *Mol. Pharmacol.*, 46, 957–963, 1994.
99. Shen, W.X., Mennerick, S., Zorumski, E.C., Covey, D.F., and Zorumski, C.F., Pregnenolone sulfate and dehydroepiandrosterone sulfate inhibit GABA-gated chloride currents in Xenopus oocytes expressing picrotoxin-insensitive GABA(A) receptors, *Neuropharmacology*, 38, 267–271, 1999.
100. Eisenmen, L.N., He. Y., Fields, C., Zorumski, C.F., and Mennerick, S., Activation dependent properties of pregnenolone sulfate inhibition of $GABA_A$ receptor-mediated current, *J. Physiol.*, 550, 679–691, 2003.

6 Anabolic Androgenic Steroids and the Brain: Novel Actions at the $GABA_A$ Receptor and on $GABA_A$ Receptor-Mediated Behaviors

Ann S. Clark, Brian L. Jones, Paul Yang, and Leslie P. Henderson

CONTENTS

0-8493-2392-4/04/$0.00+$1.50

6.1 HISTORY OF AAS DEVELOPMENT

6.1.1 Therapeutic Use of AASs

The steroid hormone, testosterone, has both androgenic and anabolic activities. The identification, isolation, and subsequent synthesis of testosterone in the 1930s opened up therapeutic avenues of treating not only hypogonadism, but also a number of other catabolic disorders. Specifically, beginning in the 1940s, continuing efforts have been made to produce synthetic derivatives of testosterone with enhanced anabolic vs. androgenic actions (for review, see References 1–3). While efforts to synthesize purely anabolic steroids have not been successful, and all anabolic androgenic steroids (AASs) retain some degree of androgenic actions (for review, see Wilson[4] and Lukas[5]), efforts to this end have led to the production of approximately 60 AAS compounds to date (for review, see Kammerer[6]). AASs have been used for decades not only to provide androgen replacement for hypogonadal men, but also for treatment of diminished growth, breast cancer, hereditary angioedema, and sequelae associated with burns, surgery, and radiation therapy (for review, see Basaria et al.[2]). Most recently, AASs have become a valuable tool in combating wasting syndromes, such as those associated with HIV/AIDS (for review, see References 2, 3, and 7). AASs that are most commonly used for therapeutic purposes today include nandrolone decanoate, oxandrolone, oxymetholone, and stanozolol (for review, see Shahidi[3]). Whereas historically AASs have received the greatest attention for their effects on metabolism and peripheral structures, recent studies in the clinic and in the laboratory have highlighted the effects of AASs on the central nervous system (CNS) and on behavior (for review, see References 2, 3, and 7).

6.1.2 Abuse of AASs

Although originally developed for clinical use, AAS administration is now predominantly one of abuse, and the medical benefits of low doses of AASs stand in sharp contrast to the potential health risks associated with the excessive doses self-administered by elite athletes (for review see References 5, 8, and 9). Steroid abuse refers to the consumption or self-administration of AASs to modify physique, weight, or athletic performance. At events sponsored by the International Olympic Committee, anabolic steroids account for more than half of all the positive tests for illicit compounds.[10] Due to the legal and ethical prohibitions on AAS use, it has been difficult to estimate the prevalence of AAS use; however, assessments made a decade ago indicate that, at that time, more than 1 million adult Americans had or were using AASs.[11] Over the past few decades, the misuse of AASs by college, Olympic, and professional athletes has become widely recognized.[12] However, recent reports highlight the fact that the more insidious growth in the abuse of these drugs is not among elite athletes, but among adolescent boys and girls.[13] Present estimations are that 4 to 12% of adolescent boys and 0.5 to 2% of adolescent girls have used AASs[14,15] (for review, see Koch[16]), and the greatest increase in AAS use over the past decade has been reported for adolescent girls.[15]

Self-administration of AASs in the human population is characterized by personal experimentation, secrecy, and self-evaluation of results. The doses and combinations of AASs used by athletes are typically in great excess (6- to 100-fold) of therapeutic doses.[17] Perhaps somewhat surprisingly, 37% of users interviewed did not know the dosage in milligrams of the AASs they were taking.[18] In some cases, even the exact composition of the AASs consumed was unknown. Heavy abusers report a stacking pattern of steroid consumption, referring to the combination of different types of steroid preparations over a period of time, referred to as a cycle.[19] The complexity of AAS self-administration has confounded interpretations of the physiological and behavioral effects of AASs in human subjects. In contrast, mechanistic information on AAS actions in the brain and on behavior has been provided by studies in animal models in which known, more simplified AAS administration paradigms are employed.

6.1.3 Characterization of AAS Chemistry

The AASs, like the endogenous androgens, have a four-ringed backbone structure with 19 carbon atoms (Figure 6.1). Modifications of this androstane backbone have been introduced to prolong the metabolic half-life and thus the efficacy of synthetic AASs. Three main classes of AASs have been described.[2,3] The first, used primarily as injectable compounds, is derived from esterification of the 17β-hydroxyl group of testosterone and includes testosterone propionate and testosterone cypionate. Esterification retards absorption and prolongs the duration of action after injection of the hormone by slowing its release into circulation. Testosterone esters can be hydrolyzed into free testosterone, reduced to 5α-dihydrotestosterone, an androgen with higher biological activity at brain androgen receptors than testosterone,[20–22] or aromatized to estrogens[21] (for review, see Kochakian and Yesalis[22]). Molecules that have been 5α-reduced cannot be metabolized into estrogens but may be metabolized into other androgenic compounds such as 3α-androstanediol.

The second class of AASs is also composed of injectable androgen esters, called 19-nor-testosterone derivatives. These compounds have, in conjunction with the addition of long side chain moieties, a substitution of a hydrogen for the methyl group at C19 (for review, see Basaria et al.[2] and Shahidi[3]) (Figure 6.1). AASs in this class include nandrolone decanoate. The substitution at C19 extends the half-life of this class of AASs beyond that contributed by esterification alone (for review, see Shahidi[3]). Nandrolone decanoate can be aromatized to 17β-estradiol, albeit only ~20% as efficiently as testosterone.[21,23] Aromatizable AASs in both class 1 and class 2 may thus have additional and significant central nervous system effects not only at the androgen receptor, but also vis-à-vis the actions of their estrogenic metabolites at brain estrogen receptors.[4]

The final class of AASs comprises those compounds that are alkylated at C17, including 17α-methyltestosterone (17α-MeT), oxymetholone, methandrostenolone, and stanozolol (Figure 6.1). Because alkylation retards metabolism by the liver, this

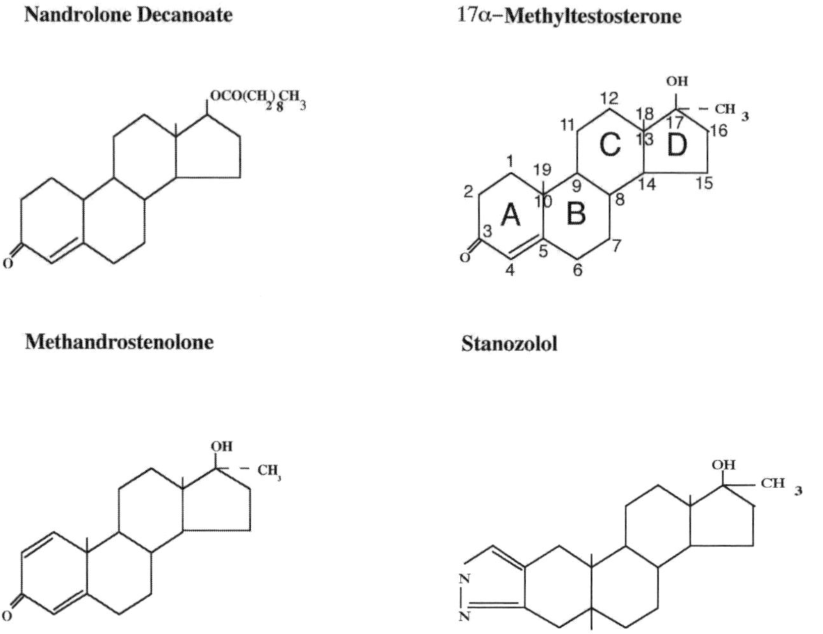

FIGURE 6.1 Chemical structures of four commonly self-administered AASs. Numbering convention for androstane backbone is shown for 17α-MeT.

group of AASs is orally active (for review, see Basaria et al.[2]). Although none of the 17α-alkylated AASs is converted into dihydrotestosterone or 17β-estradiol,[21] recent data indicate that the 17α-alkylated AASs can be converted to their respective A-ring metabolites and that peripheral effects of the 17α-alkylated AASs are mediated by both estrogen and androgen receptors.[24,25]

Over the past several decades, there has accumulated a wealth of data characterizing the actions of the endogenous neurosteroid metabolites of progesterone (e.g., allopregnanolone: 5α-pregnan-3α-ol-20-one), corticosterone (e.g., THDOC: 5α-pregnane-3α,21-diol-20-one), and testosterone (e.g., 3α-diol: 5α-androstane-3α,17β-diol) on the expression and function of γ-aminobutyric acid type A ($GABA_A$) receptors (see other chapters in this volume). Recent data have suggested that the AASs, like the neurosteroids, modulate $GABA_A$ receptor function and expression. One possibility is that the AASs are simply structural variants of the neurosteroids; however, none of the data to date support this assertion. First, with respect to structure, although the AASs share obvious similarities (e.g., the pregnane and androstane backbones) with the neurosteroids, key differences between the AASs and the neurosteroids exist. Specifically, the presence of a 3α-hydroyx group is critical for neurosteroid modulation, especially for the androstane steroids[26] (for review, see Lambert et al.[27,28]), and a C_{20} or C_{17} keto group is believed to enhance neurosteroid modulation.[26,29,30] Both moieties are notably absent on commonly abused AASs (Figure 6.1). Second, as described below, the mechanisms by which the AASs and neurosteroids induce allosteric modulation of $GABA_A$ receptor-mediated currents clearly diverge.

6.2 AAS MODULATION OF GABA$_A$ RECEPTOR FUNCTION AND EXPRESSION

6.2.1 Allosteric Modulation of GABA$_A$ Receptors

The breadth of behaviors altered in humans and animals subjected to supraphysiological levels of AASs suggests that numerous brain regions and many different signaling systems may be altered by AASs, and data demonstrate that AASs have a multiplicity of actions on serotonergic, dopaminergic, and peptidergic signaling in the central nervous system (for review, see Clark and Henderson[31]). However, the prevalence of behavioral reports demonstrating AAS effects on anxiety, aggression, and reproductive behaviors (for review, see Clark and Henderson[31]) also strongly implicate GABA$_A$ receptors as important targets of AAS action, as GABAergic transmission has been shown to play a pivotal role in expression of these behaviors.[32–37] Studies performed within the past decade have directly shown that acute exposure to AASs induces allosteric modulation of GABA$_A$ receptor function[38–41] and that chronic AAS exposure alters GABA$_A$ receptor subunit mRNA expression.[42] In addition, recent data demonstrate that the ability of the AASs to modulate the GABA$_A$ receptor depends upon subunit composition, and that the AASs act at a separate site from the neurosteroid binding sites.

6.2.1.1 GABA$_A$ Receptor Subunit Complexity

GABA$_A$ receptors (Figure 6.2) are widely expressed throughout the vertebrate central nervous system and subserve an essential role in mediating fast synaptic inhibitory transmission in the adult CNS. Moreover, the GABA$_A$ receptor is the primary molecular target for a broad range of therapeutic and abused drugs. GABA$_A$ receptors

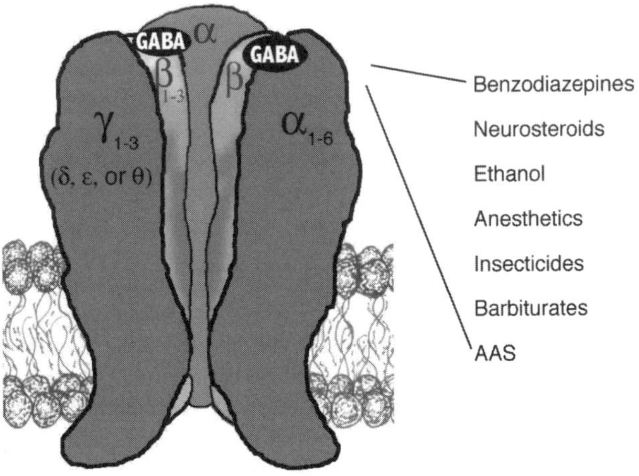

FIGURE 6.2 Schematic representation of the GABA$_A$ receptor indicating the posited arrangement of subunits within the pentamer, the heterogeneity in possible subunit composition, the location of the GABA binding sites at the extracellular interface of the α and the β subunits, and a list of some of the allosteric modulators of this receptor.

are members of the evolutionarily related superfamily of ligand-gated ion channels that includes the nicotinic acetylcholine receptor and the glycine receptor families (for review, see References 43–45) and the 5-HT_3 subclass of receptors gated by serotonin (for review, see Reference 46). The first $GABA_A$ receptor cDNAs were cloned in 1987, and 19 different members (α_1 to α_6, β_1 to β_3, γ_1 to γ_3, δ, ε, π, ρ_1 to ρ_3 and θ) have now been identified in mammals (for review, see References 47–49) with mRNA splice variants reported for the γ_2,[50–52] β_2,[53,54] β_3,[55] and α_6[56] subunits. With such extraordinary diversity of subunit transcripts, an amazing range of pentameric combinations could potentially be formed. The actual diversity of $GABA_A$ receptors expressed in neuronal tissue, however, is more limited, and studies to date indicate that most native $GABA_A$ receptors in the CNS contain two α, two β, and a γ subunit, with the most commonly expressed isoform consisting of α_1, β_2 or β_3, and γ_2 subunits.[57] Less abundant subunits, such as γ_1, γ_3, δ, or ε, may substitute for or be coexpressed with α, β, and γ_2 subunits (for review, see Barnard et al.[58]), although specific restrictions for particular subunit combinations within a single receptor have been described.[59,60]

While subunit combinations may be somewhat restricted in the CNS, there is nonetheless both tremendous developmental and regional diversity in the expression of $GABA_A$ receptor subunits. This structural diversity, in turn, endows $GABA_A$ receptors with both region-specific and developmentally regulated functional complexity, especially with respect to the ability of drugs, toxins, and hormones to induce allosteric modulation of receptor function (for review, see Mehta and Ticku[47] and Sieghart[61]). The biophysical mechanisms by which benzodiazepines, barbiturates, anesthetics, and neurosteroids allosterically modulate the $GABA_A$ receptor, as well as the importance of subunit composition for allosteric modulation by these compounds, are well characterized (see other chapters in this volume). In contrast, studies have only recently been performed to assess allosteric modulation of $GABA_A$ receptors by the AASs.

6.2.1.2 AASs Induce Allosteric Modulation of $GABA_A$ Receptors

The first studies to suggest that AASs have modulatory actions at the $GABA_A$ receptor were those provided by Bitran and colleagues.[62,63] These investigators showed that prolonged exposure (>1 week) to the AAS, testosterone propionate (TP) or methandrostenolone, left-shifted the EC_{50} for chloride ($^{36}Cl^-$) flux induced by GABA in cortical synaptosomes and induced anxiolytic effects in rats. These investigators hypothesized that the observed changes in $GABA_A$ receptor-mediated chloride flux and behavior were mediated by allosteric modulation of the $GABA_A$ receptor following metabolism of testosterone propionate or methandrostenolone to neurosteroid derivatives, specifically androsterone and 3α-androstanediol. Although the novel hypothesis that AASs may induce both functional and behavioral effects via allosteric modulation of the $GABA_A$ was attractive, the prolonged time course of these experiments did not allow these investigators to directly demonstrate allosteric modulation of $GABA_A$ receptor function. Moreover, the observed changes in chloride flux did not correlate with comparable changes in serum levels of 3α-androstanediol,[63] and no studies to date have demonstrated that methandrostenolone is metabolized to neuroactive metabolites.

Other studies have now directly demonstrated that the AASs, in the absence of metabolism, can induce allosteric modulation of the $GABA_A$ receptor. Masonis and McCarthy[38,39] showed that acute application of micromolar concentrations of either stanozolol or 17α-methyltestosterone significantly inhibited the binding of the benzodiazepine site ligand, flunitrazepam, to rat brain synaptoneurosomal membrane preparations and that stanozolol both potentiated GABA-mediated chloride flux and inhibited the flunitrazepam-mediated enhancement of this GABA-elicited chloride flux. These investigators concluded that the AASs act as direct allosteric modulators of the $GABA_A$ receptor. Masonis and McCarthy[38] also showed that while stanozolol or 17α-methyltestosterone inhibited benzodiazepine binding, the neurosteroid tetrahydroxycorticosterone (THDOC) enhanced benzodiazepine binding, suggesting distinct functional mechanisms and perhaps separate binding sites for these two classes of steroid modulators.

Results using these biochemical approaches to demonstrate that the AASs are indeed allosteric modulators of $GABA_A$ receptors have been recently confirmed and extended by direct electrophysiological assessments. Jorge-Rivera et al.[40] demonstrated that acute and direct application of 1 μM stanozolol, nandrolone, or 17α-MeT induced significant modulation of synaptic currents in two rat forebrain areas, the ventromedial nucleus of the hypothalamus (VMN) and the medial preoptic area (mPOA), in slices from prepubertal (postnatal days 3 to 14) female Sprague–Dawley rats. Taken together with the results of Masonis and McCarthy,[38,39] these data unambiguously demonstrate that the AAS can itself induce allosteric modulation of the $GABA_A$ receptor, and that metabolism to neurosteroids is not required for this modulation. It should also be noted that although concentrations of endogenous steroids are in the nanomolar range, serum levels of AASs have been estimated to reach micromolar concentrations in human subjects who self-administer AASs.[38,64] Thus, modulation induced by micromolar concentrations of AASs can be predicted to be relevant to steroid abuse conditions.

6.2.1.3 AAS- vs. Neurosteroid-Induced Allosteric Modulation

Different classes of positive allosteric modulators of the $GABA_A$ receptor elicit enhanced chloride flux through distinct biophysical mechanisms. For example, positive benzodiazepine site modulators are believed to act by increasing receptor affinity for GABA and increasing the frequency of channel openings[65–67] (for review, see Sieghart[61]). In contrast, positive neurosteroid modulators, at nanomolar to low micromolar concentrations, act to slow the rate of recovery from desensitization,[68] thus promoting late reentry into an open state and prolonging GABA-elicited macroscopic responses.[68–70] Moreover, at concentrations greater than 1 μM, positive neurosteroid modulators can activate the receptor directly in the absence of GABA (for review, see references 27, 28, and 61). Although AASs and neurosteroids share some obvious structural similarities, physiological data do not support the hypothesis that they share a common mechanism of allosteric modulation or act at the same site on the receptor. First, as mentioned above, the 3α-hydroxyl moiety common to

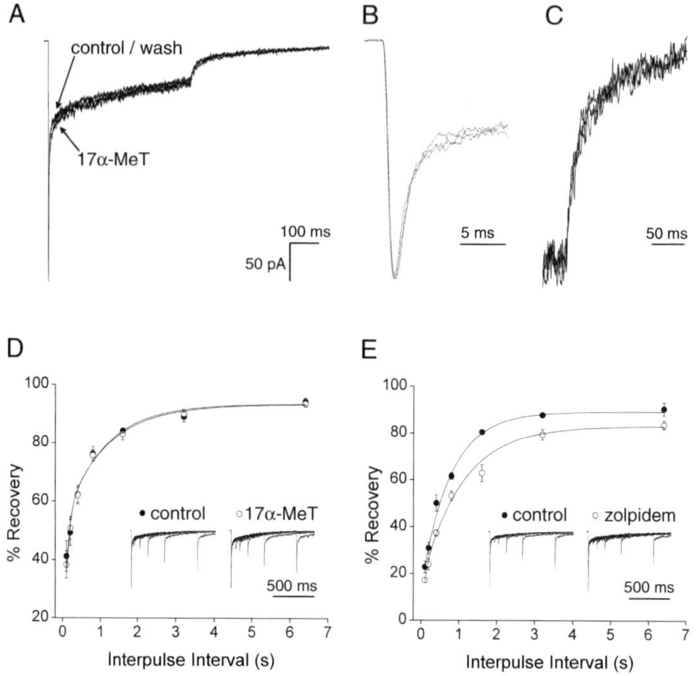

FIGURE 6.3 AASs do not modulate desensitization or recovery from desensitization for recombinant $\alpha_1\beta_3\gamma_{2L}$ $GABA_A$ receptors. (A) Currents elicited by a step application (500 msec) of 1 m*M* GABA in the presence and absence of 1 μ*M* 17α-MeT. The AASs had no appreciable effect on peak current (I_{peak}), desensitization during the step, or deactivation following the step. Currents in (B) and (C) are from the current trace in (A) and have been peak-scaled and expanded to show that there is no difference in the time course of desensitization (B) or in the deactivation following the end of the GABA step (C) in the absence and presence of 17α-MeT. (D) Comparison of the recovery time course of the second response in the paired pulse protocol for exposure to 1 μ*M* 17α-MeT (n = 10). Brief (4 msec) pulses of 1 m*M* GABA were applied with variable interpulse intervals. Percent recovery from desensitization (% Recovery), as assessed by $(I_{peak2} - onset_2)/(I_{peak1} - onset_1) \times 100$, was plotted as a function of interpulse interval and described by two exponential components with time constants τ_{r1} and τ_{r2}. Inset shows currents elicited by paired pulse paradigm from a representative experiment. (E) Percent recovery from desensitization as shown in (D) but for exposure to 1 μ*M* zolpidem (n = 4). Exposure to this benzodiazepine site agonist slowed recovery from desensitization. (From Figure 4 in Yang, P., Jones, B.L., and Henderson, L.P., *Neuropharmacology*, 43, 619, 2002. With permission.)

all active neurosteroids and the C_{17} or C_{20} keto groups (for review, see Lambert et al.[27,28]) are not present in any of the AASs shown to elicit allosteric modulation of $GABA_A$ receptors. Second, positive neurosteroid modulators have been shown to elicit their effects by enhancing desensitization, slowing recovery from desensitization, and prolonging deactivation following desensitization.[68,71] Yang et al.[41] have shown for $\alpha_1\beta_3\gamma_{2L}$ recombinant receptors that the AAS 17α-MeT had no effect on any of these three parameters (Figure 6.3). Third, unlike the neurosteroids, AASs, at concentrations as high as 10 μ*M*, had no ability to directly activate $GABA_A$ receptor-mediated chloride currents.[40,41] Finally, as noted above, Masonis and McCarthy[38] have shown that AASs inhibit binding of ^{3}H-flunitrazepam to rat brain cerebrocortical membranes, whereas the neurosteroid THDOC potentiated ^{3}H-flunitrazepam binding.

6.2.1.4 AAS- vs. Benzodiazepine-Induced Allosteric Modulation

In contrast to the neurosteroids, two lines of evidence from previous studies suggested that the AASs might be acting at a site shared by the benzodiazepines. First, Masonis and McCarthy[38] hypothesized that the ability of stanozolol and 17α-MeT to displace ^{3}H-flunitrazepam binding was consistent with a competitive interaction between the two classes of compounds. Second, the negative modulation by AASs of $GABA_A$ receptor-mediated synaptic currents in mPOA neurons[40] mirrored that induced in these same neurons by the benzodiazepine site agonist zolpidem.[72] However, recent experiments by Yang et al.[41] have directly compared benzodiazepine and AAS modulation of recombinant $\alpha_1\beta_3\gamma_{2L}$ receptors, and these data do not support the assertion that the AASs are acting at the high-affinity benzodiazepine modulatory site for this class of $GABA_A$ receptors. First, although differences in potency at the same site cannot be ruled out, the potentiation induced by 1 μ*M* 17α-MeT was significantly less than that produced by either 1 μ*M* zolpidem or 1 μ*M* diazepam. Second, zolpidem and the AAS, 17α-MeT, induced disparate effects on both deactivation and desensitization kinetics of currents elicited by ultrafast perfusion of GABA. Third, and most compelling, the high-affinity benzodiazepine site antagonist, flumazenil, inhibited potentiation induced by zolpidem, but was without effect on the potentiation induced by 17α-MeT. Finally, as described in the section below, computational modeling suggests that optimal fits of benzodiazepine and AAS potentiation are provided by two different kinetic mechanisms.

6.2.1.5 Kinetic Modeling of AAS Allosteric Modulation

Positive benzodiazepine site modulators are believed to act by increasing receptor affinity for GABA by altering its binding (k_{on}) or unbinding (k_{off}) rates[65–67] (for review, see Sieghart[61]). Consistent with these previous reports, simulations based on the seven-state model proposed by Jones and Westbrook[69,70] indicated that optimal solutions for zolpidem modulation of currents elicited from recombinant $\alpha_1\beta_3\gamma_{2L}$ receptors were provided by changes in k_{on} or k_{off} (Jones and Yang, unpublished data). In contrast, data for the modulation induced by 17α-MeT were more accurately described by changes in β_l, the rate governing the transition to the singly liganded open state[41] (Figure 6.4). The conclusion that the best computation fit for currents from this class of $GABA_A$ receptors is provided by a mechanism in which 17α-MeT acts to increase β_l is consistent with the observation that this AAS induced no potentiation of responses elicited at $\alpha_1\beta_3\gamma_{LL}$ receptors by millimolar concentrations

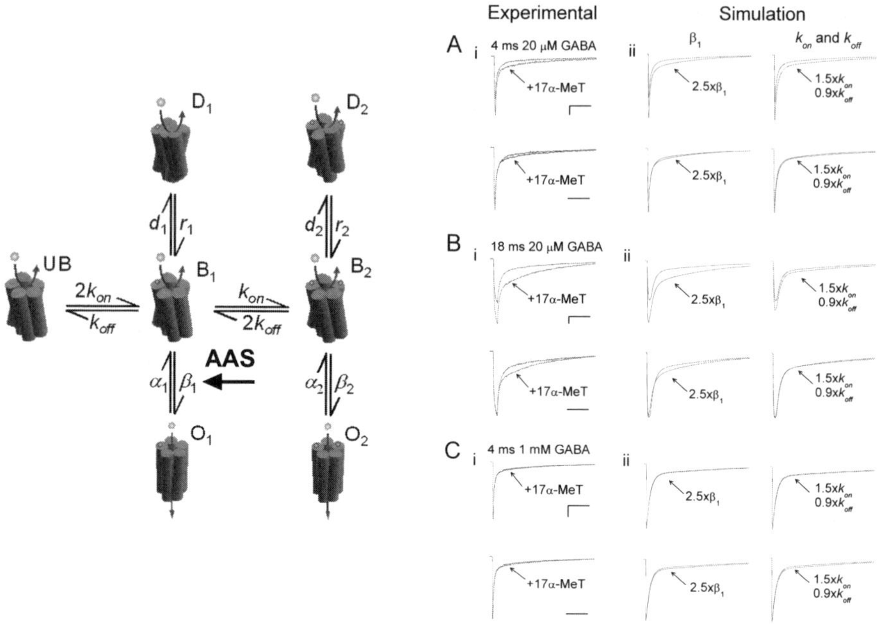

FIGURE 6.4 Kinetic modeling of AAS modulation of currents elicited from recombinant $\alpha_1\beta_3\gamma_{2L}$ $GABA_A$ receptors. The model is a minor modification of the seven-state Markov model of Jones and Westbrook[69] and assumes two equivalent agonist binding sites for the receptor. This yields one unbound state (UB) and two bound states (B_1 and B_2). From the bound states there are also transitions to singly and doubly liganded open (O_1 and O_2, respectively) and desensitized (D_1 and D_2, respectively) states. This cartoon illustrates agonist binding and hypothetical conformational changes associated with each kinetic state. The rates for the model were (in sec^{-1}): $k_{on} = 8.18 \times 10^6\ M^{-1}$, $k_{off} = 50$, $\alpha_1 = 2387$, $\alpha_2 = 523$, $\beta_1 = 874$, $\beta_2 = 2559$, $d_1 = 222$, $d_2 = 239$, $r_1 = 36$, and $r_2 = 10$. This combination of parameters gives a P_{OMax} of 0.7. Experimental data (i) and computer simulations (ii) are shown for representative responses elicited by either a brief (4 or 18 msec) pulse of 20 μ*M* GABA (A and B) or a 4-msec pulse of 1 m*M* GABA (C). In each section, immediately below each trace, are peak-scaled versions. Computer simulations in which kinetic parameters were altered indicated that the best fits to the data were obtained by both increasing k_{on} and decreasing k_{off} or by increasing β_1, the rate governing transition into the singly liganded open state (O_1). (Adapted from Figures 7 and 9 in Yang, P., Jones, B.L., and Henderson, L.P., *Neuropharmacology*, 43, 619, 2002. With permission.)

of GABA (where it is expected that activation is maximal and nearly all receptors reside in the doubly liganded states), but induced significant potentiation of responses elicited by subsaturating (20 μ*M*) concentrations of GABA.[41]

It has been shown that the ability of both the neurosteroids and the benzodiazepines to modulate $GABA_A$ receptor-mediated currents and the molecular mechanism by which this modulation occurs is dependent upon the subunit composition of the receptor. The mechanism suggested by kinetic modeling for AAS modulation of responses from $\alpha_1\beta_3\gamma_{2L}$ receptors should be interpreted with the restriction that it may not uniformly describe how these steroids modulate $GABA_A$ receptors with different subunit compositions. Specifically, current findings from the Henderson laboratory indicate that the ability of the AASs to modulate $GABA_A$ receptors varies significantly with subunit composition. This pharmacological variation most likely reflects distinct molecular mechanisms by which the AASs alter the function of different $GABA_A$ receptor isotypes.

6.2.1.6 Subunit Specificity of AAS Allosteric Modulation

The ability of the AAS, 17α-MeT, to modulate GABA-elicited currents varies significantly between α_1- and α_2-containing $GABA_A$ receptors. Specifically, as discussed above, 17α-MeT was found to have no effect on peak amplitudes, deactivation, or desensitization of currents elicited by millimolar concentrations of GABA from recombinant $\alpha_1\beta_3\gamma_{2L}$ receptors (Figure 6.5).[41] In contrast, preliminary data indicate that 17α-MeT significantly increases peak current amplitudes and prolongs deactivation for currents elicited from $\alpha_2\beta_3\gamma_{2L}$ receptors by millimolar concentrations of GABA (Figure 6.5). Consistent with data from recombinant receptors, AASs do not modulate synaptic currents recorded in cerebellar Purkinje neurons,[41] which express $\alpha_1\beta_{2/3}\gamma_2$ receptors,[57,73,74] but do potentiate synaptic currents recorded in VMN neurons[40] that express predominantly $\alpha_2\beta_3\gamma_2$ receptors.[57,73,74] Taken together, these data suggest that AASs will impart appreciably different actions on functions mediated by the forebrain, hypothalamic, and limbic structures, which express high levels of α_2-containing receptors, than on functions mediated by the cortex or cerebellum, in which α_1-containing receptors predominate.

The observation that AASs induce positive modulation of currents in VMN neurons, but negative modulation of neurons from the mPOA, was attributed to differences in γ subunit composition by Jorge-Rivera et al.,[40] based on prior studies demonstrating high levels of γ_1 subunit expression in the mPOA and contrasting high levels of γ_2 subunit expression in the VMN.[72,73,75] Although Jorge-Rivera et al.[40] indicated that substitution of the γ_1 for the γ_2 subunit in recombinant $\alpha_2\beta_3\gamma_x$ receptors could reverse AAS modulation from positive to negative, more recent data from the Henderson laboratory have not replicated these initial findings. Current data indicate that the AAS, 17α-MeT, potentiates peak current amplitudes of both $\alpha_2\beta_3\gamma_2$ and $\alpha_2\beta_3\gamma_1$ recombinant $GABA_A$ receptors (Figure 6.5). Moreover, potentiation was observed for receptors expressed from both the rat and the human γ_1 subunit cDNAs (data not shown) and for currents elicited by both high (1 m*M*) and subsaturating (20 μ*M*) concentrations of GABA (Figure 6.5).

6.2.2 Regulation of $GABA_A$ Receptor Expression by Chronic AAS Treatment

AAS-dependent changes in $GABA_A$ receptor function (Cl^- flux) as reported by Bitran et al.[63] were not noted until after 2 weeks of treatment, a time frame suggesting that chronic AAS exposure may induce changes in $GABA_A$ receptor expression. A recent study by McIntyre et al.[42] determined the effects of chronic exposure to either a high or moderate dose of 17α-MeT on steady-state levels of α_1, α_2, α_5, γ_1, γ_2, and ε $GABA_A$ receptor subunit mRNAs in the VMN, mPOA,

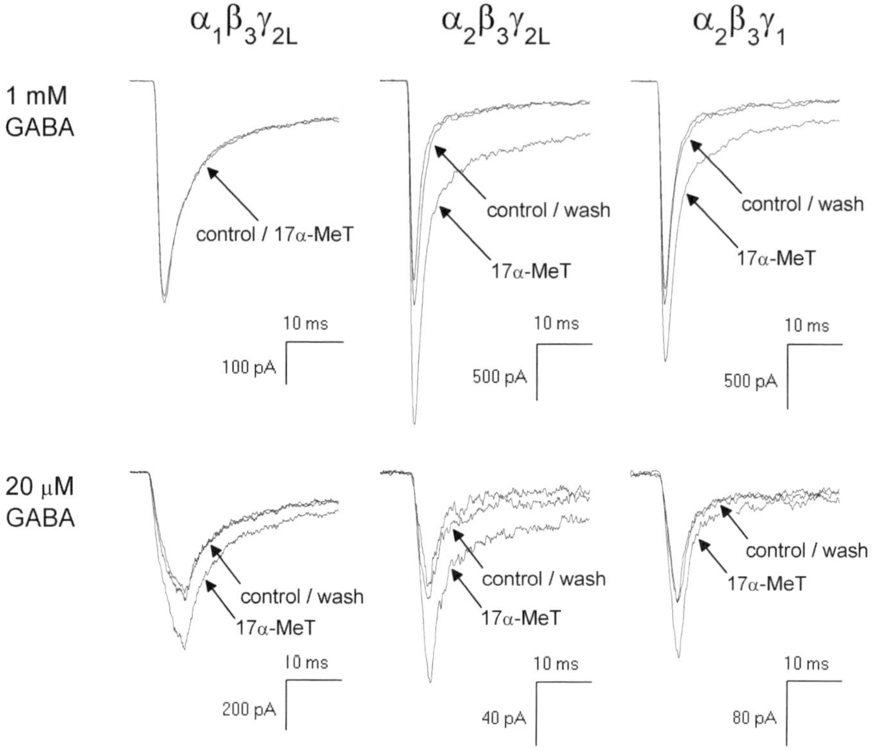

FIGURE 6.5 Alpha, but not γ, subunit composition has significant effects on AAS modulation. Representative, averaged currents elicited by ultrafast (2- to 4-msec pulse duration) perfusion of GABA at 1 m*M* (~EC_{95}) or 20 μ*M* (~EC_{15}) from recombinant receptors of the indicated subunit composition expressed in human embryonic kidney (HEK 293) cells in the presence (17α-MeT; 1 μ*M*) or absence (control/wash) of AASs (V_H = –80 mV). Recording conditions were according to Yang et al.[41] Cells were reequilibrated with 17α-MeT prior to the application of GABA for the +AAS condition. 17α-MeT reversibly and significantly (p < .05) potentiated currents elicited by 20 μ*M* GABA for receptors containing either the α_1 (see also Yang et al.[41]) or the α_2 subunit in conjunction with the β_3 and γ_{2L} subunits, and for receptors containing either the γ_{2L} or the γ_1 subunit in conjunction with the α_2 and β_3 subunits. However, with near saturating (1 m*M*) concentrations of GABA, significant potentiation was evident for receptors comprising α_2 and β_3 with *either* the γ_1 or the γ_2 subunit, but not for receptors comprising α_1, β_3, and γ_{2L} subunits.

and medial amygdala (MeA) of adult and peri-pubertal male and female mice. Significant changes in the levels of some, but not all, $GABA_A$ receptor subunit mRNAs were elicited by AAS exposure. Interestingly, when significant changes were observed, in every case but one, the effect of AAS treatment was to *decrease* the level of the specific subunit mRNA (Figure 6.6). Moreover, the effects of 17α-MeT on $GABA_A$ receptor mRNA levels depended on the dose of AAS used and the age and sex of the animals. Specifically, a greater number of significant changes in individual receptor subunit mRNA levels were observed in peri-pubertal females treated with 17α-MeT than in adults of either sex or peri-pubertal males. These

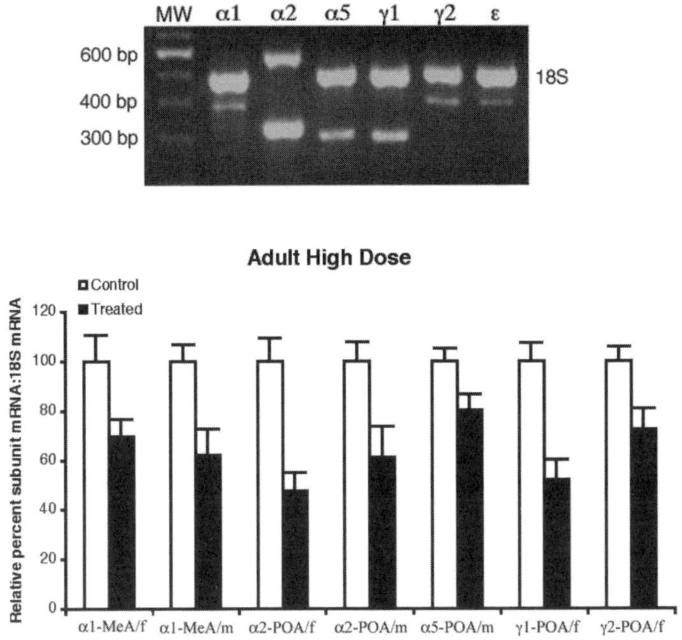

FIGURE 6.6 Chronic AAS treatment $GABA_A$ receptor subunit mRNAs in control and treated animals in the pubertal treatment group. Top: Representative agarose gel electrophoresis of the reverse transcriptase coupled with the polymerase chain reaction (RT-PCR) products detected by primers specific for the α_1, α_2, α_5, γ_1, γ_2, and ε subunit mRNAs, as well as the 18S mRNA (18S), in tissue isolated from the VMN of an adult male mouse treated with a dose of 17α-MeT (7.5 mg/kg/day for 4 weeks) that corresponds to high doses self-administered by human AAS users. Bands of expected size for PCR products specific for the α_1 (400 bp), α_2 (583 bp), α_5 (300 bp), γ_1 (290 bp), γ_2 (391 bp), and ε (392 bp) subunit mRNAs were detected. A band of 488 bp corresponding to the 18S ribosomal mRNA was also detected in each sample analyzed for products corresponding to the α_1, α_5, γ_1, γ_2, and ε subunit mRNAs. Different 18S primers were required in the multiplex reaction for the α_2 subunit mRNA since the 18S primers used above interfered with amplification of the α_2 PCR product. The product corresponding to the 18S mRNA band with this alternative set of primers produced a band of the expected size (315 bp). Molecular weight standards (MW) are shown to the left. Bottom: Cumulative data are shown for those $GABA_A$ receptor subunit mRNAs that were significantly ($p < .05$) different between animals given daily injections of 17α-MeT for 4 weeks (black bars) and oil-injected age- and sex-matched controls (white bars). Data for subunit mRNAs that did not change significantly with treatment are not shown. Individual bars indicate the relative levels of each subunit mRNA in tissue harvested from the indicated brain region (MeA, mPOA, or VMN) taken from female (f) or male (m) mice; e.g., α_2 MeA/f indicates cumulative data for α_2 subunit mRNA assessed in the MeA of female mice. Densitometric values corresponding to each subunit mRNA were first normalized to those corresponding to the 18S mRNA for that individual sample. These ratios were subsequently averaged across all animals in a given group (8 to 10 animals for each brain region, age, sex, and dosage regime), and the averaged data have been plotted with values for the same brain region in control (oil-injected) animals of comparable age and sex, the dose being set to 100%. Error bars represent SEM. (Adapted from Figures 1 and 4 in McIntyre, K.L., Porter, D.M., and Henderson, L.P., *Neuropharmacology*, 43, 634, 2002. With permission.)

results may be of particular relevance with respect to AAS abuse in human populations because the greatest increase in AAS use over the past decade has been reported for adolescent girls.[15]

AASs may elicit changes in $GABA_A$ receptor subunit mRNA expression via multiple pathways. Gonadal steroids have previously been shown to regulate $GABA_A$ receptor subunit mRNA levels,[75,76] presumably by signaling mediated by nuclear hormone receptors. Clearly AASs, like their parent steroids, could alter $GABA_A$ receptor expression via androgen receptor-mediated or, subsequent to aromatization to their A-ring metabolites,[24,25] estrogen receptor-mediated changes in gene expression. In addition, it is known that changes in neuronal activity that arise due to the actions of allosteric modulators of the $GABA_A$ receptor can also result in subsequent changes in $GABA_A$ receptor subunit expression (for review, see Mehta and Ticku[47]). Thus, AAS modulation of the $GABA_A$ receptor function may elicit both acute effects on neuronal activity and long-term effects on $GABA_A$ receptor expression. Interestingly, preliminary data from androgen receptor null (testicular feminization (*tfm*)) mice demonstrate that chronic AAS treatment alters forebrain $GABA_A$ receptor subunit mRNA levels (McIntyre and Henderson, unpublished data). While these data do not preclude a role for the androgen receptor in regulating $GABA_A$ receptor subunit expression, they do indicate that androgen receptor activation is not required for these effects of AASs.

6.3 BEHAVIORAL CORRELATES OF AAS ADMINISTRATION ON BEHAVIORS MEDIATED BY GABAERGIC TRANSMISSION

6.3.1 Anxiety

The widespread use of benzodiazepines as anxiolytic therapeutic agents (for review, see Woods et al.[77]), in concert with laboratory studies affirming the anxiolytic effects of these drugs (for review, see Wilson[36]), highlights the importance of positive allosteric modulation of $GABA_A$ receptors in alleviating anxiety. Although benzodiazepines have been known for years to elicit their anxiolytic effects via modulation of $GABA_A$ receptors, recent studies employing *in vivo* alpha subunit point mutations have demonstrated that anxiolytic actions of benzodiazepines can be ascribed solely to α_2-containing receptors (for review, see Möhler et al.[78]). Effects on anxiety are also produced by other allosteric modulators of the $GABA_A$ receptor, including the neurosteroids and ethanol (for review, see Romeo et al.[79] and Morrow et al.[80]).

AAS effects on anxiety were first reported by Bitran et al.,[62] who assessed the effects of a high dose of TP (silastic capsules) on the performance of male rats on the elevated plus maze. Animals were tested at either 6 or 14 days after implants were made. Serum levels of testosterone were 7-fold higher in rats receiving TP for 1 week and 10-fold higher in rats receiving TP for 2 weeks relative to intact controls. In spite of the continued high levels of serum testosterone, the anxiolytic effects, as determined by increased exploration of the open arms of the elevated plus maze relative to controls, were evident only in rats treated for 6 days. Fourteen days of TP was without effect. Bitran et al.[62] hypothesized that anxiolytic actions

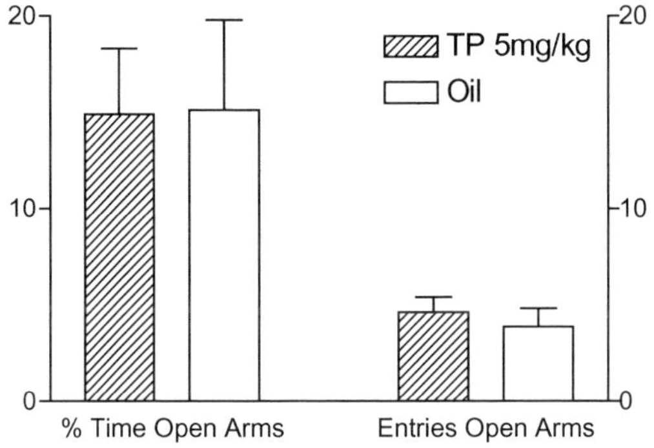

FIGURE 6.7 Gonadally intact male rats received daily subcutaneous injections of TP (5 mg/kg, n = 8) or the sesame oil vehicle (n = 8) daily for 6 days and were tested on the elevated plus maze on day 7. The percentage of time spent in the open arms and the number of entries to the open arms did not differ between the groups.

of TP arose from modulation of $GABA_A$ receptors by testosterone metabolites (namely, androsterone and androstanediol), and that the loss of anxiolytic action in rats treated for 2 weeks may have reflected the development of tolerance at the $GABA_A$ receptor; however, neither hypothesis was tested. We (Clark and Gudmundsen, unpublished data) have also evaluated plus maze performance in male rats treated with six daily subcutaneous injections of TP (5 mg/kg). As shown in Figure 6.7, rats receiving TP and rats receiving the vehicle performed similarly on the plus maze, such that the time spent in the open arms and number of entries did not differ between the groups. The basis for the discrepancy between our results and those of Bitran et al.[62] is not clear, but may reflect differences in the method of TP delivery (silastic implants vs. injections).

Evidence that acute exposure to AASs may have anxiolytic effects is provided by the study of Bing et al.[81] in which male rats that received a single injection of testosterone before testing on Vogel's conflict test accepted significantly more shocks than controls, consistent with an antianxiety action of testosterone. It is noteworthy that Bing et al.[81] observe AAS-induced changes in anxiety within 24 h of a single injection of testosterone. First, it is unlikely that anything beyond physiological levels of testosterone would be present at 24 h after a single injection, given the rapid clearance of unesterified testosterone.[4] However, if the anxiolytic effects were due to this testosterone treatment, then the narrow time frame, although not categorically excluding androgen receptor-dependent changes in gene expression, would be more consistent with a nongenomic mechanism of AAS action. Further evidence that androgens do indeed elicit rapid anxiolytic effects has been provided by a recent study that found that a single exposure to 500 μg of either testosterone, androsterone, or 3α-androstanediol reduced anxiety within 30 min.[82] The anxiolytic effects of testosterone were blocked by co-injection of the $GABA_A$ receptor antagonists, bicuculline or picrotoxin, suggesting that the ability of testosterone to reduce anxiety arose from nongenomic allosteric modulation of this neurotransmitter receptor, although secondary effects on locomotor behavior observed with bicuculline or picrotoxin complicate interpretation of experiments with these antagonists. Taken together, these studies are consistent with the hypothesis that the effects of androgenic steroids on anxiety may be regulated, in part, by allosteric modulation of $GABA_A$ receptors. However, no studies to date have assessed the effects of AASs (except TP) on anxiety behaviors. Moreover, additional studies are needed to assess the dose–response relationship of TP to the production of anxiolytic behaviors and to determine whether the AASs reach concentrations in the brain that would lead to allosteric modulation of these $GABA_A$ receptors.

6.3.2 Aggression

A role for GABA in the display of aggression is well established. As reviewed in Miczek et al.,[35] a number of studies have shown that an increase in GABAergic transmission is associated with a reduction in behavior in animal models of aggression. In particular, defensive aggression, such as that elicited by electric foot shock, is sensitive to the inhibitory effects of GABA. In contrast, there is evidence that activation of $GABA_A$ receptors (or increased GABAergic transmission) may exaggerate offensive aggressive behavior, such as the behavior displayed by experienced aggressors in a resident–intruder paradigm. There are no studies that have addressed the interactions between AASs and $GABA_A$ receptors in the modulation of aggression. There is, however, a growing literature demonstrating that specific AASs can induce offensive aggression. Moreover, other positive allosteric modulators of the $GABA_A$ receptor, including alcohol, allopregnanolone, and the benzodiazepines, produce aggression-heightening effects (for discussion, see Fish et al.[83]), suggesting that some of the effects of AASs on aggression may arise from allosteric modulation of the $GABA_A$ receptor.

A number of studies using the resident–intruder paradigm have demonstrated that individual AASs or AAS combinations elicit aggression in gonadally intact male rats,[84–87] gonadectomized male rats,[88] and adolescent male hamsters.[89,90] Based on these studies, a number of conclusions can be drawn. First, AAS effects on aggression are compound specific.[85,88] Whereas aggressive behaviors were found to be elicited by supraphysiological doses of TP, nandrolone decanoate was without effect, and stanozolol reduced or was ineffective in stimulating aggressive behaviors.[85,88] Second, the effects of AASs on aggression are dependent upon environmental stimuli that modulate the display of aggression.[85,86] Specifically, AASs may sensitize the rat to its surroundings and lower the threshold to respond to provocation with aggression.[85,86] Finally, the effects of AASs on aggression appear to be reversible, because withdrawal from AASs was accompanied by a return to pre-AAS levels of aggressive behavior.[87]

As mentioned above, no studies have examined the role of the $GABA_A$ receptors in AAS-induced aggression. Because AAS effects on aggression have been observed in the resident–intruder paradigm, and GABAergic transmission may increase the display of aggression in this context,[35] one could hypothesize that AASs may increase GABAergic transmission in brain areas important for the display of aggression. In addition, analysis of the ability of AASs to induce aggression in mice with targeted deletions of specific alpha subunits, or engineer point mutations in $GABA_A$ receptor subunit genes (for review, see Möhler et al.[78]), may provide important information on the role of GABAergic transmission in AAS-induced aggression.

6.3.3 Female Sexual Behavior

Although the role of GABAergic transmission in the display of male sexual behaviors is not clear,[91] studies utilizing the central administration of $GABA_A$ receptor agonists and antagonists have provided information about the specific brain regions involved in the GABAergic modulation of female sex behaviors (for review, see Pfaff et al.[92]). These studies have demonstrated that GABAergic transmission in the VMN, mPOA, and midbrain central gray (MCG) plays a key role in the expression of female sexual behaviors. For example, infusion of $GABA_A$ receptor agonists, including positive allosteric modulators of the $GABA_A$ receptor, directly into the hypothalamus or the MCG of hormone-primed females induces a significant facilitation of lordosis. In contrast, bicuculline infused into the VMN reduces lordosis in receptive females. These data unequivocally demonstrate that GABAergic transmission is necessary for the display of this female sexual behavior in the rat (for review, see Pfaff et al.[92]). The observations of regionally specific effects of GABAergic agonists and antagonists are complemented by findings that show regional differences in the distribution of specific subunits of the $GABA_A$ receptor[93] and by findings of region-specific modulation of GABAergic transmission by AAS.[40,41]

Chronic exposure to AASs has been found to influence the expression of female sexual behavior in a compound- and dose-dependent manner. Clark and colleagues[94–96] reported that exposure to high, but not moderate, doses of any of six individual AASs inhibited the display of lordosis in gonadally intact female rats during a 2-week treatment period. AASs also inhibited the display of lordosis in gonadectomized rats primed with estrogen, albeit the effects on gonadectomized animals were compound specific and inhibition was observed with four of the six AASs tested. Of primary importance, experiments in this study demonstrated that AAS actions at the androgen receptor were critical for the suppression of lordosis in gonadectomized subjects as the androgen receptor-specific antagonist flutamide completely reversed the inhibitory effects of 17α-methyltestosterone, stanozolol, and nandrolone decanoate.[96] Although these studies highlight the importance of the androgen receptor in mediating the inhibitory effects of chronic AAS exposure on lordosis, studies have not been performed to determine whether androgen receptor-mediated signaling alters $GABA_A$ receptor expression, and thus $GABA_A$ receptor-mediated transmission, in these animals. Moreover, these studies do not address whether AASs acutely modulate the expression of lordosis or if they modulate the expression of other female sexual behaviors (e.g., proceptive behaviors) via allosteric modulation of $GABA_A$ receptors. To this end, it would be valuable to infuse AASs centrally into specific brain regions and to determine their direct effects on the display of a range of hormone-induced female sexual behaviors.

6.4 SUMMARY

Studies reported here demonstrate that the AASs act to allosterically modulate the $GABA_A$ receptor, and that they do so via a mechanism that is distinct from that of other known allosteric modulators. The effects of AASs on $GABA_A$ receptor function vary with the subunit composition of the receptor and with the concentration of GABA. These data suggest that AASs will have significantly different effects in different brain regions and different actions on synaptic transmission vs. tonic or spillover conductances. Finally, initial studies have demonstrated that chronic exposure to the AASs, like the benzodiazepines, ethanol, and the neurosteroids, alters the levels of $GABA_A$ receptor subunit mRNAs. However, whether these changes in subunit transcripts are reflected in concomitant changes in receptor subunit protein expression, or in the physiology or pharmacology of the receptor, remains to be determined. Studies in rodents have also shown that exposure to AASs at doses that mimic levels observed with human abuse elicit significant changes in aggression, anxiety, and sexual behaviors — all behaviors mediated by GABAergic transmission and implicated as being altered in human steroid abusers. While it is tantalizing to suggest that these behavioral effects of AASs arise from the changes in $GABA_A$ receptor function and expression just noted, more studies are needed that mechanistically tie the former to the latter, and that more fully explore both the physiological and behavioral actions of these steroids.

REFERENCES

1. Blue, J.G. and Lombardo, J.A., Nutritional aspects of exercise: steroids and steroid-like compounds, *Clin. Sports Med.*, 8, 667, 1999.
2. Basaria, S., Wahlstrom, J.T., and Dobs, A.S., Clinical review 138: anabolic-androgenic steroid therapy in the treatment of chronic diseases, *J. Clin. Endocrinol. Metab.*, 86, 5108, 2001.
3. Shahidi, N.T., A review of the chemistry, biological action, and clinical applications of anabolic-androgenic steroids, *Clin. Ther.*, 23, 1355, 2001.
4. Wilson, J., Androgen abuse by athletes, *Endocr. Rev.*, 9, 181, 1988.
5. Lukas, S.E., CNS effects and abuse liability of anabolic-androgenic steroids, *Annu. Rev. Toxicol.*, 36, 333, 1996.
6. Kammerer, R.C., Drug testing and anabolic steroids, in *Anabolic Steroids in Sport and Exercise*, Yesalis, C.E., Ed., Human Kinetics, Champaign, IL, 2000, chap. 13.
7. Bahrke, M.S., Yesalis, C.E., and Wright, J.E., Psychological and behavioral effects of endogenous testosterone and anabolic-androgenic steroids, *Sports Med.* 22, 367, 1996.
8. Friedl, K.E., Effects of anabolic steroids on physical health, in *Anabolic Steroids in Sport and Exercise*, Yesalis, C.E., Ed., Human Kinetics, Champaign, IL, 2000, chap. 6.
9. Franke, W.W. and Berendonk, B., Hormonal doping and androgenization of athletes: a secret program of the German Democratic Republic government, *Clin. Chem.*, 43, 1262, 1997.
10. Bowers, L.D. and Segura, J., Anabolic steroids, athletic drug testing, and the Olympic games, *Clin. Chem.*, 42, 999, 1996.

11. Yesalis, C.E., Kennedy, N.J., Kopstein, A.N., and Bahrke, M.S., Anabolic-androgenic steroid use in the United States, *JAMA*, 270, 1217, 1993.
12. Kolata, G., With No Answers on Risks, Steroid Users Still Say 'Yes,' *New York Times*, December 2, 2002, p. 1-19.
13. NIDA Research Report Series, http://165.112.78.61/ResearchReports/Steroids/Anabolicsteroids.html; 2002.
14. Yesalis, C.E., Barsukiewicz, C.K., Kopstein, A.N., and Bahrke, M.S., Trends in anabolic-androgenic steroid use among adolescents, *Arch. Pediatr. Adolesc. Med.*, 151, 1197, 1997.
15. Bahrke, M.S., Yesalis, C.E., and Brower, K.J., Anabolic-androgenic steroid abuse and performance-enhancing drugs among adolescents, *Child Adolesc. Psychiatr. Clin. N. Am.*, 7, 821, 1998.
16. Koch, J.J., Performance-enhancing substances and their use among adolescent athletes, *Pediatr. Rev.*, 23, 310, 2002.
17. Rubinow, D.R. and Schmidt, P.J., Androgens, brain and behavior, *Am. J. Psychiatry*, 153, 974, 1996.
18. Korkia, P. and Stimson, G.V., Indications of prevalence, practice and effects of anabolic steroid use in Great Britain, *Int. J. Sports Med.*, 18, 557, 1997.
19. Kuhn, C.M., Anabolic steroids, *Recent Prog. Horm. Res.*, 57, 411, 2002.
20. Martini, L., The 5α-reduction of testosterone in the neuroendocrine structures: biochemical and physiological implications, *Endocr. Rev.*, 3, 1, 1982.
21. Winters, S., Androgens: endocrine physiology and pharmacology, *NIDA Res. Monogr.*, 102, 113, 1990.
22. Kochakian, C. and Yesalis, C.E., Anabolic-androgenic steroids: a historical perspective and definition, in *Anabolic Steroids in Sport and Exercise*, Yesalis, C.E., Ed., Human Kinetics, Champaign, IL, 2000, chap. 1.
23. Ryan, K.J., Biological aromatization of steroids, *J. Biol. Chem.*, 234, 268, 1959.
24. Papaconstantinou, C.D., Umbreit, T.H., Goering, P.L., and Brown, K.M., Effects of 17α-methyltestosterone on uterine morphology and heat shock protein expression are mediated through estrogen and androgen receptors, *J. Steroid Biochem. Mol. Biol.*, 82, 305, 2002.
25. de Gooyer, M.E. et al., Tibolone is not converted by human aromatase to 7α-methyl-17α-ethynylestradiol (7α-MEE): analyses with sensitive bioassays for estrogens and androgens with LC-MSMS, *Steroids*, 68, 235, 2003.
26. Turner, J.P. and Simmonds, M.A., Modulation of the $GABA_A$ receptor complex by steroids in slices of rat cuneate nucleus, *Br. J. Pharmacol.*, 96, 409, 1989.
27. Lambert, J.J., Belelli, D., Hill-Venning, C., and Peters, J.A., Neurosteroids and $GABA_A$ receptor function, *Trends Pharmacol. Sci.*, 16, 295, 1995.
28. Lambert, J.J., Harney, S.C., Belelli, D., and Peters, J.A., Neurosteroid modulation of recombinant and synaptic $GABA_A$ receptors, *Int. Rev. Neurobiol.*, 46, 177, 2001.
29. Gee, K.W. et al., Steroid modulation of the chloride ionophore in rat brain: structure-reactivity requirements, regional dependence and mechanisms of action, *J. Pharmacol. Exp. Ther.*, 246, 803, 1988.
30. Purdy, R.H., Morrow, A.L., Blinn, J.R., and Paul, S.M., Synthesis metabolism, and pharmacological activity of 3α-hydroxy steroids which potentiate GABA-receptor-mediated chloride uptake in rat cerebral cortical synaptoneurosomes, *J. Med. Chem.*, 33, 1572, 1990.
31. Clark, A.S. and Henderson, L.P., Behavioral and physiological responses to anabolic-androgenic steroids, *Neurosci. Biobehav. Rev.*, in press.

32. Fernandez-Guasti, A., Larsson, K., and Beyer, C., GABAergic control of masculine sexual behavior, *Pharmacol. Biochem. Behav.*, 24, 1065, 1986.
33. Ojeda, S.R. and Urbanski, H.F., Puberty in the rat, in *The Physiology of Reproduction*, Vol. 2, Knobil, E. and Neill, J.D., Eds., Raven Press Ltd., New York, 1994, chap 40.
34. McCarthy, M.M., Frank A. Beach Award. Functional significance of steroid modulation of GABAergic neurotransmission: analysis at the behavioral, cellular, and molecular levels, *Horm. Behav.*, 29, 131, 1995.
35. Miczek, K.A., Fish, E.W., De Bold, J.F., and De Almeida, R.M.M., Social and neural determinants of aggressive behavior: pharmacotherapeutic targets at serotonin, dopamine and γ-aminobutyric acid systems, *Psychopharmacology,* 163, 434, 2002.
36. Wilson, M.A., GABA physiology: modulation by benzodiazepines and hormones, *Crit. Rev. Neurosci.*, 10, 1, 1996.
37. Siegel, A., Roeling, T.A., Gregg, T.R., and Kruk, M.R., Neuropharmacology of brain-stimulation-evoked aggression, *Neurosci. Biobehav. Rev.*, 23, 359, 1999.
38. Masonis, A.E. and McCarthy, M.P., Direct effects of the anabolic/androgenic steroids, stanozolol and 17α-methyltestosterone, on benzodiazepine binding to the γ-aminobutyric acid A receptor, *Neurosci. Lett.*, 189, 35, 1995.
39. Masonis, A.E. and McCarthy, M.P., Effects of the androgenic/anabolic steroid stanozolol on $GABA_A$ receptor function: GABA-stimulated $^{36}Cl^-$ influx and [^{35}S]TBPS binding, *J. Pharmacol. Exp. Ther.*, 279, 186, 1996.
40. Jorge-Rivera, J.C., McIntyre, K.L., and Henderson, L.P., Anabolic steroids induce region- and subunit-specific rapid modulation of $GABA_A$ receptor-mediated currents in the rat forebrain, *J. Neurophysiol.*, 83, 3299, 2000.
41. Yang, P., Jones, B.L., and Henderson, L.P., Mechanisms of anabolic androgenic steroid modulation of $\alpha_1\beta_3\gamma_{2L}$ $GABA_A$ receptors, *Neuropharmacology,* 43, 619, 2002.
42. McIntyre, K.L., Porter, D.M., and Henderson, L.P., Anabolic androgenic steroids induce age-, sex- and dose-dependent changes in $GABA_A$ receptor subunit mRNAs in the mouse forebrain, *Neuropharmacology,* 43, 634, 2002.
43. Betz, H., Ligand-gated ion channels in the brain: the amino acid receptor superfamily, *Neuron*, 5, 383, 1990.
44. Boyd, R.T., The molecular biology of neuronal nicotinic acetylcholine receptors, *Crit. Rev. Toxicol.*, 27, 299, 1997.
45. Betz, H. et al., Structure and functions of inhibitory and excitatory glycine receptors, *Ann. N.Y. Acad. Sci.*, 868, 667, 1999.
46. Jackson, M.B. and Yakel, J.L., The 5-HT3 receptor channel, *Ann. Rev. Physiol.*, 57, 447, 1995.
47. Mehta, A.K. and Ticku, M.K., An update on $GABA_A$ receptors, *Brain Res. Rev.*, 29, 196, 1999.
48. Whiting, P.J., The GABA-A receptor gene family: new targets for therapeutic intervention, *Neurochem. Int.*, 34, 387, 1999.
49. Whiting, P. et al., Molecular and functional diversity of the expanding GABA-A receptor gene family, *Ann. N.Y. Acad. Sci.*, 868, 645, 1999.
50. Whiting, P., McKernan, R.M., and Iversen, L.L., Another mechanism for creating diversity in γ-aminobutyrate type A receptors: RNA splicing directs expression of two forms of γ_2 subunit, one of which contains a protein kinase C phosphorylation site, *Proc. Natl. Acad. Sci. U.S.A.*, 87, 9966, 1990.
51. Kofuji, P. et al., Generation of two forms of the γ-aminobutyric $acid_A$ receptor γ_2-subunit in mice by alternative splicing, *J. Neurochem.*, 56, 713, 1991.

52. Moss, S.J., Doherty, C.A., and Huganir, R.L., Identification of the cAMP-dependent protein kinase and protein kinase C phosphorylation sites within the major intracellular domains of the β_1,γ_2S and γ_2L subunits of the γ-aminobutyric acid type$_A$ receptor, *J. Biol. Chem.* 267, 14470, 1992.
53. Harvey, R.J., Chinchetru, M.A., and Darlison, M.G., Alternative splicing of a 51-nucleotide exon that encodes a putative protein kinase C phosphorylation site generates two forms of the chicken γ-aminobutyric acid$_A$ receptor β_2 subunit, *J. Neurochem.*, 62, 10, 1994.
54. McKinley, D.D., Lennon, D.J., and Carter, D.B., Cloning, sequence analysis and expression of two forms of mRNA coding for the human β_2 subunit of the GABA$_A$ receptor, *Mol. Brain Res.*, 28, 175, 1995.
55. Kirkness, E.F. and Fraser, C.M., A strong promoter element is located between alternative exons of a gene encoding the human γ-aminobutyric acid type A receptor β_3 subunit (GABRB3), *J. Biol. Chem.*, 268, 4420, 1993.
56. Korpi, E.R. et al., Small N-terminal deletion by splicing in cerebellar α_6 subunit abolishes GABA$_A$ receptor function, *J. Neurochem.*, 63, 1167, 1994.
57. Fritschy, J.M. and Möhler, H., GABA$_A$-receptor heterogeneity in the adult rat brain: differential regional and cellular distribution of seven major subunits, *J. Comp. Neurol.*, 359, 154, 1995.
58. Barnard, E.A. et al., International Union of Pharmacology: XV. Subtypes of γ-aminobutyric acid$_A$ receptors: classification on the basis of subunit structure and receptor function, *Pharmacol. Rev.*, 50, 291, 1998.
59. Khan, Z.U., Gutiérrez, A., and De Blas, A.L., Short and long form γ_2-subunits of the GABA$_A$/benzodiazepine receptors, *J. Neurochem.*, 63, 1466, 1994.
60. Quirk, K. et al., γ-Aminobutyric acid type A receptors in the rat brain can contain both γ_2 and γ_3 subunits, but γ_1 does not exist in combination with another γ-subunit, *Mol. Pharmacol.*, 45, 1061, 1994.
61. Sieghart, W., Structure and pharmacology of γ-aminobutyric acid$_A$ receptor subtypes, *Pharmacol. Rev.*, 47, 181, 1995.
62. Bitran, D., Kellogg, C.K., and Hilvers, R.J., Treatment with an anabolic-androgenic steroid affects anxiety-related behavior and alters the sensitivity of cortical GABA$_A$ receptors in the rat, *Horm. Behav.*, 27, 568, 1993.
63. Bitran, D., Hilvers, R.J., Frye, C.A., and Erskine, M.S., Chronic anabolic-androgenic steroid treatment affects brain GABA$_A$ receptor-gated chloride ion transport, *Life Sci.*, 58, 573, 1996.
64. Wu, F.C., Endocrine aspects of anabolic steroids, *Clin. Chem.*, 43, 1289, 1997.
65. Rogers, C.J., Twyman, R.E., and Macdonald, R.L., Benzodiazepine and β-carboline regulation of single GABA$_A$ receptor channels of mouse spinal cord neurones in culture, *J. Physiol. (Lond.)*, 475, 69, 1994.
66. Lavoie, A.M. and Twyman, R.E., Direct evidence for diazepam modulation of GABA$_A$ receptor microscopic affinity, *Neuropharmacology*, 35, 1383, 1996.
67. Mellor, J.R. and Randall, A.D., Frequency-dependent actions of benzodiazepines on GABA$_A$ receptors in cultured murine cerebellar granule cells, *J. Physiol. (Lond.)*, 503, 353, 1997.
68. Zhu, W.J. and Vicini, S., Neurosteroid prolongs GABA$_A$ channel deactivation by altering kinetics of desensitized states, *J. Neurosci.*, 17, 4022, 1997.
69. Jones, M.V. and Westbrook, G.L., Desensitized states prolong GABA$_A$ channel responses to brief agonist pulses, *Neuron*, 15, 181, 1995.
70. Jones, M.V. and Westbrook, G.L., The impact of receptor desensitization on fast synaptic transmission, *Trends Neurosci.*, 19, 96, 1996.

71. Wohlfarth, K.M., Bianchi, M.T., and Macdonald, R.L., Enhanced neurosteroid potentiation of ternary $GABA_A$ receptors containing the delta subunit, *J. Neurosci.*, 22, 1541, 2002.
72. Nett, S.T., Jorge-Rivera, J.-C., Myers, M., Clark, A.S., and Henderson, L.P., Properties and sex-specific differences of $GABA_A$ receptors in neurons expressing $\gamma 1$ subunit mRNA in the preoptic area of the rat, *J. Neurophysiol.*, 81, 192, 1999.
73. Wisden, W., Laurie, D.J., Monyer, H., and Seeburg, P.H., The distribution of 13 $GABA_A$ receptor subunit mRNAs in the rat brain. I. Telencephalon, diencephalon, mesencephalon, *J. Neurosci.*, 12, 1040, 1992.
74. Pirker, S. et al., $GABA_A$ receptors: immunocytochemical distribution of 13 subunits in the adult rat brain, *Neuroscience*, 101, 815, 2000.
75. Herbison, A.E. and Fénelon, V.S., Estrogen regulation of $GABA_A$ receptor subunit mRNA expression in preoptic area and bed nucleus of the stria terminalis of female rat brain, *J. Neurosci.*, 15, 2328, 1995.
76. Zhang, L. et al., The expression of $GABA_A$ receptor α_2 subunit is upregulated by testosterone in rat cerebral cortex, *Neurosci. Lett.*, 265, 25, 1999.
77. Woods, J.H., Katz, J.L., and Winger, G., Benzodiazepines: use, abuse and consequences, *Pharmacol. Rev.*, 44, 151, 1992.
78. Möhler, H., Fritschy, J.M., and Rudolph, U., A new benzodiazepine pharmacology. *J. Pharmacol. Exp. Ther.*, 300, 2, 2002.
79. Romeo, E. et al., Marked decrease of plasma neuroactive steroids during alcohol withdrawal, *Clin. Neuropharmacol.*, 19, 366, 1996.
80. Morrow, A.L., VanDoren, M.J., Penland, S.N., and Matthews, D.B., The role of GABAergic neuroactive steroids in ethanol action, tolerance and dependence, *Brain Res. Rev.*, 37, 98, 2001.
81. Bing, O. et al., High doses of testosterone increase anticonflict behaviour in rat, *Eur. Neuropsychopharmacol.*, 8, 321, 1998.
82. Aikey, J.L., Nyby, J.G., Anmuth, D.M., and James, P.J., Testosterone rapidly reduces anxiety in male house mice (Mus musculus), *Horm. Behav.* 42, 448, 2002.
83. Fish, E.W., De Bold, J.F., and Miczek, K.A., Aggressive behavior as a reinforcer in mice: activation by allopregnanolone, *Psychopharmacology*, 163, 459, 2002.
84. Lumia, A.R., Thorner, K.M., and McGinnis, M.Y., Effects of chronically high doses of the anabolic androgenic steroid, testosterone, on intermale aggression and sexual behavior in male rats, *Physiol. Behav.*, 55, 331, 1994.
85. Breuer, M.E., McGinnis, M.Y., Lumia, A.R., and Possidente, B.P., Aggression in male rats receiving anabolic androgenic steroids: effects of social and environmental provocation, *Horm. Behav.*, 40, 409, 2001.
86. McGinnis, M.Y., Lumia, A.R., Breuer, M.E., and Possidente, B., Physical provocation potentiates aggression in male rats receiving anabolic androgenic steroids, *Horm. Behav.*, 41, 101, 2002.
87. McGinnis, M.Y., Lumia, A.R., and Possidente, B., Effects of withdrawal from anabolic androgenic steroids on aggression in adult male rats, *Physiol. Behav.*, 75, 541, 2002.
88. Clark, A.S. and Barber, D.M., Anabolic-androgenic steroids and aggression in castrated male rats, *Physiol. Behav.*, 56, 1107, 1994.
89. Melloni, R.H., Jr. and Ferris, C.F. Adolescent anabolic steroid use and aggressive behavior in golden hamsters, *Ann. N.Y. Acad. Sci.*, 794, 372, 1996.
90. Melloni, R.H., Jr. et al., Anabolic-androgenic steroid exposure during adolescence and aggressive behavior in golden hamsters, *Physiol. Behav.*, 61, 359, 1997.

91. Agmo, A., Paredes, R.G., Sierra, L., and Garces, I., The inhibitory effects on sexual behavior and ambulatory activity of the mixed $GABA_A/GABA_B$ agonist progabide are differentially blocked by GABA receptor antagonists, *Psychopharmacology*, 129, 27, 1997.
92. Pfaff, D.W., Schwartz-Giblin, S., McCarthy, M.M., and Kow, L.-M., Cellular and molecular mechanisms of female reproductive behaviors, in *The Physiology of Reproduction*, Vol. 2, Knobil, E. and Neill, J.D., Eds., Raven Press Ltd., New York, 1994, chap 36.
93. Clark, A.S. et al., Hormone-dependent regulation of $GABA_A$ receptor γ subunit mRNAs in sexually dimorphic regions of the rat brain, *Proc. R. Soc. Lond. B*, 265, 1853, 1998.
94. Blasberg, M.E. and Clark, A.S., Anabolic-androgenic steroid effects on sexual receptivity in ovariectomized rats, *Horm. Behav.*, 32, 201, 1997.
95. Clark, A.S., Blasberg, M.E., and Brandling-Bennett, E.M., Stanozolol, oxymetholone, and testosterone cypionate effects on the rat estrous cycle, *Physiol. Behav.*, 63, 287, 1998.
96. Blasberg, M.E., Robinson, S., Henderson, L.P., and Clark, A.S., Inhibition of estrogen-induced sexual receptivity by androgens, *Horm. Behav.*, 34, 283, 1998.

7 Withdrawal Effects of a Neuroactive Steroid as a Model of PMS: Synaptic Physiology to Behavior

Sheryl S. Smith

CONTENTS

0-8493-2392-4/04/$0.00+$1.50

ABBREVIATIONS

3α,5α-THP 3α-OH-5α-pregnan-20-one (allopregnanolone)
3α,5β-THP 3α-OH-5β-pregnan-20-one (pregnanolone)
BDZ Benzodiazepine
CNS Central nervous system
EPSP Excitatory postsynaptic potential
GABAR $GABA_A$ receptor
gGABA GABA-gated conductance
IPSP Inhibitory postsynaptic potential
LZM Lorazepam
mIPSC Miniature inhibitory postsynaptic current
P Progesterone
PKC Protein kinase C
PMS Premenstrual syndrome
PPI Paired pulse inhibition
PS_{cond} Conditioning population spike
PS_{test} Test population spike
PTZ Pentylenetetrazol
sIPSC Spontaneous inhibitory postsynaptic current
τ Tau, decay time constant
TBPS t-Butylbicyclophosphorothionate
THIP 4,5,6,7-Tetrahydroisoxazolo[5,4-c]pyridin-3-ol (gaboxadol)
TTX Tetrodotoxin

7.1 INTRODUCTION

As a potent GABA-modulatory steroid, 3α,5α-THP (3α-OH-5α-pregnan-20-one or allopregnanolone) significantly increases GABA-gated current recorded from cultured neurons or acutely isolated hippocampal pyramidal cells at concentrations (10 to 30 n*M*) in the physiological range.[1] Similar properties have been reported for its isomer 3α,5β-THP.[1] (For a complete discussion of these effects and subunit specificity, see Chapters 2 to 5.) Because this steroid is easily converted from the ovarian/adrenal steroid progesterone within the central nervous system (CNS) or periphery, via two enzymatic conversions (see Chapter 1), its levels increase during the luteal phase of the menstrual cycle in the human and the afternoon of proestrus in the rodent, as well as during pregnancy. In addition, it is increased by stressful stimuli, when it can be formed *de novo* from cholesterol in the brain (see Chapter 1), and thus is classified both as a neurosteroid and as a neuroactive steroid. The acute effects of this steroid have been well characterized since the mid-1980s, when its modulatory effect was first identified. These include direct effects of 3α,5α-THP on the $GABA_A$ receptor (GABAR) in cultured neurons and recombinant systems,[1–3] as well as effects of the parent compound progesterone, which were shown to potentiate GABA-mediated inhibition *in vivo* via conversion to 3α,5α-THP within the CNS.[4,5] However, under physiological conditions, neurons are exposed to the steroid for sustained periods of time (hours to weeks to months), and abrupt declines in steroid levels occur in a cyclic fashion. Thus, the possibility exists that homeostatic mechanisms may be brought into play to alter GABAergic inhibition under these more sustained fluctuations in CNS levels of 3α,5α-THP across naturally occurring cycles. Alterations in GABA function have been reported following chronic administration of most known GABA modulators, including benzodiazepines (BDZs),[6–11] barbiturates, and alcohol.[12,13] Notably, tolerance to these compounds develops,[14,15] as does cross-tolerance to related compounds,[7,16,17] in many cases as a result of subunit substitution in expressed receptor isoforms.[11,18] In addition, most known GABA-modulatory drugs[19] and GABA itself exhibit withdrawal properties.[20] That is, abrupt discontinuation following sustained administration of these agents produces a state of hyperexcitability, characterized by increased anxiety and seizure susceptibility.[19,21] Recently, a number of investigators have explored the possibility that, similar to other GABA-modulatory drugs, endogenous steroids such as 3α,5α-THP also exhibit withdrawal properties. The repercussions of altered excitability during the withdrawal phase for these steroids during the late luteal phase or the postpartum period may have relevance for altered mood or seizure susceptibility reported for certain individuals during these periods, i.e., premenstrual syndrome (see Chapter 13), postpartum dysphoria, and catamenial epilepsy (see Chapter 14). In addition, fluctuations in brain neurosteroid levels are also reported for certain rodent models of stress, such as the socially isolated mouse.[22] This review will focus on the effects of steroid withdrawal that are related to altered function of the GABAR.

7.2 STEROID WITHDRAWAL AND BEHAVIORAL EXCITABILITY

Steroid withdrawal has been investigated with the use of a progesterone-filled capsule implanted subcutaneously in adult female rats for 3 weeks. This paradigm results in brain levels of both progesterone and 3α,5α-THP in the physiological range.[23] Both intermittent and sustained administrations of the parent compound result in significant withdrawal effects examined 8 to 24 h after capsule removal (progesterone withdrawal).[24,25] At this time, progesterone withdrawal results in an increase in anxiety (see Figure 7.1B), assessed using a variety of animal models, including the defensive burying paradigm,[26] light:dark transitions,[26] the elevated plus maze,[25] and the acoustic startle response.[27] The anxiogenic effect observed after withdrawal from progesterone is blocked by indomethacin,[26] which prevents conversion of the parent compound to 3α,5α-THP,[28] suggesting that this GABA-modulatory steroid is producing the withdrawal effect. In addition, these withdrawal effects recover within 3 days of cessation of steroid administration, suggesting that they are transient. The use of a pseudopregnancy model results in increased ovarian release of P, and thus elevates levels of 3α,5α-THP in the circulation and CNS.[29] Under these conditions, administration of a 5α-reductase blocker produces a selective withdrawal from 3α,5α-THP, in the presence of elevated levels of P, and also results in significant increases in anxiety, assessed as the number of open arm entries in the elevated plus maze.[29]

In addition to producing increased anxiety, 3α,5α-THP withdrawal also results in a significant increase in seizure susceptibility,[23,24,30–35] assessed by the duration and severity of seizures induced by kainic acid, pentylenetetrazol (PTZ), picrotoxin,

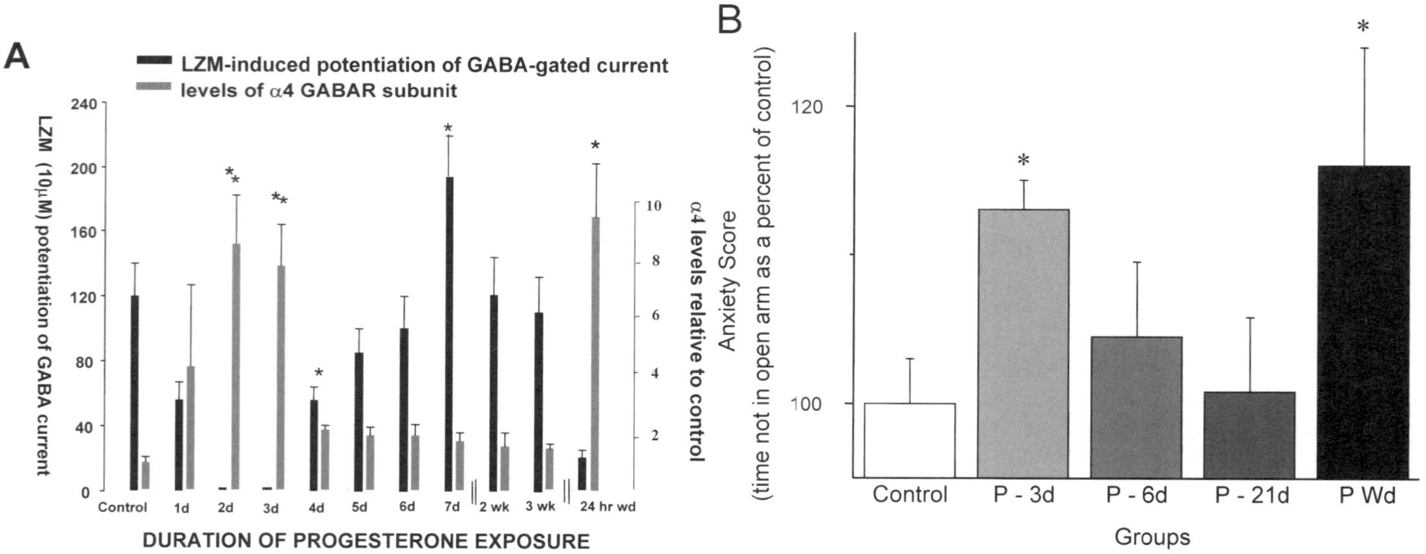

FIGURE 7.1 Time course of progesterone effects on $GABA_A$ receptor (GABAR) subunit composition, pharmacology, and anxiety response. (A) Time course of changes in α4 GABAR subunit expression and benzodiazepine sensitivity. Levels of α4 expression in hippocampus were detected with Western blot techniques on a daily or weekly basis during sustained exposure to progesterone (and thus 3α,5α-THP), via subcutaneous capsule, across a 3-week period. Levels of α4 expression increased significantly by 2 to 3 days of progesterone exposure compared to control; then values recovered to control levels across the remainder of the exposure period until progesterone withdrawal (24 h after capsule removal), when significant increases in α4 expression were again observed ($p < .001$). Concomitant pharmacological evaluation of benzodiazepine (LZM) potentiation of GABA-gated current evaluated in acutely isolated CA1 hippocampus pyramidal cells from the same hippocampal tissue revealed a LZM insensitivity closely correlated with the changes in α4 expression. (B) Anxiety score varies across the progesterone exposure period. Assessed as the amount of time not in the open arm as a percent of control, anxiety score was significantly increased in association with α4 upregulation by 3 days of progesterone exposure, as well as during the progesterone withdrawal period, with control levels of anxiety response noted during chronic (>4 days) steroid exposure. (From Gulinello, M. et al., *Brain Res.*, 910, 55–66, 2001. Printed with permission, Elsevier Science.)

or a β-carboline BDZ inverse agonist. In one study, pregnant rats administered a 5α-reductase blocker to induce withdrawal from 3α,5α-THP also exhibited an increase in seizure duration,[30] demonstrating direct withdrawal effects of the neurosteroid. In another study, acute neurosteroid withdrawal from a single exposure was shown to precipitate handling-induced convulsions in a strain of seizure-prone mice.[36] Thus, the behavioral excitability produced by progesterone withdrawal is due to withdrawal from its GABA-modulatory metabolite and comparable to that resulting from withdrawal from other GABA-modulatory drugs such as ethanol, where withdrawal syndromes are also characterized by increased anxiety and seizure susceptibility.[19] These findings may have relevance for catamenial epilepsy,[37–39] a condition where seizure-prone women experience exacerbation of their condition at selective time points across the menstrual cycle (see Chapter 14): during the midcycle peak and late luteal phase decline in hormones, the latter similar to the progesterone withdrawal paradigm.

There are gender differences in the capacity of progesterone to exhibit withdrawal properties that are selective for the type of anxiety model employed. The elevated plus maze is a model of risk-taking behavior that assesses the frequency of open arm exploration.[40] Progesterone withdrawal produces a decrease in open arm entries in males as in females using this paradigm,[25] suggesting that males are susceptible to increases in anxiety associated with risk-taking parameters. However, use of another model of anxiety, the acoustic startle response, reveals withdrawal properties of progesterone only in females — not in males.[27] This apparent gender difference may be secondary to an increased baseline startle response in males compared to females, rather than a lack of steroid withdrawal effect. The baseline acoustic startle response is a brainstem reflex when employed without fear conditioning.[41] The rapid startle response to a random nonhabituated auditory stimulus is a function of the auditory pathway, pontine relay nuclei, and motor neurons, with modification possible by limbic structures.[41] As such, it has been used as a measure of anxiety and behavioral excitability in both rodents and humans under conditions of stress[42,43] or anxiety and following withdrawal from drugs such as ethanol[44,45] or in individuals at risk for alcoholism. As a model of behavioral excitability, it can be differentially altered by local activation/inhibition of dorsal and ventral hippocampus,[46] and is attenuated by drugs such as ethanol[47] and benzodiazepines.[14] Thus, the behavioral excitability produced by progesterone withdrawal is similar to that produced by other GABA-modulatory compounds.

7.3 STEROID WITHDRAWAL AND GABAR PHARMACOLOGY

Initial studies investigating progesterone withdrawal have been focused on events in CA1 hippocampus, a limbic region involved in synaptic plasticity,[48] with importance for regulation of emotional response in addition to memory consolidation and spatial memory. In the control state, CA1 pyramidal cells predominantly express α1-, α2-, α3-, and α5-containing GABAR,[49] which are potently modulated by classic broad-spectrum BDZs such as lorazepam (LZM).[50] The most striking finding following progesterone withdrawal is a two- to threefold increase in expression of the α4 subunit[24] (see Figure 7.1A), a BDZ-insensitive subunit that is relatively underexpressed in this region.[49] Levels of both the protein and mRNA were observed in CA1 hippocampus, suggesting the possibility that steroid withdrawal induces transcriptional events. This change is accompanied by a nearly complete insensitivity to the GABA-modulatory effects of LZM[24] (see Figure 7.1A), without any change in peak GABA-gated current, assessed using whole-cell patch clamp recording procedures on pyramidal cells acutely isolated from CA1 hippocampus. The fact that the peak current is unchanged suggests a subunit substitution rather than increased expression of additional α4-containing GABAR beyond the normally expressed GABAR population. This effect is seen as early as 8 h after cessation of progesterone treatment, peaks at 24 h, and recovers within 72 h of withdrawal from the steroid. This type of cross-tolerance to other classes of GABA-modulatory agents has been reported for chronic ethanol[16] and BDZ treatment[7] regimens. This withdrawal effect of progesterone is due to its GABA-modulatory metabolite, 3α,5α-THP, as prevention of progesterone metabolism with a 5α-reductase blocker precludes the observed effect, and direct withdrawal from 3α,5α-THP produces a similar BDZ insensitivity.[51] In addition, the ovarian steroid 17β-estradiol has a permissive effect upon progesterone withdrawal,[51] allowing a 10-fold lower concentration of the steroid to result in withdrawal effects. A rigorous pharmacological assessment of GABA-gated current recorded from isolated hippocampal neurons following progesterone withdrawal reveals a profile characteristic of recombinant α4-containing GABAR expressed in oocytes.[52,53] Specifically, neuronal responses to BDZ antagonists, such as flumazenil, are converted to BDZ agonist-like effects, as are responses to the BDZ partial inverse agonist, RO15-4513. In addition to these pharmacological alterations, the decay of GABA-gated current is markedly accelerated after neurosteroid withdrawal.[24] The role of α4-containing GABAR in mediating these effects has been verified by suppressing α4 expression using intraventricular administration of antisense oligonucleotide, a treatment that prevents both the BDZ insensitivity and the change in decay kinetics.[24] Thus, increased expression of the α4 subunit produces alterations in both the pharmacology and kinetics of the resultant GABAR expressed in hippocampus.

7.4 STEROID WITHDRAWAL: OTHER BRAIN REGIONS AND GABA MODULATORS

More recently, progesterone withdrawal has been shown to increase α4 expression in the amygdala,[27] a CNS site pivotally important for the generation of anxiety, suggesting that similar pharmacological and physiological changes may occur in this region, as observed in CA1 hippocampus.[24] Neurosteroid withdrawal has also been shown to increase α4 expression in cerebellar granule cells,[54,55] an effect dependent upon 3α,5α-THP.[54] A similar BDZ insensitivity developed, in conjunction with increased response to a β-carboline BDZ inverse agonist. Similar increases in α4 expression are also induced by withdrawal from, but not by long-term exposure to, ganaxolone, a synthetic analog of the neuroactive steroid

3α,5α-THP, assessed using the same *in vitro* system.[56] In contrast, chronic exposure to oral contraceptives that result in reduced brain levels of 3α,5α-THP decrease γ2 expression in the cortex,[57] but do not alter α4 expression.

The effect of steroid withdrawal on α4 expression is similar to the reported effects of other nonsteroid GABA modulators, such as BDZs[7,10] and ethanol,[12,13] which have been shown to increase α4 expression after chronic administration or withdrawal following sustained exposure. In addition, decreases in α1 expression have been reported after chronic BDZ administration.[11,18] The chronic intermittent ethanol administration paradigm produces increased α4 expression in association with an increased negative modulatory response to a β-carboline.[58] Chronic exposure to ethanol has also been shown to increase α4 expression[13] and produces a pharmacological profile similar to that reported for neurosteroid withdrawal, i.e., with insensitivity to BDZ agonists and conversion of a BDZ antagonist to a BDZ agonist.[16] Thus, these examples of increased α4 expression are all a result of prolonged exposure and subsequent withdrawal from a GABA modulator, which are additionally correlated with increased behavioral excitability. As such, this suggests a common mechanism that triggers increased expression of GABAR containing the α4 subunit.

7.5 CROSS-TOLERANCE WITH OTHER GABA MODULATORS

Withdrawal from progesterone produces a relative insensitivity to the GABA-modulatory effects of 3α,5α-THP, assessed using whole-cell patch clamp procedures in pyramidal cells acutely isolated from CA1 hippocampus.[24] However, other studies suggest that withdrawal from this neurosteroid produces an increased response to the anticonvulsant effect of the steroid on PTZ-generated seizures,[59] similar to the increased anticonvulsant effect of the steroid reported after ethanol withdrawal,[60,61] when endogenous neuroactive steroid levels are decreased.[62] This discrepancy may reflect selectivity of effect mediated by specific GABAR isoforms or brain regions responsible for particular behavioral effects. In fact, although withdrawal from ethanol increases the anticonvulsant effect of neurosteroids,[61] it decreases the anxiolytic effect of this compound.[63] In addition, ethanol increases the ability of 3α,5α-THP to increase t-butylbicyclophosphorothionate (TBPS) binding,[17] while GABA withdrawal also results in increased response to progesterone assessed using the paired pulse inhibition (PPI) paradigm.[20] Differential development of tolerance may explain these apparently discrepant results, as tolerance does not develop to the anticonvulsant effects of 3α,5β-THP,[64] but does develop in response to the anxiolytic and sedative effects of 3α,5α-THP,[65] assessed in a pseudopregnancy model, when levels of the steroid are elevated for 11 days. In addition to subunit composition, alterations in the phosphorylation state of the receptor[66–68] also determine the steroid sensitivity of the GABAR. Co-administration of progesterone also enhances the withdrawal effects of GABA[69] and BDZs,[70] while preventing the development of tolerance and attenuating the withdrawal effects resulting from chronic BDZ administration.[71]

7.6 STEROID WITHDRAWAL: HYPOTHALAMIC NEURONS

In the hypothalamus, the effects of hormone withdrawal may have very different implications. During pregnancy, activity in oxytocin-containing neurons is depressed due to extensive GABAergic innervation of these neurons and the influence of high ambient 3α,5α-THP levels on spontaneous inhibitory postsynaptic current (sIPSC), which results in prolongation of the decay time.[72] Additionally, 3α,5α-THP prevents suppression of sIPSCs by oxytocin, an effect mediated by protein kinase C (PKC)-induced phosphorylation, thus further ensuring that premature labor does not occur.[67] At the time of parturition, however, a number of changes in GABAergic inhibition occur that tip the balance of neuronal excitability in the positive direction. GABAR subunit composition in oxytocin neurons within the supraoptic nuclei is altered during pregnancy, with an increase in the α2:α1 subunit ratio occurring at the time of parturition.[72–74] Consistent with previous reports of GABAR subunit effects on the kinetics of GABA-gated current,[75] the increase in α2-containing GABAR is associated with a marked increase in the decay time constant (τ) for GABAergic monoquantal sIPSCs. In fact, this subunit switch is the mediating factor in promoting the increase in τ, as antisense-induced suppression of α2 expression prevents the change in τ.[72] However, oxytocin-induced activation of PKC produces a relative neurosteroid insensitivity that is independent of receptor isoform.[68] During parturition and lactation, this steroid insensitivity renders oxytocin neurons with a net acceleration in decay time, as compared with this value during pregnancy, a time of elevated 3α,5α-THP levels. Initial observations during early pregnancy also demonstrate an increase in the frequency of these monoquantal events, a finding consistent with other reports of increases in GABAergic innervation of these neurons during pregnancy. In order to determine the net GABAergic inhibition of this circuit, the authors calculated the synaptic current densities, by integrating synaptic current in the presence of 3α,5α-THP across time, and expressed this value with respect to the cell capacitance, a quantitative measure of the cell membrane surface area and thus cell size. Because swelling of oxytocin cell bodies occurs during pregnancy and lactation,[76] cell capacitance measurements also increase across the period of lactation. Assessed in this way, the total synaptic inhibition is markedly decreased, beginning in late pregnancy and continuing throughout the lactation period, compared to that determined for virgin control animals. Such a change would permit increases in excitability of the oxytocin neurons, whose bursting discharge would release oxytocin to promote the uterine contractility necessary for the birth process and subsequent lactation period.

7.7 CHRONIC NEUROSTEROID EXPOSURE

A number of studies have suggested that chronic exposure to neuroactive steroids, such as 3α,5α-THP, distinct from a withdrawal effect, results in altered GABAR pharmacology, which in some cases is associated with subunit changes.[77,78] These changes are comparable to those reported for other GABA-modulatory drugs such as the BDZs[7] and ethanol.[13] Micromolar concentrations of 3α,5α-THP have been shown to decrease GABA-gated current recorded from neocortical neurons in culture after 5 days of exposure.[77] In addition, modulation of GABA-gated current by barbiturates and 3α,5α-THP is also significantly decreased by this long-term steroid exposure.[77,79] These changes are accompanied by decreases in the β subunits generally and a decrease in the α3 subunit specifically.[80] Although basal binding is unaltered by this treatment, heterologous uncoupling between GABA, barbiturate, neurosteroid sites, and the BDZ site are observed as a decrease in E_{max} values.[78,81] Heterologous uncoupling produced by 3α,5β-THP has an EC_{50} of 1.7 μ*M* and is blocked by the GABA site antagonist SR-95531, whereas homologous uncoupling produced by this steroid is resistant to this antagonist, suggesting two distinct sites of action for the two effects of the steroid. These effects of chronic steroid administration may be dependent upon the developmental stage, as in one study,[82] which utilized developmental age 19 neurons, where chronic (5 day) neurosteroid exposure was shown to decrease α4 expression, with recovery to control levels upon withdrawal of the steroid.

Other studies have demonstrated significant alteration in GABAR subunits, which are differentially expressed in subzones of the hippocampal formation by 2 to 3 days' exposure to progesterone.[83] In addition to exerting effects on GABAR, a number of novel effects of 3α,5α-THP have also been reported, including induction of cytoarchitectural regression in fetal hippocampal neurons.[84]

7.8 TIME COURSE OF STEROID-INDUCED CHANGES IN α4 SUBUNIT EXPRESSION

The time course of changes in α4 GABAR subunit expression has been examined across 3 weeks of continuous exposure to progesterone, administered via s.c. implant to female rats. Under these conditions, α4 GABAR subunit levels in the hippocampus increase by two- to threefold after 48 to 72 h of hormone exposure.[85] This increase is well correlated with insensitivity to BDZ modulation of GABA-gated current assessed using whole-cell patch clamp procedures on pyramidal cells acutely dissociated from CA1 hippocampus (see Figure 7.1A). That the GABA-modulatory 3α,5α-THP is the mediating factor is suggested by studies showing that blockade of progesterone conversion prevents the effect,[85] and in fact, α4 upregulation can be observed after direct injection of 3α,5α-THP or its active isomer 3α,5β-THP (10 mg/kg, i.p.) across the same time period. The increase in α4 levels is also correlated with a significant acceleration in the decay time constant, τ, for GABA-gated current. In the absence of alterations in E_{max} for peak GABA-gated current, this change would yield a net decrease in total charge transfer, and reflect a neuron with increased excitability, as demonstrated for steroid withdrawal.[24] Consistent with this, increases in anxiety are also observed after 48 to 72 h of hormone exposure,[85] assessed as the number of open arm entries on the elevated plus maze (see Figure 7.1B), suggesting a correlation between increases in the α4 subunit and anxiety-like behavior. In contrast, by 4 to 5 days of hormone exposure, α4 levels return to control levels and remain at this steady-state level until removal of the hormone (withdrawal) when α4 upregulation is again observed in conjunction with BDZ insensitivity and decreases in GABA decay time. This pattern of changes may be relevant for the human, where increases in 3α,5α-THP and progesterone are elevated across the 11 to 12 days of the luteal phase. Altered mood in susceptible women has been reported not only during the time of hormone withdrawal during the late luteal phase[86] (i.e., PMS), but also during the early days of the luteal phase.[87–90] Such a complex pattern of α4 expression may reflect a homeostatic response to prolonged inhibition, with the initial increase in α4-containing GABAR an attempt to decrease inhibition via a faster-decaying receptor, and the late decrease related to the development of at least partial tolerance to the neurosteroid. The increase in α4 expression during neurosteroid withdrawal may represent a rebound from steroid-induced suppression of expression during the later stages of the chronic paradigm.

7.9 PHARMACOLOGY OF BEHAVIORAL CHANGES ASSOCIATED WITH CHRONIC STEROID EXPOSURE AND WITHDRAWAL

Both 48-h exposure to 3α,5α/β-THP and its withdrawal result in increased expression of the α4 subunit in association with increases in anxiogenic responses.[24,85] Because expression of α4-containing GABAR yields a distinctive pharmacological profile,[52,53] both LZM and the BDZ antagonist flumazenil have been tested for their effects on the anxiety generated by 48-h neurosteroid exposure and withdrawal. Under control conditions, LZM produces a significant anxiolytic effect on open arm time and frequency of entries using the elevated plus maze, while flumazenil produces no significant effect on these parameters.[25] Use of both hormonal paradigms, which increase α4 expression, results in an anxiety response that is relatively unresponsive to LZM, but markedly attenuated by flumazenil[25,91] in a manner consistent with upregulation of α4βγ2 GABAR in relevant limbic structures. The sedative effects of a BDZ are also markedly attenuated following progesterone withdrawal.[92] These pharmacological changes are largely observed in both males and females, suggesting a robust effect of neurosteroid treatment on GABAR subunit plasticity independent of gender differences, which is reflected behaviorally as well as by *in vitro* measures of GABAR function. Interestingly, a similar BDZ insensitivity has been reported for women with premenstrual dysphoric disorder[88] (see Chapter 13), suggesting possible relevance of the short-term steroid paradigm with this clinical syndrome. However, flumazenil induces a greater panic response in women with this disorder compared to asymptomatic controls,[93] implicating additional mechanisms.

7.10 α4 SUBUNIT EXPRESSION AND INHIBITORY TONE

The results reported from a number of different studies suggest that increased expression of α4-containing GABAR is associated with reduced inhibitory tone, resulting in greater excitability in affected CNS areas, for which α4 subunits may substitute for other normally expressed α subunits. First, a number of studies have demonstrated increased levels of this subunit following kindling paradigms[94] when BDZ insensitivity has also been reported.[95] α4 upregulation is also seen[94] in seizure states[96] and drug-induced tremor.[97] α4 subunit upregulation also accompanies the behavioral excitability associated with chronic treatment and withdrawal from ethanol[12] or BDZs.[7,10] Recent studies[98] suggest that α4-containing GABAR channels display shorter mean open times than α2-containing GABAR, and our preliminary data[99] suggest that following neuroactive steroid exposure, GABAR pharmacologically characterized as α4-containing decay more quickly than BDZ-responsive GABA-gated currents recorded under control conditions. Conversely, states that decrease expression of the α4 subunit are characterized by increases in GABAR mediated Cl^- flux.[100] In addition, the seizure susceptibility resulting from progesterone withdrawal is prevented when α4 expression is prevented with the use of antisense technology.[24] Taken together, these results suggest that increased expression of the α4 subunit, as a substitution for the normally expressed GABAR α subunit pool, produces a state of hyperexcitability. In light of this, the effects of progesterone withdrawal have been tested directly on hippocampal circuit excitability using a model paradigm, paired pulse inhibition.

7.11 A MODEL OF HIPPOCAMPAL EXCITABILITY: PAIRED PULSE INHIBITION

As a model of hippocampal excitability, PPI tests the percentage change in neuronal response to the second of paired electrical stimuli delivered with a 10-msec interpulse interval. Under such conditions, it has been used to evaluate hippocampal excitability[101] as a function of GABAR response in kindling and drug withdrawal paradigms. Using this paradigm, the population response to the second (test) of two paired stimuli (PS_{test}) is routinely of smaller amplitude than the response to the initial (conditioning) stimulus (PS_{cond})[101] (see Figure 7.2). The degree of this inhibition is significantly decreased (40 to 60%, $p < .01$) after progesterone withdrawal,[102] compared to the control (see Figure 7.2), as is the initial population spike (PS) amplitude, while the threshold for eliciting a PS is also significantly increased in the withdrawal state. The pharmacology of this response is also consistent with increased expression of functional α4-containing GABAR in the CA1 hippocampus.[53] In control slices, PPI is strongly potentiated (PPI > 70%) after bath application of either the BDZ LZM (10 μ*M*) or the barbiturate pentobarbital (10 μ*M*). However, the robust potentiation of hippocampal PPI observed following administration of LZM under control conditions is completely prevented after progesterone withdrawal. In contrast, pentobarbital (10 μ*M*) is able to significantly ($p < .05$) potentiate PPI after progesterone withdrawal.[102]

Antisense-induced suppression of expression of α4 subunit levels prevents the reduction in PPI normally observed following hormone withdrawal[102] across a range of stimulus intensities (3 to 10 T) (see Figure 7.2), suggesting that upregulation of GABAR containing the α4 subunit reduces inhibition in CA1 hippocampal circuits. Under reduced levels of α4 subunit expression, progesterone withdrawal results in an average PPI of 61.95 ± 5.97% at a 5x threshold stimulus intensity, similar to the original control value of PPI (65.18 ± 8.28%). In contrast, missense treatment fails to prevent the reduction in PPI normally observed after progesterone withdrawal, compared to the control (20 to 30% PPI).

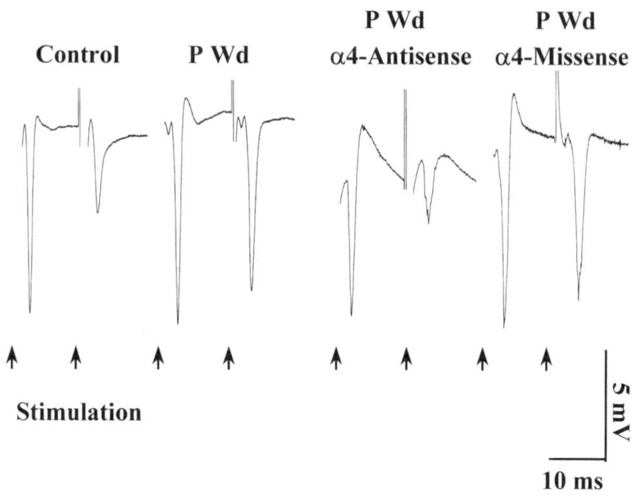

FIGURE 7.2 Progesterone withdrawal increases hippocampal circuit excitability in a paired pulse inhibition paradigm: an effect dependent upon increased α4 subunit expression. This study made use of the PPI paradigm, which, as a model of circuit excitability, tests the population response of CA1 hippocampal pyramidal cells to the second of two paired stimuli, delivered 10 msec apart. Under control conditions, the representative paired population spike traces reveal a significantly reduced response to the second stimulus compared to the first, an effect resulting from feedback inhibitory control of pyramidal cell excitability. Progesterone withdrawal resulted in a marked attenuation of PPI, suggesting impairment of this inhibitory mechanism. This alteration in circuit excitability following progesterone withdrawal is dependent upon upregulation of α4 expression, as the reduction in PPI normally observed after progesterone withdrawal was prevented by the prior intraventricular administration of GABAR α4 antisense oligonucleotide to suppress α4 expression. In contrast, administration of GABAR α4 missense oligonucleotide did not prevent the decreased PPI produced by progesterone withdrawal (n = 5 to 13 slices/group). (From Hsu, F.C. and Smith, S.S., *J. Neurophysiol.*, 89, 186–198, 2003. Printed with permission, American Physiological Society.)

The results from this study suggest that PPI, one measure of circuit excitability, is significantly attenuated in the CA1 hippocampus following withdrawal from progesterone. Further, the upregulation of α4-containing GABAR shown to result from hormone withdrawal[24] produces a decreased level of inhibitory control of pyramidal cell excitability.

7.12 PPI IN CA1 HIPPOCAMPUS: CIRCUIT PROPERTIES

Inhibitory pathways contributing to PPI in the hippocampal slice have been well characterized and can involve both feed-forward and feed-back inhibitory circuits.[103–105] Feed-forward inhibition is the result of activity from interneurons in the lacunosum-moleculare, with peak inhibitory postsynaptic potentials (IPSPs) that coincide with the rising phase of the pyramidal cell excitatory postsynaptic potential (EPSP), or from interneurons in the alveus-oriens, with peak IPSPs that coincide with the decay phase of the pyramidal cell EPSP.[103] A second mechanism of

GABAergic control of pyramidal cell excitability is through increases in GABA-gated conductance (gGABA), which are shorter latency events than either the GABAergic IPSC or IPSP.[103] Both of these mechanisms may have been attenuated under conditions of progesterone withdrawal, when increases in the amplitude of the PS response to the conditioning pulse are seen in conjunction with a significantly lower threshold for activation of this short-latency response to the first stimulus, compared with control slices.

7.13 RECURRENT INHIBITION AND PPI

The temporal properties of recurrent inhibition are most relevant as a mechanism explaining the observed differences in the degree of PPI across experimental groups. Stimulation of the Schaeffer collateral pathway elicits an action potential in the pyramidal cells with latencies in the range of ~2 to 5 msec in the stimulation paradigm employed. This pyramidal cell then activates interneurons, such as bistratified cells, with a latency of 6.3 msec.[103] Across a range of stimulation intensities, the resulting recurrent IPSP assessed in the pyramidal cell would then occur from 6 to 28 msec following its response to the first stimulus.[103] This time course fits well with the 10-msec interpulse interval used in the PPI paradigm for the current study, which demonstrates a reduced PS amplitude in response to the test pulse. Thus, the decrease in PPI observed following progesterone withdrawal may be due to a reduction in both GABAergic feed-forward and recurrent inhibition of pyramidal cell activity directly as a result of altered subunit composition, suggesting that both classes of inhibitory inputs known to regulate pyramidal cell excitability are impaired in the withdrawal state.

7.14 PPI AND NEURONAL EXCITABILITY

PPI has been used as a model paradigm to assess hippocampal circuit excitability in different seizure models that employ electrical kindling[106–108] as well as chemical kindling.[109–113] The PPI paradigm has also been used to evaluate hippocampal excitability after GABA withdrawal,[114] ethanol withdrawal,[104,115] and chronic BDZ administration,[116] where it is decreased primarily as a result of attenuated feed-forward inhibition.[117] In the majority of these studies, the PPI in CA1 hippocampus is reduced, indicating in many cases impairment of recurrent inhibition,[104,106,115,118–120] although in some kindling paradigms, increased synaptic GABAR density is reported[121] and PPI of dentate gyrus granule cells is robust.[110] Thus, the decrease in PPI observed following progesterone withdrawal compares well with most of these other known states of behavioral excitability.

7.15 ALTERATIONS IN SYNAPTIC CURRENT AFTER 48-H 3α,5β-THP TREATMENT

GABAR in the hippocampus and cerebellum have been well characterized in terms of their subcellular localization[122] with regard to synaptic input. Many GABAR subunit subtypes, including the $\alpha 2$, are believed to be localized exclusively to the synapse. Others, such as the δ and $\alpha 5$, are believed to reside exclusively in the extrasynaptic GABAR population, while a subset, including the $\alpha 1$, $\alpha 6$, and $\beta 1$–3, may reside in either domain. Within the synaptic GABAR population,[122] $\alpha 1$-containing GABAR have been localized to somatic and proximal dendrites on the CA1 hippocampal pyramidal cell, while the $\alpha 2$-containing GABAR are present on the axon-initiating segment, where inhibitory input would control action potential generation. Specific targeting of these GABAR subtypes by steroids would thus yield distinct effects on the net output of the cell.

In order to determine whether hormone-induced upregulation of the $\alpha 4$ subunit results in a change in synaptic current, analysis of unitary synaptic events is required. Under conditions where action potentials are suppressed with tetrodotoxin (TTX), the recorded miniature inhibitory postsynaptic currents (mIPSCs) are believed to reflect the postsynaptic quantal response from a single vesicle at one synapse.[123] The decay time constant of these unitary events thus reflects the kinetics of postsynaptic GABAR clusters, whereas compound events occur in response to the asynchronous release of the transmitter at multiple synapses and are not useful for estimates of postsynaptic GABAR kinetic properties.

Acutely applied, neurosteroids produce significant prolongation of the decay time of both evoked IPSCs and mIPSCs, assessed using whole-cell patch clamp techniques,[73,124,125] although the phosphorylation state has been shown to determine sensitivity of synaptic GABAR to modulation by steroids.[66,68]

In contrast, *in vivo* exposure to the neurosteroid 3α,5β-THP for 48 h results in a significant, 30% faster ($p < .05$) mIPSC decay time constant (τ_w), weighted for the relative contribution of fast and slow components.[126] Antisense-induced suppression of $\alpha 4$ expression allows comparison of conditions favoring high (missense) vs. low (antisense) levels of $\alpha 4$ expression following 48-h neurosteroid exposure.[126] Using this approach, conditions of high $\alpha 4$ expression are associated with a greater fraction of mIPSCs best fit with a biexponential decay, as compared to conditions of low $\alpha 4$ expression (42.3 vs. 16.3%, respectively) (see Figure 7.3). The distribution of values for the fast component of τ (τ_{fast}) exhibits two peaks with values of 0.54 $\pm$ 0.007 msec (75%) and 1.08 $\pm$ 0.04 msec (25%) (mean = 0.67 $\pm$ 0.03 msec, $p <$ .05) under conditions of high $\alpha 4$ expression, with the fast component accounting for 62% of the total current. In contrast, τ_{fast} is significantly slower (1.14 $\pm$ 0.06 msec) and accounts for a smaller fraction of the total current (47.0%) under conditions of low $\alpha 4$ expression ($p < .05$). Values for τ_{slow} are not significantly different between high and low $\alpha 4$ expression groups.

From the total population of mIPSCs sampled after 48 h of 3α,5β-THP treatment (both mono- and biexponential decays), the total percentage of current decaying with a faster rate than control currents under conditions of $\alpha 4$ upregulation is 47%, thus suggesting that approximately half of the synaptic GABAR clusters exhibit faster rates of deactivation following short-term neurosteroid exposure.

This conclusion is further supported by pharmacological data[126] from 3α,5β-THP-treated animals demonstrating a bimodal distribution of τ for individual mIPSCs following bath application of the $\alpha 4$-selective benzodiazepine agonist RO15-4513,[53] with a shift to slower values. Of the individual τ values, 40 to 50% were

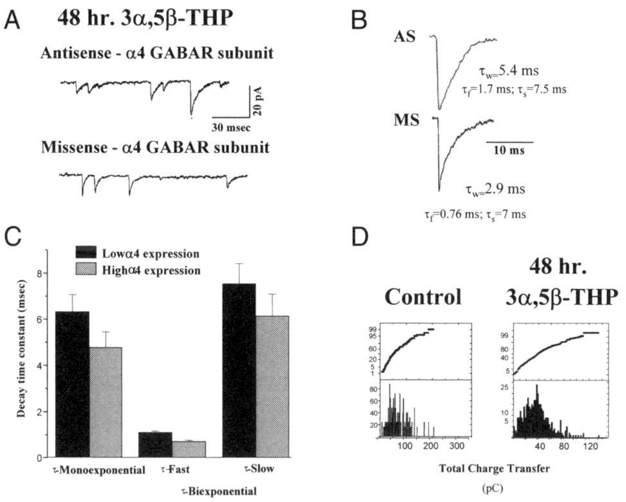

FIGURE 7.3 Suppression of α4 $GABA_A$ receptor subunit expression alters the kinetics of miniature inhibitory postsynaptic currents following neurosteroid exposure. For this study, animals were administered 3α,5β-THP over a 48-h period (10 mg/kg, i.p.), in conjunction with α4 antisense oligonucleotide administered intraventricularly to suppress α4 subunit expression. Miniature inhibitory postsynaptic currents were recorded with whole-cell patch clamp recording techniques from CA1 hippocampal pyramidal cells at near-physiological temperature (35˚C) using the slice preparation. Representative traces (A) and averaged, scaled mIPSCs (B) from both α4 antisense (AS; upper panel, low α4) and α4 missense-treated (MS; lower panel, high α4) animals during the 48-h 3α,5β-THP administration paradigm (n = 30, AS; n = 34, MS). (C) Summary diagram: Conditions of high α4 expression resulted in an accelerated τ_{fast} for mIPSCs compared to values recorded under conditions of low α4 expression, but no significant change in τ_{slow} (n = 489, AS; n = 1269, MS). Significant, 30% decreases in τ_{mono} are also observed under conditions of high α4 expression (n = 80 to 100 mIPSCs/cell, 4 to 8 cells/animal, 8 to 12 rats/group). Similar decreases in τ_{fast} are observed when mIPSCs recorded from hormone-treated animals (48 h of 3α,5β-THP) are compared with control (data not shown), resulting in a reduction in total charge transfer following hormone treatment. (D) The histogram and probability plot of values for total charge transfer are presented for representative mIPSCs recorded from hippocampal slices from 10 to 12 rats following 48 h of exposure to 3α,5β-THP or control conditions. This reduction in total charge transfer would reduce inhibition and thus lead to increased excitability. (From Hsu, F.C. et al., *J. Neurophysiol.*, 89, 1929–1940, 2003. Printed with permission, American Physiological Society.)

also shifted to slower values after bath application of the BDZ agonist LZM.[52] Because the α4 subunit is insensitive to BDZs, this suggests that the number of GABAR synapses containing α4 subunits is equivalent to those that do not after 48 h of administration of 3α,5β-THP. Similar effects have been reported following ethanol withdrawal, when α4 expression is also increased, with acceleration of τ_{fast} for mIPSCs recorded from CA1 pyramidal cells in association with a pharmacological profile characteristic of α4-containing GABAR.[127]

In contrast to these effects of chronic treatment with neurosteroids and ethanol, chronic BDZ treatment decreases GABA response[6] primarily by decreasing the amplitude of inhibitory potentials[8,128] or, in some cases, producing populations of silent synapses,[9] rather than through an effect on current decay kinetics.

7.16 CURRENT DEACTIVATION

The mIPSC decay time constant is an approximate measure of the deactivation rate of synaptically localized GABAR, given that the GABA released at synapses on the pyramidal cell soma, the locus of most inhibitory activity,[129] is quickly (<1 msec) removed from the synaptic cleft.[130,131] The most striking effect of neurosteroid exposure is to accelerate τ_{fast}, an effect blocked by prior administration of antisense oligonucleotide to prevent α4 subunit upregulation, while τ_{slow} is either unchanged or, in some cases, prolonged in comparison to control values. The fast component of decay is thought to represent the duration of the initial burst of channel openings gated by agonist,[132,133] while τ_{slow} more likely represents final channel closing and unbinding of the ligand.[132] The acceleration in τ_{fast} after 48 h of 3α,5β-THP exposure is consistent with recent findings demonstrating a shorter mean open time for α4-containing GABAR, assessed using fluctuation analysis.[98] In fact, rates of GABAR deactivation are known to be influenced by the expression of particular GABAR α subunits; α_1-containing GABAR deactivate with a time constant sixfold faster than α_2-containing GABAR, an outcome demonstrated by using both excised membrane patches containing recombinant receptors and synaptic current recording.[134,135]

In addition to producing decreases in τ, 48-h exposure to 3α,5β-THP also increases the percentage of currents decaying biexponentially from 16 to 42% (see Figure 7.3). Interestingly, in addition to reflecting a change in GABAR subunit composition, this phenomenon may be a result of postreceptor mechanisms such as receptor phosphorylation. Increases in phosphorylation[136] produce a number of changes that are strikingly similar to those we observe after short-term hormone treatment, including (1) an increase in currents with a biexponential decay, (2) an acceleration in τ[137], and (3) a decrease in mIPSC amplitude. These intriguing similarities suggest that alterations in phosphorylation state may also play a role in mediating the changes in synaptic current that we observe following neurosteroid exposure. The decrease in τ observed after neurosteroid exposure is associated with a reduction in total charge transfer (total integrated current, see Figure 7.3D) and would thus be expected to produce less inhibition. This effect is consistent with the increase in excitability observed at both the circuit and behavioral levels following steroid exposure/withdrawal.[138]

7.17 THE GABAR Δ SUBUNIT

The α4 subunit can coexpress with either γ2 or δ subunits.[139] Alterations in synaptic current (see above) are due to expression of α4βγ2 GABAR because the presence of a γ2 subunit is necessary for tethering at the synapse.[140] However, in addition to increasing α4 expression, progesterone withdrawal also increases expression of the δ subunit, resulting in formation of receptors containing both α4 and δ subunits (see Figure 7.4), as demonstrated using co-immunoprecipitation techniques.[144]

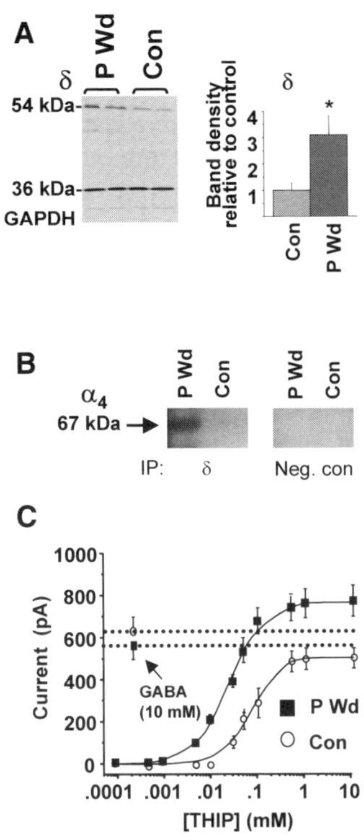

FIGURE 7.4 Progesterone withdrawal increases $\alpha_4\beta\delta$ $GABA_A$ receptors. (A) Left: A Western blot showing increased expression of the δ subunit (54 kDa), but not a control protein (GAPDH, 36 kDa) following progesterone withdrawal (P Wd) compared to control (Con). Right: Mean values (n = 20 to 25, performed in triplicate). (B) Co-assembly of α_4 and δ $GABA_A$ receptor subunits. Following immunoprecipitation using protein A beads coupled to antibodies for the δ subunit (IP: δ) or a cytosolic protein (IP: Neg. control), membranes were probed with digoxygenin-labeled anti-α_4 on a Western blot (n = 4, performed twice). A prominent 67-kDa band was detected following P Wd, but was barely detectable under control conditions. (C) THIP pharmacology. The maximum current produced by THIP compared to 10 m*M* GABA (THIP:GABA ratio) was 1.41 after P Wd (THIP $EC_{50} = 39 \pm 2$ μ*M*) vs. 0.95 in control neurons (THIP $EC_{50} = 81 \pm 6$ μ*M*), assessed using patch clamp recording in neurons from CA1 hippocampus (n = 10 to 15). This effect is consistent with upregulation of α4βδ GABAR following P Wd. *, $p < .05$. (From Sundstrom-Poromaa, I. et al., *Nat. Neurosci.,* 5, 721–722, 2002. Printed with permission, Nature Publishing Group.)

α4βδ GABAR exhibit distinctive biophysical and pharmacological properties[142–145] compared to the major native GABAR isoforms. They deactivate more quickly and desensitize less than α1β3γ2 GABAR (see Chapter 4). In addition, saturating concentrations of the GABA partial agonist 4,5,6,7-tetrahydroisoxazolo[5,4-c]pyridin-3-ol (THIP) ($THIP_{max}$) generates a maximum current greater than saturating concentrations of GABA ($GABA_{max}$) only at the α4βδ isoform.[146,147] This increase in the $THIP_{max}$:$GABA_{max}$ response ratio (1.25 to 1.4) is in striking contrast to the majority of native $GABA_A$-R isoforms (α1–3, β1–3, γ2), for which the

$THIP_{max}$:$GABA_{max}$ ratio is significantly less than 1 (~0.7).[146–148] In isolated pyramidal neurons, THIP produces a maximum response greater than GABA only after progesterone withdrawal (see Figure 7.4C), suggesting incorporation of these subunits into functional receptors.[141] Our recent studies[149] also demonstrate that THIP, at a dose subthreshold for its sedative effect, produces anxiolytic effects only after progesterone withdrawal, but not in control animals, assessed using the acoustic startle response, suggesting that this expression of α4βδ GABAR is behaviorally relevant. Pharmacologically, α4βδ GABAR are also BDZ insensitive but respond more to pentobarbital than α1βγ2 GABAR,[147] as also evidenced when α4βδ GABAR are increased following progesterone withdrawal using both *in vitro*[24] and behavioral[25,149] measures.

This increase in δ expression distinguishes progesterone withdrawal from other withdrawal paradigms, such as the chronic intermittent ethanol protocol, where α4 expression is increased but δ expression is decreased.[127] However, other excitability states, including both human and rodent models of temporal lobe epilepsy, also result in increased expression of both α4 and δ GABAR subunits in dentate gyrus granule cells.[96]

7.18 ETHANOL EFFECTS ON α4βΔ GABAR

One unique pharmacological property of α4βδ GABAR is a striking sensitivity to modulation by low concentrations of ethanol (1 to 3 m*M*), as assessed by using recombinant receptors expressed in oocytes and tested with two electrode voltage clamp techniques.[141] In general, most known effects of the sedative, anxiolytic drug ethanol are at much higher concentrations (20 to 100 m*M*) when it acts via a binding pocket on the M1–M2 segment of the α subunit[151] (see Chapter 10). Under conditions of increased α4βδ GABAR expression produced by progesterone withdrawal, however, significant (60 to 80%) potentiation of GABA-gated current is produced by these low concentrations of ethanol, an effect prevented when α4 expression is suppressed pharmacologically.[144] Similar low concentrations of ethanol produce anxiolytic effects in progesterone-withdrawn, but not control, rats using the acoustic startle response as a measure of anxiety, suggesting a behaviorally relevant effect. Interestingly, higher doses of ethanol are ineffective in both the electrophysiology and behavioral paradigms.[144] These results suggest that under conditions of progesterone withdrawal, low-dose ethanol may be anxiolytic, but higher doses not as sedative, leading to reinforcing conditions favoring increased ethanol consumption, which has been reported for women with premenstrual syndrome.[151] Interestingly, ethanol withdrawal also increases neurosteroid responsiveness,[61] suggesting a reciprocal relationship between the two GABA-modulatory compounds.

In addition to producing a greater response at α4βδ GABAR, ethanol administered across a 2-h period prevents the increased expression of the α4 subunit in progesterone-withdrawn rats.[141] These results suggest that α4 expression is dependent upon the time course and the presence of GABA modulators, and that this expression can be rapidly altered. In fact, BDZ administration has also been shown to prevent the increase in α4 expression produced by ethanol withdrawal.[152]

7.19 EXTRASYNAPTIC GABAR

α4βδ GABAR are thought to be extrasynaptic.[153–155] These receptor isoforms[156] contribute to tonic current recorded from CA1 hippocampal pyramidal cells after 48-h neurosteroid treatment, when an increase in THIP-evoked current is seen, but the bicuculline-sensitive tonic current is not altered following hormone treatment. Thus, these results suggest that α4βδ GABAR replace the existing pool of extrasynaptic receptors, perhaps including α5βγ2.

An extrasynaptic site of action for α4-containing GABAR would alter the degree of tonic inhibition following increased activity of GABAergic afferents to the cell, when transmitter spillover has been reported to occur,[157] and extrasynaptic GABAR populations would have access to the ligand. Extrasynaptic GABAR may act as a resistive shunt to limit excitability of the neuron, as demonstrated recently in the dentate gyrus[158] and cerebellum.[159] Under these conditions, replacement of the ambient extrasynaptic GABAR population in the CA1 hippocampus by α4-containing GABAR would result in current with faster decay kinetics, thus decreasing the total charge transfer after progesterone withdrawal, an outcome that would decrease inhibitory input and thereby decrease PPI.

7.20 NEUROSTEROID EFFECTS ON EXTRASYNAPTIC GABAR

Under control conditions, the dentate gyrus expresses the highest levels of α4βδ GABAR in the hippocampal formation.[49,154,155] Recently, these receptor isoforms have been identified as the most responsive to modulation by neuroactive steroids such as THDOC[147,148,160,162] (see Chapter 4). Therefore, the dentate gyrus may be a region of high neurosteroid sensitivity. Consistent with this is recent data demonstrating a tonic current highly responsive to low, physiological concentrations (10 n*M*) of THDOC,[162] as assessed by recording of the holding current with whole-cell patch clamp techniques, and displacement in response to bath application of the steroid. In contrast, a similar paradigm recording from the dentate gyrus of transgenic mice lacking the δ subunit was unresponsive to this concentration of steroid. Thus, the α4βδ $GABA_A$ receptor may be a primary target of GABA-modulatory steroids, and possibly a major site of change following steroid withdrawal.

7.21 CONCLUSIONS

In conclusion, the GABA-modulatory steroid 3α,5α/β-THP exhibits withdrawal properties similar to those of most conventional GABA-modulatory drugs, increasing anxiety and seizure susceptibility following chronic exposure to its parent compound progesterone. These changes are accompanied by increases in hippocampal circuit excitability, most probably due to attenuation of both feed-forward and feed-back inhibitory input to the pyramidal cell in the CA1 hippocampus. This alteration in excitability state is dependent upon a subunit switch involving upregulation of hippocampal GABAR expressing the α4 subunit, which confers unique pharmacological properties to the receptor and results in less total charge transfer, to produce less inhibitory tone. Across the time course of exposure to this steroid, the pattern of α4 expression is complex, increasing by 48 h after the onset of exposure, recovering to control levels by 5 days, and increasing again during withdrawal following cessation of steroid exposure, suggesting a series of compensatory mechanisms after prolonged inhibition and the development of tolerance to the steroid. In all cases, increased α4 expression is closely correlated with insensitivity to BDZ modulation and increased anxiety. However, α4 expression is differentially targeted to synaptic and extrasynaptic pools of receptors: 48 h of neurosteroid exposure results in a synaptic current with a faster decay kinetics and a pharmacological profile characteristic of α4-containing GABAR. In contrast, progesterone withdrawal results in minimal effects on synaptic current following increased expression of both α4βγ2 and α4βδ GABAR, an extrasynaptic isoform that may be a major target of neurosteroid action. The pharmacological profile of these new isoforms is reflected in the anxiolytic efficacy of GABA-modulatory drugs, suggesting that alterations in GABAR subunit composition may play a role in altering behavior during naturally occurring fluctuations in this steroid across the cycle. These results may be relevant for alterations in mood and seizure susceptibility associated with the premenstrual, postpartum, and postmenopausal periods, when alterations in BDZ and ethanol responses have also been reported.

ACKNOWLEDGMENTS

This work is supported by NIH grants DA09618 and AA12958. The author is grateful to Qi Hua Gong, Xinshe Li, Fu-Chun Hsu, and Maria Gulinello for contributing to the work described herein.

REFERENCES

1. Majewska, M.D., Harrison, N.L., Schwartz, R.D., Barker, J.L., and Paul, S.M., Steroid hormone metabolites are barbiturate-like modulators of the GABA receptor, *Science*, 232, 1004, 1986.
2. Gee, K.W., Chang, W.-C., Brinton, R.E., and McEwen, B.S., GABA-dependent modulation of the Cl– ionophore by steroids in rat brain, *Eur. J. Pharmacol.*, 136, 419, 1987.
3. Callachan, H., Cottrell, G.A., Hather, N.Y., Lambert, J.J., Nooney, J.M., and Peters, J.A., Modulation of the GABAA receptor by progesterone metabolites, *Proc. R. Soc. Lond. B Biol. Sci.*, 231, 359, 1987.
4. Smith, S.S., Waterhouse, B.D., and Woodward, D.J., Progesterone alters GABA and glutamate responsiveness: a possible mechanism for its anxiolytic action, *Brain Res.*, 400, 353, 1987.
5. Smith, S.S., Waterhouse, B.D., and Woodward, D.J., Locally applied progesterone metabolites alter neuronal responsiveness in the cerebellum, *Brain Res. Bull.*, 18, 739, 1987.
6. Gallager, D.W., Lakoski, J.M., Gonsalves, S.F., and Rauch, S.L., Chronic benzodiazepine treatment decreases postsynaptic GABA sensitivity, *Nature*, 308, 74, 1984.

7. Holt, R.A., Bateson, A.N., and Martin, I.L., Chronic treatment with diazepam or abecarnil differently affects the expression of $GABA_A$ receptor subunit mRNAs in the rat cortex, *Neuropharmacology*, 35, 1457, 1996.
8. Zeng, X. and Tietz, E.I., Depression of early and late monosynaptic inhibitory postsynaptic potential in hippocampal CA1 neurons following prolonged benzodiazepine administration: role of a reduction in Cl–driving force, *Synapse*, 25, 125, 1997.
9. Poisbeau, P., Williams, S.R., and Mody, I., Silent $GABA_A$ synapses during flurazepam withdrawal are region-specific in the hippocampal formation, *J. Neurosci.*, 17, 3467, 1997.
10. Follesa, P., Cagetti, E., Mancuso, L., Biggio, F., Manca, A., Maciocco, E., Massa, F., Desole, M., Carta, M., Busonero, F., Sanna, E., and Biggio, G., Increase in expression of the GABA(A) receptor alpha(4) subunit gene induced by withdrawal of, but not by long-term treatment with, benzodiazepine full or partial agonists, *Brain Res. Mol. Brain Res.*, 92, 138, 2001.
11. Chen, S., Huang, X., Zeng, X.J., Sieghart, W., and Tietz, E.I., Benzodiazepine-mediated regulation of alpha-1, alpha-2, beta 1-3 and gamma-2 GABA(A) receptor subunit proteins in the rat brain hippocampus and cortex, *Neuroscience*, 93, 33, 1999.
12. Mahmoudi, M., Kang, M.H., Tillakaratne, N., Tobin, A.J., and Olsen, R.W., Chronic intermittent ethanol treatment in rats increases $GABA_A$ receptor a4 subunit expression: possible relevance to alcohol dependence, *J. Neurochem.*, 68, 2485, 1997.
13. Devaud, L.L., Fritschy, J.-M., Sieghar, W., and Morrow, A.L., Bidirectional alterations of $GABA_A$ receptor subunit peptide levels in rat cortex during chronic ethanol consumption and withdrawal, *J. Neurochem.*, 69, 126, 1997.
14. Davis, M. and Gallagher, D.W., Continuous slow release of low levels of diazepam produces tolerance to its depressant and anxiolytic effects on the startle reflex, *Eur. J. Pharmacol.*, 150, 23, 1988.
15. Zeng, X.J. and Tietz, E.I., Benzodiazepine tolerance at GABAergic synapses on hippocampal CA1 pyramidal cells, *Synapse*, 31, 263, 1999.
16. Buck, K. and Harris, R., BDZ agonist and inverse agonist actions on GABA-A receptor-operated chloride channels. II. Chronic effects of ethanol, *J. Exp. Pharmacol. Exp. Ther.*, 253, 713, 1990.
17. Mehta, A.K. and Ticku, M.K., Chronic ethanol administration alters the modulatory effect of 5alpha-pregnan-3alpha-ol-20-one on the binding characteristics of various radioligands of GABA-A receptors, *Brain Res.*, 805, 88, 1998.
18. Tietz, E.I., Huang, X., Weng, X., Rosenberg, H.C., and Chiu, T.H., Expression of alpha-1, alpha-5, and gamma-2 GABA-A receptor subunit mRNAs measured *in situ* in rat hippocampus and cortex following chronic flurazepam administration, *J. Mol. Neurosci.*, 4, 277, 1993.
19. Ueno, S., Harris, R.A., Messing, R.O., Sanchez-Perez, A.M., Hodge, C.W., McMahon, T., Wang, D., Mehmert, K.K., Kelley, S.P., Haywood, A., Olive, M.F., Buck, K.J., Hood, H.M., Blednov, Y., Findlay, G., and Maxcia, M.P., Alcohol actions on GABA(A) receptors: from protein structure to mouse behavior, *Alcohol. Clin. Exp. Res.*, 25, 76S, 2001.
20. Calixto, E., Lopez-Colome, A.M., Casasola, C., Montiel, T., Bargas, J., and Brailowsky, S., Neocortical hyperexcitability after GABA withdrawal in vitro, *Epilepsy Res.*, 39, 13, 2000.
21. File, S.E., The history of benzodiazepine dependence: a review of animal studies, *Neurosci. Biobehav. Rev.*, 14, 135, 1990.

22. Dong, E., Matsumoto, K., Uzunova, V., Sugaya, I., Takahata, H., Nomura, H., Watanabe, H., Costa, A., and Guidotti, A., Brain 5alpha-dihydroprogesterone and allopregnanolone synthesis in a mouse model of protracted social isolation, *Proc. Natl. Acad. Sci.*, 98, 2849, 2001.
23. Moran, M.H. and Smith, S.S., Progesterone withdrawal I: pro-convulsant effects, *Brain Res.*, 807, 91, 1998.
24. Smith, S.S., Gong, Q.H., Hsu, F.C., Markowitz, R.S., Ffrench-Mullen, J.M.H., and Li, X., $GABA_A$ receptor α4 subunit suppression prevents withdrawal properties of an endogenous steroid, *Nature*, 392, 926, 1998.
25. Gulinello, M., Gong, Q.H., and Smith, S.S., Progesterone withdrawal increases the alpha-4 subunit of the GABA(A) receptor in male rats in association with anxiety and altered pharmacology: a comparison with female rats, *Neuropharmacology*, 43, 701, 2002.
26. Gallo, M.A. and Smith, S.S., Progesterone withdrawal decreases latency to and increases duration of electrified prod burial: a possible rat model of PMS anxiety, *Pharmacol. Biochem. Behav.*, 46, 897, 1993.
27. Gulinello, M., Orman, R., and Smith, S.S., Sex differences in anxiety, sensorimotor gating and expression of the alpha-4 subunit of the GABA-A receptor in the amygdala after progesterone withdrawal, *Eur. J. Neurosci.*, 17, 1, 2003.
28. Compagnone, N.A. and Mellon, S.H., Neurosteroids: biosynthesis and function of these novel neuromodulators, *Front. Neuroendocrinol.*, 21, 1, 2000.
29. Smith, S.S., Gong, Q.H., Li, X., Moran, M.H., Bitran, D., Frye, C.A., and Hsu, F.-C., Withdrawal from 3α-OH-5α-pregnan-20-one using a pseudopregnancy model alters the kinetics of hippocampal $GABA_A$-gated current and increases the $GABA_A$ receptor a4 subunit in association with increased anxiety, *J. Neurosci.*, 18, 5275, 1998.
30. Frye, C.A. and Bayon, L.E., Seizure activity is increased in endocrine states characterized by decline in endogenous levels of the neurosteroid 3 alpha,5 alpha-THP, *Neuroendocrinology*, 68, 272, 1998.
31. Frye, C.A. and Bayon, L.E., Cyclic withdrawal from endogenous and exogenous progesterone increases kainic acid and perforant pathway induced seizures, *Pharmacol. Biochem. Behav.*, 62, 315, 1999.
32. Reddy, D., Kim, H., and Rogawski, M., Neurosteroid withdrawal model of perimenstrual catamenial epilepsy, *Epilepsia*, 42, 328, 2001.
33. Smith, S.S., Gong, Q.H., Hsu, F.C., Markowitz, R.S., Ffrench-Mullen, J.M.H., and Li, X., $GABA_A$ receptor a4 subunit suppression prevents withdrawal properties of an endogenous steroid, *Nature*, 392, 926, 1998.
34. Frye, C.A. and Bayon, L.E., Cyclic withdrawal from endogenous and exogenous progesterone increases kainic acid and perforant pathway induced seizures, *Pharmacol. Biochem. Behav.*, 62, 315, 1999.
35. Frye, C.A. and Bayon, L.E., Seizure activity is increased in endocrine states characterized by decline in endogenous levels of the neurosteroid 3 alpha,5 alpha-THP, *Neuroendocrinology*, 68, 272, 1998.
36. Reilly, M.T., Crabbe, J.C., Rustay, N.R., and Finn, D.A., Acute neuroactive steroid withdrawal in withdrawal seizure-prone and withdrawal seizure-resistant mice, *Pharmacol. Biochem. Behav.*, 67, 709, 2000.
37. Herzog, A.G., Progesterone therapy in women with complex partial and secondary generalized seizures, *Neurology*, 45, 1660, 1995.
38. Herzog, A., Klein, P., and Ransil, B., Three patterns of catamenial epilepsy, *Epilepsia*, 38, 1082, 1997.

39. Backstrom, T., Epileptic seizures in women related to plasma estrogen and progesterone during the menstrual cycle, *Acta Neurol. Scand.*, 54, 321, 1976.
40. Pellow, S., Chopin, P., File, S.E., and Briley, M., Validation of open:closed arm entries in an elevated plus maze as a measure of anxiety in the rat, *J. Neurosci. Methods*, 14, 149, 1995.
41. Davis, M., Gendelman, D.S., Tischler, M.D., and Gendelman, P.M., A primary acoustic startle circuit: lesion and stimulation studies, *J. Neurosci.*, 2, 791, 1982.
42. Garrick, T., Morrow, N., Shalev, A.Y., and Eth, S., Stress-induced enhancement of auditory startle: an animal model of posttraumatic stress disorder, *Psychiatry*, 64, 346, 2001.
43. Metzger, L.J., Orr, S.P., Berry, N.J., Ahern, C.E., Lasko, N.B., and Pitman, R.K., Physiologic reactivity to startling tones in women with posttraumatic stress disorder, *J. Abnorm. Psychol.*, 108, 347, 1999.
44. van Erp, A.M. and Miczek, K.A., Persistent suppression of ethanol self-administration by brief social stress in rats and increased startle response as index of withdrawal, *Physiol. Behav.*, 73, 301, 2001.
45. Krystal, J.H., Webb, E., Grillon, C., Cooney, N., Casal, L., Morgan, C.A., III, Southwick, S.M., Davis, M., and Charney, D.S., Evidence of acoustic startle hyperreflexia in recently detoxified early onset male alcoholics: modulation by yohimbine and m-chlorophenylpiperazine (mCPP), *Psychopharmacology*, 131, 207, 1997.
46. Bitran, D., Dugan, M., Renda, P., Ellis, R., and Foley, M., Anxiolytic effects of the neuroactive steroid pregnanolone (3alpha-OH-5beta-pregnan-20-one) after microinjection in the dorsal hippocampus and lateal septum, *Brain Res.*, 850, 217, 1999.
47. Grillon, C., Sinha, R., and O'Malley, S.S., Effects of ethanol on the acoustic startle reflex in humans, *Psychopharmacology*, 114, 167, 1994.
48. Anagnostaras, S.G., Gale, G.D., and Fanselow, M.S., Hippocampus and contextual fear conditioning: recent controversies and advances, *Hippocampus*, 11, 8, 2001.
49. Wisden, W., Laurie, D.J., Monyer, H., and Seeburg, P., The distribution of 13 GABA-A receptor subunit mRNAs in the rat brain. I. Telencephalon, diencephalon, mesencephalon, *J. Neurosci.*, 12, 1040, 1992.
50. Hevers, W. and Luddens, H., The diversity of $GABA_A$ receptors: pharmacological and electrophysiological properties of $GABA_A$ channel subtypes, *Mol. Neurobiol.*, 18, 35, 1998.
51. Costa, A.-M.N., Spense, K.T., Smith, S.S., and Ffrench-Mullen, J.M.H., Withdrawal from the endogenous steroid progesterone results in $GABA_A$ currents insensitive to benzodiazepine modulation in rat CA1 hippocampus, *J. Neurophysiol.*, 74, 464, 1995.
52. Wisden, W., Laurie, D.J., Monyer, H., and Seeburg, P., Cloning, pharmacological characteristis and expression pattern of the rat $GABA_A$ receptor a4 subunit, *FEBS Lett.*, 289, 227, 1991.
53. Wafford, K.A., Thompson, S.A., Sikela, J., Wilcox, A.S., and Whiting, P.J., Functional characterization of human GABA receptors containing the a4 subunit, *Mol. Pharmacol.*, 50, 670, 1996.
54. Follesa, P., Serra, M., Cagetti, E., Pisu, M., Porta, S., Floris, S., Massa, F., Sanna, E., and Biggio, G., Allopregnanolone synthesis in cerebellar granulse cells: roles in regulation of GABA(A) receptor expression and function during progesterone treatment and withdrawal, *Mol. Pharmacol.*, 57, 1262, 2000.
55. Concas, A., Follesa, P., Barbaccia, M.L., Purdy, R.H., and Biggio, G., Physiological modulation of GABA(A) receptor plasticity by progesterone metabolites, *Eur. J. Pharmacol.*, 375, 225, 1999.

56. Mascia, M.P., Biggio, F., Mancuso, L., Cabras, S., Cocco, P.L., Gorini, G., Manca, A., Marra, C., Purdy, R.H., Follesa, P., and Biggio, G., Changes in GABA(A) receptor gene expression induced by withdrawal of, but not by long-term exposure to, ganaxolone in cultured rat cerebellar granule cells, *J. Pharmacol. Exp. Ther.*, 303, 1014, 2002.
57. Follesa, P., Porcu, P., Sogliano, C., Cinus, M., Biggio, F., Mancuso, L., Mostallino, M.C., Paoletti, A.M., Purdy, R.H., Biggio, G., and Concas, A., Changes in GABA-A receptor gamma-2 subunit gene expression induced by long-term administration of oral contraceptives in rats, *Neuropharmacology*, 42, 325, 2002.
58. Kang, M.H., Spigelman, I., and Olsen, R.W., Alteration in the sensitivity of GABA(A) receptors to allosteric modulatory drugs in rat hippocampus after chronic intermittent ethanol treatment, *Alcohol. Clin. Exp. Res.*, 22, 2165, 1998.
59. Reddy, D.S. and Rogawski, M.A., Enhanced anticonvulsant activity of neuroactive steroids in a rat model of catamenial epilepsy, *Epilepsia*, 42, 337, 2001.
60. Devaud, L.L., Purdy, R.H., and Morrow, A.L., The neurosteroid, 3a-OH-5a-pregnan-20-one, protects against bicuculline-induced seizures during ethanol withdrawal in rats, *Alcohol. Clin. Exp. Res.*, 19, 350, 1995.
61. Devaud, L.L., Purdy, R.H., Finn, D.A., and Morrow, A.L., Sensitization of gamma-aminobutyric acid A receptors to neuroactive steroids in rats during ethanol withdrawal, *J. Pharmacol. Exp. Ther.*, 278, 510, 1996.
62. Romeo, E., Brancati, A., DeLorenzo, A., Fucci, P., Furnari, C., Pompili, E., Sasso, G.F., Spalleta, G., Troisi, A., and Pasini, A., Marked decrease of plasma neuroactive steroids during alcohol withdrawal, *Clin. Neuropharmacol.*, 19, 366, 1996.
63. Finn, D.A., Gallaher, E.J., and Crabbe, J.C., Differential change in neuroactive steroid sensitivity during ethanol withdrawal, *J. Pharmacol. Exp. Ther.*, 292, 394, 2000.
64. Kokate, T.G., Yamaguchi, S., Pannell, L.K., Rajamani, U., Carroll, D.M., Grossman, A.B., and Rogawski, M.A., Lack of anticonvulsant tolerance to the neuroactive steroid pregnanolone in mice, *J. Pharmacol. Exp. Ther.*, 287, 553, 1998.
65. Marshall, F.H., Stratton, S.C., Mullings, J., Ford, E., Worton, S.P., Oakley, N.R., and Hagan, R.M., Development of tolerance in mice to the sedative effects of the neuroactive steroid minaxolone following chronic exposure, *Pharmacol. Biochem. Behav.*, 58, 1, 1997.
66. Fancsik, A., Linn, D.M., and Tasker, J.G., Neurosteroid modulation of GABA IPSCs is phosphorylation dependent, *J. Neurosci.*, 20, 3067, 2000.
67. Brussaard, A.B., Wossink, J., Lodder, J.C., and Kits, K.S., Progesterone-metabolite prevents protein kinase C-dependent modulation of g-aminobutyric acid type A receptors in oxytocin neurons, *Proc. Natl. Acad. Sci. U.S.A.*, 97, 3625, 2000.
68. Koksma, J.J., van Kesteren, R.E., Rosahl, T.W., Zwart, R., Smit, A.B., Luddens, H., and Brussaard, A.B., Oxytocin regulates neurosteroid modulation of GABA(A) receptors in supraoptic nucleus around parturition, *J. Neurosci.*, 23, 788, 2003.
69. Calixto, E., Montiel, T., Lemini, C., and Brailowsky, S., Allopregnanolone potentiates a GABA-withdrawal syndrome in the rat cerebral cortex, *Neurosci. Lett.*, 195, 73, 1995.
70. Pesce, M.E., Acevedo, X., Pinardi, G., and Miranda, H.F., Progesterone modulation of diazepam withdrawal syndrome in mice, *Pharmacol. Toxicol.*, 79, 331, 1996.
71. Reddy, D.S. and Kulkarni, S.K., Neurosteroid coadministration prevents development of tolerance and augments recovery from benzodiazepine withdrawal anxiety and hyperactivity in mice, *Methods Find. Exp. Clin. Pharmacol.*, 19, 395, 1997.

72. Brussaard, A.B., Kits, K.S., Baker, R.E., Willems, W.P., Leyting-Vermuelen, J.W., Voom, P., Smit, A.B., Bicknell, R.J., and Herbison, A.E., Plasticity in fast synaptic inhibition of adult oxytocin neurons by switch in GABA(A) receptor subunit expression, *Neuron*, 19, 1103, 1997.
73. Brussaard, A.B., Devay, P., Leyting-Vermuelen, J.W., and Kits, K.S., Changes in properties and neurosteroid regulation of GABAergic synapses in the supraoptic nucleus during the mammalian female reproductive cycle, *J. Physiol.*, 516, 513, 1999.
74. Brussaard, A.B., Devay, P., Leyting-Vermuelen, J.W., and Kits, K.S., Changes in properties and neurosteroid regulation of GABAergic synapses in the supraoptic nucleus during the mammalian female reproductive cycle, *J. Physiol.*, 516, 513, 1999.
75. Lavole, A.M., Tingey, J.J., Harrison, N.L., Pritchett, D.B., and Twyman, R.E., Activation and deactivation rates of recombinant $GABA_A$ receptor channels are dependent on a-subunit isoform, *Biophys. J.*, 73, 1, 1997.
76. Hatton, G.I., Function-related plasticity in hypothalamus, *Annu. Rev. Neurosci.*, 20, 375, 1997.
77. Yu, R., Hay, M., and Ticku, M.K., Chronic neurosteroid treatment attenuates single cell $GABA_A$ response and its potentiation by modulators in cortical neurons, *Brain Res.*, 706, 160, 1996.
78. Friedman, L.K., Gibbs, T.T., and Farb, D.H., Gamma-aminobutyric acidA receptor regulation: heterologous uncoupling of modulatory site interactions induced by chronic steroid, barbiturate, benzodiazepine, or GABA treatment in culture, *Brain Res.*, 707, 100, 1996.
79. Yu, R. and Ticku, M.K., Chronic neurosteroid treatment decreases the efficacy of benzodiazepine ligands and neurosteroids at the $GABA_A$ receptor complex in mammalian cortical neurons, *J. Pharmacol. Exp. Ther.*, 275, 784, 1995.
80. Yu, R., Follesa, P., and Ticku, M.K., Down-regulation of the GABA receptor subunit mRNA levels in mammalian cultured cortical neurons following chronic neurosteroid treatment, *Brain Res. Mol. Brain Res.*, 41, 163, 1996.
81. Yu, R. and Ticku, M.K., Chronic neurosteroid treatment produces functional heterologous uncoupling at the gamma-aminobutyric acid type A/benzodiazepine receptor complex in mammalian cortical neurons, *Mol. Pharmacol.*, 47, 603, 1995.
82. Grobin, A.C. and Morrow, A.L., 3alpha-Hydroxy-5alpha-pregnan-20-one exposure reduces GABA(A) receptor alpha4 subunit mRNA levels, *Eur. J. Pharmacol.*, 409, R1, 2000.
83. Weiland, N.G. and Orchinik, M., Specific subunit mRNAs of the $GABA_A$ receptor are regulated by progesterone in subfields of the hippocampus, *Brain Res. Mol. Brain Res.*, 32, 271, 1995.
84. Brinton, R.D., The neurosteroid 3a-hydroxy-5a-pregnan-20-one induces cytoarchitectural regression in cultured fetal hippocampal neurons, *J. Neurosci.*, 14, 2763, 1994.
85. Gulinello, M., Gong, Q.H., Li, X., and Smith, S.S., Short-term exposure to a neuroactive steroid increases a4 $GABA_A$ receptor subunit levels in association with increased anxiety, *Brain Res.*, 910, 55, 2001.
86. Rapkin, A., Morgan, M., Goldman, L., Brann, D., Simone, D., and Mahesh, V., Progesterone metabolite allopregnanolone in women with premenstrual syndrome, *Obstet. Gynecol.*, 90, 709, 1997.
87. Schmidt, P., Purdy R.H., Moore, P.H., Jr., Paul, S., and Rubinow, D., Circulating levels of anxiolytic steroids in the luteal phase in women with premenstrual syndrome and in control subjects, *J. Clin. Endocrinol. Metab.*, 79, 1256, 1994.

88. Sundstrom, I., Nyberg, S., and Backstrom, T., Patients with premenstrual syndrome have reduced sensitivity to midazolam compared to control subjects, *Neuropsychopharmacology*, 17, 370, 1997.
89. Redei, E. and Freeman, E.W., Daily plasma estradiol and progesterone levels over the menstrual cycle and their relation to premenstrual symptoms, *Psychoneuroendocrinology*, 20, 259, 1995.
90. Schmidt, P.J., Nieman, L.K., Grover, G.N., Muller, K.L., Meriam, G.R., and Rubinow, D.R., Lack of effect of induced menses on symptoms in women with premenstrual syndrome, *N. Engl. J. Med.*, 324, 1174, 1991.
91. Gulinello, M. and Smith, S.S., Anxiogenic effects of neurosteroid exposure: sex differences and altered GABA-A receptor pharmacology in adult rats, *J. Pharmacol. Exp. Ther.*, 305, 541, 2003.
92. Moran, M., Goldberg, M., and Smith, S., Progesterone withdrawal II: insensitivity to the sedative effects of a benzodiazepine, *Brain Res.*, 807, 91, 1998.
93. Le Melledo, J.M., Van Driel, M., Coupland, N.J., Lott, P., and Jhangri, G.S., Response to flumazenil in women with premenstrual dysphoric disorder, *Am. J. Psychiatry*, 157, 821, 2000.
94. Clark, M., Sensitivity of the rat hippocampal GABA(A) receptor alpha-4 subunit to electroshock seizures, *Neurosci. Lett.*, 250, 17, 1998.
95. Kapur, J., Hippocampal neurons express GABA-A receptor insensitive to diazepam in hyperexcitable conditions, *Epilepsia*, 41, S86, 2000.
96. Brooks-Kayal, A.R., Shumate, M.D., Jin, H., Rikhter, T.Y., and Coulter, D.A., Selective changes in single cell GABA(A) receptor subunit expression and function in temporal lobe epilepsy, *Nat. Med.*, 4, 1166, 1998.
97. Frostholm, A., Evans, J., Cummings, S., and Rotter, A., Harmaline-induced changes in gamma aminobutyric acid(A) receptor subunit mRNA expression in murine olivocerebellar nuclei, *Brain Res. Mol. Brain Res.*, 85, 200, 2000.
98. Maric, D., Maric, I., Wen, X., Fritschy, J.-M., Sieghart, W., Barker, J.L., and Serafini, R., GABA-A receptor subunit composition and functional properties of Cl– channels with differential sensitivity to zolpidem in embryonic rat hippocampal cells, *J. Neurosci.*, 19, 4921, 1999.
99. Smith, S., Wieland, S., and Gong, Q.H., Withdrawal from the reproductive steroid progesterone alters deactivation and desensitization kinetics of GABA-gated current in hippocampal neurons, *Society for Neuroscience Abstract*, 2000.
100. Papadeas, S., Grobin, A., and Morrow, A., Chronic ethanol consumption differentially alters GABA(A) receptor alpha1 and alpha4 subunit peptide expression and GABA(A) receptor-mediated 36 Cl(–) uptake in mesocorticolimbic regions of rat brain, *Alcohol. Clin. Exp. Res.*, 25, 1270, 2001.
101. Davies, C.H., Davies, S.N., and Collingridge, G.L., Paired-pulse depression of monosynaptic GABA-mediated inhibitory postsynaptic responses in rat hippocampus, *J. Physiol.*, 424, 513, 1990.
102. Hsu F.-C. and Smith, S.S., Progesterone withdrawal reduces paired-pulse inhibition in rat hippocampus: dependence on GABA-A receptor alpha-4 upregulation, *J. Neurophysiol.*, 89, 186, 2003.
103. Karnup, S. and Stelzer, A., Temporal overlap of excitatory and inhibitory afferent input in guinea-pig CA1 pyramidal cells, *J. Physiol.*, 516, 485, 1999.
104. Rogers, C.J. and Hunter, B.E., Chronic ethanol treatment reduces inhibition in CA1 neurons of the rat hippocampus, *Brain Res. Bull.*, 28, 587, 1992.
105. Turner, D.A., Feed-forward inhibitory potentials and excitatory interactions in guinea-pig hippocampal pyramidal cells, *J. Physiol.*, 422, 333, 1990.

106. Kamphuis, W., Gorter, J.A., Wadman, W.J., and Lopes da Silva, F.H., Hippocampal kindling leads to different changes in paired-pulse depression of local evoked field potentials in CA1 area and in fascia dentata, *Neurosci. Lett.*, 141, 101, 1992.
107. Sloviter, R.S., Decreased hippocampal inhibition and a selective loss of interneurons in experimental epilepsy, *Science*, 235, 73, 1987.
108. Zhao, D. and Leung, L.S., Effects of hippocampal kindling on paired-pulse response CA1 in vitro, *Brain Res.*, 564, 220, 1991.
109. Barkal, E., Grossman, Y., and Gutnick, M., Long-term changes in neocortical activity after chemical kindling with systemic pentylenetetrazol: an *in vitro* study, *J. Neurophysiol.*, 72, 72, 1994.
110. Bragin, A., Mody, I., Wilson, C.L., and Engel, J., Local generation of fast ripples in epileptic brain, *J. Neurosci.*, 22, 2012, 2002.
111. Fathollahi, Y., Motamedi, F., Semnanian, S., and Zardoshti, M., Examination of persistent effects of repeated administration of pentylenetetrazol on rat hippocampal CA1: evidence from *in vitro* study on hippocampal slices, *Brain Res.*, 758, 92, 1997.
112. Psarropoulou, C., Matsokis, N., Angelatou, F., and Kostopoulos, G., Pentylenetetrazol-induced seizures decrease gamma-aminobutyric acid-mediated recurrent inhibition and enhance adenosine-mediated depression, *Epilepsia*, 35, 12, 1994.
113. Corda, M.G., Giorgi, O., Longoni, B., Orlandi, M., and Biggio, G., Decrease in the function of the g-aminobutyric acid-coupled chloride channel produced by the repeated administration of pentylenetetrazol to rats, *J. Neurochem.*, 55, 1216, 1990.
114. Garcia-Ugalde, G., Galarraga, E., Bargas, J., and Brailowsky, S., Hyperexcitability of hippocampal CA1 region in brain slices after GABA withdrawal, *Neurosci. Lett.*, 147, 229, 1992.
115. Kang, M., Spigelman, I., Sapp, D.W., and Olsen, R.W., Persistent reduction of GABA(A) receptor-mediated inhibition in rat hippocampus after chronic intermittent ethanol treatment, *Brain Res.*, 709, 221, 1996.
116. Xie, X.-H. and Tietz, E.I., Chronic benzodiazepine treatment of rats induces reduction of paired-pulse inhibition in CA1 region of *in vitro* hippocampus, *Brain Res.*, 561, 69, 1991.
117. Zeng, X. and Tietz, E.I., Impairment of feed-forward inhibition in CA1 region of hippocampus after chronic benzodiazepine treatment, *Neurosci. Lett.*, 173, 40, 1994.
118. Denslow, M., Eid, T., Du, F., Schwarcz, R., Lothman, E., and Steward, O., Disruption of inhibition in area CA1 of hippocampus in a rat model of temporal lobe epilepsy, *J. Neurophysiol.*, 86, 2231, 2001.
119. Doherty, J. and Dingledine, R., Reduced excitatory drive onto interneurons in the dentate gyrus after status epilepticus, *J. Neurosci.*, 21, 2048, 2001.
120. Kapur, J., Michelson, H.B., Buterbaugh, G.G., and Lothman, E.W., Evidence for a chronic loss of inhibition in the hippocampus after kindling: electrophysiological studies, *Epilepsy Res.*, 4, 90, 1989.
121. Nusser, Z., Hajos, N., Somogyi, P., and Mody, I., Increased number of synaptic GABA(A) receptors underlies potentiation at hippocampal inhibitory synapses, *Nature*, 395, 172, 1998.
122. Nusser, Z., Sieghart, W., Benke, D., Fritschy, J.-M., and Somogyi, P., Differential synaptic localization of two major g-aminobutyric acid type A receptor a subunits on hippocampal pyramidal cells, *Proc. Natl. Acad. Sci. U.S.A.*, 93, 11939, 1996.
123. Mody, I., De Konick, Y., Pace, J.R., and Soltesz, I., Bridging the cleft at GABA synapses in the brain, *Trends Neurosci.*, 17, 517, 1994.
124. Harrison, N.L., Vicini, S., and Barker, J.L., A steroid anesthetic prolongs inhibitory postsynaptic currents in cultured rat hippocampal neurons, *J. Neurosci.*, 7, 604, 1987.

125. Cooper, E.J., Johnston, G.A., and Edwards, F.A., Effects of a naturally occurring neurosteroid on $GABA_A$ IPSCs during development in rat hippocampal or cerebellar slices, *J. Physiol.*, 521, 437, 1999.
126. Hsu, F.-C., Waldeck, R., Faber, D.S., and Smith, S.S., Neurosteroid effects on GABAergic synaptic plasticity in hippocampus, *J. Neurophysiol.*, 89, 1929, 2003.
127. Cagetti, E., Liang, J., Spigelman, I., and Olsen, R.W., Withdrawal from chronic intermittent ethanol treatment changes subunit composition, reduces synaptic function, and decreases behavioral responses to positive allosteric modulators of GABA-A receptors, *Mol. Pharmacol.*, 63, 53, 2003.
128. Zeng, X., Xie, X.-H., and Tietz, E.I., Reduction of GABA-mediated inhibitory postsynaptic potentials in hippocampal CA1 pyramidal neurons following oral flurazepam administration, *Neuroscience*, 66, 87, 1995.
129. Soltesz, I., Smetters, D.K., and Mody, I., Tonic inhibition originates from synapses close to the soma, *Neuron*, 14, 1273, 1995.
130. Maconochie, D.J., Zempel, J.M., and Steinbach, J.H., How quickly can $GABA_A$ receptors open? *Neuron*, 12, 61, 1994.
131. Williams, S.R., Buhl, E.H., and Mody, I., The dynamics of synchronized neurotransmitter release determined from compound spontaneous IPSCs in rat dentate granule neurones in vitro, *J. Physiol.*, 510, 477, 1998.
132. Jones, M. and Westbrook, G.L., The impact of receptor desensitization on fast synaptic transmission, *Trends Neurosci.*, 19, 96, 1996.
133. McLellan, A.M. and Twyman, R., Receptor system response kinetics reveal functional subtypes of native murine and recombinant human GABA(A) receptors, *J. Physiol.*, 515, 711, 2001.
134. Vicini, S., Ferguson, C., Prybylowski, K., Kralic, J., Morrow, A.L., and Homanics, G.E., GABA-A receptor alpha-1 subunit deletion prevents developmental changes of inhibitory synaptic currents in cerebellar neurons, *J. Neurosci.*, 21, 3009, 2001.
135. Vicini, S., Ferguson, C., Prybylowski, K., Kralic, J., Morrow, A.L., and Homanics, G.E., GABA-A receptor alpha-1 subunit deletion prevents developmental changes of inhibitory synaptic currents in cerebellar neurons, *J. Neurosci.*, 21, 3009, 2001.
136. Poisbeau, P., Feltz, P., and Schlichter, R., Modulation of synaptic $GABA_A$ receptor function by PKA and PKC in adult hippocampal neurons, *J. Physiol.*, 500, 475, 1996.
137. Jones, M.V. and Westbrook, G.L., Shaping of IPSCs by endogenous calcineurin activity, *J. Neurosci.*, 17, 7626, 1997.
138. Mtchedlishvili, Z., Bertram, E., and Kapur, J., Diminished allopregnanolone enhancement of GABA(A) receptor currents in a rat model of chronic temporal lobe epilepsy, *J. Physiol.*, 537, 453, 2001.
139. Sur, C., Farrar, S., Kerby, J., Whiting, P.J., Atack, J.R., and McKernan, R.M., Preferential coassembly of alpha-4 and delta subunits of the GABA-A receptor in rat thalamus, *Mol. Pharmacol.*, 56, 110, 1999.
140. Essrich, C., Lorez, M., Benson, J.A., Fritschy, J.M., and Luscher, B., Postsynaptic clustering of major GABA-A receptor subtypes requires the gamma-2 subunit and gephyrin, *Nat. Neurosci.*, 1, 541, 998.
141. Sundstrom-Poromaa, I., Smith, D.H., Gong, Q., Sabado, T.N., Li, X., Light, A., Wiedmann, M., Williams, K., and Smith, S., Hormonally regulated alpha(4)beta(2)delta GABA(A) receptors are a target for alcohol, *Nat. Neurosci.*, 5, 721, 2002.
142. Saxena, N., Neelands T.R., and Macdonald, R.L., Contrasting actions on lanthanum on different recombinant gamma-aminobutyric acid receptor isoforms expressed in L929 fibroblasts, *Mol. Pharmacol.*, 51, 328, 1997.

143. Bianchi, M.T., Haas, K.F., and Macdonald, R.L., Structural determinants of fast desensitization and desensitization-deactivation coupling in GABAa receptors, *J. Neurosci.*, 21, 1127, 2001.
144. Haas, K.F. and Macdonald, R.L., $GABA_A$ receptor subunit gamma2 and delta subtypes confer unique kinetic properties on recombinant $GABA_A$ receptor currents in mouse fibroblasts, *J. Physiol.*, 514, 27, 1999.
145. Saxena, N.C. and Macdonald, R.L., Assembly of $GABA_A$ receptor subunits: role of the delta subunit, *J. Neurosci.*, 14, 7077, 1994.
146. Adkins, C.E., Pillai, G.V., Kerby, J., Bonnert, T.P., Haldon, C., McKernan, R.M., Gonzalez, J.E., Oades, K., Whiting, P.J., and Simpson, P.B., $GABA_A$ receptors characterized by FRET-derived measurements of membrane potential, *J. Biol. Chem.*, 276, 38934, 2001.
147. Brown, N., Kerby, J., Bonnert, T.P., Whiting, P.J., and Wafford, K.A., Pharmacological characterization of a novel cell line expressing human alpha(4)beta(3)delta GABA(A) receptors, *Br. J. Pharmacol.*, 136, 965, 2002.
148. Ebert, B., Thompson, S.A., Saqunatsou, K., McKernan, R., Krogsgaard-Larsen, P., and Wafford, K., Differences in agonist/antagonist binding affinity and receptor transduction using recombinant human gamma-aminobutyric acid type A receptors, *Mol. Pharmacol.*, 52, 1150, 1997.
149. Gulinello, M., Gong, Q.H., and Smith, S.S., Progesterone withdrawal increases the anxiolytic actions of gaboxadol: role of alpha4betadelta GABA(A) receptors, *Neuroreport*, 14, 43, 2003.
150. Mihic, S.J., Ye, Q., Wick, M.J., Koltchine, V.V., Krasowski, M.D., Finn, S.E., Mascia, M.O., Valenzuela, C.F., Hanson, K.K., Greenblatt, E.P., Harris, R.A., and Harrison, N.L., Sites of alcohol and volatile anaesthetic action on GABA(A) and glycine receptors, *Nature*, 389, 385, 1997.
151. Tobin, M.B., Schmidt, P.J., and Rubinow, D.R., Reported alcohol use in women with premenstrual syndrome, *Am. J. Psychiatry*, 151, 1503, 1994.
152. Follesa, P., Mancuso, L., Biggio, F., Mostallino, M.C., Manca, A., Mascia, M., Busonero, F., Talani, G., Sanna, E., and Biggio, G., Gamma-hydroxybutyric acid and diazepam antagonize a rapid increase in GABA(A) receptors alpha(4) subunit mRNA abundance induced by ethanol withdrawal in cerebellar granule cells, *Mol. Pharmacol.*, 63, 896, 2003.
153. Nusser, Z., Sieghart, W., and Somogyi, P., Segregation of different GABA(A) receptors to synaptic and extrasynaptic membranes of cerebellar granule cells, *J. Neurosci.*, 18, 1693, 1998.
154. Peng, Z., Hauer, B., Mihalek, R.M., Homanics, G.E., Sieghart, W., Olsen, R.W., and Houser, C.R., GABA(A) receptor changes in delta subunit-deficient mice: altered expression of alpha4 and gamma2 subunits in the forebrain, *J. Comp. Neurol.*, 446, 179, 2002.
155. Wei, W., Mody, I., Zhang, N., Peng, Z., and Houser, C.R., Overspill dependent activation and anatomical localization of delta subunit containing GABA-A receptors in the mouse dentate gyrus, *Society for Neuroscience Abstract*, 2002.
156. Smith, S.S. and Shen, H., Evoked current recorded in CA1 hippocampus after neurosteroid exposure exhibits the pharmacological properties of alpha4beta delta GABA-A receptors, *Society for Neuroscience Abstract*, 2002.
157. Isaacson, J., Solis, J., and Nicoll, R., Local and diffuse synaptic actions of GABA in the hippocampus, *Neuron*, 10, 165, 1993.
158. Nusser, Z. and Mody, I., Selective modulation of tonic and phasic inhibitions in dentate gyrus granule cells, *J. Neurophysiol.*, 87, 2624, 2002.

159. Brickley, S.G., Revilla, V., Cull-Candy, S.G., Wisden, W., and Farrant, M., Adaptive regulation of neuronal excitability by a voltage-independent potassium conductance, *Nature*, 409, 88, 2001.
160. Wohlfarth, K.M., Bianchi, M.T., and Macdonald, R.L., Enhanced neurosteroid potentiation of ternary GABA(A) receptors containing the delta subunit, *J. Neurosci.*, 22, 1541, 2002.
161. Belelli, D., Casula, A., Ling, A., and Lambert, J.J., The influence of subunit composition on the interaction of neurosteroids with $GABA_A$ receptors, *Neuropharmacology*, 43, 651, 2002.
162. Stell, B.M., Brickley, S., Farrant, M., Tang, C.Y., and Mody, I., Neurosteroids enhance the tonic conductance generated by delta subunit containing GABA-A receptors, *Society for Neuroscience Abstract*, 2002.

8 GABA, Estrogen, and Sex Differences in the Brain

Margaret M. McCarthy, Joseph L. Nunez, and Tara S. Perrot-Sinal

CONTENTS

8.1 INTRODUCTION

This chapter stretches the boundaries in a discussion of neurosteroids and the $GABA_A$ receptor: first in that the focus is on estradiol, which, despite being synthesized in the brain, is not considered a classic neurosteroid; and second, in that the primary target of regulation is not the $GABA_A$ receptor per se, but rather the consequences of receptor activation to cell function. Nonetheless, as a demonstration of how profoundly important the consequences of steroid modulation of the GABA system can be to neural functioning, this is as good an example as any. In the adult brain, there has been considerable evidence of estradiol modulation of the GABA system to impact on reproductive physiology and behavior, as well as cognitive functions.[1,2] Recently, the fundamental importance of GABA to brain development has become increasingly evident as our appreciation for this simple molecule with complex actions continues to grow. Most important is the detection of GABA's dual nature,

0-8493-2392-4/04/$0.00+$1.50

starting life as an excitatory agent prior to becoming the major source of inhibition in the adult brain. Superimposed on this new appreciation we can add the influence of estradiol, which co-opts the GABA system at a time when its actions are excitatory to exert a masculinizing influence on the developing brain. In this chapter we will review sexual differentiation of the brain and the role of GABA in this process. We will also discuss the recent development of a novel animal model of pediatric brain damage in which GABA is an excitotoxin and its actions are enhanced by estradiol. A mechanistic understanding of estradiol-mediated GABA action serves as an excellent model system for brain development in general, and specifically may provide insight into the etiology of sex differences in neurological disorders and diseases of mental health.

8.2 SEXUAL DIFFERENTIATION OF THE BRAIN

For embryonic mammals, one's ultimate fate as either male or female is a function of whether one's genetic profile will dictate the development of the bipotential gonad to become a testis or an ovary. Once gonadal sex is determined, the survival of either the Wolffian or Mullerian duct system will establish the appropriate genital tract, and circulating hormones will finalize the development of the genitalia and other developmentally established secondary sex characteristics. Among these is the differentiation of the brain into a male or female phenotype. In rodents, we have a considerable understanding of the nature and degree of sex differences in the brain, and the variety of sexually dimorphic physiological and behavioral end points that are a consequence of the differentiation process,[3] but we have a very limited understanding of the cellular and molecular mechanisms establishing these sex differences. Central to this process are the gonadal steroids, which are themselves dimorphic in profile and are the perpetrators of the differentiation process. In the rat, gestation takes 21 to 22 days. Gonadal sex is determined as early as embryonic day 13. Beginning around embryonic day 18, circulating gonadal steroid levels diverge in males and females with the onset of testicular synthesis and secretion of testosterone in males. Steroid levels remain dimorphic for several days after birth until becoming undetectable in both sexes for the duration of the juvenile period. The developmental time course for sexual differentiation is presented schematically in Figure 8.1. In the rodent, the aromatization of testosterone to its bioactive metabolite, estradiol, is a requisite step in the masculinization process.[4] Aromatization of testosterone to estradiol occurs in neurons, the estradiol binds to its cognate receptor, which is a ligand-activated transcription factor, something magical happens, and a male brain develops. Determining what that something magical is has proven a major challenge to the field of neuroendocrinology. Recent work in our laboratory points to sex differences in GABA action as a principal divergence point on the pathway to a male vs. female brain.

8.3 γ-AMINOBUTYRIC ACID (GABA)

GABA is synthesized by the rate-limiting enzyme glutamic acid decarboxylase (GAD), which exists in two forms that differ slightly in size, cellular localization,

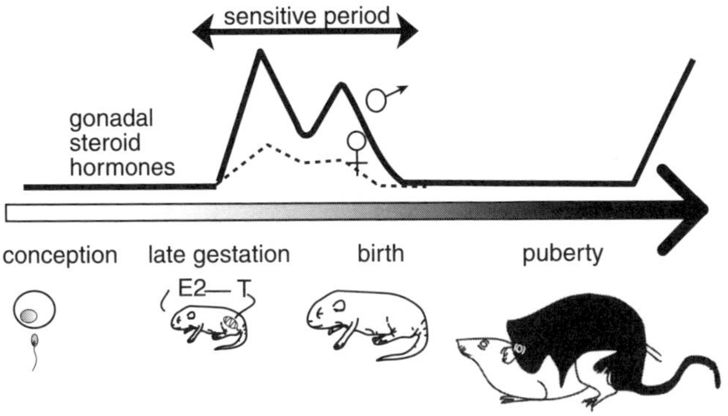

FIGURE 8.1 Sexual differentiation of the brain. Sex differences in the brain are established during a perinatal sensitive period. This developmental window is operationally defined as beginning with the onset of sexually dimorphic hormone exposure, which occurs around embryonic day 18 in the rat with testicular secretion of testosterone in male fetuses. The termination of the sensitive period varies for different end points and is defined by a lack of effect of exogenous hormone treatment to females. These organizational actions of hormones, in particular the aromatized product of testosterone, estradiol, will dictate the adult activational response to gonadal hormones. Adult sexual behavior is stereotypic for males and females and activated by gonadal hormones, but determined by the organizational effects of hormones during the perinatal sensitive period.

and regulation and are known as GAD_{65} and GAD_{67}.[5–7] In the adult brain, GABA acts via one of three receptor subtypes, $GABA_A$, $GABA_B$, or $GABA_C$, and the major action is to either hyperpolarize membranes or alter presynaptic release of other transmitters, resulting in inhibition of neuronal excitation. Nearly every cell in the brain possesses GABA receptors of some sort, and the vital importance of the inhibition imposed by this system is evident in the rapid induction of seizures or seizure-like activity when GABA receptors are antagonized.

8.3.1 Sex Differences in the GABA System during the Perinatal Sensitive Period

Similar to many neurotransmitters, GABA is modulated by gonadal steroids at multiple levels, including the expression and activity of the receptor as well as the rate-limiting enzyme for synthesis of the ligand (for review, see McCarthy et al.[8]). Given the pronounced sex differences in gonadal steroid levels during the perinatal sensitive period, we explored the potential for a sex difference by quantifying mRNA for each form of GAD in discrete brain regions on the day of birth and postnatal day 15. These time points were chosen to reflect the middle of the perinatal sensitive period and postsensitive period, respectively. We found that GAD levels were nearly twice as high in numerous steroid-concentrating regions of the male brain, including the arcuate nucleus (ARC), dorsomedial nucleus (DMN), and CA1 region of the hippocampus.[9] We further used HPLC to quantify GABA levels in the same regions in which GAD had been measured, with the addition of the ventromedial nucleus

(VMN). GABA levels reflected GAD mRNA levels, with concentrations being significantly higher in males than females in the ARC, DMN, CA1, and the VMN. Sex differences are apparent on the day of birth but are absent as early as postnatal day 5. Treatment with testosterone on the day of birth significantly increased GABA in females on postnatal day 1, compared to oil vehicle-injected controls, confirming that this is a steroid-mediated sex difference. Regardless of hormonal status, GABA levels increased developmentally in all brain areas examined.[10] Notably, neither GAD mRNA nor GABA levels were found to be sexually dimorphic in the medial preoptic nucleus (MPN), a steroid receptor-containing brain region that is sexually dimorphic, integral to the control of male copulatory behavior and contains substantial levels of GAD. This observation suggests either that GABA may not be a critical player in the sexual differentiation of the MPN or that the time course of action of GABA in the preoptic area is distinct.

While sex differences in the level of GABA are of obvious importance, a difference in the quantity or quality of the $GABA_A$ receptor would be of equal interest. The $GABA_A$ receptor is a heteromeric pentamer, and variability in subunit composition confers differential kinetics and binding profiles on the channel. The expression of particular subunits varies substantially across development, and we found two to be of particular interest. The $\alpha 1$ and $\alpha 2$ subunits exhibit a developmental switch in several brain regions, with $\alpha 2$ being very high early in development but gradually replaced by $\alpha 1$.[11] This switch occurs in the same cells, suggesting $\alpha 1$ is replacing $\alpha 2$. The time course of the switch is around postnatal days 4 to 6, coinciding with the end of the sensitive period for sexual differentiation. This combination of attributes compelled us to examine the expression of the $\alpha 1$ and $\alpha 2$ subunits for sex differences during the perinatal sensitive period and beyond. The results replicated and extended the previous finding of a developmental switch in $\alpha 1$ and $\alpha 2$ levels, and demonstrated several new and unique patterns of subunit expression in different brain regions. However, there was no evidence for sex differences in the amounts of $\alpha 1$ or $\alpha 2$ or the timing of the switch in expression.[12] A limitation of this study was the focus on two of 16 potential subunits, leaving open the question of whether other changes went undetected. However, the binding of the $GABA_A$ agonist muscimol in distinct nuclei of male and female brains was also assessed across development. There was again no evidence for sex differences in receptor binding,[13] suggesting that changes in levels of the ligand, and not the receptor, constitute the steroid-responsive component of the GABA system during the perinatal sensitive period.

8.3.2 GABA Is Excitatory in the Developing Brain

GABA in the adult brain is the major regulator of neuronal inhibition, but during development is postulated to be the primary source of excitatory drive.[14,15] GABA is one of the earliest transmitters detected in rat brain and is found long before there is synaptic activity in most brain areas. The $GABA_A$ receptor is a chloride ionophore that is selectively permeable to Cl^- and bicarbonate ions. Chloride will flow through the channel in either direction, depending on its concentration gradient. In adulthood, activation of the $GABA_A$ receptor is associated with a hyperpolarizing (inhibitory) effect since the resting anion gradient favors net Cl^- influx at resting membrane potential. Hence, opening of the anion-selective $GABA_A$ ionophore results in accumulated negative charge intracellularly. The increased intracellular negative charge results in hyperpolarization of the membrane, a process that represents the principal mechanism of synaptic inhibition. However, in recent years it has become apparent that variation in the transmembrane anionic gradient can alter the properties of the cell such that activation of the $GABA_A$ receptor can result in a depolarizing or excitatory effect. Select cells in the adult brain retain a depolarizing response to GABA, for instance, dorsal root ganglion cells and some types of glia. More importantly, during development it appears that most, if not all, neurons first respond to GABA with depolarization and gradually develop a hyperpolarizing response as they mature. The excitatory action of GABA in the developing brain has been convincingly demonstrated in dispersed neuronal cultures for the hypothalamus,[16] hippocampus,[17] neocortex,[18,19] cerebellum,[20] and spinal cord.[21]

An obvious question is: How does GABA switch from being excitatory to inhibitory? The transmembrane concentrations for cations such as sodium, potassium, and calcium do not vary significantly between classes of excitable cells, and the intracellular bicarbonate concentration is kept close to the extracellular concentration by the homeostatic mechanisms that control intracellular pH. In contrast, the transmembrane Cl^- concentration gradient varies widely, and this is the basis for differential effects of $GABA_A$ receptor activation across development (reviewed in McCarthy[22]). Variability in intracellular Cl^- concentration is achieved by differential expression of ionic pumps moving ions into and out of cells (see Figure 8.2). In cells where GABA is hyperpolarizing (i.e., inhibitory), electroneutral Cl^- cotransporters pump Cl^- out of the cell so that the equilibrium potential for GABA is kept negative relative to the resting membrane potential. For example, in pyramidal neurons of the hippocampus, the resting membrane potential is around –55 mV and the equilibrium potential for GABA is –60 mV. The K-Cl cotransporter KCC2 is the principal pump that keeps intracellular Cl^- concentrations low in neurons and is widely expressed in the adult brain.[23,24] The presence of KCC2 has been functionally linked to the hyperpolarizing action of GABA, and blocking KCC2 expression with antisense oligonucleotides produces a marked positive shift in the reversal potential of $GABA_A$ responses in mature hippocampal neurons.[25] In contrast, in cells where GABA is depolarizing (i.e., excitatory), NKCC1, a Na-K-2Cl exchange cotransporter, pumps Cl^- into the neurons. Intracellular Cl^- accumulation then leads to an equilibrium potential for Cl^- that is positive to the resting membrane potential. The NKCC1 isoform of the Na-K-2Cl cotransporters is heavily expressed in the brain and developmentally regulated, being high in the first postnatal week and declining thereafter.[26] Elevated NKCC1 appears to be the primary mechanism mediating excitatory GABA early in development, whereas KCC2 is the primary determinant of inhibitory GABA. An additional level of cellular specificity is achieved by the expression of an inwardly rectifying conductive Cl^- channel, ClC-2, which is expressed only in neurons with outward Cl^- transport and activated at membrane potentials more negative than the equilibrium potential for Cl^-. Thus, Cl^- can flow out of, but not in through, ClC-2. This transporter prevents the passive accumulation of Cl^-, such as happens after synaptic activity in the adult, whereas the KCC2 and

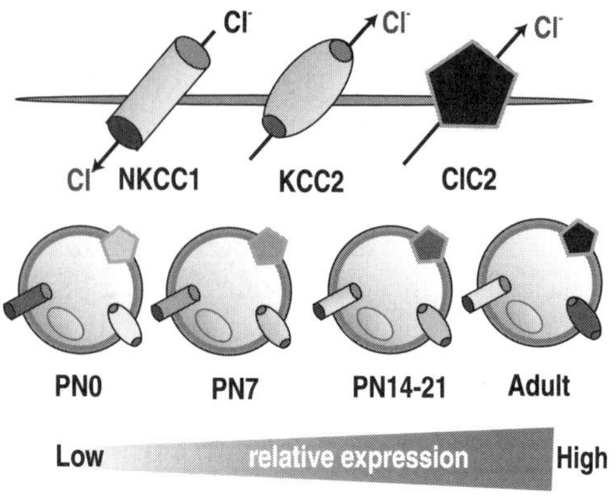

FIGURE 8.2 Chloride cotransporters. The transmembrane concentration gradient of chloride is the primary determinant of a depolarizing vs. hyperpolarizing effect of $GABA_A$ receptor activation. Developmentally, a depolarizing response gradually gives way to hyperpolarization as neurons mature. This change in membrane excitability following $GABA_A$ receptor activation is dictated by changes in the expression and activity of two electroneutral chlorides, potassium and sodium cotransporters. The cotransporter NKCC1 transports chloride into cells and is expressed at high levels in immature neurons but gradually declines as development progresses. In contrast, KCC2, which transports chloride out of cells, is found at very low levels in immature neurons but increases in expression as neurons mature. The inwardly rectifying chloride channel, ClC2, transports chloride out of cells as well and is expressed in mature adult neurons. (Shown as Color Figure 8.2, following page 74.)

NKCC1 channels are required for active extrusion and accumulation of Cl^-, respectively. Nonetheless, if ClC-2 is transiently transfected into neurons with high intracellular Cl^-, such as dorsal root ganglion neurons, GABA is no longer depolarizing in these neurons but rather becomes hyperpolarizing due to shunting inhibition and changes in membrane resistance.[27] By analogy, ClC-2 expression would be predicted to be low in the early postnatal brain but increase progressively beginning around PN 6 to 10, and this is indeed the case in the neocortex.[28] Taken together, these studies demonstrate the critical role played by these three Cl^- transporters in determining whether a particular cell's response to GABA is excitatory or inhibitory.

An additional question frequently raised in discussions of excitatory GABA action during development is: What is the source of the GABA? Synaptic transmission would seem the obvious answer, but in many regions where GABA is excitatory, there is little to no synaptic activity. Evidence that GABA is indeed released is the detection of an early slow current (ESC) in immature hippocampal neurons that is blocked by $GABA_A$ antagonism.[29] Ambient GABA that is released in a nonvesicular, non-calcium-dependent fashion mediates the current, but it remains unknown how the GABA is released and what variables might regulate it. Our previous observation of sex differences in GABA concentration in some brain regions suggests regional specificity, but does not address the source, except that it appears to involve traditional synthesis via GAD. Related to the issue of early GABA action in the absence of synapses is the ability of GABA to reverse its own action. Exogenous administration of muscimol to cultured cortical neurons produced the predicted rise in intracellular calcium, which activates as yet unidentified signal transduction pathways that lead to gene transcription. In particular, the synthesis of KCC2 was increased post-$GABA_A$ receptor activation, decreasing the intracellular chloride concentration and thereby driving GABA toward a hyperpolarizing response.[30] This has been suggested as a mechanism by which GABA prevents overexcitation and promotes synaptic maturation. Preventing overexcitation by GABA when there appears to be no source of inhibition may also be accomplished by the ability of glutamate to antagonize excitatory GABA.[31] Despite these compensatory mechanisms, the potential for GABA to become excitotoxic exists and may be of considerable physiological significance under pathological conditions such as brain injury.

8.3.3 GABA as Both a Neurotrophic and Neurotoxic Agent during Development

The autoregulatory effect of GABA on its own action and the apparent maturation of synapses further add to the concept of GABA as a trophic agent early in development. Central to this trophic action is the influx of calcium, but elevated intracellular calcium is also a principal trigger for neuronal death, both necrotic and apoptotic.[32–34] Excitotoxicity is a global term to characterize the common mediator of neuronal damage following a variety of insults, including hypoxia-ischemia, blunt trauma, and aging-related dementia such as Alzheimer's. In each case it is overexcitation produced by glutamate, whether acutely as in injury or chronically during degenerative disorders, that is the ultimate purveyor of damage. Consequent to the overexcitation is an influx of extracellular calcium and the potential for activation of signal transduction pathways that lead to programmed cell death or apoptosis.[34] The primary source of the increased calcium is influx through the NMDA receptor following activation by glutamate. However, additional routes for calcium entry into the cell include calcium-permeable AMPA receptors and voltage-gated calcium channels found on the membrane. Calcium entering the cell via these routes can also be damaging if in excess, but is generally not considered the primary route leading to cell death.[35]

Brain damage in infants is of considerable clinical interest given the lifelong consequences and the high rates of risk for asphyxia associated with birth. A preferred animal model of early brain damage is the PN7 rat pup, which shows an enhanced sensitivity to glutamate excitation compared to both older and younger animals and is considered analogous to the newborn human.[36–38] The hippocampus is particularly sensitive to excitotoxicity following hypoxia-ischemia. The lack of damage induced by glutamate in the early postnatal period is due to the immature state of the hippocampal glutamatergic system at birth, as the three primary receptor types of the glutamatergic system are not fully functional. Moreover, the kinetic properties of the channel only allow for limited activation and accompanying calcium influx at this age.[39–41] Premature infants are at high risk for intracranial hemorrhage, asphyxia, and the damaging effects of subsequent hypoxia-ischemia (reviewed in McDonald and Johnston[42]). Yet, little attention has been given to the development of an animal model for human prenatal brain damage.

Following a hypoxic event, extracellular glutamate increases due to reversal of the glutamate transporters, as well as release from degenerating necrotic cells.[38] However, there is also a release of substantial amounts of GABA,[43] and this is true in the neonatal brain too.[44] If the 7-day-old rat pup is considered analogous to the newborn human, the newborn rat pup can be considered analogous to the fetal human and a model for premature infants. The observations that excessive release of GABA follows an ischemic event and that $GABA_A$ receptor activation induces calcium influx in the neonatal period led us to hypothesize that excessive activation of the $GABA_A$ receptor early in development would be excitotoxic, and this excitotoxicity would be the result of increased calcium. We have used exogenous administration of muscimol as a novel model of prenatal brain damage mediated by excitoxic GABA to test this hypothesis.

Consistent with the published literature, we also found the newborn rat pup to be relatively impervious to glutamate excitoxicity, although females exhibit damage in the dentate gyrus.[45] However, administration of exogenous muscimol to newborn rat pups, a time when activation of $GABA_A$ receptors is excitatory, increases cell death in the hippocampus proper and the dentate gryus.[46,47] The effects are region specific, persistent, and greater in males. With the use of primary hippocampal cultures, results observed *in vitro* parallel those *in vivo* and argue against a peripherally mediated effect. Furthermore, the damaging effects of muscimol were either completely prevented or attenuated by pretreatment with the L-type calcium channel blocker diltiazem, both *in vivo* and *in vitro*. This alternate pathway for calcium influx is functional on the day of birth, prior to maturation of NMDA and AMPA/KA receptors.[48–50] Calcium entry via voltage-sensitive calcium channels has been documented to be excitotoxic,[51,52] but see Sattler et al.[35] Blockade of the L-type VSCC following neonatal models of glutamate-induced excitotoxicity is partially neuroprotective.[53] Since ischemia in newborn rats induces GABA release,[43] we propose that treatment with muscimol mimics the cell death cascade induced by hypoxia-ischemia in premature human infants.

Of interest to the current discussion are observations in animal models as well as humans that males are more sensitive than females to early brain injury.[54,55] We found that in our model of prenatal brain damage, $GABA_A$-mediated excitotoxity is greater in males than females. The same insult produced a slightly greater loss in males of neurons in the hippocampus proper (8% more) and a substantially greater loss of neurons in the dentate gyrus (43% more) than in females. Evidence for continued cell death in the hippocampus of males indicates that the disparity between males and females may increase as development progresses. More importantly, when animals were tested as juveniles (PN 21) on a simplified version of the Morris water maze, there were deficits in both sexes, but they were significantly more severe in males.[46] Thus males exhibit greater cell death and more functional impairment than females in response to $GABA_A$ receptor activation early in life. Sex differences in exposure to estradiol in neonatal development may be an important basis for the increased sensitivity in males given that newborn males have significantly higher levels of estradiol in the brain than females.[56]

In our model, muscimol was administered on the day of birth and 1 day after, yet dying cells were still evident up to 7 days later. This persistence of cell death for 1 week after insult is consistent with reports documenting elevated levels of cell death for at least 7 days following neonatal hypoxia-ischemia. Excitotoxic cell death in neonates occurs over a continuum, including both apoptosis and necrosis.[57,58] A careful examination of time course of hippocampal cell loss after muscimol exposure using an *in vitro* system indicated an early period of cell death that appears to be necrotic and lasts up to 24 to 48 h, followed by a period of prolonged death similar to apoptosis. When examined *in vivo*, dying neurons were observed 6 days after muscimol treatment in both the male and female rat hippocampus proper, but in the dentate gyrus dying cells were found only in the male. It is possible that cell death persists for an extended period of time due to the lack of appropriate connections being formed within the developing hippocampus.

When ascertaining the basis of a developmental sex difference, hormones are the primary suspects, and in the rodent, estradiol is the primary mediator of most sex differences, with males having higher intracerebral levels. Exogenous administration of estradiol to females elevated the level of cell death following muscimol to that of males, and estradiol given to males further increased damage. When examined *in vitro*, a complex pattern of estradiol action emerged in which the early, and presumably necrotic, component of cell death after muscimol exposure was actually attenuated by estradiol, but the later apoptotic component was markedly enhanced.[59] Figure 8.3 schematically illustrates this mixed action of estradiol, which is consistent with published reports and may reflect an early free-radical scavenging effect followed by a transcriptionally mediated change in some survival factor or cell death-related proteins.[60] Alternatively, estradiol may be acting to alter the properties of GABA action more directly, so that it becomes more depolarizing or allows for greater calcium influx.

In summary, based on these observations, we propose that the excitatory drive through the $GABA_A$ receptor in early development may be an important contributor to some forms of brain damage in premature infants. Understanding the basis of the

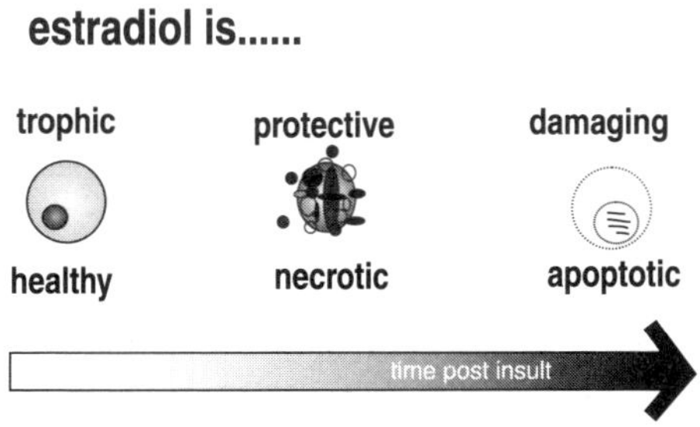

FIGURE 8.3 Estradiol exerts mixed effects on cell death induced by $GABA_A$ receptor activation. In a novel model of pediatric brain damage, we have found that excessive activation of the $GABA_A$ receptor and consequent calcium influx via L-type voltage-gated calcium channels is excitotoxic. Estradiol treatment exacerbates the damage induced by $GABA_A$ receptor activation, but it has a biphasic effect, being protective against the early necrotic phase of cell death but increasing the later apoptotic phase. The precise mechanism(s) for this differential effect is unknown but may involve changes in the chloride cotransporters that determine a depolarizing vs. hyperpolarizing effect of $GABA_A$ receptor activation.

sex difference in sensitivity to GABA-mediated excitoxicity will be essential for therapeutic interventions that are appropriate to the gender of the individual.

8.3.4 Estradiol Enhances Excitatory GABA

The principal effector of depolarizng GABA action is opening of L-type calcium channels and influx of calcium. Monitoring changes in the frequency, magnitude, and duration of intracellular calcium transients provides an accurate picture of the effects of excitatory GABA. This end is achieved with the use of dispersed cultured neurons loaded with the ratiometric calcium indicator fura-2, applying the $GABA_A$ agonist muscimol, and measuring free $[Ca^{2+}]_i$. Analysis of cultured hypothalamic neurons at two different developmental ages reveals that in neurons roughly equivalent in age to embryonic day 19, 85% of neurons respond to muscimol with a rise in $[Ca^{2+}]_i$. In hypothalamic neurons pretreated with estradiol for 2 days, muscimol-induced calcium flux was double the magnitude and the onset of response was hastened by almost 2 sec. Moreover, the mean time required for estradiol-treated neurons to lower $[Ca^{2+}]_i$ after muscimol tended to be longer. In hypothalamic neurons the developmental equivalent of day of birth, the magnitude of the muscimol-induced Ca^{2+} transient and the percentage of cells responding were substantially diminished, consistent with the gradually waning excitatory effect of GABA as development progresses. However, estradiol pretreatment again significantly increased the magnitude of calcium influx and, more importantly, maintained the responsiveness of neurons to muscimol in a significantly greater number of neurons (72% for estradiol vs. 35% for control). Importantly, it was confirmed that the source of increased intracellular free calcium was influx via L-type voltage-gated calcium channels, and not via a novel mechanism induced by estradiol. Thus, prior treatment with estradiol increases the likelihood that hypothalamic cells will respond to GABA with excitation, and in cells that do respond, the rise in $[Ca^{2+}]_i$ is faster, larger, and tends to be of longer duration.[61]

Recently we have extended the analysis of estradiol modulation of GABA-mediated calcium influx to cultured hippocampal neurons. A similar preparation to that used for hypothalamic neurons was employed with the exception that neurons were cultured at a later stage of development, beginning on embryonic day 18 and imaged on what would be considered the developmental equivalent of the day of birth. Muscimol exerted a depolarizing effect in the majority of hippocampal neurons imaged at this age, as evidenced by calcium influx. The source of the increased intracellular free calcium was again confirmed to be via the L-type calcium channel, and the magnitude of the calcium transient was again increased by pretreatment of cultured neurons with estradiol. Representative traces and color-enhanced images of calcium influx into cultured hippocampal neurons are presented in Figure 8.4. The ability of estradiol to enhance calcium influx in hippocampal neurons depolarized by GABA could be the causal variable in sex differences in hippocampal damage following excitatory GABA, as well as the increased cell death observed after estradiol pretreatment *in vivo* and *in vitro*.

When considering the mechanism by which estradiol enhances GABA-mediated calcium influx, the fact that the source of calcium is not novel, but rather the

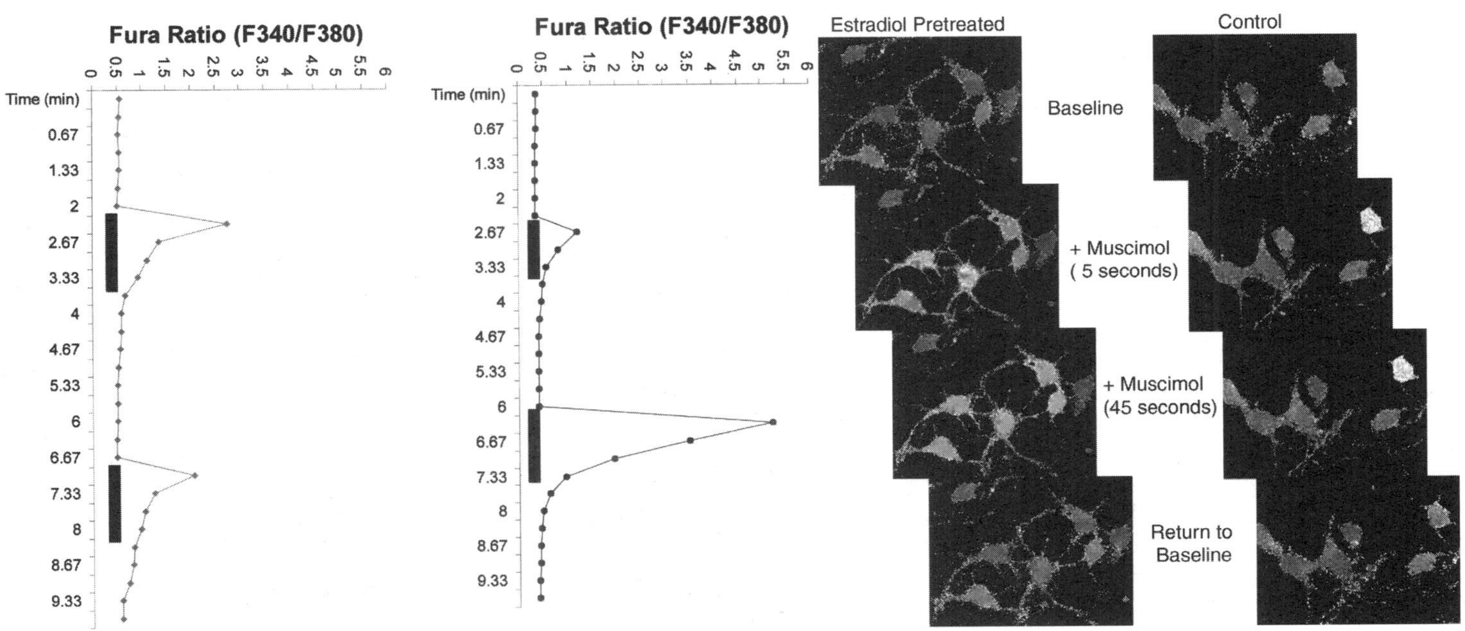

FIGURE 8.4 Estradiol increases calcium influx induced by depolarizing GABA. Intracellular free calcium was monitored with the use of the calcium-sensitive dye Fura-AM. Application of 10 μ*M* glutamate induces a moderate influx of calcium (green bar, left panel, top) compared to equimolar muscimol (red bars). In cells that are pretreated with estradiol (left panel, top), the influx of calcium is about twice the magnitude seen in cells that were not pretreated with estradiol (left panel, bottom). The photomicrograph in the right panel illustrates the time course of calcium flux in estradiol pretreated and control cells, with the lighter color indicating greater activation of Fura by free calcium. (Shown as Color Figure 8.4, following page 74.)

same as that for GABA effects independent of estradiol, leads to logical and testable predictions. Several features of our results are consistent with the hypothesis that estradiol exerts its effects on GABA-induced Ca^{2+} influx by altering the chloride gradient that maintains the depolarizing action of GABA. Estradiol not only decreased the latency to an excitatory response and increased peak cytosolic Ca^{2+} levels, but also *maintained* an excitatory response to muscimol as development proceeded. These results suggest that estradiol-exposed neurons maintained a chloride reversal potential characteristic of neonates, allowing a depolarizing response to $GABA_A$ receptor activation. Given that the excitatory nature of GABA requires changes in intracellular [Cl^-], an assessment of the three transporters responsible for regulating cellular chloride levels can provide an indirect measure of the potential for excitatory vs. inhibitory GABA. To date there has been little attention given to the potential for highly local regional variations in both the magnitude of the excitatory action and the precise timing of the change from excitatory to inhibitory GABA. This is in part due to the challenging technical aspects of measuring depolarizing GABA responses, which requires use of a gramicidin-perforated patch clamp.[14] Comparison across studies suggests there are regional differences, with the hypothalamus being on a relatively fast track for switching to hyperpolarizing GABA responses compared to later-maturing regions, such as the cortex or hippocampus.[22] However, transgenic mice expressing GFP in LHRH neurons of the preoptic area have allowed for an analysis of GABA responses selectively in these neurons, revealing an extraordinarily late switch from depolarizing to hyperpolarizing occurring at puberty[62] or not at all; in other words, GABA is always depolarizing.[63] Thus the potential exists for substantial variation between brain regions and cell types in the timing of the depolarizing/hyperpolarizing hand-off.

An indirect way to assess changes in GABA action is to monitor changes in the expression or activity of the Cl^- cotransporters that are responsible for maintaining the transmembrane concentration gradient. Quantification of both the mRNA and protein levels for NKCC1 indicates marginally but significantly higher levels in the male hypothalamus than in females on embryonic day 20 and day of birth (Figure 8.5). This electroneutral cotransporter maintains a high intracellular Cl^- concentration by transporting in one Na^+ and one K^+ ion along with two Cl^- ions. No reliable sex differences were detected in KCC2, which transports chloride out of the cell and increases in activity with advancing development. The relatively small magnitude of the sex difference in NKCC1 suggests that there may be other factors at play. Quantification of protein and mRNA unfortunately provides no information on the activity of the cotransporters, which may be a more critical regulatory variable than the actual amount of cotransporters.

In summary, the ability of estradiol to enhance GABA-mediated depolarization-induced calcium influx via L-type voltage-gated calcium channels has been established for both hypothalamic and hippocampal neurons. What remains is to establish the precise mechanism by which estradiol is achieving this end. Increased synthesis of the cotransporter that maintains high intracellular chloride induced by estradiol is one plausible and likely mechanism, but the potential for additional and undiscovered mechanisms of estradiol action exist.

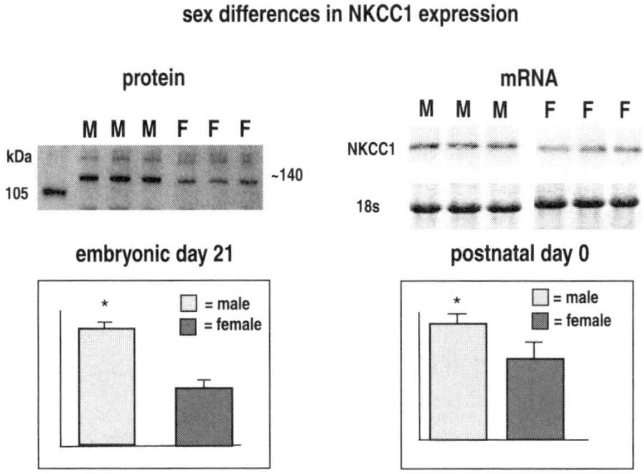

FIGURE 8.5 NKCC1 levels are higher in the hypothalamus of neonatal males. The cotransporter NKCC1 transports chloride into cells and increases the ability of $GABA_A$ receptor activation to induce a depolarizing excitatory response. Analysis of NKCC1 protein levels by Western blot of gestational day 21 fetuses indicates higher levels in the hypothalamus of males than in females. Analysis of mRNA levels by Northern blot on the day of birth indicates that the sex difference persists to this developmental time point. (Shown as Color Figure 8.5, following page 74.)

8.3.5 $GABA_A$ Receptor Activation Increases Phosphorylation of CREB in Males but Decreases It in Females

Establishing the ce*lular consequences of sex differences in GABA content and depolarizing GABA action is a requisite step in determining the mechanisms of estradiol-mediated sexual differentiation of the brain. The transcription factor cyclic AMP response element binding (CREB) protein is a well-established target for activation by calcium-induced kinases.[64] Phosphorylation of critical serine residues on the CREB protein results in its activation and induction of gene transcription. Antibodies that selectively detect the phosphorylated form of CREB, referred to as pCREB, allow for the identification of individual cells in particular brain regions that have experienced an influx of calcium and activation of signal transduction pathways leading to CREB-mediated transcription. Exogenous administration of a bolus of muscimol to newborn male and female pups and quantification of pCREB-expressing neurons 1 h later allowed for an *in vivo* assay of excitatory GABA as well as a comparison between the sexes. The induction of pCREB following muscimol was dramatic in several steroid-concentrating regions of the brain, including the MPN, VMN, ARC, and CA1 field of the hippocampus. There was no change in pCREB in the CA3 field or in certain subnuclei of the thalamus. More importantly, the dramatic increase in pCREB-expressing neurons after $GABA_A$ receptor activation was seen consistently only in male brains. In newborn females, exogenous muscimol resulted in a decrease in the number of neurons expressing pCREB in all of the same regions in which it was seen to increase in males, with the exception of the arcuate nucleus, where there was an increase in females that was of the same magnitude of that seen in males.[65] Pretreatment of pups with the L-type calcium channel blocker nimodipine further highlighted the sex difference in responsiveness to excitatory GABA. In males, preventing calcium influx exerted lawful and predictable effects; i.e., the phosphorylation of CREB was blocked. However, in females, the apparent dephosphorylation of CREB observed was in some cases enhanced by pretreatment with nimodipine, in some regions prevented, and still in others left unaltered.[66] Thus two important conclusions can be reached. First is that there is a divergence in the responsiveness of males and females to $GABA_A$ receptor activation at birth such that in males there appears to be activation of kinases leading to the increased phosphorylation of CREB. In females, however, the response downstream to $GABA_A$ receptor opening appears to be the activation of phosphatases and the subsequent dephosphorylation and hence deactivation of CREB. This represents a genuine divergence point in the signal transduction pathways activated in males vs. females and may represent the critical juncture in responsiveness leading to brain sexual phenotype (Figure 8.6). The second important conclusion is that the cellular

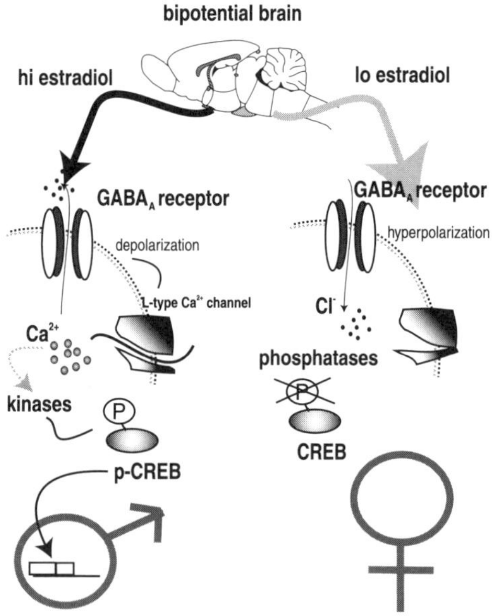

FIGURE 8.6 GABA action is a primary divergence point in sexual differentiation of the brain: activation of $GABA_A$ receptors results in either hyperpolarization or depolarization of the membrane as a function of the transmembrane chloride concentration gradient. Estradiol, which masculinizes the brain during development, appears to increase the cotransporter that regulates high intracellular chloride. This results in a greater depolarizing response to GABA in a larger number of cells and for a longer developmental period. Depolarizing GABA and the consequent calcium influx activates an as of yet unidentified signal transduction pathway that leads to the phosphorylation of the transcription factor CREB. In the absence of high levels of estradiol, as seen in the female brain, GABA reverts to hyperpolarizing sooner; opening of the $GABA_A$ receptor results in the dephosphorylation of CREB and via unknown pathways will ultimately produce a feminized brain phenotype.

processes leading to a male brain are revealing themselves in a straightforward manner, whereas those underlying the female brain appear to be far more complex and, given the lack of a clear hormonal trigger, likely to be far harder to elucidate.

8.3.6 $GABA_A$ Receptor Activation Mediates Sexually Dimorphic Astrocyte Morphology

A focus on neurons is natural when considering the neurotransmitter GABA, which is made exclusively in neurons and acts at postsynaptic receptors. However, astrocytes express fully functional $GABA_A$ receptors and respond to exogenous administration of $GABA_A$ agonists and antagonists with changes in morphology. An appreciation for the fundamental importance of glia, and in particular astrocytes, is continuing to grow, and it has been long established that astrocytes in neuroendocrine brain regions exhibit substantial plasticity in response to steroids and other physiologically relevant stimuli.

A hallmark of astrocyte maturation is increased stellation and synthesis of glial fibrillary acidic protein (GFAP), an intermediate filament protein that is readily visible by immunocytochemistry.[67] Astrocytes have traditionally been viewed as secondary to neurons, acting as support cells or scavengers of metabolic waste and excess neurotransmitters. As appreciation for the importance of astrocytes has grown, so has awareness that these cells behave in a regionally specific manner. Astrocytes of one brain region express unique complements of signaling molecules that are exclusively relevant to local neurons.[68–70] Also apparent is that hormonal modulation of astrocytes is also region specific. One of the first and most prominent examples of hormonal modulation of astrocyte morphology is in the arcuate nucleus of the adult female rat. The amount of surface area covered by astrocytic processes changes dramatically across the estrous cycle, and this modification regulates the density of inhibitory GABAergic synapses.[71] Evidence suggests that this "phasic synaptic remodeling" involves physical changes in astrocyte morphology and is central to the control of the luteinizing hormone (LH) surge at proestrus.[72] Naturally occurring changes in astrocyte morphology across the cycle are mimicked by exogenous estradiol administration to ovariectomized rats,[73,74] implicating this gonadal steroid as the mediator of this process. Estradiol is also central to the sexual differentiation of arcuate astrocytes in the developing brain. Beginning as early as the day of birth and continuing throughout life, males exhibit more stellate astrocytes with longer processes and increased branching. Administration of testosterone or estradiol, but not the nonaromatizable androgen dihydrotestosterone, to females results in an astrocyte morphology indistinguishable from that of gonadally intact males.[75–77]

Astrocytes throughout the brain are exquisitely sensitive to changes in their surroundings, and they are careful monitors of the local environment, generally not extending beyond a 50-μm radius. They express a host of functional neurotransmitter receptors, including $GABA_A$ receptors.[78–80] In mature neurons, the reversal potential for chloride is negative to the resting membrane potential, and opening of the $GABA_A$ receptor constitutes the basis of most synaptic inhibition. However, in astrocytes, the reversal potential for chloride is positive to the resting membrane potential, and activation of these $GABA_A$ receptors has the opposite effect, depolarization of the astrocytic membrane, a situation highly analogous to that observed for immature neurons. In acutely isolated astrocytes, activation of the $GABA_A$ receptor and subsequent membrane depolarization is of sufficient magnitude to activate voltage-dependent Ca^{2+} channels, resulting in a transient increase in intracellular Ca^{2+} concentrations.[78] Study of acutely isolated brain slices further supports GABA's ability to elicit increases in the internal concentration of Ca^{2+} within astrocytes,[81] and *in vivo* observations suggest that neuronal activity regulates astrocyte function and physiology. In astrocytic-neuronal co-cultures, astrocyte differentiation is blocked by the addition of bicuculline, a $GABA_A$ receptor antagonist. Moreover, when astrocytes are cultured in the absence of neurons GABA and muscimol (both induce differentiation within 24 h), this response is blocked by bicuculline. Baclofen, a $GABA_B$ receptor agonist, has no effect, suggesting that GABA mediates astrocyte differentiation exclusively through its ionotropic receptor.[82]

Both GABA release and $GABA_A$ receptor activation are detected in hypothalamic neurons during late embryonic and early postnatal development. As previously mentioned, the neonatal GABAergic system is sensitive to the hormonal milieu such that males and masculinized females exhibit higher levels of both GAD mRNA and GABA in various loci, including the arcuate nucleus.[83,84] During the same developmental period, astrocytes in the arcuate nucleus are becoming differentiated and sexually dimorphic such that arcuate astrocytes in masculinized brains are morphologically more complex (increased number of processes with an increased length and more frequent branching) than arcuate astrocytes in feminized brains. To date, the presence of estrogen receptors in astrocytes of the neonatal rat arcuate has not been demonstrated (reviewed in Mong and McCarthy[85]); leading to the question of how estradiol mediates changes in arcuate astrocyte morphology. Based on the observation that estrogen receptors are present in the neurons of the neonatal arcuate,[86] including GABAergic neurons,[87] and given that neonatal males exhibit higher levels of GABA, we hypothesized that gonadal steroid-induced astrocyte differentiation is mediated through neuronally released GABA. An *in vivo* analysis indicates that GABAergic signaling is a necessary component of gonadal steroid-induced astrocyte differentiation. When GAD_{65} and GAD_{67} protein levels in the neonatal hypothalamus are reduced with intrahypothalamic administration of oligonucleotides antisense to GAD mRNA, the gonadal steroid-induced masculinization of astrocytes in the arcuate is completely blocked during the first few days of life. Conversely, muscimol administered to females induces astrocyte differentiation in the absence of exogenous steroids. Thus, steroid-induced astrocyte differentiation in the neonatal arcuate of masculinized animals requires elevated GABA signaling.[88] Given that GABA is synthesized only in neurons and its synthesis is increased by estradiol, it must be acting as a diffusible factor inducing the differentiation of neighboring astrocytes (see Figure 8.7). It remains unknown how the released GABA is directing the increase in astrocytic process length and process number. It is also unclear whether this phenomenon is exclusive to the arcuate nucleus or also occurs in other sexually dimorphic brain regions. Astrocytes in the neonatal preoptic area exhibit a sexual dimorphism in morphology similar to that seen in the arcuate nucleus,[89] with males having more complex cells. In this region, prostaglandin E2 is partially involved in the sexual differentiation of astrocytes, but the potential for a GABAergic

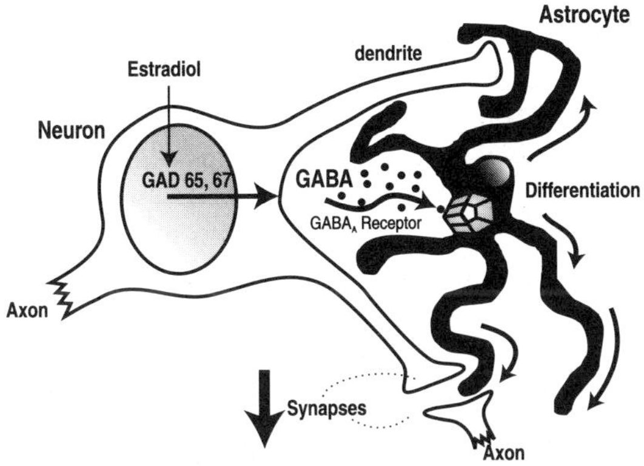

FIGURE 8.7 GABA mediates estradiol-induced sex differences in astrocyte morphology. Astrocytes of the neonatal arcuate nucleus are sexually dimorphic, with males having more complex cells with longer processes that branch more frequently than those of females. Estradiol is the mediator of this sex difference and acts by increasing GABA synthesis in adjacent neurons that release the GABA, which then acts on astrocytes to induce their differentiation.

component has not been explored. There are no other reports of sexually dimorphic astrocyte populations in the brain, but this is more likely the result of no one having looked, rather than not existing.

8.4 SUMMARY AND FUTURE PERSPECTIVES

As is apparent from this review, GABA is centrally involved in the process of sexual differentiation of the brain. Both the amount and the outcome of GABA action are modulated by estradiol, the principal hormone mediating masculinization in the rodent brain. Males possess more GABA in many brain regions, and this GABA exerts a greater depolarizing effect for a longer period of time developmentally. Activation of $GABA_A$ receptors leads to the activation of the transcription factor CREB in males and may lead to its active deactivation in females. When the activation of $GABA_A$ receptors becomes excessive, as occurs after hypoxia or injury, excitoxity can ensue and cause substantial cell death in the hippocampus. This damage is worse in males and may be a consequence of estradiol-induced enhancement of depolarizing GABA action. However, when $GABA_A$ receptor activation is not excessive, it is often trophic. This includes an action on a nonneuronal cell type, astrocytes of the arcuate nucleus. These cells are sexually dimorphic, being more complex in males, and this complexity is mediated by GABA. Thus, GABA action in the developing brain is potently modulated by steroids, in particular estradiol, and has a myriad of effects that exert profound and permanent influences on the developing brain.

Traditionally steroid hormones act via their ligand-dependent transcription factor receptors to interact directly with DNA and alter gene expression. An awareness of nontraditional mechanisms of estradiol action has increased enormously in the past decade. Distinguishing the route of estradiol effects on the GABA system has not been a topic of investigation, but parsimony would suggest that simple induction of gene transcription could account for all of the effects currently observed. As few as two genes could be modulated to alter the sexual differentiation process: (1) GAD, resulting in synthesis of more GABA, and (2) NKCC1, to alter the transmembrane chloride gradient so that $GABA_A$ receptor activation is more depolarizing. No change in the amount or subunit composition of the $GABA_A$ receptor is required or has been detected, although the possibility of alterations to the receptor cannot be ruled out. However, this model does require that estradiol act in both the presynaptic GABA-synthesizing cell and the postsynaptic $GABA_A$ receptor-expressing cell. Given the dense distribution of estrogen receptors within sexually dimorphic brain regions, this is a feasible scenario but remains to be confirmed.

One thing that has become clear in attempting to elucidate the cellular and molecular mechanisms of estradiol-mediating sexual differentiation of the brain is that regional variation is paramount. How is this regional specificity established and what role does GABA play? Differences in the timing and magnitude of the depolarizing-to-hyperpolarizing shift of GABA action are one potential source of variation, although determining this will be technically challenging. Cellular diversity could also exist in the signal transduction pathways activated downstream to calcium influx, a possibility that also exists as an intrinsic difference between the two sexes. Sex differences may also compound each other. Initial differences in GABA action, linked to preexisting differences in signal transduction pathways and the establishment of sexually dimorphic synaptic patterns, could assure divergence between the two sexes so that a genuinely distinct brain phenotype is established for males vs. females. Establishing a role for GABA is surely just the first step in what will no doubt prove to be a complex process involving numerous and, as of yet, unpredictable factors.

REFERENCES

1. McCarthy, M.M., Pfaff, D.W., and Schwartz-Giblin, S., The role of steroid modulation of amino acid transmitters in the regulation of female reproduction, *Am. Zool.*, 33, 275–284, 1993.
2. Murphy, D.D., Cole, N.B., Greenberger, V., and Segal, M., Estradiol increases dendritic spine density by reducing GABA neurotransmission in hippocampal neurons, *J. Neurosci.*, 18, 2550–2559, 1998.
3. Simerly, R.B., Development of sexually dimorphic forebrain pathways, in *Sexual Differentiation of the Brain*, Matsumoto, A., Ed., CRC Press, Boca Ratan, FL, 2000, pp. 175–202.
4. Naftolin, F., Ryan, K.J., Davies, I.J., Reddy, V.V., Flores, F., Petro, Z., Kuhn, M., White, R.J., Takaoka, Y., and Wolin, L., The formation of estrogens by central neuroendocrine tissues, *Recent Prog. Horm. Res.*, 31, 295–319, 1975.
5. Erlander, M.G. and Tobin, A.J., The structural and functional heterogeneity of glutamic acid decarboxylase: a review, *Neurochem. Res.*, 16, 215–226, 1991.
6. Esclapez, M., Tillakaratne, N.J.K., Kaufman, D.L., Tobin, A.J., and Houser, C.R., Comparative localization of two forms of glutamic acid decarboxylase and their mRNAs in rat brain supports the concept of functional differences between the forms, *J. Neurosci.*, 14, 1834–1855, 1994.

7. Kaufman, D.L., Houser, C.R., and Tobin, A.J., Two forms of the γ-aminobutyric acid synthetic enzyme glutamate decarboxylase have distinct intraneuronal distributions and cofactor interactions, *J. Neurochem.*, 56, 720–723, 1991.
8. McCarthy, M.M., Davis, A.M., and Mong, J.A., Excitatory neurotransmission and sexual differentiation of the brain, *Brain Res. Bull.*, 44, 487–495, 1997.
9. Davis, A.M., Grattan, D.R., Selmanoff, M.K., and McCarthy, M.M., Sex differences in glutamic acid decarboxylase mRNA in neonatal rat brain: implications for sexual differentiation, *Horm. Behav.*, 30, 538–552, 1996.
10. Davis, A.M., Ward, S.C., Selmanoff, M., Herbison, A.E., and McCarthy, M.M., Developmental sex differences in amino acid neurotransmitter levels in hypothalamic and limbic areas of rat brain, *Neuroscience*, 90, 1471–1482, 1999.
11. Fritschy, J.-M., Paysan, J., Enna, A., and Mohler, H., Switch in expression of rat GABA-A receptor subtypes during postnatal development: an immunohistochemical study, *J. Neurosci.*, 14, 5302–5324, 1994.
12. Davis, A.M., Penshuck, S., Fritschy, J.-M., and McCarthy, M.M., Developmental switch in the expression of $GABA_A$ receptor subunits α1 and α2 in the hypothalamus and limbic system of the rat, *Dev. Brain Res.*, 119, 127–138, 2000.
13. Davis, A.M. and McCarthy, M.M., Developmental increase in [3H]muscimol binding to the $GABA_A$ receptor in the rat excludes the ventromedial nucleus of the hypothalamus, *Neurosci. Lett.*, 288, 223–227, 2000.
14. Owens, D.F., Boyce, L.H., Davis, M.B.E., and Kriefstein, A.R., Excitatory GABA responses in embryonic and neonatal cortical slices demonstrated by gramicidin perforated-patch recordings and calcium imaging, *J. Neurosci.*, 16, 6414–6423, 1996.
15. Ben-Ari, Y., Tseeb, V., Raggozzino, D., Khazipov, R., and Gaiasa, J.L., Gamma-aminobutyric acid (GABA): a fast excitatory transmitter which may regulate the development of hippocampal neurones in early postnatal life, *Prog. Brain Res.*, 102, 261–273, 1994.
16. Obrietan, K. and van den Pol, A.N., GABA neurotransmission in the hypothalamus: developmental reversal from Ca2+ elevating to depressing, *J. Neurosci.*, 15, 5065–5077, 1995.
17. Cherubini, E., Rovira, C., Gaiarsa, J.L., Corradetti, R. and Ben-Ari, Y., GABA mediated excitation in immature rat CA3 hippocampal neurons, *Int. J. Dev. Neurosci.*, 8, 481–490, 1990.
18. Obata, K., Excitatory and trophic actions of GABA and related substances in newborn mice and organotypic cerebellar culture, *Dev. Neurosci.*, 19, 117–119, 1997.
19. van den Pol, A.N., Excitatory actions of GABA after neuronal trauma, *J. Neurosci.*, 16, 4283–4292, 1996.
20. Rego, A., Lambert, J.J., and Nicholls, D.G., Developmental profile of excitatory GABA(A) responses in cultured rat cerebellar granule cells, *Neuroreport*, 12, 477–482, 2001.
21. Rohrbough, J. and Spitzer, N.C., Regulation of intracellular Cl^- levels by Na^+-dependent Cl^- cotransport distinguishes depolarizing from hyperpolarizing $GABA_A$ receptor-mediated responses in spinal neurons, *J. Neurosci.*, 16, 82–91, 1996.
22. McCarthy, M.M., Auger, A.P., and Perrot-Sinal, T.S., Getting excited about GABA and sex differences in the brain, *Trends Neurosci.*, 25, 307–312, 2002.
23. Clayton, G.H., Owens, G.C., Wolff, J.S., and Smith, R.L., Ontogeny of cation-Cl– cotransporter expression in rat neocortex, *Dev. Brain Res.*, 109, 281–292, 1998.
24. Kanaka, C., Ohno, K., Okabe, A., Kuriyama, K., Itoh, T., Fukuda, A., and Sato, K., The differential expression patterns of messenger RNAs encoding K-Cl cotransporters (KCC1,2) and Na-K-2Cl cotransporter (NKCC1) in the rat nervous system, *Neuroscience*, 104, 933–946, 2001.

25. Rivera, C., Voipio, J., Payne, J.A., Ruusuvuori, E., Lahtinen, H., Lamsa, K., Pirvola, U., and Saarma, M., The K+/Cl– co-transporter KCC2 renders GABA hyperpolarizing during neuronal maturation, *Nature*, 397, 251–255, 1999.
26. Plotkin, M.D., Kaplan, M.R., Peterson, L.N., Gullans, S.R., Herbert, S.C., and Delpire, E., Expression of the Na-K-2Cl cotransporter BSC2 in the nervous system, *Am. J. Physiol.*, 272, C173–C183, 1997.
27. Staley, K., Smith, R., Schaack, J., Wilcox, C., and Jentsch, T.J., Alteration of GABA-A receptor function following gene transfer of the CLC-2 chloride channel, *Neuron*, 17, 543–551, 1996.
28. Clayton, G.H., Staley, K.J., Wilcox, C.L., Owens, G.C., and Smith, R.L., Developmental expression of CLC-2 in the rat nervous system, *Dev. Brain Res.*, 108, 307–318, 1998.
29. Demarque, M., Represa, A., Becq, H., Khalilov, I., Ben-Ari, Y., and Aniksztejn, L., Paracrine intracellular communication by a Ca2+- and SNARE-independent release of GABA and glutamate prior to synapse formation, *Neuron*, 36, 1051–1061, 2002.
30. Ganguly, K., Schinder, A.F., Wong, S.T., and Poo, M., GABA itself promotes the developmental switch of neuronal GABAergic responses from excitation to inhibition, *Cell*, 105, 521–532, 2001.
31. van den Pol, A.N., Gao, X.-B., Patrylo, P.R., Ghosh, P.K., and Obrietan, K., Glutamate inhibits GABA excitatory activity in developing neurons, *J. Neurosci.*, 18, 10749–10761, 1998.
32. Nakamura, T., Minamisawa, H., Katayama, Y., Ueda, M., Terashi, A., Nakamura, K., and Kudo, Y., Increased intracellular Ca^{2+} concentration in the hippocampal CA1 area during global ischemia and reperfusion in the rat: a possible cause of delayed neuronal death, *Neuroscience*, 88, 57–67, 1999.
33. Simon, R.P., Griffiths, T., Evans, M.C., Swan, J.H., and Meldrum, B.S., Calcium overload in selectively vulnerable neurons of the hippocampus during and after ischemia: an electron microscopic study in the rat, *J. Cereb. Blood Flow Metab.*, 4, 350–361, 1984.
34. Choi, D.W., Excitotoxic cell death, *J. Neurobiol.*, 23, 1261–1276, 1992.
35. Sattler, R., Charlton, M.P., Hafner, M., and Tymianski, M., Distinct influx pathways, not calcium load, determine neuronal vulnerability to calcium neurotoxicity, *J. Neurochem.*, 71, 2349–2364, 1998.
36. McDonald, J.W., Silverstein, F.S., and Johnston, M.V., Neurotoxicity of N-methyl-D-aspartate is markedly enhanced in developing rat central nervous system, *Brain Res.*, 459, 200–203, 1988.
37. McDonald, J.W., Trescher, W.H., and Johnston, M.V., Susceptibility of brain to AMPA induced excitotoxicity transiently peaks during early postnatal development, *Brain Res.*, 583, 54–70, 1992.
38. Vannuci, R.C., Brucklacher, R.M., and Vannucci, S.J., CSF glutamate during hypoxia-ischemia in the immature rat, *Dev. Brain Res.*, 118, 147–151, 1999.
39. Bickler, P.E., Gallego, S.M., and Hansen, B.M., Developmental changes in intracellular calcium regulation in rat cerebral cortex during hypoxia, *J. Cereb. Blood Flow Metab.*, 13, 811–819, 1993.
40. Cheng, C., Fass, D.M., and Reynolds, I.J., Emergence of excitotoxicity in cultured forebrain neurons coincides with larger glutamate-stimulated [Ca^{2+}] increases and NMDA receptor mRNA levels, *Brain Res.*, 849, 97–108, 1999.
41. Marks, J.D., Bindokas, V.P., and Zhang, X.M., Maturation vulnerability to excitoxicity: intracellular mechnisms in cultured postnatal hippocampal neurons, *Dev. Brain Res.*, 124, 101–116, 2000.

42. McDonald, J.W. and Johnston, M.V., Physiological and pathophysiological roles of excitatory amino acids during central nervous system development, *Brain Res. Rev.*, 15, 41–70, 1990.
43. Andine, P., Orwar, O., Jacobson, I., Sandberg, M., and Hagberg, H., Changes in extracellular amino acids and spontaneous neuronal activity during ischemia and extended reflow in the CA1 of the rat hippocampus, *J. Neurochem.*, 57, 222–229, 1991.
44. Cataltepe, O., Towfighi, J., and Vannucci, R.C., Cerebrospinal fluid concentrations of glutamate and GABA during perinatal cerebral hypoxia-ischemia and seizures, *Brain Res.*, 709, 326–330, 1996.
45. Hilton, G.D., Nunez, J.L., and McCarthy, M.M., Sex differences in response to kainic acid and estradiol in the hippocampus of newborn rats, *Neuroscience*, 117, 383–391, 2003.
46. Nunez, J.L., Alt, J., and McCarthy, M.M., A novel model for prenatal brain damage. II. Long-term deficits in hippocampal cell number and hippocampal dependent behavior following neonatal $GABA_A$ receptor activation, *Exp. Neurol.*, 181, 270–280, 2003.
47. Nunez, J.L., Alt, J., and McCarthy, M.M., A new model for prenatal brain damage. I. $GABA_A$ receptor activation induces cell death in developing rat hippocampus, *Exp. Neurol.*, 181, 258–269, 2003.
48. Jones, O.T., Bernstein, G.M., Jones, E.J., Jugloff, D.G., Law, M., Wong, W., and Mills, L.R., N-Type clacium channels in the developing rat hippocampus: subunit, complex and regional expression, *J. Neurosci.*, 17, 6152–6164, 1997.
49. Jones, S.W., Overview of voltage-dependent calcium channels, *J. Bioenerg. Biomembr.*, 30, 299–312, 1998.
50. Chameau, P., Lucas, P., Melliti, K., Bournaud, R., and Shimahara, T., Development of multiple calcium channel types in cultured mouse hippocampal neurons, *Neuroscience*, 90, 383–388, 1999.
51. Choi, D.W., Calcium-mediated neurotoxicity: relationship to specific channel types and role in ischemic damage, *Trends Neurosci.*, 14, 515–519, 1988.
52. Siesjo, B.K., Calcium in the brain under physiological and pathological conditions, *Eur. Neurol.*, 30, 3–9, 1990.
53. Gunn, A.J., Williams, C.E., Millard, C.E., Tan, W.K.M., and Gluckman, P.D., Flunarizine, a calcium channel anatagonist, is partially prophylactically neuroprotective in hypoxic-ischemic encephalopathy in the fetal sheep, *Ped. Res.*, 35, 657–663, 1994.
54. Simon, N.P., Long-term neurodevelopmental outcome of asphyxiated newborns, *Clin. Perinatol.*, 26, 767–778, 1999.
55. Naeye, R.L., Burt, L.S., Wright, D.L., Blanc, W.A., and Tatter, D., Neonatal mortality, the male disadvantage, *Pediatrics*, 48, 902–906, 1971.
56. Rhoda, J., Corbier, P., and Roffi, J., Gonadal steroid concentrations in serum and hypothalamus of the rat at birth: aromatization of testosterone to 17β-estradiol, *Endocrinology*, 114, 1754–1760, 1984.
57. Nakajima, W., Ishida, A., Lange, M.S., Gabrielson, K.L., Wilson, M.A., Martin, L.J., Blue, M.E., and Johnston, M.V., Apoptosis has a prolonged role in the neurodegeneration after hypoxic ischemia in the newborn rat, *J. Neurosci.*, 20, 7994–8004, 2000.
58. Portera-Cailliau, C., Price, D.J., and Martin, L.J., Excitotoxic neuronal death in immature brain is an apoptosis-necrosis morphological continuum, *J. Comp. Neurol.*, 378, 70–87, 1997.
59. Nunez, J.L. and McCarthy, M.M., Estradiol exacerbates hippocampal damage in a model of preterm brain injury, *Endocrinology*, 2003, 144, 2350–2359, 2003.

60. Harms, C., Lautenschlager, M., Bergk, A., Katchanov, J., Freyer, D., Kapinya, K., Herwig, U., Megow, D., Weber, J. R., and Hortnagl, H., Differential mechanisms of neuroprotection by 17 β-estradiol in apoptotic versus necrotic neurodegeneration, *J. Neurosci.*, 21, 2600–2609, 2001.
61. Perrot-Sinal, T.S., Davis, A.M., Gregerson, K.A., Kao, J.P.Y., and McCarthy, M.M., Estradiol enhances excitatory gamma-aminobutyric acid-mediated calcium signaling in neonatal hypothalamic neurons, *Endocrinology*, 143, 2238–2243, 2001.
62. Han, S.K., Abraham, I.M., and Herbison, A.E., Effect of GABA on GnRH neurons switches from depolarization to hyperpolarization at puberty in the female mouse, *Endocrinology*, 143, 1459–1466, 2002.
63. DeFazio, R.A., Heger, S., Ojeda, S.R., and Moenter, S.M., Activation of A-type gamma-aminobutyric acid receptors excites gonadotropin-releasing hormone neurons, *Mol. Endocrinology*, 16, 2872–2891, 2002.
64. Sheng, M., McFadden, G., and Greenberg, M.E., Membrane depolarization and calcium induced c-fos transcription via phosphorylation of transcription factor CREB, *Neuron*, 4, 571–582, 1990.
65. Auger, A.P., Perrot-Sinal, T.S., and McCarthy, M.M., Excitatory versus inhibitory GABA as a divergence point in steroid-mediated sexual differentiation of the brain, *Proc. Natl. Acad. Sci. U.S.A.*, 98, 8059–8064, 2001.
66. Perrot-Sinal, T.S., Auger, A.P., and McCarthy, M.M., Excitatory GABA-induced pCREB in developing brain is mediated by L-type Ca+2 channels and dependent on age, sex and brain region, *Neuroscience*, 116, 995–1003, 2003.
67. Eng, L.F., GFAP: the major protein of glial intermediate filaments in differentiated astrocytes, *J. Neurosci.*, 8, 203–214, 1985.
68. Gooday, D., Retinal axons in Xenopus laevis recognize differences between tectal and diencephalic glial cells in vitro, *Cell Tissue Res.*, 259, 595–598, 1990.
69. Le Roux, P.D. and Reh, T.A., Regional differences in glial-derived factors that promote dendritic outgrowth from mouse cortical neurons in vitro, *J. Neurosci.*, 14, 4639–4655, 1994.
70. Le Roux, P.D. and Reh, T.A., Independent regulation of primary dendritic and axonal growth by maturing astrocytes in vitro, *Neurosci. Lett.*, 198, 5–8, 1995.
71. Parducz, A., Perez, J., and Garcia-Segura, L.M., Estradiol induces plasticity of GABAergic synapses in the hypothalamus, *Neuroscience*, 53, 395–401, 1993.
72. Garcia-Segura, L.M., Cardona-Gomez, G.P., Trejo, J.L., Fernandez-Galaz, M.C., and Chowen, J.A., Glial cells are involved in organizational and activational effects of sex hormones in the brain, in *Sexual Differentiation of the Brain*, Matsumoto, A., Ed., CRC Press, Boca Raton, FL, 2000, pp. 83–93.
73. Garcia-Segura, L.M., Canas, B., Parducz, A., Rougon, G., Theodosis, D., Naftolin, F., and Torres-Aleman, I., Estradiol promotion of changes in the morphology of astroglia growing in culture depends on the expression of polysialic acid of neural membranes, *Glia*, 13, 209–216, 1995.
74. Garcia-Segura, L.M., Olmos, G., Robbins, R.J., Hernandez, P., Meyer, J.H., and Naftolin, F., Estradiol induces rapid remodelling of plasma membranes in developing rat cerebrocortical neurons in culture, *Brain Res.*, 498, 339–343, 1989.
75. Mong, J.A., Glaser, E., and McCarthy, M.M., Gonadal steroids promote glial differentiation and alter neuronal morphology in the developing hypothalamus in a regionally specific manner, *J. Neurosci.*, 19, 1464–1472, 1999.
76. Mong, J.A., Kurzweil, R.L., Davis, A.M., Rocca, M.S., and McCarthy, M.M., Evidence for sexual differentiation of glia in rat brain, *Horm. Behav.*, 30, 553–562, 1996.

77. Mong, J.A. and McCarthy, M.M., Ontogeny of sexually dimorphic astrocytes in the neonatal rat arcuate, *Dev. Brain Res.*, 139, 151–158, 2002.
78. Fraser, D.D., Duffy, S., Angelides, K.J., Perez-Velazquez, J.L., Kettenmann, H., and MacVicar, B.A., $GABA_A$/benzodiazepine receptors in acutely isolated hippocampal astrocytes, *J. Neurosci.*, 15, 2720–2732, 1995.
79. Fraser, D.D., Mudrick-Donnon, L.A., and MacVicar, B.A., Astrocytic GABA receptors, *Glia*, 11, 83–93, 1994.
80. Poulter, M.O. and Brown, L.A., Transient expression of $GABA_A$ receptor subunit mRNAs in the cellular processes of cultured cortical neurons and glia, *Brain Res. Mol. Brain Res.*, 69, 44–52, 1999.
81. Kang, J., Jiang, L., Goldman, S.A., and Nedergaard, M., Astrocyte-mediated potentiation of inhibitory synaptic transmission, *Nat. Neurosci.*, 1, 683–692, 1998.
82. Matsutani, S. and Yamamoto, N., Neuronal regulation of astrocyte morphology in vitro is mediated by GABAergic signaling, *Glia*, 20, 1–9, 1997.
83. Davis, A.M., Grattan, D.R., Selmanoff, M., and McCarthy, M.M., Sex differences in glutamic acid decarboxylase mRNA in neonatal rat brain: implications for sexual differentiation, *Horm. Behav.*, 30, 538–552, 1996.
84. Davis, A.M., Ward, S.C., Selmanoff, M., Herbison, A.E., and McCarthy, M.M., Developmental sex differences in amino acid neurotransmitter levels in hypothalamic and limbic areas of rat brain, *Neuroscience*, 90, 1471–1482, 1999.
85. Mong, J.A. and McCarthy, M.M., Steroid-induced developmental plasticity in hypothalamic astrocytes: implications for synaptic patterning, *J. Neurobiol.*, 40, 602–619, 1999.
86. Shughrue, P.J., Lane, M.V., and Merchenthaler, I., Comparative distribution of estrogen receptor-alpha and beta mRNA in the rat central nervous system, *J. Comp. Neurol.*, 388, 507–525, 1997.
87. Leranth, C., Sakamoto, H., Maclusky, N.J., Shanabrough, M., and Naftolin, F., Estrogen responsive cells in the arcuate nucleus of the rat contain glutamic acid decarboxylase (GAD): an electron microscopic immunocytochemical study, *Brain Res.*, 331, 376–381, 1985.
88. Mong, J.A., Nunez, J.L., and McCarthy, M.M., GABA mediates steroid-induced astrocyte differentiation in the neonatal rat hypothalamus, *J. Neuroendocrinol.*, 14, 1–16, 2002.
89. Amateau, S.K. and McCarthy, M.M., Sexual differentiation of astrocyte morphology in the developing rat preoptic area, *J. Neuroendocrinol.*, 14, 904–910, 2002.

9 FGIN-1-X: A Series of Novel Specific and Efficacious Ligands for the Peripheral Benzodiazepine Receptor

Ethan M. Skowronski-Lutz and Daniel Bitran

CONTENTS

ABSTRACT

Synthesized in 1992, the FGIN-1 derivatives (2-aryl-indole-3-acetamides) were the first series of compounds demonstrated to have a high efficacy in stimulating

0-8493-2392-4/04/$0.00+$1.50

steroidogenesis while also showing a high and selective binding affinity for the peripheral benzodiazepine receptor (PBR), a binding site located on the outer mitochondrial membrane. The PBR is a poorly characterized binding site in peripheral organs and the brain. Its function, while multifaceted, is strongly associated with substrate-based regulation of neurosteroidogenesis via control of cholesterol transport across the mitochondrial membranes. FGIN-1-X compounds, most notably FGIN-1-27 and FGIN-1-44, show significant anxiolytic, anticonvulsant, antiamnestic, and antidepressant activity in rats and mice. In this manner the FGIN-1 derivatives demonstrate a pharmacological profile very similar to that of the nongenomically active neurosteroids such as 3α-OH-5α-pregnan-20-one (allopregnanolone), 5α-pregnan, 21-3α-diol-20-one (allotetrahydrodeoxycorticosterone (THDOC)), and pregnenolone sulfate — products of FGIN-1-X-induced neurosteroidogenesis. FGIN-1-27 has also been demonstrated to inhibit cell proliferation and potentiate apoptosis in three separate lines of colorectal cancer and to potentiate induction of cellular necrosis in cancerous fibroblasts by tumor necrosis factor α (TNFα). This review examines the current literature pertaining to FIGN-1 derivatives and examines the proposed physiological roles of the PBR *in vivo*. The evidence suggests that FGIN-1 derivatives may have significant clinical potential in treating a variety of psychological and nonpsychological disorders, and that further research into this potential is warranted. Furthermore, the unique pharmacological profile of FGIN-1 derivatives gives it critical importance in the further elucidation of the physiological role of the PBR.

9.1 INTRODUCTION

The 2-aryl-indole-3-acetamide series of compounds were the first agonists discovered that bind to the peripheral benzodiazepine receptor exclusively and with a high affinity, and that strongly promote PBR-mediated steroidogenesis, making them highly specific and efficacious ligands.[1] The first accounts of these drugs (hereafter referred to as FGIN-1-X compounds because their discovery was made under the auspices of the now defunct Fidia Georgetown Institute of Neuroscience) were published in 1992 and 1993.[1,2] Since then, researchers have used the FGIN-1-X series of compounds to characterize the PBR and further elucidate its roles in the body. Despite the unique pharmacological profile of FGIN-1-X compounds and their ability to reveal a great deal about the important, but poorly understood, PBR, there is, to date, no literature review of FGIN-1-X compounds. This paper investigates the FGIN-1-X series of compounds by review of all available literature pertaining directly to FGIN-1 derivatives grounded in an understanding of the complexities and difficulties surrounding research into the PBR.

9.2 PHARMACOLOGY AND STRUCTURE–ACTIVITY STUDIES OF FGIN-1 COMPOUNDS

Based on the structure of the imidazopyridines alpidem and zolpidem, but lacking a nitrogen atom at the ring fusion position (Figure 9.1A), FGIN-1 derivatives consist of a 2-aryl-indole-3-acetamide core and are prepared by a Fischer indole synthesis. While only three papers have attempted to characterize the structure–activity relationship of FGIN-1-X and FGIN-1-X-related compounds,[1–3] a substantial number of chemical variants have been studied with respect to the structure–activity relationships governing their binding affinity to numerous receptors; their *in vitro* and *in vivo* efficacy, as judged by mediation of steroidogenic activity; and their behavioral effects in rats, using a number of different animal models of anxiolytic drug effects. FGIN-1 derivatives with the following substitutions have been studied: (1) substitutions of straight alkyl chains ranging from CH_3 to $C_{10}H_{21}$ at the amide nitrogen (R^1 and R^2 in Figure 9.1A), (2) hydrogen and halogen substitutions to the C5 position and para-position of the benzene rings (X and Z, respectively, in Figure 9.1A), and (3) replacement of the aryl moiety with a benzofuran.[1,3] Complete descriptions of the chemical and pharmacological properties of all the derivatives tested are not

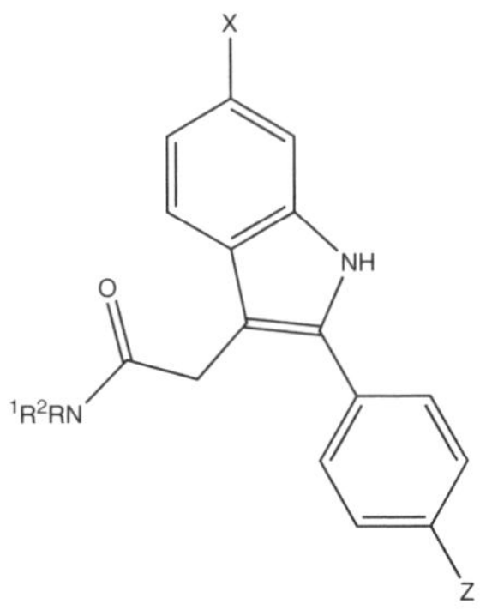

FGIN-1 Derivatives

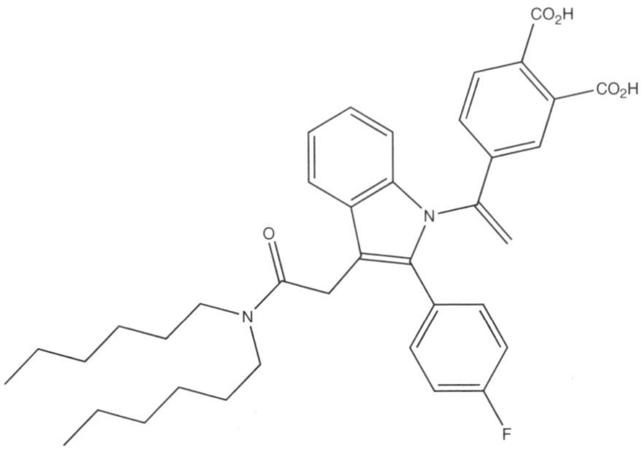

FGIN-1-44

FIGURE 9.1 (A) Structure of FGIN-1 derivatives — FGIN-1-27 and FGIN-1-43. FGIN-1-27 is comprised of an n-hexyl group at R1 and R2, a hydrogen at X, and a fluorine at Z; FGIN-1-43 is comprised of an n-hexyl group at R1 and R2, and a chlorine at X and Z. (B) FGIN-1-44 — this is a FGIN-1-27 molecule with a benzofuran group substitution to the indole nitrogen.

available, however, as only a subset of the FGIN-1 derivatives bound to the PBR with sufficient affinity to warrant further testing.

The binding affinity of FGIN-1 and its analogues to the PBR, as measured by displacement of [^{3}H]PK 11195 and [^{3}H]Ro 5-4864 from cerebellar astroglia, is evidently dependent on three main variables: the presence or absence of the two alkyl groups on the amide nitrogen, the chain length of the alkyl groups, and the halogenation of the benzene rings.[2] Halogenation of both rings significantly improved the binding affinity of the ligand, as did the presence of the larger alkyl groups on the amide nitrogen, six carbons being the ideal length. Substitution of hexyl groups for methyl groups altered binding affinity by a 30- to 70-fold increase. A model FGIN-1 derivative, FGIN-1-27, displays many of these properties (Figure 9.1). FGIN-1-43 displayed the highest binding affinity, but is not deemed a model FGIN-1 derivative, as it suffered from poor steroidogenic performance *in vivo*, presumably because it cannot penetrate the blood–brain barrier due to low lipophilicity (Figure 9.1). The FGIN-1 derivatives with the highest binding affinities bound in the nanomolar range. All other substitutions displayed no marked effect on binding affinity of the ligand for the PBR. Extensive structure–activity tests performed on 2-aryl-substituted benzofuran 3-acetamide and 3-aryl-3-pyrrolylpropanamides produced results that largely parallel those of the FGIN-1 structure–activity relationship tests.[3,4]

The pharmacological specificity of FGIN-1 derivatives has been extensively documented.[1] Several FGIN-1 derivatives, in concentrations up to 1 μM, failed to displace [^{3}H]flumazenil, [^{3}H]zolpidem, [^{3}H]TBPS, or [^{3}H]GABA from their relative binding sites on the $GABA_A$ receptor complex. In concentrations up to and exceeding 1 μM, FGIN-1 derivatives reportedly failed to displace [^{3}H]baclofen from $GABA_B$ receptors; [^{3}H]spiperone from dopamine and 5-HT_1 receptors; [^{3}H]ketanserin from 5-HT_2 receptors; [^{3}H]glycine, [^{3}H]dibenzocyclohepteneimine, or [^{3}H]-amino-3-hydroxy-5-methyl-4-isoxazoleproprionic acid from glutamate NMDA receptors; [^{3}H]naloxone or [^{3}H]3-(3-hydroxylphenyl)-N-(1-propyl)piperidine from opiate and σ receptors; [^{3}H]3-L-365,260 from cholecystokinin receptors; [^{125}I]pindolol from β-adrenergic receptors; and [^{3}H]CP-55,940 from canabinoid receptors. Together, these experiments demonstrate the significant selectivity of FGIN-1 compounds for the PBR receptor, with a binding affinity difference of nearly three orders of magnitude for the best-performing FGIN-1 derivatives.

9.3 LOCALIZATION AND FUNCTION OF THE PERIPHERAL BENZODIAZEPINE RECEPTOR

The significance of FGIN-1 compounds is made salient in a consideration of the putative physiological functions mediated by the PBR. The mammalian body possesses two distinct types of benzodiazepine (BDZ) receptors, often referred to as central and peripheral benzodiazepine receptors. The central benzodiazepine receptor (CBR) is by far the more studied and better understood of the two. It is located primarily at the interface of the α and γ subunits of the hetero-oligomeric $GABA_A$ receptor,[5] and thus is almost exclusively localized to neurons in the central nervous system (CNS).[6,7] Stimulation of the CBR by an agonist increases GABA binding affinity and thereby facilitates GABAergic inhibition of synaptic currents in most associated cells. This property is believed to be at the root of the actions of most prescription benzodiazepine drugs, namely, sedation, anxiolysis, and anticonvulsant activity.[8,9] The peripheral benzodiazepine receptor is not as well understood and appears to possess a greater diversity of physiological functions and cellular and subcellular locales than the CBR. The PBR was originally discovered in peripheral tissues,[10] hence its name, yet it also exists in significant numbers in the CNS.[11,12] The exact functions of the PBR on either the tissue or cellular level are not yet known, and its physiological role in the body appears to be complex and significantly varied, further obfuscating the therapeutic potential of drugs that target the PBR.

The dearth of knowledge regarding the PBR and its physiological role is not simply due to lack of research. Though far more attention has been paid to the CBR, the site of action for many commonly used therapeutic agents, our understanding of the PBR has been additionally hindered by two greater concerns. The first is that the subcellular localization of the PBR and its functional role on the cellular level are multifaceted. While the PBR's most common location is understood to be in the outer mitochondrial membrane,[13–15] where it is believed to regulate steroidogenesis,[16–18] it is also found in significant quantities in nonmitochondrial compartments such as the nuclear envelope[19,20] and cellular membrane.[19,21,22] Correspondingly, in addition to its role in steroidogenesis, the PBR has been implicated in regulation of mitochondrial respiration,[23–25] gliosis,[26,27] inflammatory response,[28,29] nerve regeneration,[30] insulin secretion,[31] Ca^{2+} influx,[23,32,33] anion transport,[23,34–36] apoptosis[37,38] and the formation of the mitochondrial permeability pore (MPT), a relatively large protein channel that is thought to be regulated by the PBR to allow for cholesterol transport across the inner mitochondrial membrane.[39–42]

The second major obstacle to a clear understanding of the functional consequences of PBR activation lies in the fact that researchers have been relatively bereft of the highly specific drugs necessary for this task. Until very recently, the only drugs that bound with any significant affinity to the PBR were the isoquinoline carboxamides (most notably PK 11195 and PK 14105), the imidazopyridines (alpidem and its congeners), and 4′-chlorodiazepam (4′CD), also known as Ro 5-4864. The isoquinoline carboxamides were limited in their utility because PK 11195, by most accounts, possesses no appreciable agonist effects, as measured by the ligand's potentiation of mitochondrial steroidogenesis of pregnenolone and its derivatives in cells to which the ligands bind.[43] A handful of research has demonstrated that PK 11195 induced some steroidogenesis, but these effects are presumed to be due to a very low level of intrinsic agonist efficacy on the part of PK 11195 that is normally concealed by endogenous agonists.[44–46] If PK 11195 does possess any *de facto* agonist properties, they are very weak with respect to the PBRs tested thus far and inappropriate for a great deal of research into the normal actions of the PBR. Problems with PK 14105 stem from both a failure to elicit notable steroidogenic effects and irreversible binding to the PBR.[47] Ro 5-4864 and some of the imidazopyridines (alpidem mainly) are potent agonists of the PBR with a high (n*M*) binding affinity, but they are also nonspecific and possess a lower (μ*M*) affinity for the chloride ionophore itself, where Ro 5-4864 exerts antagonistic effects,[45,48] and alpidem and many of its congeners exert agonistic effects.[49] As a result, any *in vivo* and, in some cases, *in vitro* work using these ligands is indeterminately skewed, precluding proper study of PBR function.

9.4 LOCALIZATION OF THE PERIPHERAL BENZODIAZEPINE RECEPTOR

Most PBR localization studies have been performed with either [^{3}H]PK 11195 or [^{3}H]Ro 5-4864. While the significant difference in binding affinity of Ro 5-4864 for PBR vs. the GABA-gated chloride ionophore provides a fair margin of error in the interpretation of many localization studies, caution is still advised until studies that use Ro 5-4864 are performed with more specific ligands, such as FGIN-1-27, PK 11195, or the highly specific, fluorescent probe 7-nitro-2,1,3-benzoxadiazol-4-yl, recently developed by Kozikowski and coworkers.[50] The existence of so-called low-affinity binding sites for PBR agonists, however, does not seem to be subject to this particular criticism,[51–54] and their existence has even been implicated through the use of FGIN-1-27.[55]

The gross anatomical localization of the PBR is characteristically more diverse than that of the CBR. The CBR is found almost solely in the CNS, and the highest densities, in ascending order, are found in the spinal cord, medulla oblongata-pons, striatum, hippocampus, midbrain, cerebellum, hypothalamus, and cerebral cortex.[6,7] The PBR, as characterized by [^{3}H]PK 11195, [^{3}H]PK 14105, and [^{3}H]Ro 5-4864, is found in both the CNS and in peripheral tissues. In peripheral tissues, PBRs are found in their highest densities in the adrenal cortex[56,57] and also in high densities in the heart,[58] smooth muscle,[59] kidney and lung,[10] testis and uterus,[57] blood platelets,[60] red blood cells,[61] monocytes,[62] macrophages,[63] and many immune system organs.[64,65] Areas of the CNS displaying the highest level of radioligand binding for PBR include, in no particular order, the choroid plexus, pineal gland, olfactory bulb, olfactory nerve, frontal cortex, cerebellum, and hippocampus.[11,27,66,67] PBR expression in these areas is typically associated with glial and ependymal cells.[67,68] Some neuronal localization of the PBR has been evidenced, but appears to be far less common.[35,66,69,70]

9.5 INVOLVEMENT OF THE PERIPHERAL BENZODIAZEPINE RECEPTOR IN STEROIDOGENESIS

Apparently, in most tissues and in the brain especially, the receptive portion of the PBR is located in the outer membrane of mitochondria. This location coincides with the PBR's best-documented role: mediation of steroidogenesis. Occupation of the mitochondrial PBR (MBR) seems to facilitate the influx of cholesterol through the mitochondrial membranes into the inner cistae of the mitochondria, where it becomes available for conversion into the steroid precursor pregnenolone by the cytochrome P450scc enzyme, localized on the inner mitochondrial membrane.[71–73] The aqueous filling between the outer and inner mitochondrial membranes appears to deter lipophillic cholesterol molecules from entering the inner mitochondrial space, decreasing the availability of cholesterol, and probably many other molecules, to enzymes located in the inner mitochondrial membrane and matrix.[74] It is not cytochrome P450scc activity per se that is rate limiting in mitochondrial steroidogenesis, but rather substrate availability, which is determined by the rate of cholesterol transport to the inner mitochondrial space.[75] Thus, in theory, the MBR controls the rate of steroidogenesis via its control of cholesterol influx.[72,75,76]

The pregnenolone produced during neurosteroidogenesis is almost immediately processed into a number of nongenomically active, neuromodulatory steroids, including 3α-OH-5α-pregnanane-20-one (allopregnanolone) and allotetrahydrodeoxycorticosterone (alloTHDOC), which allosterically potentiate $GABA_A$ receptor function, and pregnenolone sulfate, which allosterically attenuates $GABA_A$ receptor function and is a precursor to allopregnanolone.[77–82] The specific mechanisms regulating pregnenolone's conversion into its various processing products are poorly understood. Nonetheless, pregnenolone synthesis or administration seems to consistently result in a net positive allosteric modulation of $GABA_A$ receptors in the brain.[43,83,84]

9.6 *IN VIVO* AND *IN VITRO* STEROIDOGENIC EFFICACY OF FGIN-1 COMPOUNDS

The capacity of FGIN-1 derivatives to positively modulate steroidogenesis in cultured C6-2B glioma cells is very closely correlated to the binding affinity of the compounds for the PBR.[85] FGIN-1-27, initially the most efficacious ligand *in vivo,* by a small margin, increased pregnenolone synthesis nearly 2.5-fold with an EC_{50} of ~3.3 n*M*, higher than Ro 5-4864, but slightly lower than diazepam binding inhibitor (DBI). The steroidogenic effects of FGIN-1 derivatives are subject to complete antagonism by both PK 11195 and the P450scc inhibitor α-aminoglutethimide, strongly supporting the conclusions that the PBR is the site of action for FGIN-1 derivatives and that FGIN-1-induced steroidogenesis is a result of *de novo* synthesis in mitochondria. FGIN-1-27 also potentiated steroidogenesis in at least one nonmammalian system. Snyder and Van Antwerpen[86] used FGIN-1-27 to induce the *de novo* synthesis of ecdysteroids, steroids critical to maturation in many insects, *in vivo* in the prothoracic glands (PGs) of the insect *Manduca sexta.* The demonstration of FGIN-1-27-induced steroidogenesis in an arthropod not only suggests the evolutionary conservation of a PBR-like protein, but also demonstrates that the steroidogenic efficacy of FGIN-1-27 is maintained in evolutionarily disparate biological systems.

Unfortunately, the steroidogenic potency of FGIN-1-27 in *in vivo* studies in mammalian systems is nearly one tenth its potency in some *in vitro* tests.[85] This difference has been attributed to poor bioavailability, presumably due to the low aqueous solubility of the compound. In an attempt to overcome this shortcoming, a novel FGIN-1 derivative was made by addition of a 1,2,4-benzenetricarboxylic acid group to the indole nitrogen of FGIN-1-27 (Figure 9.1B), producing FGIN-1-44. This increases the solubility of the compound in blood, thereby increasing its bioavailability and greatly increasing its *in vivo* potency. Interestingly, FGIN-1-44 is prevented from stimulating the PBR receptor itself due to interference by the 1,2,4-benzenetricarboxylic acid radical group. However, *in vivo* FGIN-1-44 is hydrolyzed by an unknown enzyme to FGIN-1-27, which in turn stimulates the PBR. This was confirmed by the near absence of FGIN-1-44 from brain and plasma of treated rats after approximately 4 h and the corresponding appearance of FGIN-1-27 in the brain and plasma of the same animals. The utility of this addition stems not only from the increased potency of the drug *in vivo*, but also from the option of safer intravenous administration at high doses, due to the greater aqueous solubility of the compound.

Some disagreement exists as to the relative efficacy of FGIN-1-27 in the brain vs. the periphery of the body. Korneyev et al.[43] examined circulating pregnenolone levels in blood from the brain and periphery and saw a significant increase in steroid levels only in the brain, implying low PBR-mediated steroidogenic potential in peripheral tissues. Similar studies carried out by Bitran et al.[84] measured the levels of the circulating pregnenolone metabolite allopregnanolone and demonstrated a parallel increase in steroidogenesis in both the brain and periphery. It is important that this discrepancy be addressed, as periphery-derived steroids can have profound effects on animal metabolism.[87–89] However, a major difference in methodologies, the use of trilostane, may reconcile the differing results. Trilostane, which does not readily cross the blood–brain barrier, prevents conversion of pregnenolone into its metabolites, allowing researchers to determine the rate of steroidogenesis by measurement of a single variable. Because Korneyev et al.[43] only injected trilostane into the brain, it is quite possible that significant amounts of peripherally produced pregnenolone were processed into various pregnenolone metabolites. As only pregnenolone levels were measured, the rate of steroidogenesis in the periphery may have been misjudged, masking the steroidogenic effect of FGIN-1-27 in that context. Bitran et al.,[84] on the other hand, did not use trillostane pretreatment, and they directly measured the levels of allopregnanolone in both the brain and periphery,[84] which may have given more accurate measures of steroidogenesis in the periphery. This seems to indicate that FGIN-1-27 does not suffer reduced efficacy in the periphery of the body relative to the brain. It remains for future research to resolve this issue.

9.7 EFFECTS OF FGIN COMPOUNDS ON APOPTOSIS AND CELL PROLIFERATION

The implication that the PBR may partially regulate the mitochondrial permeability pore[39–42] suggests, according to some lines of research, that the PBR may also regulate the induction of apoptosis, which may involve formation of the mitochondrial permeability pore.[38,55,90–92] Indeed, the PBR has been detected in elevated numbers in some cancers,[19,93–96] and the application of PBR ligands *in vitro* in micromolar concentrations inhibited cell proliferation in cultures of astrocytoma,[97] melanoma,[98] breast cancer,[99,100] and testis cancer.[21] In nanomolar concentrations PK 11195 and Ro 5-4864 have been found to increase DNA synthesis and cellular proliferation in C6 glioma cells,[101] suggesting that the PBR may be involved in normal cell proliferation and that the apoptotic effect seen upon application of PBR ligands' micromolar quantities may be taking advantage of a natural mechanism to prevent excess cell proliferation.

Maaser et al.[55] demonstrated that FGIN-1-27, as well as PK 11195 and Ro 5-4864, inhibited cell proliferation, dose dependently, in three separate lines of human colorectal cancer. All three ligands also increased caspase-3 activity, a key element in the apoptotic signaling pathway,[102] and induced cell cycle arrest in the G_1/G_0

phase.[55] The last observation is in contrast to the observed induction of cell cycle arrest at the G_1/S and G_2/M phases in breast cancer[99] and at the G_2/M phase in lung cancer[103] and melanoma cells,[98] implying that there may be tissue-specific transduction of PBR-mediated signals. FGIN-1-53, a steroidogenically inert indole acetamide, did not noticeably affect any of the three colorectal cancer lines, supporting the conclusion that FGIN-1-27's pro-apoptotic effects were mediated by the potentiation of steroidogenesis.[55]

FGIN-1-27 also potentiated, but was not sufficient to cause, cell death induced by tumor necrosis factor α in L929 fibroblasts.[38] FGIN-1-27's potentiation of cellular necrosis was antagonized by a combination of cyclosporin (CyA), an inhibitor of mitochondrial permeability pore formation, and aristolochic acid (ArA), a phospholipase A_2 inhibitor. The authors interpreted these results as evidence that FGIN-1-27 is capable of inducing the mitochondrial permeability pore in L929 fibroblasts and that it is by this mechanism that FGIN-1-27 potentiated TNFα-induced cellular necrosis.[38]

9.8 ANXIOLYTIC-LIKE EFFECTS OF FGIN-1 COMPOUNDS

Given that FGIN-1-27 results in the synthesis of many neurosteroids with potent positive allosteric effects on the $GABA_A$ receptor,[77,79,80,104–106] it is not surprising that anxiolysis, characteristic of $GABA_A$ receptor stimulation, is also a characteristic effect of FGIN-1-27 *in vivo*. Numerous FGIN-1 derivatives, foremost among them FGIN-1-27 and FGIN-1-44, have significant and powerful anxiolytic-like properties in the plus-maze, Vogel conflict test, and passive avoidance paradigm where the derivatives displayed properties very similar to those of Ro 5-4864 and alloTHDOC.[1–3,43,84,85,107,108] The anxiolytic-like effects of FGIN-1-27 were present even after adrenalectomy and castration of the rodents, implying that the behavioral effects are at least not dependent upon peripheral sources of neurosteroidogenesis.[85,108] Intracranial co-infusion of PK 11195 prevented the anxiolytic-like effects of intrahippocampal FGIN-1-27 administration, demonstrating the pharmacological specificity of FGIN-1-27-induced effects to the neurosteroidogenesis that results from PBR activation.[84] Moreover, pretreatment with 4-MA (a 5α-reductase inhibitor) or picrotoxin (a $GABA_A$ receptor-gated chloride ionophore blocker) prevented the anxiolytic-like effect of FGIN-1-27, showing that subsequent to PBR stimulation, the formation of allopregnanolone and its action at the $GABA_A$ receptor complex was responsible for the FGIN-1-27-induced anxiolytic effect.[84]

FGIN-1-27's anxiolytic-like properties may also mediate its effects on pentobarbital-induced sleep in socially isolated mice.[109] Socially isolated mice display a shorter duration of pentobarbital-induced sleep, relative to group-housed mice. The sleep period of isolated mice is normalized (i.e., prolonged) by administration of FGIN-1-27. PK 11195 antagonizes this effect, presumably as a result of the displacement of FGIN-1-27 from the PBR; however, PK 11195 alone lacks any intrinsic effects. FGIN-1-27, on the other hand, has no effects on the duration of pentobarbital-induced sleep in group-housed mice. In contrast to the lack of effect of PK 11195 on pentobarbital-induced sleep duration in isolated mice, PK 11195 decreases sleep duration in group-housed mice. The latter effect is antagonized by FGIN-1-27 co-administration. This evidence supports the existence of an endogenous ligand that positively modulates the PBR and the presence of an endogenous modulator at the PBR of group-housed (nonstressed) but not socially isolated (stressed) mice. This is a matter that deserves more attention.

9.9 ANTICONVULSANT EFFECTS OF FGIN-1 COMPOUNDS

FGIN-1-27 has been shown to delay the onset of isoniazid- and metrazol-induced convulsions in rats in much the same way as THDOC.[1,107] FGIN-1-27 does not have an effect on bicuculline-induced seizures, which are readily antagonized by diazepam and alpidem, but not by THDOC. This evidence further supports the supposition that the effects of FGIN-1-27 are indirectly mediated through the $GABA_A$ receptor via neurosteroidogenesis. Blocking of pentylenetetrazol-induced seizures in mice by FGIN-1-27 has also been demonstrated.[110] The potency of THDOC in antagonizing pentylenetetrazol-induced seizures is increased during periods of diazepam withdrawal in mice.[107] As THDOC is a pregnenolone metabolite, it would stand to reason that the anticonvulsant potency of FGIN-1-27 would also be increased during diazepam withdrawal. Surprisingly, FGIN-1-27 had no effects on seizure latency when administered to mice withdrawing from diazepam.[110] One explanation is that chronic diazepam administration may have resulted in the down-regulation of CBR and PBR and, as a consequence of PBR down-regulation, a decrease in potential neurosteroid production in response to FGIN-1-27 administration. In turn, the decrease in neurosteroidogenesis may have also resulted in an increased sensitivity to the less abundant neurosteroids, explaining the enhanced effects of THDOC during diazepam withdrawal. Unfortunately, there have been no studies seeking to resolve this issue, though it may have important implications for the treatment of benzodiazepine withdrawal. Not only would co-tolerance to FGIN-1-27 potentially limit or eliminate its use for treating benzodiazepine withdrawal, but if the sensitivity of the $GABA_A$ receptor to neurosteroid positive modulators is increased due to prolonged benzodiazepine administration, this may contribute to the decreasing therapeutic index of benzodiazepines and other anxiolytics. Such a scenario might bode well for the clinical use of FGIN-1 derivatives; if one mechanism were to down-regulate or desensitize PBRs in response to chronic FGIN-1-X administration, and another mechanism were to up-regulate or sensitize $GABA_A$ receptors in response to decreased neurosteroid production due to PBR down-regulation, then the two mechanisms may antagonize one another and at least partially maintain both the long-term potency and efficacy of a FGIN-1-X or similar compound. This is highly conjectural, of course, but an interesting prospect all the same and deserves further attention.

9.10 ANTIDEPRESSANT EFFECTS OF FGIN-1 COMPOUNDS

Implications that changes in neurosteroid levels may be involved in some emotional disorders raise the intriguing possibility that FGIN-1 derivatives may help us understand

and treat some of these disorders. In addition to the anxiolytic and sedative effects of many endogenous neurosteroids, low levels of allopregnanolone in cerebrospinal fluid (CSF) have been demonstrated in patients with major unipolar depression[111–113]; treatment with selective serotonin reuptake inhibitors (SSRIs) provides beneficial effects and restores normal circulating levels of allopregnanolone.[111] A potent antidepressant effect of allopregnanolone was observed on the time of immobility in a forced-swim test in mice.[114,115] Similar effects resulted from Ro 5-4864 and FGIN-1-27 administration. Indeed, the antidepressant effects of the PBR ligands were equivalent to those of fluoxetine and metyrapone.[114] FGIN-1-27 has also been demonstrated to potentiate the antidepressant effect of ethanol in the Porsolt forced-swim test.[116]

9.11 LEARNING AND MEMORY

Evidence exists that neurosteroids may play an important role in human and animal cognition. Oral consumption of a solution containing dehydroepiandrosterone sulfate improved memory on a win-shift water escape task in aged mice.[117] Intracerebroventricular administration of pregnenolone and pregnenolone sulfate in male mice has been observed to enhance both memory and learning in active and passive avoidance tasks.[118–120] Da Cunha et al.[121] demonstrated that memory retention in a step-down inhibitory avoidance task is at least partially mediated by centrally located PBR and $GABA_A$ receptors. Administration of neurosteroids to adult males suffering from Alzheimer's has also proven beneficial,[122] which complements other research demonstrating abnormally low levels of diazepam binding inhibitor, a putative endogenous ligand for PBRs in the brain, in Alzheimer's patients.[123]

Together, these findings suggest that FGIN-1 derivatives may also exhibit similar pro-cognitive effects. Few studies have investigated FGIN-1 derivatives' effects on memory and learning. PK 11195 has been shown to induce memory deficits in the taste aversion conditioning paradigm when injected directly into the right hemisphere of the lobus parolfactorius (LPO) of newborn chicks.[124] FGIN-1-27 did not noticeably affect taste aversion conditioning when similarly injected, although it did antagonize the memory deficits induced by PK 11195 administration. Neither drug produced any observable motor impairment. FGIN-1-27 was also found to antagonize amnesia induced by the noncompetitive NMDA antagonist dizocilpine in rats trained to retain a passive avoidance response or to retrieve a learned task in radial and water maze tests.[125] Dizocilpine-induced amnesia was also prevented by administration of pregnenolone sulfate, THDOC, or allopregnanolone.

It is important to point out that not all behavioral effects elicited by allopregnanolone administration are mimicked by FGIN-1-27 administration. Whereas allopregnanolone administration produced catalepsy in mice,[126,127] FGIN-1-27 failed to induce catalepsy, even though it was found to potentiate catalepsy when co-administered with serotonergic agents such as fluoxetine.[128] The authors suggested that FGIN-1-27's failure to induce catalepsy on its own may be due to an inability of the doses used to stimulate sufficient levels of neurosteroids. Although the increase in neurosteroids that resulted from serotonergic drugs was found to occur independently of FGIN-1-27,[114,128–130] no investigation was made to determine if neurosteroidogenesis following the co-administration of the serotonergic agents and FGIN-1-27 increased linearly, or if a synergistic or even partially antagonistic relationship existed between the steroidogenic effects of these compounds.

9.12 SIDE EFFECTS

Perhaps the most remarkable characteristic of FGIN-1 derivatives is their highly favorable therapeutic index and apparent lack of side effects. With the exception of one paper, none of the authors reviewed reported any adverse affects as a result of FGIN-1-X administration.[1–3,43,85,107,108,110,114,115,124,125,128] In the one paper that did report side effects,[1] sedation, ataxia, and ultimately death due to respiratory failure followed intravenous administration of FGIN-1-27 to rats, but only when very high doses (2.3 to 11.5 μmol/kg and 13.8 to 23 μmol/kg for sedative and lethal doses, respectively) — far exceeding doses necessary to induce anxiolysis and other desired behavioral effects — were administered. The lack of reported side effects is not entirely surprising; many neurosteroids have long been characterized by a very high therapeutic index.[80,131] The relative disuse of neurosteroids clinically has largely been attributed to their low aqueous solubility and generally unfavorable pharmacokinetic profiles,[80] a shortcoming that FGIN-1 derivatives may be able to overcome. Nonetheless, the exact functions of neuromodulatory steroids *in vivo* are not known, and investigations into the long-term effects of neurosteroid administration and the possibility of subtle side effects affecting cognition or endocrine system regulation need to be carried out in order to obtain a better picture of FGIN-1 derivatives' suitability for clinical use.

9.13 OTHER SELECTIVE PBR LIGANDS

Since the synthesis of the first FGIN-1 derivatives, three new series of PBR ligands boasting selectivity for the PBR equal to or greater than that of the FGIN-1 derivatives have been described: the 2-phenyl-imidazol[1,2-a]pyridine derivatives (CB 34, CB 50, and CB 54),[132] N-(4-chloro-2-phenoxyphenyl)-N-(2-isopropoxybenzyl)acetamide (DAA1097) and N-(2,5-dimethoxybenzyl)-N-(5-fluoro-2-phenoxyphenyl)acetamide (DAA1106),[133,134] and the 3-aryl-3-pyrrol-ylpropanamides.[4] The 2-phenyl-imidazol[1,2-a]pyridine derivatives increased brain and plasma content of the neurosteroids pregnenolone, progesterone, allopregnanolone, and alloTHDOC in adrenalectomized-orchiectomized rats, and administration of the derivative CB 34 was marked by significant anticonflict effect in the Vogel test, paralleling the anxiolytic behavior of FGIN-1 derivatives.[132] DAA1097 and DAA1106 have not been specifically tested for steroidogenic efficacy, but have been demonstrated to elicit anxiolytic effects when administered to rats in the light–dark exploration and elevated plus maze tests,[133] as would be expected from a FGIN-1-like compound.

While one might expect the effects of FGIN-1 derivatives to differ only quantitatively from the effects of similarly selective and potent PBR ligands, this is not necessarily the case. Guillon et al.[4] recently reported that substitution of a pyrrol ring for FGIN-1's indole ring produced a compound with a high specificity for the PBR, yet it proved to be both sedative and nonanxiolytic in mice in the light–dark transition test. While the lack of an anxiolytic effect of this putative specific PBR ligand may be attributed to differences in testing paradigms, test species (mice as opposed to rats), or a possible lack of steroidogenic effect of the ligands (their ability to potentiate steroidogenesis *in vivo* or *in vitro* was not tested), the sedative action of the ligands remains unexplained. Until further tests are performed and the PBR is conclusively determined to be the 3-aryl-3-pyrrol-ylpropanamides' site of action *in vivo*, nothing can be said with certainty about the nature of their effects. As a result, the possibility that FGIN-1 derivatives' and other PBR ligands' effects may be mediated by more complex mechanisms than was previously thought precludes reliable prediction of the effects of any PBR ligands that have not been extensively tested. For instance, despite the similarities of DAA1106 to FGIN-1-27, it would currently be inadvisable to assume that both perform similarly in a wide range of tests. Exploration of the unique pharmacological profile of the 3-aryl-3-pyrrol-ylpropanamides is unquestionably called for and may reveal important characteristics of PBR function.

9.14 SUMMARY

FGIN-1 derivatives have proven effective as anxiolytics, anticonvulsants, antiamnetics, and, to a lesser extent, antidepressants, as well as antiproliferatives in some cancers. Unlike benzodiazepines, FGIN-1-X administration does not appear to produce motor impairment or other adverse side effects, even at relatively high therapeutic doses. The density of PBRs in the human brain is also four to five times higher than in the rat brain, even higher than the $GABA_A$ benzodiazepine recognition site, implying that FGIN-1 derivatives may be even more efficacious in humans than in the rodent models used thus far. Because of this, FGIN-1-27 and FGIN-1-44 may have significant clinical potential, effectively acting as a *de facto* delivery system for pregnenolone metabolites that otherwise suffer from low bioavailability. Our humble understanding of the PBR's physiological roles may mean that FGIN-1-X compounds induce significant, yet undetected, side effects, especially following chronic administration, but even the inadvertent discovery of adverse side effects stemming from FGIN-1-X administration would be beneficial, as it would further elucidate the physiological role of the PBR.

The capacity of FGIN-1 derivatives to specifically bind to and positively modulate the PBR also makes them critical to the further characterization of both PBR and neurosteroid function *in vivo*. However, a substantial gap exists in the characterization of FGIN-1 derivatives' capacity to interact with the PBR: FGIN-1 derivatives have only been concretely demonstrated to be efficacious in the stimulation of neurosteroidogenesis via the mitochondrial localization of the PBR. At least one study found that more than 50% of the PBRs in rat monocytes and T-4 lymphocytes are not compartmentalized in mitochondria,[135] and PBRs have also been found in red blood cells, which lack mitochondria.[61] Yet the only nonsteroidogenic function that has been attributed to a FGIN-1 derivative thus far is the induction of the MPT in cancerous L929 fibroblasts.[38] Whether FGIN-1 is able to potentiate non-steroidogenic PBR-mediated functions is unclear, and FGIN-1 derivatives have not been demonstrated to bind to nonmitochondrially localized PBRs. Clearly then, not only should the behavioral effects of FGIN-1 derivatives be further characterized, but the full variety of cellular and subcellular effects of FGIN-1 derivatives should also be viewed as a promising area of research.

REFERENCES

1. Romeo, E., Auta, J., Kozikowski, A.P., Ma, D., Papadopoulos, V., Puia, G., Costa, E., and Guidotti, A., 2-Aryl-3-indoleacetamides (FGIN-1): a new class of potent and specific ligands for the mitochondrial DBI receptor (MDR), *J. Pharmacol. Exp. Ther.*, 262, 971–978, 1992.
2. Kozikowski, A.P., Ma, D., Brewer, J., Sun, S., Costa, E., Romeo, E., and Guidotti, A., Chemistry, binding affinities, and behavioral properties of a new class of antineophobic mitochondrial DBI receptor complex (mDRC) ligands, *J. Med. Chem.*, 36, 2908–2920, 1993.
3. Liao, Y., Kozikowski, A.P., Guidotti, A., and Costa, E., Synthesis and pharmacological evaluation of benzofuran-acetamides as "antineophobic" mitochondrial DBI receptor complex ligands, *Bioorg. Med. Chem. Lett.*, 8, 2099–2102, 1998.
4. Guillon, J., Boulouard, M., Lelong, V., Dallemagne, P., Rault, S., and Jarry, C., Synthesis and preliminary behavioral evaluation of mice of new 3-aryl-3-pyrol-1-ylpropanimides, analogues of FGIN-1-27 and FGIN-1-43, *Pharmacy Pharmacol.*, 53, 1561–1568, 2001.
5. DeLorey, T.M. and Olsen, R.W., $GABA_A$ receptor structure and function, *J. Biol. Chem.*, 267, 16747–16750, 1992.
6. Mohler, H. and Okada, T., Benzodiazepine receptors: demonstration in the central nervous system, *Science*, 198, 849–851, 1977.
7. Squires, R. and Braestrup, C., Benzodiazepine receptors in rat brain, *Nature*, 266, 732–734, 1977.
8. Tallman, J.F., Paul, S.M., Skolnick, P., and Gallager, D.W., Receptors for the age of anxiety: pharmacology of the benzodiazepines, *Science*, 207, 274–281, 1980.
9. Paul, S.M., Marangos, P.J., and Skolnick, P., The benzodiazepine-GABA-chloride ionophore receptor complex: common site of minor tranquilizer action, *Biol. Psychiatry*, 16, 213–229, 1981.
10. Braestrup, C. and Squires, R.F., Specific benzodiazepine receptor in rat brain characterized by high affinity [^{3}H]diazepam binding, *Proc. Natl. Acad. Sci. U.S.A.*, 74, 3805–3809, 1977.
11. Marangos, P.J., Patel, J., Boulenger, J.P., and Clark-Rosenberg, R., Characterizations of the peripheral-type benzodiazepine binding sites in brain using [^{3}H] Ro 5-4864, *Mol. Pharmacol.*, 22, 26–32, 1982.
12. Verma, A. and Snyder, S.H., Peripheral type benzodiazepine receptors, *Annu. Rev. Pharmacol. Toxicol.*, 29, 307–322, 1989.
13. Anholt, R.R.H., Pedersen, P.L., De Souza, E.B., and Snyder, S.H., The peripheral-type benzodiazepine receptor: localization to the mitochondrial outer membrane, *J. Biol. Chem.*, 261, 576–583, 1986.
14. Papadopoulos, V., Berkovich, A., Krueger, K.E., Costa, E., and Guidotti, A., Diazepam binding inhibitor (DBI) and its processing products stimulate mitochondrial steroid biosynthesis via an interaction with mitochondrial benzodiazepine receptors, *Endocrinology*, 129, 1481–1488, 1991.
15. Sprengel, R., Werner, P., Seeburg, P.H., Mukhin, A.G., Santi, M.R., Grayson, D.R., Guidotti, A., and Krueger, K.E., Molecular cloning and expression of cDNA encoding a peripheral-type benzodiazepine receptor, *J. Biol. Chem.*, 264, 20415–20421, 1989.

16. Papadopoulos, V., Peripheral-type benzodiazepine/diazepam binding inhibitor receptor: biological role in steroidogenic cell function, *Endocr. Rev.*, 14, 222–240, 1993.
17. Mukhin, A.G., Papadopoulos, V., Costa, E., and Krueger, K.E., Mitochondrial benzodiazepine receptors regulate steroid biosynthesis, *Proc. Natl. Acad. Sci. U.S.A.*, 86, 9813–9816, 1989.
18. Cascio, C., Guarneri, P., Li, H., Brown, R.C., Amri, H., Boujrad, N., Kotoula, M., Vidic, B., Drieu, K., and Papadopoulos, V., Peripheral-type benzodiazepine receptor: role in the regulation of steroid and neurosteroid biosynthesis, in *Neurosteroids: A New Regulatory Function in the Nervous System*, Balieu, E.-E., Robel, P., and Schumacher, M., Eds., Humana Press, Totowa, NJ, 1999, pp. 75–96.
19. Brown, R.C., Cascio, C., and Papadopoulos, V., Neurosteroid biosynthesis in cell lines from human brain: regulation of dehydroepiandrosterone formation by oxidative stress and amyloid peptide, *J. Neurochem.*, 74, 847–859, 2000.
20. Hardwick, M., Fertikh, D., Culty, M., Li, H., Vidic, B., and Papadopoulos, V., Peripheral-type benzodiazepine receptor (PBR) in human breast cancer: correlation of breast cancer cell aggressive phenotype with PBR expression, nuclear localization, and PBR-mediated cell proliferation and nuclear transport of cholesterol, *Cancer Res.*, 59, 831–842, 1999.
21. Garnier, M., Boujrad, N., Oke, B.O., Brown, A.S., Riond, J., Ferrara, P., Shoyab, M., Suarez-Ouain, C.A., and Papadopoulos, V., Diazepam binding inhibitor is a paracrine/autocrine regulator of Leydig cell proliferation and steroidogenesis: action via peripheral-type benzodiazepine receptor and independent mechanisms, *Endocrinology*, 132, 444–458, 1993.
22. Oke, B.O., Suarez-Ouain, C.A., Riond, J., Ferrara, P., and Papadopoulos, V., Cell surface localization of the peripheral-type benzodiazepine receptor (PBR) in adrenal cortex, *Mol. Cell. Endocrinol.*, 87, R1–R6, 1992.
23. Drugan, R.C., Peripheral benzodiazepine receptors: molecular pharmacology to possible physiological significance in stress-induced hypertension, *Clin. Neuropharmacol.*, 19, 475–496, 1996.
24. Hirsch, J.D., Beyer, C.F., Malkowitz, L., Beer, B., and Blume, A.J., Mitochondrial benzodiazepine receptors mediate inhibition of mitochondrial respiratory control, *Mol. Pharmacol.*, 35, 157–163, 1989.
25. Moreno-Sanchez, R., Hogue, B.A., Bravo, C., Newman, A.H., Basile, A.S., and Chiang, P.K., Inhibition of substrate oxidation in mitochondria by the peripheral-type benzodiazepine receptor ligand AHN 086, *Biochem. Pharmacol.*, 41, 1479–1484, 1991.
26. Kuhlmann, A.C. and Guilarter, T.R., Cellular and subcellular localization of peripheral benzodiazepine receptors after trimethylin neurotoxicity, *J. Neurochem.*, 74, 1694–1704, 2000.
27. Schoemaker, H., Morelli, M., Deshmukh, P., and Yamamura, H.I., Specific high affinity saturable binding of [^{3}H] RO5-4864 to benzodiazepine binding sites in rat cerebral cortex, *Eur. J. Pharmacol.*, 71, 173–175, 1981.
28. Guilarte, T.R., Kuhlmann, A.G., O'Callaghan, J.P., and Miceli, R.C., Enhanced expression of peripheral benzodiazepine receptors in trimethylin-exposed rat brain: a biomarker of neurotoxicity, *Neurotoxicology*, 16, 441–450, 1995.
29. Torres, S.R., Frode, T.S., Nardi, G.M., Vita, N., Reeb, R., Ferrara, P., Ribeiro-do-Valle, R.M., and Farges, R.C., Anti-inflammatory effects of peripheral benzodiazepine receptor ligands in two mouse models of inflammation, *Eur. J. Pharmacol.*, 408, 199–211, 2000.

30. Lacor, P., Gandolfo, P., Tonon, M.C., Brault, E., Dalibert, I., Schumacher, M., Benavides, J., and Ferzaz, B., Regulation of the expression of peripheral benzodiazepine receptors and their endogenous ligands during rat sciatic nerve degeneration and regeneration: a role of PBR in neurosteroidogenesis, *Brain Res.*, 516, 295–303, 1999.
31. Marchetti, P., Trincavelli, L., Giannarelli, R., Giusti, L., Coppelli, A., Martini, C., Navalesi, R., and Lucacchini, A., Characterization of peripheral benzodiazepine receptors in purified large mammal pancreatic islets, *Biochem. Pharmacol.*, 51, 1437–1442, 1996.
32. Mestre, M., Carriot, T., Berlin, C., Uzan, A., Renault, C., Dubroeucq, M.C., Gueremy, C., and Le Fur, G., Electrophysiological and pharmacological characterization of peripheral benzodiazepine receptors in guinea pig heart preparation, *Life Sci.*, 35, 953–962, 1984.
33. Bisserbe, J.C., Patel, J., and Eskay, R.L., Evidence that the peripheral-type benzodiezapine receptor ligand Ro 5-4864 inhibits beta-endorphin release from AtT-20 cells by blockade of voltage-dependent calcium channels, *J. Neurochem.*, 47, 1419–1424, 1986.
34. Basile, A.S., Lueddens, H.W., and Skolnick, P., Regulation of renal peripheral benzodiazepine receptors by anion transport inhibitors, *Life Sci.*, 42, 715–726, 1988.
35. Basile, A.S., Bolger, G.T., Lueddens, H.W.M., and Skolnick, P., Electrophysiological actions of Ro 5-4864 on cerebellar Purkinje neurons: evidence for peripheral benzodiazepine receptor-mediated depression, *J. Pharmacol. Exp. Ther.*, 248, 463–469, 1989.
36. Leuddens, H.W. and Skolnick, P., 'Peripheral-type' benzodiazepine receptors in the kidney: regulation of radioligand binding by anions and DIDS, *Eur. J. Pharmacol.*, 133, 205–214, 1987.
37. Bono, F., Lamarche, I., Prabonnaud, V., Le Fur, G., and Herbert, J.M., Peripheral benzodiazepine receptor agonists exhibit potent antiapoptic activities, *Biochem. Biophys. Res. Commun.*, 265, 457–461, 1999.
38. Pastorino, J.G., Simbula, G., Yamamoto, K., Glascott, P.A., Jr., Rothman, R.J., and Farber, J.L., The cytotoxicity of tumor necrosis factor depends on induction of the mitochondrial permeability transition, *J. Biol. Chem.*, 271, 29792–29798, 1996.
39. Zonov, D.B., Mitochondrial damage as a source of diseases and aging: a strategy of how to fight these, *Biochem. Biophys. Acta*, 1275, 10–15, 1996.
40. Fennell, D.A., Corbo, M., Pallaska, A., and Cotter, F.E., Bcl-2 resistant mitochondrial toxicity mediated by the isoquinoline carboxamides PK 11195 involves de novo generation of reactive oxygen species, *Br. J. Cancer*, 84, 1397–1404, 2001.
41. Gunter, T.E., Gunter, K.K., Sheu, S., and Gavin, C.E., Mitochondrial calcium transport: physiological and pathological relevance, *Am. J. Physiol.*, 267, C313–C319, 1994.
42. Zoratti, M. and Szabo, I., The mitochondrial permeability transition, *Biochim. Biophys. Acta*, 1241, 139–176, 1995.
43. Korneyev, A., Pan, B.S., Polo, A., Romeo, E., Guidotti, A., and Costa, E., Stimulation of brain pregnenolone synthesis by mitochondrial diazepam binding inhibitor receptor ligands in vivo, *J. Neurochem.*, 61, 1515–1524, 1993.
44. Rago, L., Saano, V., Auvinen, T., Adojaan, A., Holma, R., and Airaksinen, M.M., The effect of chronic treatment with peripheral benzodiazepine receptor ligands on behavior and GABAA/benzodiazepine receptors in rat, *Naunym Schmiedeberg Arch. Pharmacol.*, 346, 432–436, 1992.
45. Benavides, J.J., Guilloux, F., Allam, D.E., Uzan, A., Mizoule, J., Renault, C., Dubroeucq, C., Gueremy, C., and Le Fur, G., Opposite effects of an agonist, Ro 5-4864, and an antagonist PK 11195 of the peripheral type benzodiazepine binding sites on audiogenic seizures in DBA-2J mice, *Life Sci.*, 34, 2613–2620, 1984.

46. Pellow, S. and File, S.E., Behavioral actions of Ro 5-4864: a peripheral-type benzodiazepine? *Life Sci.*, 35, 229–240, 1984.
47. Skowronski, R., Fanestil, D.D., and Beaumont, K., Photoaffinity labeling of peripheral-type benzodiazepine receptors in rat kidney mitochondria with [^{3}H] PK 14105, *Eur. J. Pharmacol.*, 148, 187–193, 1988.
48. Puia, G., Santi, M.R., Vicini, S., Pritchett, D.B., Seeburg, P.H., and Costa, E., Differences in the negative allosteric modulation of γ-aminobutyric acid receptors elicited by 4'-chlordiazepam and by β-carboline-3-carboxylate ester: a study with natural and reconstituted receptors, *Proc. Natl. Acad. Sci. U.S.A.*, 86, 7275–7279, 1989.
49. Puia, G., Vicini, S., Seeburg, P.H., and Costa, E., Influence of recombinant γ-aminobutyric acid-A receptor subunit composition on the action of allosteric modulators of γ-aminobutyric acid-gated Cl– currents, *Mol. Pharmacol.*, 39, 691–696, 1991.
50. Kozikowski, A.P., Kotoula, M., Ma, D., Boujrad, N., Tuckmantel, W., and Papadopoulos, V., Synthesis and biology of a 7-nitro-2,1,3-benzoaxdiazol-4-yl derivative of 2-phenylindole-3-acetamide: a fluorescent probe for the peripheral-type benzodiazepine receptor, *J. Med. Chem.*, 40, 2435–2439, 1997.
51. File, S.E. and Pellow, S., The effects of PK 11195, a ligand for benzodiazepine binding sites, in animal tests of anxiety and stress, *Pharmacol. Biochem. Behav.*, 23, 737–741, 1985.
52. Kunert-Radek, J., Stepien, H., and Pawlikowski, M., Inhibition of rat pituitary tumor cell proliferation by benzodiazepines in vitro, *Neuroendocrinology*, 59, 92–96, 1994.
53. Pawlikowski, M., Kunert-Radek, J., Radek, A., and Stepien, H., Inhibition of cell proliferation of human gliomas by benzodiazepines in vitro, *Acta Neurol. Scand.*, 77, 231–233, 1988.
54. Calogero, A.E., Kamilaris, T.C., Bernardini, R., Johnson, E.O., Chrousos, G.P., and Gold, P.W., Effects of peripheral benzodiazepine receptor ligands on hypothalamic-pituitary-adrenal axis function in the rat, *J. Pharmacol. Exp. Ther.*, 253, 729–737, 1990.
55. Maaser, K., Hopfner, M., Jansen, A., Weisinger, G., Gavish, M., Kozikowski, A.P., Weizman, A., Carayon, P., Riecken, E.O., Zeitz, M., and Scherubl, H., Specific ligands of the peripheral benzodiazepine receptor induce apoptosis and cell cycle arrest in human colorectal cancer cells, *Br. J. Cancer*, 85, 1771–1780, 2001.
56. Benavides, J., Malgouris, C., Imbault, F., Begassat, F., Uzan, A., Renault, C., Dubroeucq, M.C., Gueremy, C., and Le Fur, G., “Peripheral type” benzodiazepine binding sites in rat adrenals: binding studies with [^{3}H] PK 11195 and autoradiographic localization, *Arch. Int. Pharmacodyn. Ther.*, 266, 38–49, 1983.
57. DeSouza, E.B., Anholt, R.R., Murphy, K.M., Snyder, S.H., and Kuhar, M.J., Peripheral-type benzodiazepine receptors in endocrine organs: autoradiographic localization in rat pituitary, adrenal, and testis, *Endocrinology*, 116, 567–573, 1985.
58. Davies, L.P. and Huston, V., Peripheral benzodiazepine binding sites in heart and their interaction with dipyridamole, *Eur. J. Pharmacol.*, 73, 209–211, 1981.
59. French, J.F. and Matlib, M.A., Identification of a high-affinity peripheral-type benzodiazepine binding site in rat aortic smooth muscle membranes, *J. Pharmacol. Exp. Ther.*, 247, 23–28, 1988.
60. Wang, J.K., Morgan, J.I., and Spector, S., Benzodiazepines that bind at peripheral sites inhibit cell proliferation, *Proc. Natl. Acad. Sci. U.S.A.*, 81, 753–756, 1984.
61. Olson, J.M., Ciliax, B.J., Mancini, W.R., and Young, A.B., Presence of peripheral-type benzodiazepine binding sites on human erythrocyte membranes, *Eur. J. Pharmacol.*, 152, 47–53, 1988.

62. Ruff, M.R., Pert, C.B., Weber, R.J., Wahl, L.M., Wahl, S.M., and Paul, S.M., Benzodiazepine receptor-mediated chemotaxis of human monocytes, *Science*, 229, 1281–1283, 1985.
63. Zavala, F., Haumont, J., and Lenfant, M., Interaction of benzodiazepines with mouse macrophages, *Eur. J. Pharmacol.*, 106, 561–566, 1984.
64. Benavides, J., Dubois, A., Dennis, T., Hamel, E., and Scatton, B., Omega 3 (peripheral type benzodiazepine binding) site distribution in the rat immune system: an autoradiographic study with the photoaffinity ligand [^{3}H] PK 14105, *J. Pharmacol. Exp. Ther.*, 249, 333–339, 1989.
65. Drugan, R.C. and Holmes, P.V., Central and peripheral benzodiazepine receptors: involvement in an organism's response to physical and psychological stress, *Neurosci. Biobehav. Rev.*, 15, 277–298, 1991.
66. Doble, A., Malgouris, C., Daniel, M., Daniel, N., Imbault, F., Basbaum, A., Uzan, A., Gueremy, C., and Le Fur, G., Labeling of peripheral-type benzodiazepine binding sites in human brain with [^{3}H] PK 11195: anatomical and subcellular distribution, *Brain Res. Bull.*, 18, 49–61, 1987.
67. Benavides, J., Quartenoret, D., Imbault, F., Malgouris, C., Uzan, A., Renault, C., Dubroeucq, M.C., Gueremy, C., and Le Fur, G., Labeling of "peripheral-type" benzodiazepine binding sites in the rat brain by using [^{3}H] PK 11195, an isoquinoline carboximide derivative: kinetic studies and autoradiographic localization, *J. Neurochem.*, 41, 1744–1750, 1983.
68. Bender, A.S. and Hertz, L., Binding of [^{3}H] Ro 5-4864 in primary cultures of astrocytes, *Brain Res.*, 341, 41–49, 1985.
69. Basile, A.S. and Skolnick, P., Subcellular localization of "peripheral-type" binding sites for benzodiazepines in rat brain, *J. Neurochem.*, 46, 305–308, 1986.
70. Viliger, J., Specific [3H] Ro 5-4864 binding to rat spinal cord membranes: evidence for peripheral-type benzodiazepine recognition sites, *Neurosci. Lett.*, 46, 267–270, 1984.
71. Hall, P.F., Cellular organization of steroidogenesis, *Int. Rev. Cytol.*, 86, 53–95, 1984.
72. Besman, M.J., Yanagibashi, K., Lee, T., Kawamura, M., Hall, P.F., and Shiveley, J.E., Identification of des-(Gly-Ile)-endozepine as an effector of corticotropin-dependent adrenal steroidogenesis: stimulation of cholesterol delivery is mediated by the peripheral benzodiazepine receptor, *Proc. Natl. Acad. Sci. U.S.A.*, 86, 4897–4901, 1989.
73. Papadopoulos, V., Amri, H., Boujard, N., Cascio, C., Culty, M., Garnier, M., Hardwick, M., Li, H., Vidic, B., Brown, A.S., Reversa, J.L., Bernassau, J.M., and Drieu, K., Peripheral benzodiazepine receptor in cholesterol transport and steroidogenesis, *Steroids*, 62, 21–28, 1997.
74. Yanagibashi, K., Ohno, Y., Nakamichi, N., Matsui, T., Hayashida, K., Takamura, M., Yamada, K., Tou, S., and Kawamura, M., Peripheral-type benzodiazepine receptors are involved in the regulation of cholesterol side chain cleavage in adrenocortical mitochondria, *J. Biochem. (Tokyo)*, 106, 1026–1029, 1989.
75. Krueger, K.E. and Papadopoulos, V., Peripheral-type benzodiazepine receptors mediate translocation of cholesterol from outer to inner mitochondrial membranes in adrenocortical cells, *J. Biol. Chem.*, 265, 15015–15022, 1990.
76. Yanagibashi, K., Ohno, Y., Nakamichi, N., Matsui, T., Hayashida, K., Takamura, M., Yamada, K., Tou, S., and Kawamura, M., Diazepam potentiates the corticodogenic response of bovine adrenal fasciculata cells to dibutyryl cyclic AMP, *Jpn. J. Pharmacol.*, 51, 347–355, 1989.

77. Morrow, A.L., Suzdak, P.D., and Paul, S.M., Steroid hormone metabolites potentiate GABA receptor-mediated chloride ion influx with nanomolar potency, *Eur. J. Pharmacol.*, 142, 483–485, 1987.
78. Majewska, M.D. and Vaupel, D.B., Steroid control of uterine motility via gamma-aminobutyric acidA receptors in the rabbit: a novel mechanism? *J. Endocrinol.*, 131, 427–434, 1991.
79. Lan, N.C. and Gee, K.W., Neuroactive steroid actions at the $GABA_A$ receptor, *Horm. Behav.*, 28, 537–544, 1994.
80. Paul, S.M. and Purdy, R.H., Neuroactive steroids, *FASEB J.*, 6, 2311–2322, 1992.
81. Barnea, A., Hajibeigi, A., Trant, J.M., and Mason, J.I., Expression of steroid metabolizing enzymes by aggregating fetal brain cells in culture: a model for developmental regulation of the progesterone 5alpha-reductase pathway, *Endocrinology*, 127, 500–502, 1990.
82. Melacagni, R.C., Celotti, F., Ballabio, M., Castano, P., Massarelli, R., Poletti, A., and Martini, L., 5α-Reductase activity in isolated and cultured neuronal and glial cells of the rat, *Brain Res.*, 516, 229–236, 1990.
83. Purdy, R.H., Morrow, A.L., Blinn, J.R., and Paul, S.M., Synthesis, metabolism, and pharmacological activity of 3α-hydroxy steroids which potentiate GABA-receptor-mediated chloride ion uptake in rat cerebral cortical synaptoneurosomes, *J. Med. Chem.*, 33, 1572–1581, 1990.
84. Bitran, D., Foley, M., Audette, D., Leslie, N., and Frye, C.A., Activation of peripheral mitochondrial benzodiazepine receptors in the hippocampus stimulates allopregnanolone synthesis and produces anxiolytic-like effects in the rat, *Psychopharmacology*, 151, 64–71, 2000.
85. Romeo, E., Cavallaro, S., Korneyev, A., Kozikowski, A.P., Ma, D., Polo, A., Costa, E., and Guidotti, A., Stimulation of brain steroidogenesis by 2-aryl-indole-3-acetamide derivatives acting at the mitochondrial diazepam-binding inhibitor receptor complex, *J. Pharmacol. Exp. Ther.*, 267, 462–471, 1993.
86. Snyder, M. and Antwerpen, R.V., Evidence for a diazepam-binding inhibitor (DBI) benzodiazepine receptor-like mechanism in ecdysteroidogenesis by the insect prothoracic gland, *Cell Tissue Res.*, 294, 161–168, 1998.
87. Cutolo, M., Seriolo, B., Villagio, B., Pizzorni, C., Craviotto, C., and Sulli, A., Androgens and estrogens modulate the immune and inflammatory responses in rheumatoid arthritis, *Ann. N.Y. Acad. Sci.*, 966, 131–142, 2002.
88. Castagnetta, L., Granata, O.M., Traina, A., Cocciadiferro, L., Saetta, A., Stefano, R., Cutolo, M., and Carruba, G., A role for sex steroids in autoimmune diseases: a working hypothesis and supporting data, *Ann. N.Y. Acad. Sci.*, 966, 193–203, 2002.
89. Goldstein, J.S. and Sites, C.K., Selective modulation of sex steroids, *Ageing Res. Rev.*, 1, 17–28, 2002.
90. Pastorino, J.G., Snyder, J.W., Serroni, A., Hoek, J.B., and Farber, J.L., Cyclosporin and carnitine prevent anoxic death of cultured hepatocytes by inhibiting the mitochondrial permeability transition, *J. Biol. Chem.*, 268, 13791–13798, 1993.
91. Pastorino, J.G., Snyder, J.W., Hoek, J.B., and Farber, J.L., Ca2+ depletion prevents anoxic death of hepatocytes by inhibiting mitochondrial permeability transition, *Am. J. Physiol.*, 268, C676–C685, 1995.
92. Snyder, J.W., Pastorino, J.G., Attie, A., and Farber, J.L., Protection by cyclosporin A of cultured hepatocytes from the toxic consequences of the loss of mitochondrial energization produced by 1-methyl-4-phenylpyridinium, *Biochem. Pharmacol.*, 44, 833–835, 1992.

93. Cornu, P., Benavides, J., Scatton, B., Hauw, J.J., and Philippson, J., Increase in omega-3 tumors: a quantitative autoradiographic study, *Acta Neurochim.*, 119, 146–152, 1992.
94. Miettinen, H., Kononen, J., Haapasalo, H., Helen, P., Sallinen, P., Harjuntausta, T., Helin, H., and Alho, H., Expression of peripheral-type benzodiazepine receptor and diazepam binding inhibitor in human astrocytomas: relationship to cell proliferation, *Cancer Res.*, 15, 2691–2695, 1995.
95. Starosta-Rubenstein, S., Ciliax, B.J., Penney, J.B., McKeever, P., and Young, A.B., Imaging of a glioma using peripheral benzodiazepine receptor ligands, *Proc. Natl. Acad. Sci. U.S.A.*, 84, 891–895, 1987.
96. Zisterer, D.M. and Williams, D.C., Peripheral-type benzodiazepine receptors, *Gen. Pharmacol.*, 29, 305–314, 1997.
97. Neary, J.T., Jorgensen, S.L., Oracion, A.M., Bruce, J.H., and Norenburg, M.D., Inhibition of growth factor-induced DNA synthesis in astrocytes by ligands of peripheral-type benzodiazepine receptors, *Brain Res.*, 675, 27–30, 1995.
98. Landau, M., Weizman, A., Zoref-Shani, E., Beery, E., Wasseman, L., Landau, O., Gavish, M., Brenner, S., and Nordenberg, J., Antiproliferative and differentiating effects of benzodiazepine receptor ligands on B16 melanoma cells, *Biochem. Pharmacol.*, 56, 1029–1034, 1998.
99. Carmel, I., Fares, F.A., Leschiner, S., Scherubl, H., Weisinger, G., and Gavish, M., Peripheral-type benzodiazepine receptors in the regulation of proliferation of MCF-7 human breast carcinoma cell line, *Biochem. Pharmacol.*, 58, 273–278, 1999.
100. Beinlich, A., Strohmeir, R., Kaufmann, M., and Kuhl, H., Specific binding of benzodiazepines to human breast cancer cell lines, *Life Sci.*, 65, 2099–2108, 1999.
101. Ikezaki, K. and Black, K.L., Stimulation of cell growth and DNA synthesis by peripheral benzodiazepine receptor agonist, *Cancer Lett.*, 49, 115–120, 1990.
102. Nunez, G., Benedict, M.A., Hu, Y., and Inohara, N., Caspases: the proteases of the apoptotic pathway, *Oncogene*, 17, 3237–3245, 1998.
103. Camins, A., Diez-Fernandez, C., Fujadas, E., Camarasa, J., and Escubedo, E., A new aspect of the antiproliferative action of peripheral-type benzodiazepine receptor ligands, *Eur. J. Pharmacol.*, 272, 289–292, 1995.
104. Reddy, D.S. and Kulkarni, S.K., Reversal of benzodiazepine inverse agonist FG 7142-induced anxiety syndrome by neurosteroids in mice, *Methods Find. Exp. Clin. Pharmacol.*, 19, 665–681, 1997.
105. Reddy, D.S. and Kulkarni, S.K., Differential anxiolytic effects of neurosteroids in the mirrored chamber behavior test in mice, *Brain Res.*, 752, 61–71, 1997.
106. Baulieu, E.-E., Neurosteroids: a novel function of the brain, *Psychoneuroendocrinology*, 23, 963–987, 1998.
107. Auta, J., Romeo, E., Kozikowski, A., Ma, D., Costa, E., and Guidotti, A., Participation of mitochondrial diazepam binding inhibitor receptors in the anticonflict, antineophobic and anticonvulsant action of 2-aryl-3-indoleacetamide and imidazopyridine derivatives, *J. Pharmacol. Exp. Ther.*, 265, 649–656, 1993.
108. Costa, E., Auta, J., Guidotti, A., Korneyev, A., and Romeo, E., The pharmacology of neurosteroidogenesis, *J. Steroid Biochem.*, 49, 385–389, 1994.
109. Dong, E., Matsumoto, K., and Watanabe, H., Involvement of peripheral type of benzodiazepine receptor in social isolation stress-induced decrease in pentobarbital sleep in mice, *Life Sci.*, 65, 1561–1568, 1999.
110. Tsuda, M., Suzuki, T., and Misawa, M., Subsensitivity to mitochondrial diazepam binding inhibitor receptor agonist FGIN 1-27-induced antiseizure effect in diazepam-withdrawn mice, *Life Sci.*, 62, PL213–PL217, 1998.

111. Uzanov, V., Sheline, Y., Davis, J.M., Rasmusson, A., Uzanov, D.P., Costa, E., and Guidotti, A., Increase in cerebrospinal fluid content of neurosteroids in patients with unipolar major depression who are receiving fluoxetine or fluvoxamine, *Proc. Natl. Acad. Sci. U.S.A.*, 95, 3239–3244, 1998.
112. Strohle, A., Romeo, E., Hermann, B., Pasini, A., Spalletta, G., di Michele, F., Holsboer, F., and Rupprecht, R., Concentrations of 3α-reduced neuroactive steroids and their precursors in plasma of patients with major depression and after clinical recovery, *Biol. Psychiatry*, 45, 274–277, 1999.
113. Romeo, E., Strohle, A., Spalletta, G., di Michele, F., Hermann, B., Holsboer, F., Pasini, A., and Rupprecht, R., Effects of antidepressant treatment on neuroactive steroids in major depression, *Am. J. Psychiatry*, 155, 910–913, 1998.
114. Khisti, R.T. and Chopde, C.T., Serotonergic agents modulate antidepressant-like effect of the neurosteroid 3α-hydroxy-5α-pregnan-20-one in mice, *Brain Res.*, 865, 291–300, 2000.
115. Khisti, R.H., Chopde, C.T., and Jain, S.P., Antidepressant-like effect of the neurosteroid 3α-hydroxy-5α-pregnan-20-one in mice forced swim test, *Pharmacol. Biochem. Behav.*, 67, 137–143, 2000.
116. Hirani, K., Khisti, R.H., and Chopde, C.T., Behavioral action of ethanol in Porsolt's forced swim test: modulation by 3alpha-hydroxy-5alpha-pregnan-20-one, *Neuropharmacology*, 43, 1339–1350, 2003.
117. Markowski, M., Ungeheuer, M., Bitran, D., and Locurto, C., Memory-enhancing effects of DHEAS in aged mice on a win-shift water escape task, *Physiol. Behav.*, 72, 521–525, 2001.
118. Flood, J.F. and Roberts, E., Dehydroepiandrosterone sulfate improves memory in aging mice, *Brain Res.*, 448, 178–181, 1988.
119. Flood, J.F., Smith, G.E., and Roberts, E., Dehydroepiandrosterone and its sulfate enhance memory retention in mice, *Brain Res.*, 447, 269–278, 1988.
120. Flood, J., Morley, J., and Roberts, E., Memory-enhancing effects in male mice of pregnenolone and steroids metabolically derived from it, *Neurobiology*, 89, 1567–1572, 1992.
121. Da Cunha, C., Huang, C.H., Walz, R., Dias, M., Koya, R., Bianchin, M., Pereira, M.E., Izquierdo, I., and Medina, J.H., Memory facilitation by post-training intraperitoneal, intracerebroventricular, and intra-amygdala injection of Ro 5-4864, *Brain Res.*, 554, 133–136, 1991.
122. Roberts, E. and Fitten, L.J., Serum steroid levels in two old men with Alzheimer's disease (AD), before, during, and after oral administration of dehydroepiandrosterone (DHA): pregnenolone synthesis may be rate limiting in aging, in *The Biological Role of Dehydroepiandrosterone (DHEA)*, Kalimi, M. and Regelson, W., Eds., De Gruyter, Berlin, 1990, pp. 43–63.
123. Edgar, P.F., Schonberger, S.J., Dean, B., Faull, R.L., Kydd, R., and Cooper, G.J., A comparative proteome analysis of hippocampal tissue from schizophrenic and Alzheimer's disease individuals, *Mol. Psychiatry*, 4, 173–178, 1999.
124. Freeman, F.M. and Young, I.G., The mitochondrial benzodiazepine receptor and avoidance learning in the day-old chick, *Pharmacol. Biochem. Behav.*, 67, 355–362, 2000.
125. Romeo, E., Cheney, D.L., Zivkovic, I., Costa, E., and Guidotti, A., Mitochondrial diazepam-binding inhibitor receptor complex agonists antagonize dizolcipine amnesia: putative role for allopregnanolone, *J. Pharmacol. Exp. Ther.*, 270, 89–96, 1994.

126. Mandhane, S.N., Khisti, R.T., and Chopde, C.T., Adenosinergic modulation of 3α-hydroxy-5α-pregnan-20-one induced catalepsy in mice, *Psychopharmacology*, 144, 398–404, 1999.
127. Khisti, R.T., Mandhane, S.N., and Chopde, C.T., The neurosteroid 3α-hydroxy-5α-pregnan-20-one induces catalepsy in mice, *Neurosci. Lett.*, 251, 85–88, 1998.
128. Deshpande, L.S., Khisti, R.T., and Chopde, C.T., Cataleptic effect of neurosteroid 3α-hydroxy-5α-pregnan-20-one in mice: modulation by serotonergic agents, *Brain Res.*, 898, 13–26, 2001.
129. Uzanov, D.P., Cooper, T.B., Costa, E., and Guidotti, A., Fluoxetine-elicited changes in brain neurosteroid content measured by negative ion mass fragmentography, *Proc. Natl. Acad. Sci. U.S.A.*, 93, 12599–12604, 1996.
130. Griffin, L.D. and Mellon, S.H., Selective serotonin reuptake inhibitors directly alter activity of neurosteroidogenic enzymes, *Proc. Natl. Acad. Sci. U.S.A.*, 96, 13512–13517, 1999.
131. Gyermek, L. and Soyka, L.F., Steroid anesthetics, *Anesthesiology*, 42, 331–344, 1975.
132. Serra, M., Madau, P., Chessa, M.F., Caddeo, M., Sanna, E., Trapani, G., Franco, M., Liso, G., Purdy, R.H., Barbaccia, M.L., and Biggio, G., 2-Phenyl-imidazo [1,2-a] pyridine derivatives as ligands for peripheral benzodiazepine receptors: stimulation of neurosteroid synthesis and anticonflict action in rats, *Br. J. Pharmacol.*, 127, 177–187, 1999.
133. Okuyama, S., Chaki, S., Yoshikawa, R., Ogawa, S., Suzuki, Y., Okubo, T., Nakazato, A., Nagamine, M., and Tomisawa, K., Neuropharmacological profile of peripheral benzodiazepine receptor agonists, DAA1097 and DAA1106, *Life Sci.*, 64, 1455–1464, 1999.
134. Chaki, S., Funakoshi, T., Yoshikawa, R., Okuyama, S., Okubo, T., Nakazato, A., Nagamine, M., and Tomisawa, K., Binding characteristics of [3H]DAA1106, a novel and selective ligand for peripheral benzodiazepine receptors, *Eur. J. Pharmacol.*, 371, 197–204, 1999.
135. Cahard, D., Canat, X., Carayon, P., Rogue, C., Casellas, P., and Le Fur, G., Subcellular localization of peripheral benzodiazepine receptors on human leukocytes, *Lab. Invest.*, 70, 23–28, 1994.

10 GABAergic Neuroactive Steroids Modulate Selective Ethanol Actions: Mechanisms and Significance

A. Leslie Morrow, Rahul Khisti, Sayaka Tokunaga, Janelle R. McDaniel, and Douglas B. Matthews

CONTENTS

10.1 ETHANOL INTERACTIONS WITH $GABA_A$ RECEPTORS

Ethanol is anxiolytic, sedative-hypnotic, and anticonvulsant[1,2] and causes impairments in cognitive processing.[3,4] These effects of acute ethanol administration are remarkably similar to the effects of $3\alpha,5\alpha$-reduced neurosteroids as well as benzodiazepines and barbiturates — all modulators of $GABA_A$ receptors. Research using $GABA_A$ receptor antagonists and inverse agonists has identified this primary inhibitory neurotransmitter system as a key site that mediates the behavioral effects of ethanol. A number of the behavioral effects of ethanol are enhanced by $GABA_A$ receptor agonists and attenuated by antagonists or inverse agonists,[5] the latter reducing signs of ethanol

0-8493-2392-4/04/$0.00+$1.50

intoxication in mice and rats[6–8] and blocking the anxiolytic effects of ethanol,[9–11] while $GABA_A$ receptor agonists potentiate ethanol-induced anxiolysis.[12]

Further research shows that there is also a functional relationship between ethanol action and $GABA_A$ receptor activity. Physiologically relevant concentrations of ethanol (20 to 60 m*M*) potentiate muscimol-stimulated Cl^- uptake in cerebral cortical synaptoneurosomes of rats,[13,14] mouse cerebellar microsacs,[15] and cultured spinal cord neurons.[16] Similar to ethanol-induced behavioral effects, ethanol potentiation of $GABA_A$ receptor-mediated Cl^- influx is blocked by Ro15-4513 and $GABA_A$ receptor antagonists.[14,17,18]

However, *in vivo* studies examining ethanol interactions with $GABA_A$ receptors give very mixed results, suggesting that complex mechanisms are involved in the actions of ethanol. Acute ethanol administration potentiates GABA-mediated inhibition *in vivo*, but only in specific brain regions or cell types.[19–26] Systemic ethanol administration potentiates GABA-mediated inhibition in the cortex, medial septum, and inferior colliculus, but not in the lateral septum or hippocampus.[27–32] Recent studies show that ethanol has direct, consistent, and low-dose effects on both presynaptic and postsynaptic sites in central amygdala neurons.[33] Ethanol (44 m*M*) had effects on both evoked and spontaneous $GABA_A$ receptor-mediated responses using whole-cell recordings. Ethanol modulation of $GABA_A$ receptor function has also been reported in dorsal root ganglion cells.[34] In outside-out patches from dorsal root ganglion cells, ethanol increased mean open time, frequency of opening, and burst duration, while decreasing mean closed time of single $GABA_A$ receptors when it was co-applied with 1 μM GABA.[35] Other studies indicate that direct interaction of ethanol with neuronal $GABA_A$ receptors using patch clamp recording techniques is absent or minimal at pharmacologically relevant ethanol concentrations.[21,36,37]

Ethanol potentiation of GABA responses at remarkably low concentrations (1 to 5 m*M*) has recently been demonstrated in recombinant $GABA_A$ receptors containing $\alpha 4\delta$ subunits in oocytes.[38] Furthermore, this sensitivity is recapitulated *in vivo* in progesterone-withdrawn animals that exhibit an increase in the expression of this receptor subtype in the hippocampus. This receptor subtype is also particularly sensitive to neuroactive steroids.[39] These findings may be highly significant since $GABA_A$ δ subunit-containing receptors are thought to be primarily extrasynaptic in the brain[40] and may account for the majority of tonic GABAergic inhibition.[41,42] It is interesting to consider the possibility that these receptors account for the potentiating effects of ethanol on GABA responses in Cl^- flux assays[43,44] and extracellular recordings of GABAergic neurons.[45] Notably, many other recombinant $GABA_A$ receptors are insensitive to low ethanol concentrations (<50 m*M*) in any expression system,[46] including chimeric GABA/glycine receptors that respond to alcohols at high millimolar concentrations.[47] However, neurosteroids enhance the sensitivity of $\alpha 1\beta 2\gamma 2$ receptors to low ethanol concentrations.[48]

Multiple mechanisms may influence the sensitivity of $GABA_A$ receptors to ethanol. In some studies, certain experimental conditions have been manipulated to uncover an effect of ethanol. Inclusion of intracellular ATP,[49] cold shock treatment of hippocampal slices,[50] or selective blockade of $GABA_B$ receptors by CPG-35348[51] has revealed ethanol potentiation of GABA-mediated inhibition in the hippocampus. Ethanol also potentiates GABA-mediated inhibition in hippocampal CA1 pyramidal cells following proximal, but not distal stimulation.[52] Other results suggest that ethanol potentiation of GABAergic IPSPs is indirect, due to inhibition of NMDA-mediated EPSPs and subsequent Ca^{++}-dependent modulation of $GABA_A$ receptors.[37]

β-Adrenergic receptor stimulation and subsequent elevation of cAMP and protein kinase A (PKA) activity are required to detect ethanol potentiation of GABA in cerebellar Purkinje cells.[53,54] Furthermore, elevation of protein kinase C (PKC) in hippocampal slices increased the sensitivity of CA1 neurons to ethanol.[52] However, it remains unclear if manipulations of PKA or PKC unmask a nonphysiological sensitivity to ethanol or reinstate the natural phosphorylation state of $GABA_A$ receptors. Nonetheless, ethanol appears to have some indirect actions on $GABA_A$ receptors and may involve distinct intermediates that are required for ethanol enhancement of GABA action in the central nervous system (CNS).

10.2 3α,5α-THP INFLUENCES ETHANOL ACTIONS

We have made a series of discoveries that led us to propose that neurosteroids play an important role in the complex interactions between ethanol and $GABA_A$ receptors. Systemic ethanol administration dramatically elevates both plasma and cerebral cortical 3α,5α-THP levels in male and female rats.[55] In male rats, circulating 3α,5α-THP levels increased 600% above those of handling-habituated saline-injected control rats. Cerebral cortical 3α,5α-THP levels were elevated approximately 700% above those of controls. Female rats were tested during the estrous phase of the estrous cycle when both circulating and cerebral cortical 3α,5α-THP levels are substantially higher than those of male rats. Ethanol increased circulating 3α,5α-THP levels by 157% and cerebral cortical 3α,5α-THP levels by 234% in female rats. These increases in brain 3α,5α-THP levels are sufficient to enhance GABA-mediated neurotransmission in the brain[56,57] and produce anxiolytic and anticonvulsant effects.[58,59] Similar effects of ethanol on 3α,5α-THP and 3α,5α-THDOC were observed in Sardinian alcohol-preferring rats.[60] It is conceivable that brain concentrations of 3α,5α-THP could modulate the behavioral effects of ethanol discussed above.

The effects of ethanol on cerebral cortical 3α,5α-THP levels in male rats are dose and time dependent.[61] The effect of ethanol on cortical 3α,5α-THP levels is also biphasic, similar to the biphasic effects of ethanol on $GABA_A$ receptor-mediated chloride uptake in cerebral cortical synaptoneurosomes.[14] The time course of the effect of acute ethanol administration suggests that 3α,5α-THP contributes to effects of systemic ethanol administration within 20 and 120 min of administration and reaches a plateau within 40 and 80 min of ethanol injection. The maximal effect was a $341 \pm 14\%$ increase in 3α,5α-THP levels above control.[61]

Ethanol-induced elevations in progesterone and corticosterone levels suggest a nonspecific activation of steroid biosynthesis or adrenal release. If pregnane steroid biosynthesis were activated nonspecifically by ethanol, an increase in plasma or brain levels of dehydroepiandrosterone (DHEA) or its sulfated congener, DHEAS, would be predicted. However, systemic ethanol administration (2 g/kg, i.p.) did not alter DHEA or DHEAS levels in several brain regions.[62] The lack of effect on ethanol on DHEA and DHEAS levels is consistent with the idea that androstane steroids are differentially regulated than pregnane steroids. Furthermore, these data emphasize the specificity of the effects of ethanol on 3α,5α-THP and 3α,5α-THDOC levels.

Since brain concentrations of 3α,5α-THP above 5 n*M* would be expected to potentiate GABA receptor-mediated chloride flux,[56,57] we hypothesized that ethanol induction of 3α,5α-THP in the brain could contribute to the pharmacologic effects of ethanol *in vivo*. When the specific steroid biosynthetic inhibitor finasteride[63] was administered preceding ethanol administration to block subsequent 3α,5α-THP synthesis, the increase in cortical 3α,5α-THP levels induced by moderate ethanol doses was partially reversed. We used this tool to probe the role of ethanol-induced 3α,5α-THP in specific ethanol-induced behaviors.

Ethanol has anticonvulsant effects in humans and rodents. Finasteride administration prior to ethanol completely prevented the anticonvulsant effect of ethanol at 40 min.[61] Moreover, finasteride did not alter seizure thresholds 10 min after ethanol administration, a time when ethanol lacks an anticonvulsant effect. These data support the supposition that ethanol induction of 3α,5α-THP levels in the brain mediates or modulates the anticonvulsant effects of ethanol via $GABA_A$ receptors.

To examine if 3α,5α-THP contributes to ethanol-induced effects on extracellular electrophysiological recordings, we investigated the spontaneous firing rates of medial septum/diagonal band (MS/DB) of Broca neurons.[61] In this region, ethanol significantly reduces spontaneous firing rates 40 to 50%, and this effect was observed 15 min after ethanol injection and sustained for up to 55 min. This time course corresponds closely to the time course of ethanol-induced elevation in 3α,5α-THP levels in the cerebral cortex, but finasteride pretreatment also completely blocked the effect of ethanol on spontaneous firing rate.[61] In addition, the effect of exogenous 3α,5α-THP was not altered by finasteride pretreatment, suggesting that finasteride pretreatment does not alter the ability of MS/DB neurons to respond to GABA-mediated inhibition. Therefore, it is likely that finasteride pretreatment prevents the formation of 3α,5α-THP that normally mediates or modulates the effect of ethanol on MS/DB neurons.

Several behavioral effects of ethanol can be observed within minutes of systemic ethanol administration, preceding ethanol-induced elevations of cortical 3α,5α-THP levels. For example, ethanol produces muscle relaxation and motor incoordination that can be investigated by measurement of the Majchrowicz Intoxication Scale[2] or the aerial righting reflex.[64] Finasteride did not block the effect of ethanol in either of these measures, and significantly enhanced the effect of ethanol at some time points.[62] Hence, it appears that neurosteroids do *not* contribute to the motor-incoordinating effects of ethanol. Studies on the effects of adrenalectomy on ethanol-induced motor incoordination are needed to support this conclusion.

10.3 3α,5α-THP CONTRIBUTES TO THE SEDATIVE-HYPNOTIC EFFECTS OF ETHANOL

Ethanol has profound effects on sleep and loss-of-righting reflex (LORR). Furthermore, alcoholics experience more sleep-related disorders than nonalcoholics.[65,66] Common problems experienced by alcoholics are difficulty falling asleep, decreased total sleep time, obstructive sleep apnea, narcolepsy, and other parasomnias.[67] Loss-of-righting reflex is used as a crude index to study the sedative-hypnotic actions of ethanol and other sedative-hypnotic drugs.

Ethanol-induced LORR in rodents is modified by a large number of neurotransmitters and neuromodulators, including GABA, glutamate, and adenosine. Sedative-hypnotic effects produced by high doses of ethanol are potentially mediated by $GABA_A$ receptors by either direct or indirect actions. These effects can be enhanced by GABA agonists and blocked by GABA receptor antagonists.[68] Several studies show that administration of high doses of 3α,5α-THP or its precursors to rodents induces a LORR and anesthesia.[69,70] Interestingly, VanDoren et al.[61] showed that waking cerebral cortical levels of 3α,5α-THP correlate with duration of ethanol-induced LORR.[61] This study suggested a role of neuroactive steroids, including 3α,5α-THP, in the sedative-hypnotic actions of ethanol.

Systemic administration of the biosynthetic inhibitor finasteride (5α-reductase inhibitor types 1 and 2) reduced ethanol-induced elevation of 3α,5α-THP levels in rat cortex[61] when a moderate ethanol dose (2 g/kg) was investigated. However, prior administration of finasteride to rats did not alter cerebral cortical 3α,5α-THP levels at a hypnotic ethanol dose (3.5 g/kg) and had no effect on the duration of ethanol-induced LORR (Figure 10.1A and C). Since finasteride failed to modify cortical

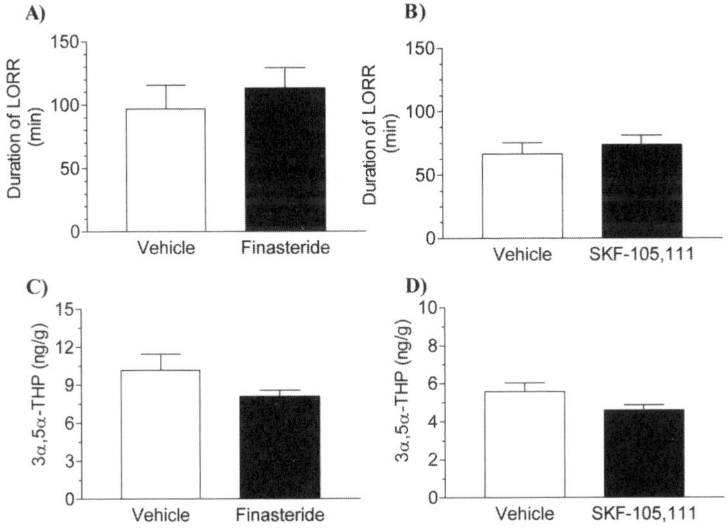

FIGURE 10.1 Finasteride and SKF-105,111 do not alter ethanol-induced LORR or cerebral cortical 3α,5α-THP levels in rats. Male rats were pretreated with vehicle, finasteride (25 mg/kg, s.c., –4 and –1.5 h) or SKF-105,111 (50 mg/kg, i.p., –15 min), before ethanol (20% v/v in saline). Rats were either decapitated 60 min (finasteride) or 40 min (SKF-105,111) after ethanol (2 g/kg, i.p.) injection, and cerebral cortical 3α,5α-THP levels were measured. The duration of LORR was recorded in rats injected with hypnotic dose (3.5 g/kg) of ethanol. Data represent the mean ± SEM of the effect of finasteride (A) and SKF-105,111 (B) on the duration of LORR (min) and the effect of finasteride (C) and SKF-105,111 (D) on cerebral cortical 3α,5α-THP levels (ng/g) in 10 to 14 rats per group.

3α,5α-THP levels, the effect of SKF-105,111 (5α-reductase type 1 inhibitor) on cortical 3α,5α-THP levels and the duration of righting reflex were tested in male rats. Reports on the effects of SKF-105,111 on ethanol-induced sleep time are controversial. Guidotti et al.[71] showed that SKF-105,111 reduced ethanol-induced LORR in mice, but Atzori et al.[72] presented evidence that SKF-105,111 does not modulate ethanol-induced LORR in mice. In our study, prior administration of SKF-105,111 to rats neither modulated the ethanol-induced LORR nor modified ethanol-induced cerebral cortical 3α,5α-THP levels (Figure 10.1B and D). The observed effects in rats could be due to the reduction of progesterone to the potent GABAergic modulator, 3α-hydroxy-4-pregnen-20-one (3α-HP), in the presence of 5α-reductase inhibitors. Since finasteride increases endogenous 3α-HP levels,[73] ethanol could theoretically increase 3α-HP under conditions where 5α-reductase activity is inhibited. Since our antibody completely cross-reacts with this steroid, it is possible that 3α-HP production explains the lack of finasteride and SKF 105,111 inhibition of both the apparent 3α,5α-THP levels and the duration of LORR. Alternately, hypnotic doses of ethanol may alter normal interactions of these inhibitors with the 5α-reductase enzyme, preventing their effects on the biosynthesis of 3α,5α-THP in the brain. Finally, the lack of effect of finasteride and SKF-105,111 at hypnotic ethanol doses may indicate novel pharmacokinetic interactions at high ethanol concentrations.

Previous studies have shown that plasma vasopressin contributes to activation of the hypothalamic pituitary adrenal (HPA) axis after systemic injection of ethanol to rats.[74] Plasma and brain content of vasopressin increases significantly after ethanol administration in rodents.[74,75] Furthermore, ethanol-induced elevation in vasopressin is responsible for tolerance to the motor-incoordinating effects of ethanol.[76] Tolerance to the hypnotic effect of ethanol in mice is prolonged by intracerebroventricular or subcutaneous injections of vasopressin.[76] Furthermore, systemic injection of vasopressin V1 receptor antagonist [d(CH(2))(5)Tyr(Me)-AVP], commonly known as the Manning's compound, alleviates ethanol-induced increases in hypothalamic corticotrophin-releasing factor (CRF), plasma adrenocorticotrophic hormone (ACTH), vasopressin, and corticosterone.[74] In view of these reports, we investigated the effect of Manning's compound on ethanol-induced LORR. The duration of LORR was not

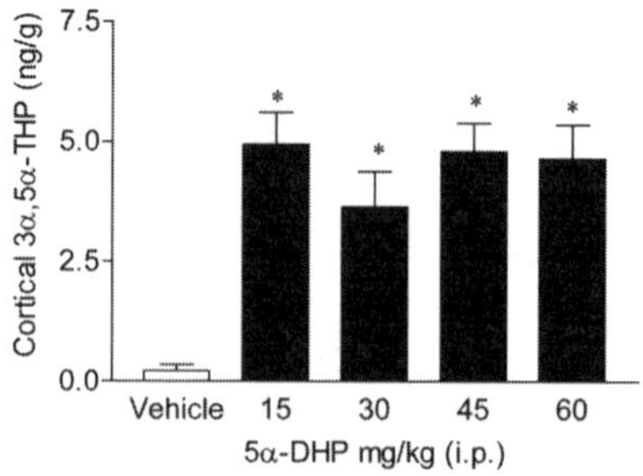

FIGURE 10.2 5α-DHP administration increases cortical 3α,5α-THP content in the rat. Male rats were injected with vehicle or 5α-DHP (15, 30, 45, or 60 mg/kg, i.p.) and decapitated after 60 min, and cerebral cortical 3α,5α-THP levels were measured. *, $p < .01$ compared with vehicle-treated rats.

TABLE 10.1
Systemic Administration of Vasopressin (V1) Receptor Antagonist Does Not Reduce Ethanol-Induced LORR, Plasma Progesterone, and Plasma Corticosterone Levels

Treatments	Duration of LORR (min ± SEM)	Waking Rat Plasma Progesterone (ng/ml)	Waking Rat Plasma Corticosterone (ng/ml)
Vehicle	80.33 ± 14.85	77.1 ± 2.6	305.6 ± 21.8
V1 antagonist			
50 μg/kg	78.78 ± 7.34	77.7 ± 2.6	333.9 ± 48.7
100 μg/kg	86.78 ± 11.14	75.8 ± 2.5	340.3 ± 22.8

Note: Rats (*n* = *10*) were injected with vasopressin receptor antagonist (50 or 100 μg/kg, i.p.) 15 min before ethanol (3.5 g/kg, i.p.) was injected, and the duration of LORR was recorded. After awakening, rats were euthanized by decapitation and plasma progesterone and corticosterone levels were measured.

affected by systemic administration of Manning's compound (50 or 100 μg/kg, i.p.) prior to administration of the hypnotic dose of ethanol. Moreover, waking plasma concentrations of progesterone and corticosterone were not modified (Table 10.1). These results suggest that ethanol stimulates steroidogenesis by pathways independent of vasopressin V1 receptors.

In addition to these biosynthetic inhibitors of the steroid metabolic pathway and vasopressin receptor antagonist, we studied the effect of adrenalectomy in male rats to identify the source of ethanol-induced elevation of neuroactive steroids. Injection of ethanol to adrenalectomized (ADX) rats reduced the total duration of loss-of-righting reflex by 90% in ADX compared to sham-adrenalec-

TABLE 10.2
Vasopressin Receptor Antagonist Does Not Alter the Effect of Moderate Doses of Ethanol (2 g/kg) on Cerebral Cortical 3α,5α-THP, Plasma Progesterone, and Plasma Corticosterone Levels

	3α,5α-THP (ng/g)		Progesterone (ng/ml)		Corticosterone (ng/ml)	
Treatments	Saline	EtOH	Saline	EtOH	Saline	EtOH
Vehicle	0.4 ± 0.9	2.4 ± 0.3[a]	13.5 ± 5.6	72.2 ± 3.5[a]	35.6 ± 13.6	324.7 ± 50.6[a]
V1 antagonist	0.2 ± 0.1	2.2 ± 0.3[a]	12.8 ± 4.4	75.5 ± 2.8[a]	64.3 ± 36.5	296.9 ± 49.8[a]

Note: Rats (*n* = *10*) were injected with vasopressin receptor antagonist (10 μg/kg, i.p.) 15 min before ethanol (EtOH, 2 g/kg, i.p.) was injected, and cerebral cortical levels of 3α,5α-THP (ng/g) and plasma progesterone levels (ng/ml) were measured after 60 min.

[a] $p < .05$ compared with saline-injected rats.

TABLE 10.3
Adrenalectomy Diminishes Ethanol-Induced LORR and Waking Rat Cerebral Cortical 3α,5α-THP Levels in Rats

	Duration of LORR (min ± SEM)	Waking Rat Cerebral Cortical 3α,5α-THP (ng/g ± SEM)
SHAM ADX	135.17 ± 24.47	9.0 ± 0.9
ADX	59.71 ± 16.92[a]	2.5 ± 0.2[a]

Note: ADX and SHAM ADX rats were injected with ethanol (3.5 g/kg, 20% v/v, i.p.), and the duration of loss-of-righting reflex and cortical 3α,5α-THP levels were measured after awakening.

[a] $p < .01$ as compared to sham ADX rats.

tomized (SHAM ADX) rats. Furthermore, waking cerebral cortical 3α,5α-THP levels of ADX rats were remarkably lower than those of the SHAM ADX rats (Table 10.3). Systemic administration of progesterone or 3α,5α-THP is known to potentiate ethanol-induced sleep (LORR).[77,78] Thus, it is clear that ethanol-induced increases in cerebral cortical 3α,5α-THP, mediated through release of adrenal steroids, contributes to the hypnotic effect of ethanol and adrenalectomy causes a loss of this effect.

Adrenalectomy removes many steroids and steroid precursors from circulation as well as 3α,5α-THP. To determine if the loss of 3α,5α-THP is specifically required for the effect of adrenalectomy on ethanol-induced LORR, we investigated if replacement of the immediate precursor of 3α,5α-THP would reverse the effects of adrenalectomy. It is widely known that 3α,5α-THP is biosynthesized only by successive metabolism of progesterone and 5α-DHP by the enzymes 5α-reductase and 3α-hydroxysteroid dehydrogenase, respectively. Systemic administration of 5α-DHP increases 3α,5α-THP levels in the brain, and high doses of 5α-DHP induce LORR by increasing brain content of 3α,5α-THP.[79] We found that systemic (i.p.) administration of 5α-DHP increased cortical 3α,5α-THP levels (Figure 10.2) without altering plasma progesterone or corticosterone levels. Thus, it was hypothesized that if 3α,5α-THP contributes to the hypnotic effects of ethanol, administration of subhypnotic doses of its immediate precursor, 5α-DHP, would reverse the effect of adrenalectomy on ethanol-induced increases in 3α,5α-THP levels and LORR. In agreement with this hypothesis, pretreatment with 5α-DHP to ADX rats not only restored the ethanol-induced increases in the duration of LORR, but also restored ethanol-induced cerebral cortical 3α,5α-THP levels (Figure 10.3). Since 5α-DHP is not a precursor of 3α-HP, it is unlikely that 3α-HP contributes to 3α,5α-THP-like levels in this experiment. Hence, these results show that ethanol elevates cortical levels of the neurosteroids by release of adrenal neuroactive steroid precursors. These precursors are clearly converted to neuroactive steroids in extra-adrenal organs, most likely the brain, and contribute to the hypnotic effects of ethanol.

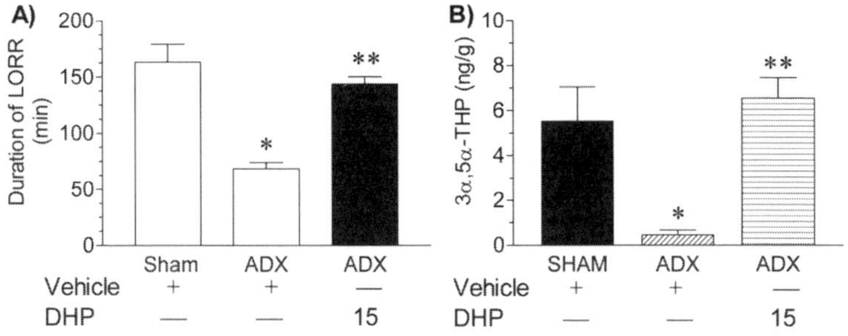

FIGURE 10.3 Systemic 5α-DHP administration to ADX rats restores the duration of ethanol-induced LORR and ethanol-induced levels of 3α,5α-THP. Male SHAM ADX or ADX rats were injected with 5α-DHP (15 mg/kg, i.p.) or vehicle 15 min before injection of ethanol (20% v/v in saline). Cerebral cortical levels of 3α,5α-THP (ng/g) were measured 40 min after ethanol (2 g/kg) injection, and the duration of LORR (min) was recorded individually in rats injected with the hypnotic dose (3.5 g/kg) of ethanol. Data represent mean ± SEM of the duration of ethanol-induced LORR (min) (A) and cerebral cortical 3α,5α-THP levels (ng/g) (B) in 10 to 16 rats per group. *, $p < .01$ compared with ethanol-pretreated SHAM ADX rats; **, $p < .05$ compared to vehicle-injected ADX rats. (Adapted from Khisti et al., *Brain Res.*, 980, 255, 2003.)

10.4 3α,5α-THP CONTRIBUTES TO ETHANOL IMPAIRMENT OF HIPPOCAMPAL FUNCTION AND SPATIAL LEARNING?

The hippocampus and related structures (e.g., the medial septum) are sensitive to ethanol's neurophysiological effects. For example, acute ethanol administration inhibits the spontaneous neural activity of both the medial septal/diagonal band of Broca neurons[80] and hippocampal pyramidal neurons.[81,82] However, similar doses of ethanol do not inhibit the spontaneous neural activity of lateral septal neurons.[80] Furthermore, it has been shown that locally applied ethanol via electro-osmosis potentiates GABA-mediated inhibition in the medial septum[80] but does not potentiate GABA-mediated inhibition in the lateral septum.[83] However, ethanol potentiation of GABA-mediated inhibition via electro-osmosis often requires preexposure to ethanol, suggesting that a direct ethanol–$GABA_A$ receptor interaction is not likely in these brain regions. In support of this suggestion, others have shown that ethanol does not directly interact with hippocampal $GABA_A$ receptors,[26,31] except following specific experimental conditions such as inclusion of intracellular ATP,[49] cold shock treatment,[50] or blockade of $GABA_B$ receptors.[51] Therefore, it appears likely that ethanol potentiation of GABA-mediated inhibition in the hippocampus via $GABA_A$ receptors involves a neurobiological intermediary.

The effect of acute ethanol exposure on the use of spatial information has been studied for many years. For example, acute ethanol administration makes animals less attentive to environmental changes and decreases spontaneous alterations in elevated radial arm mazes.[84] Furthermore, it has been repeatedly demonstrated that ethanol impairs spatial learning and memory in both rats and mice. For example, acute ethanol administration impairs the use of previously learned spatial reference memory,[3,85] the learning of new tasks that are dependent on spatial information,[86,87] spatial contextual learning,[88] and spatial working memory.[28,89,90] However, acute ethanol administration *does not* produce global memory impairments. For example, acute ethanol administration does not alter the learning of a cue-based (nonspatial) task,[3,88] and in some specific tasks, it can facilitate the use of cue (nonspatial) information.[91]

One neurophysiological function of single hippocampal pyramidal neurons is their sensitivity to spatial (i.e., place) information, termed spatial specificity. If ethanol alters the use of spatial information by altering hippocampal function, then similar doses of ethanol should alter the spatial specificity of hippocampal place cells. The first report of ethanol altering the spatial specificity of place cells demonstrated that a 1.0 g/kg i.p. injection of ethanol decreased the number of place units recorded from awake freely behaving rabbits.[92] More recently, it was shown that a higher dose of ethanol, 2.0 g/kg i.p., produces dramatic effects on the spatial specificity of place cells in freely behaving rats. The spatial specificity of each place cell recorded was significantly reduced 30 min after the ethanol injection compared to pre-ethanol levels. Furthermore, each neuron's spatial specificity completely recovered 24 h after ethanol administration.[93] The effect of ethanol on spatial specificity has been shown to be due to a selective (in that hippocampal interneurons are not affected), and rather dramatic, reduction in the firing rate of hippocampal pyramidal place cells.[90,94]

Recently, we have demonstrated that acute administration of 2.0 g/kg ethanol (the dose producing maximal impairments in spatial memory but no impairment in nonspatial memory[3,91]) increases hippocampal 3α,5α-THP levels by ~300%.[95] Since 3α,5α-THP directly enhances $GABA_A$ receptor responses determined by *in vitro* hippocampal neurophysiology,[96] we hypothesized that ethanol-induced increases in hippocampal 3α,5α-THP levels modulate ethanol's *in vivo* electrophysiological effects in the hippocampal system. Acute ethanol administration produced a dose-dependent inhibition of spontaneously active hippocampal pyramidal neurons over the 60-min recording session.[82] Importantly, preadministration of 50 mg/kg finasteride completely blocked ethanol-induced neural inhibition (Figure 10.4). This suggests that an ethanol-induced increase in 3α,5α-THP is one mechanism by which acute ethanol administration alters hippocampal function.

3α,5α-THP concentrations in the brain increase following many different types of stress, including swim stress, mild foot shock, or handling of naïve animals.[97–99] Similar stress conditions also impair performance on spatial working memory tasks.[100–102] Furthermore, intracerebroventricular infusion[103] or subcutaneous administration of 3α,5α-THP can impair spatial working memory.[104] It should be noted that the effects of 3α,5α-THP on performance in tasks requiring the use of spatial information are strikingly similar to the effects produced by acute ethanol administration. Therefore, we investigated if acute ethanol administration and acute 3α,5α-THP administration produced similar memory impairments, measured as the mean increase in escape latency when animals were tested in the Morris water maze.[105] We found that acute ethanol and 3α,5α-THP administration produce similar impairments in spatial memory, but neither drug produces impairments in nonspatial memory (Figure 10.5A and B).

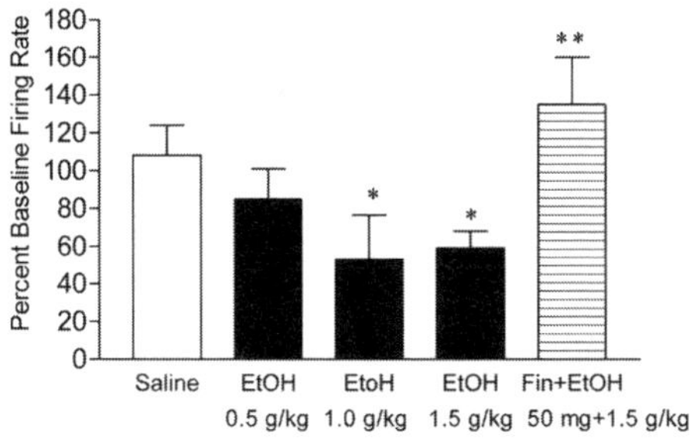

FIGURE 10.4 Blockade of neurosteroid biosynthesis by finasteride pretreatment reverses ethanol-induced inhibition of spontaneously active hippocampal pyramidal neurons. Acute ethanol administration inhibits the spontaneous firing rate of single hippocampal pyramidal neurons recorded in rats. However, pretreatment with 50 mg/kg finasteride completely blocks ethanol's effect. Error bars denote SEM; *, $p < .05$ compared to saline control values; n = 5 neurons per condition (one-way ANOVA, F = 3.73, $p < .03$). (Adapted from Tokunaga, S. et al., *Brain Res.*, 967, 273, 2003.)

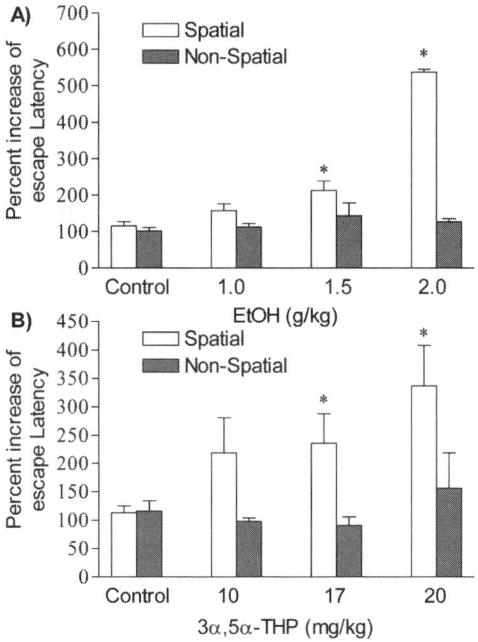

FIGURE 10.5 (A) Acute ethanol administration impairs spatial memory but does not impair nonspatial memory. Animals were trained in the Morris water maze, either the spatial or nonspatial task, prior to testing after either an ethanol injection or a saline injection. Error bars denote SEM; n = 9 to 10 animals per condition. (B) Acute 3α,5α-THP administration impairs spatial memory but does not impair nonspatial memory. Animals were trained in the Morris water maze, either the spatial or nonspatial task, prior to testing after either a 3α,5α-THP injection or a beta-cyclodextrin vehicle injection. Error bars denote SEM; n = 8 to 10 animals per condition. (Adapted from Matthews, D.B. et al., *Alcohol. Clin. Exp. Res.*, 26, 1747, 2002.)

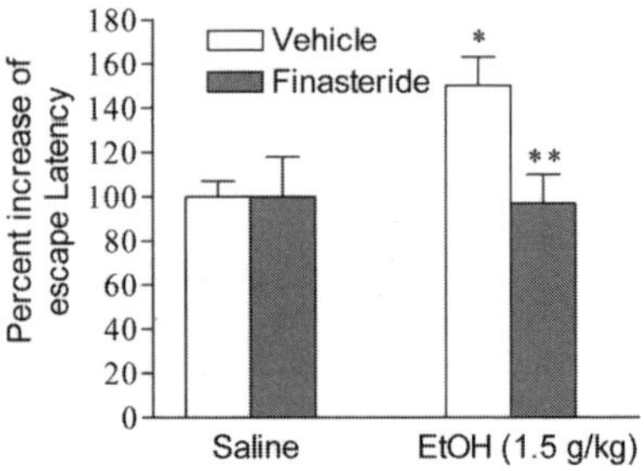

FIGURE 10.6 Blockade of neurosteroid biosynthesis by finasteride pretreatment reverses ethanol-induced spatial memory impairments. Animals were trained in the Morris water maze using the standard spatial version of the task and then pretreated with either beta-cyclodextrin vehicle or 50 mg/kg finasteride prior to testing after either a saline or a 1.5 g/kg ethanol injection (planned *t*-test, saline vs. 1.5 g/kg ethanol, t = 2.54, df = 14; finasteride + 1.5 g/kg ethanol vs. 1.5 g/kg ethanol, t = 2.32, df = 14). Error bars denote SEM; n = 8 per condition; *, $p < .05$.

Recently, we have investigated if finasteride preadministration alters ethanol-induced spatial memory deficits. Briefly, adult male rats were trained the spatial version of the Morris water maze. Following training, animals were pretreated with either 50 mg/kg finasteride or vehicle, and then given an ethanol challenge (1.5 g/kg) or a saline control challenge 30 min prior to having their spatial memory tested. As expected from previous studies, the 1.5 g/kg ethanol challenge impaired spatial memory as evidenced by an increase in mean escape latency. Moreover, pretreatment with finasteride completely blocked the spatial memory impairment (Figure 10.6). This demonstrates that spatial memory impairment due to a moderate ethanol dose is dependent upon steroidogenesis. Hence, the effects of ethanol on hippocampal place cell activity and spatial memory appear to be dependent upon the synthesis of GABAergic neurosteroids, suggesting that neurosteroids mediate the cognitive-impairing effects of ethanol.

10.5 MECHANISMS OF ETHANOL INDUCTION OF GABAERGIC NEUROSTEROIDS

Acute systemic administration of ethanol stimulates the HPA axis and elevates CRF, ACTH, plasma progesterone, and corticosterone levels.[74,106,107] Since progesterone and corticosterone are biosynthetic precursors of neurosteroids 3α,5α-THP and 3α,5α-THDOC, respectively, it was hypothesized that elevations in plasma neuroactive steroids would contribute to ethanol-induced elevation of 3α,5α-THP. However, levels of plasma progesterone and corticosterone do not correlate ($r = 0$ for progesterone, $r = 0.18$ corticosterone) with the cerebral cortical levels of 3α,5α-THP.[61] Thus, ethanol-induced elevation of 3α,5α-THP levels in the cerebral cortex may involve mechanisms that are distinct from the regulation of plasma steroid levels.

This possibility was tested by measuring cerebral cortical 3α,5α-THP and plasma progesterone and plasma corticosterone levels in adrenalectomized and

TABLE 10.4
Adrenalectomy Reduces Basal and Ethanol-Induced Cerebral Cortical 3α,5α-THP, Plasma Progesterone, and Plasma Corticosterone Levels

	3α,5α-THP (ng/g)		Progesterone (ng/ml)		Corticosterone (ng/ml)	
	Saline	EtOH	Saline	EtOH	Saline	EtOH
SHAM	3.8 ± 1.4	16.8 ± 7.2[a]	2.2 ± 0.7	68.0 ± 17.1[a]	25.0 ± 0.7	415.0 ± 10.0[a]
ADX	1.4 ± 0.4	2.0 ± 1.7[b]	0.7 ± 0.0[a]	0.8 ± 0.2[b]	25.0 ± 0.0	55.0 ± 0.2[b]

Note: Cerebral cortical levels of 3α,5α-THP (ng/g) and plasma progesterone and corticosterone levels (ng/ml) were measured 40 min after ethanol (EtOH, 2 g/kg, i.p.) administration.

[a] $p < .05$ compared with saline-injected SHAM ADX rats.
[b] $p < .05$ compared to ethanol-injected SHAM ADX rats.

sham-operated male rats. Cerebral cortical levels of 3α,5α-THP were remarkably reduced (87.90%) in adrenalectomized rats compared to in sham rats (Table 10.4). Similarly, the plasma progesterone and corticosterone levels were reduced by 98.85 and 98.40% in adrenalectomized rats (Table 10.4). These results suggest that ethanol acts as an acute stress, activates the HPA axis, and induces the release of 3α,5α-THP or progesterone into the circulation, elevating brain 3α,5α-THP levels. It appears that intact adrenals are required for ethanol induction of brain neurosteroids, possibly due to the loss of steroid precursors since replacement with 5α-DHP completely reversed the effect of adrenalectomy on ethanol-induced elevations of 3α,5α-THP (Figure 10.3).

It was recently shown that vasopressin receptor mediates activation of the HPA axis by acute ethanol (intragastic route) stress in rats.[74] Systemic administration of Manning's compound, before ethanol administration, significantly reduced levels of hypothalamic CRH, plasma ACTH, and corticosterone level.[74] It is possible that ethanol-induced elevation of neuroactive steroids is also mediated through similar activation of the HPA axis after ethanol administration. However, when we injected rats with Manning's compound (10 μg/kg, i.p.) before ethanol administration, there was no difference in the cerebral cortical levels of 3α,5α-THP and plasma levels of progesterone or corticosterone (Table 10.2). This difference may be due to a difference in the source of Manning's peptide or the route of ethanol injections. Future studies should study the effect of the V1 receptor antagonist by directly injecting in the brain. It is long known that systemic administration of dexamethasone suppresses the HPA axis. Dexamethasone reduces acute ethanol withdrawal-induced elevations in cortisol in patients.[108] Further, acute systemic dexamethasone administration is known to reverse anxiolytic,[109] sedative,[110] hypothermic,[111] and hypertensive[112] actions of ethanol in rats. In light of these reports, we investigated the effect of dexamethasone pretreatment on ethanol-induced elevations of plasma progesterone, plasma corticosterone, and cerebral cortical 3α,5α-THP. Systemic administration of

TABLE 10.5
Dexamethasone Does Not Alter Ethanol-Induced Cerebral Cortical 3α,5α-THP, Plasma Progesterone, and Plasma Corticosterone Levels

	3α,5α-THP (ng/g)		Progesterone (ng/ml)		Corticosterone (ng/ml)	
Treatments	Saline	EtOH	Saline	EtOH	Saline	EtOH
Vehicle	1.7 ± 0.5	4.4 ± 0.6[a]	47.6 ± 19.9	99.1 ± 3.7[a]	41.5 ± 5.1	729.8 ± 43.9[a]
Dexamethasone	1.5 ± 0.2	4.5 ± 0.3[a]	30.2 ± 14.2	95.2 ± 5.3[a]	44.9 ± 8.3	733.6 ± 51.7[a]

Note: Rats (*n* = *10*) were injected with dexamaethasone (15 mg/kg, i.p.) 15 min before ethanol (EtOH, 2 g/kg, i.p.) was injected, and cerebral cortical levels of 3α,5α-THP (ng/g) and plasma progesterone levels (ng/ml) were measured after 60 min.

[a] $p < .05$ compared with saline-injected rats.

dexamethasone (15 mg/kg, i.p.) prior to ethanol (2 g/kg, i.p.) injection failed to alter ethanol-induced elevation of plasma and cortical levels of neuroactive steroids (Table 10.5). These results suggest that ethanol-induced elevation of plasma neuroactive steroids may not require activation of the HPA axis to induce activation of neuroactive steroid biosynthesis or adrenal release. Ethanol could increase 3α,5α-THP levels via several mechanisms, including modulation of biosynthesis, catabolism, or release of endogenous steroid pools. Further studies are needed to elucidate this mechanism of ethanol action.

As shown in Figure 10.7, ethanol can modulate the neuroactive steroid biosynthetic pathway at many potential sites. Similar to acute stress, ethanol increases CRF in the hypothalamus, which subsequently increases ACTH hormone release from the pituitary into the circulation.[75] Elevated ACTH levels or ethanol may increase influx of cholesterol through the mitochondrial membrane by increasing the endogenous diazepam binding inhibitor (DBI) peptide and steroidogenic acute regulating protein (StAR) responsible for controlling steroid synthesis.[113] The increase in cholesterol might lead to increases in a number of neuroactive steroids, including pregnenolone, progesterone, 3α,5α-THP, and 3α,5α-THDOC. Increases in the levels of these GABAergic neuroactive steroids contribute to sedative-hypnotic, antidepressant, cognitive, anticonvulsant, and anxiolytic effects of ethanol. Several possible mechanisms for the ethanol-induced increase in brain GABAergic neurosteroids have been explored.[114] Systemic pretreatment with the biosynthetic inhibitors finasteride, indomethacin, and trilostane either moderately or completely fail to inhibit the ethanol-induced increase in cortical 3α,5α-THP content. The type II 5α-reductase inhibitor finasteride was the only enzyme inhibitor that reduced the ethanol-induced increase in cerebral cortical 3α,5α-THP at nonhypnotic doses of ethanol. The 3α-hydroxysteroid dehydrogenase (3α-HSD) inhibitor indomethacin, paradoxically, increased cerebral cortical concentrations of 3α,5α-THP, and the 3β-hydroxysteroidoxidoreducatse/isomerase inhibitor

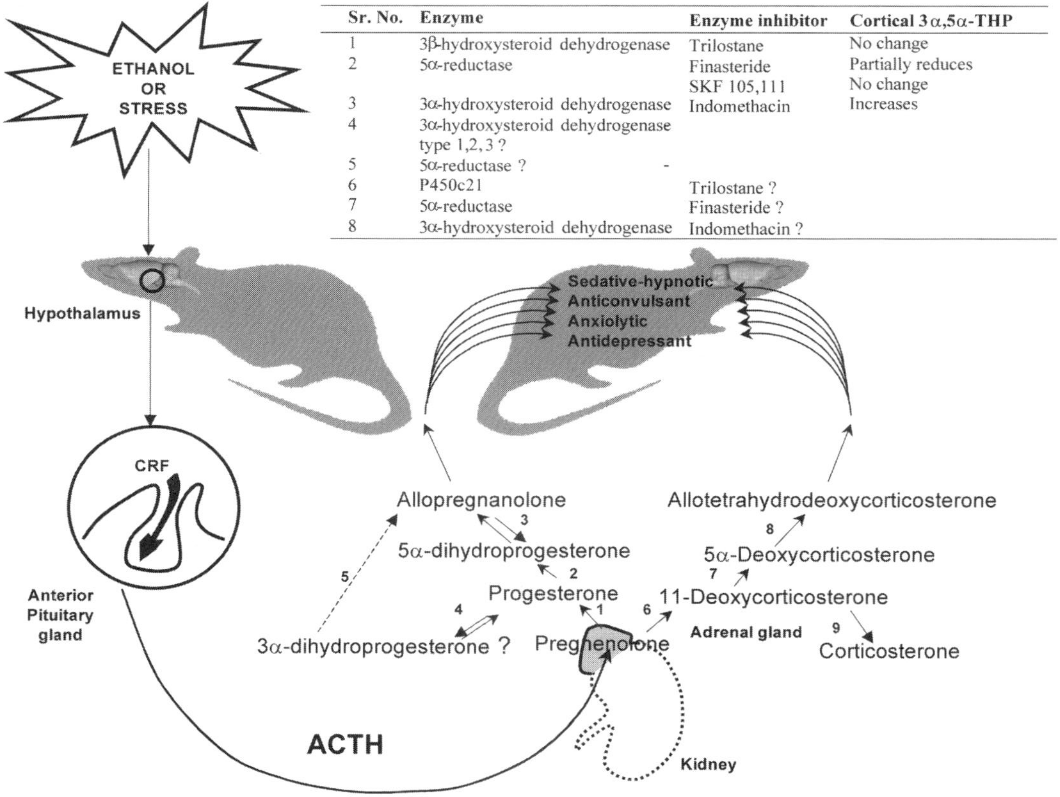

Sr. No.	Enzyme	Enzyme inhibitor	Cortical 3α,5α-THP
1	3β-hydroxysteroid dehydrogenase	Trilostane	No change
2	5α-reductase	Finasteride	Partially reduces
		SKF 105,111	No change
3	3α-hydroxysteroid dehydrogenase	Indomethacin	Increases
4	3α-hydroxysteroid dehydrogenase type 1,2,3 ?		
5	5α-reductase ?	-	
6	P450c21	Trilostane ?	
7	5α-reductase	Finasteride ?	
8	3α-hydroxysteroid dehydrogenase	Indomethacin ?	

FIGURE 10.7 Biosynthetic pathways involved in ethanol-induced increases in neuroactive steroids in rats. Systemic administration of ethanol can activate the HPA axis and elevate hypothalamic CRF content in the brain that in turn releases ACTH from the pituitary gland into the plasma. Increased plasma ACTH or ethanol may subsequently increase DBI and StAR levels in the adrenals, resulting in the transport of cholesterol to the inner mitochondrial membrane for the biosynthesis of neuroactive steroids (rate-limiting step in biosynthesis). The elevated levels of GABAergic neuroactive steroids contribute to sedative-hypnotic, anxiolytic, anticonvulsant, antidepressant, and cognitive effects of ethanol. The effects of various biosynthesis inhibitors on ethanol-induced neurosteroidogenesis are indicated in the inset table. The serial numbers represent enzymes in the biosynthetic pathway for neurosteroidogenesis. The dotted lines represent possible pathways for conversion of steroids.

trilostane did not have an effect on cerebral cortical 3α,5α-THP levels.[61] Similarly, systemic pretreatment with PK 11195, an antagonist at DBI receptors, and anti-alarmin, an antagonist at the CRF receptor and NPI-113, a partial agonist of neurosteroid sites on $GABA_A$ receptors, did not prevent ethanol-induced increases in 3α,5α-THP levels in the cortex.[114]

The lack of effect of neurosteroid biosynthesis inhibitors on ethanol-induced increases in 3α,5α-THP levels may indicate that ethanol-induced elevations in brain levels of neurosteroids are not dependent on *de novo* biosynthesis in the brain. However, the possibility exists that ethanol has pharmacokinetic interactions with steroid biosynthesis inhibitors and interferes with their inhibitory activity *in vivo*. Further studies are needed to determine if ethanol directly alters the activity of neurosteroid biosynthetic enzymes in the brain, as this mechanism may contribute to ethanol-induced elevations in neurosteroid concentrations.

10.6 THE ROLE OF NEUROSTEROIDS IN ETHANOL TOLERANCE, DEPENDENCE, AND WITHDRAWAL

Prolonged or repeated ethanol intake leads to the development of tolerance, demonstrated by a reduced behavioral response to ethanol at a particular cellular concentration. Tolerance results in a reduction in the anxiolytic, motor-incoordinating, and sedative-hypnotic effects of ethanol and a requirement for higher ethanol doses to elicit pharmacologic effects.[115–117] There is some evidence indicating that neurosteroids may play a role in the development or maintenance of ethanol tolerance.

Although acute systemic ethanol administration elevates 3α,5α-THP levels in the brain, chronic ethanol consumption does not alter steady-state 3α,5α-THP levels.[118] Indeed, ethanol-dependent male rats show decreased levels of 3α,5α-THP in the cerebral cortex compared to pair-fed controls, while ethanol-withdrawn animals showed a return of cortical 3α,5α-THP to baseline levels.[118] Furthermore, ethanol-dependent and -withdrawn animals given a moderate ethanol challenge showed reduced levels of ethanol-induced 3α,5α-THP in the cerebral cortex compared to ethanol-naïve rats.[62] Hence, tolerance to ethanol may involve a loss of ethanol induction of 3α,5α-THP and a corresponding reduction in related pharmacological effects of ethanol. Tolerance to the anticonvulsant[119] and hypnotic[120] effects of ethanol are well established. Ongoing studies in our lab examining the relationship between the development of ethanol tolerance and the loss of ethanol induction of cortical 3α,5α-THP levels will elucidate this relationship.

While ethanol tolerance has been demonstrated in many preparations, tolerance to the cognitive-impairing effects of ethanol has not been documented. Recently, we have been investigating if chronic binge ethanol administration produces tolerance to the cognitive-impairing effects of ethanol. Beginning on postnatal day 30 animals receive a hypnotic dose of ethanol (5.0 g/kg) every other day for 20 days; animals are trained in the standard spatial version of the Morris water maze on the days they are not administered ethanol. After training and the ethanol exposure period, animals' spatial memory was tested following either a 2.0 g/kg ethanol or saline challenge. Chronic binge ethanol administration produced profound tolerance to the cognitive-impairing effects of ethanol. Furthermore, in a separate experiment, we found that this ethanol exposure paradigm produced tolerance to ethanol-induced increases in hippocampal 3α,5α-THP levels (unpublished data). These data suggest that the loss of ethanol-induced increases in neurosteroids is involved in tolerance to the cognitive-impairing effects of ethanol.

Tolerance to ethanol can be environmentally dependent.[121] If alcohol administration is repeatedly paired with a specific environment, tolerance can be demonstrated in that environment, but is absent if the animal is moved to a novel environment. Stress-induced elevation of 3α,5α-THP levels may contribute to this response, since novelty itself is a potent stressor.[55] In this model, stress activation of the HPA axis may elevate 3α,5α-THP levels and reverse tolerance by enhancing the effect of ethanol. Furthermore, tolerance can be maintained by vasopressin activation of the HPA axis.[122] We propose that prolonged ethanol exposure reduces the ability of ethanol (and vasopressin) to induce 3α,5α-THP concentrations in the brain, and the loss of this response could contribute to the reduced ethanol response (tolerance). This mechanism could further explain the important role of the HPA axis in tolerance development and the ability of vasopressin to maintain ethanol tolerance in the absence of ethanol.

Alterations in $GABA_A$ receptor function and expression may also contribute to the development of ethanol tolerance.[123] Chronic ethanol administration differentially alters the expression of various $GABA_A$ receptor subunits, suggesting that alterations in the subunit expression of $GABA_A$ receptors may account for alterations in $GABA_A$ receptor function.[124–129] These effects of ethanol on $GABA_A$ receptor function and gene expression could be mediated, in part, by 3α,5α-THP. Therefore, the interplay of the neuromodulating and genomic effects of 3α,5α-THP may play a key role in adaptations to ethanol.

Changes in $GABA_A$ receptor function are thought to underlie some of the consequences of ethanol dependence. In particular, the increase in anxiety and the decrease in seizure threshold seen in dependent animals and humans during ethanol withdrawal may be mediated by changes in GABAergic neurotransmission that are brought about by chronic exposure to a $GABA_A$ receptor-potentiating drug.[123] Supporting this supposition, administration of 3α,5α-THP blocks the decrease in bicuculline-induced seizure threshold seen in ethanol-withdrawn male and female rats at doses of 3α,5α-THP that do not increase seizure threshold in pair-fed control rats.[59] Indeed, there is a dramatic sensitization to the anticonvulsant effects of 3α,5α-THP and 3α,5α-THDOC in ethanol-dependent rats.[130,131] This behavioral sensitization to neurosteroids is accompanied by an increase in sensitivity of $GABA_A$ receptors to 3α,5α-THP and 3α,5α-THDOC in cerebral cortical synaptoneurosomes using a Cl^- flux assay[131] and is not due to different endogenous levels of 3α,5α-THP in withdrawn vs. control rats.[118] Furthermore, neurosteroid sensitization during withdrawal is accompanied by an increase in $GABA_A$ receptor α_4, γ_1, β_2, and β_3 receptor subunit mRNA and peptides. These subunits have been shown to maintain or increase sensitivity of $GABA_A$ receptors to neurosteroids in recombinant expression systems.[39,132,133]

The relationship between neurosteroid levels and the sensitivity of $GABA_A$ receptors to neurosteroids remains controversial. Although one study found no change in endogenous 3α,5α-THP levels in the cortex or plasma of ethanol-withdrawn rats,[118] other studies have reported decreases in neurosteroid levels in ethanol-withdrawing mice and humans. Plasma 3α,5α-THP levels were decreased in ethanol-withdrawn B6 and D2 mice, relative to controls.[134] Interestingly, the B6 mice showed a decrease in 3α,5α-THP levels of 15%, whereas there was a 50% decrease in D2 mice. The greater reduction in endogenous neurosteroid in withdrawn D2 mice could contribute to the greater withdrawal severity experienced by those animals. In alcoholics, decreased levels of both 3α,5α-THP and 3α,5α-THDOC have been reported during the early phase of ethanol withdrawal.[135] This early phase of ethanol withdrawal is characterized by increases in anxiety and depression that can be measured by psychological tests. Later in withdrawal, when the anxiety and depression disappear, neurosteroid levels also return to normal. This study suggests that reduced levels of endogenous neurosteroids contribute to the negative consequences of ethanol withdrawal.

Withdrawal from progesterone and associated decreases in 3α,5α-THP result in behavioral changes that mimic the signs of ethanol withdrawal. Specifically, progesterone-withdrawn female rats exhibit increased anxiety,[136] increased sensitivity to convulsant drugs that act on $GABA_A$ receptors,[137] and decreased sensitivity to the sedative effects of benzodiazepines.[138] Whole-cell patch clamp electrophysiology in dissociated neurons from CA1 of the hippocampus showed that the behavioral changes were accompanied by changes in the function of $GABA_A$ receptors associated with increases in α4 subunit peptide and mRNA during progesterone withdrawal.[137,139] Furthermore, blocking this increase by infusion of antisense α4 oligonucleotides into the hippocampus during progesterone withdrawal eliminates the increase in seizure sensitivity,[137] behavioral tolerance to benzodiazepines,[140] and the change in electrophysiological properties of $GABA_A$ receptors.[137] Therefore, in this model, it seems that withdrawal from exposure to progesterone and 3α,5α-THP causes a change in $GABA_A$ receptor subunit expression that results in the changes in $GABA_A$ receptor function that are responsible for the behavioral signs of withdrawal.

Ethanol- and progesterone-withdrawing animals exhibit similar withdrawal signs, such as increased anxiety, increased seizure susceptibility, and decreased sensitivity to benzodiazepines. These changes occur at the same time as increases in α4 subunit mRNA[141] and peptide levels[142] in the cerebral cortices of ethanol-dependent rats. These adaptations are associated with alterations in $GABA_A$ receptor function, including reduced sensitivity to GABA and benzodiazepines and increased sensitivity to inverse agonists.[123] Therefore, it is possible that changes in endogenous neurosteroid levels that occur during chronic ethanol exposure or withdrawal contribute to the changes in $GABA_A$ receptor function that underlie ethanol dependence. However, ethanol withdrawal is associated with other changes in $GABA_A$ receptor expression and function that were not reported after progesterone withdrawal, including changes in α1–3, β2/3, and γ1–2 subunit expression and sensitization to 3α,5α-THP.[131,142] Moreover, ethanol-dependent and -withdrawn animals are tolerant to ethanol, and progesterone-withdrawn rats are more sensitive to ethanol.[38] Therefore, further studies are needed to determine the role of neurosteroids in ethanol-induced adaptations of $GABA_A$ receptors.

10.7 CONCLUSIONS AND FUTURE DIRECTIONS

We propose that the data reviewed in this chapter represent compelling evidence that 3α,5α-THP produced in response to systemic ethanol administration contributes to some, but not all, effects of ethanol associated with modulation of $GABA_A$ receptors. These studies demonstrate an essential correlation between the time course of ethanol-induced 3α,5α-THP production in the brain and specific effects of ethanol. Furthermore, the anticonvulsant, cognitive-impairing, and inhibitory effects of ethanol on spontaneous neural activity in two brain regions were completely prevented by a key inhibitor of steroid biosynthesis. These data are consistent with the hypothesis that 3α,5α-THP contributes to ethanol action, and this interaction may represent an important mechanism of ethanol action.

Many intriguing questions in this area remain unresolved. Where is 3α,5α-THP synthesized and released in the brain and does ethanol influence these processes? Does ethanol directly alter the biosynthesis of endogenous 3α,5α- reduced steroids in the brain or peripheral steroidogenic organs? Do neurosteroids alter the sensitivity of $GABA_A$ receptors to ethanol — by either genomic or nongenomic mechanisms? Do GABAergic neuroactive steroids contribute to ethanol actions in humans? Clearly, further studies of the neurosteroid intermediaries that are involved in ethanol action may lead to important advances in the field and may potentially lead to the development of novel therapeutics.

REFERENCES

1. Frye, G.D., Chapin, R.E., Vogel, R.A., Mailman, R.B., Kilts, C.D., Mueller, R.A., and Breese, G.R., Effects of acute and chronic 1,3-butanediol treatment on central nervous system function: a comparison with ethanol, *J. Pharmacol. Exp. Ther.*, 216, 306, 1981.
2. Majchrowicz, E., Induction of physical dependence upon ethanol and the associated behavioral changes, *Psychopharmacologia*, 43, 245, 1975.
3. Matthews, D.B., Simson, P.E., and Best, P.J., Acute ethanol impairs spatial memory but not stimulus/response memory in the rat, *Alcohol. Clin. Exp. Res.*, 19, 902, 1995.
4. Givens, B.S., Low doses of ethanol impair spatial working memory and reduce hippocampal theta, *Alcohol. Clin. Exp. Res.*, 19, 763, 1995.
5. Lister, R.G. and Linnoila, M., Alcohol, the chloride ionophore and endogenous ligands for benzodiazepine receptors, *Neuropharmacology*, 30, 1435, 1991.
6. Hoffman, P.L., Tabakoff, B., Szabo, G., Suzdak, P.D., and Paul, S.M., Effect of an imidazobenzodiazepine, Ro 15-4513, on the incoordination and hypothermia produced by ethanol and pentobarbital, *Life Sci.*, 41, 611, 1987.
7. Lister, R.G., The benzodiazepine receptor inverse agonists FG 7142 and Ro15-4513 both reverse some of the behavioral effects of ethanol in a holeboard test, *Life Sci.*, 41, 1481, 1987.
8. Suzdak, P.D., Glowa, J.R., Crawley, J.N., Schwartz, R.D., Skolnick, P., and Paul, S.M., A selective imidazobenzodiazepine antagonist of ethanol in the rat, *Science*, 234, 1243, 1986.

9. Becker, H.C. and Anton, R.F., Valproate potentiates and picrotoxin antagonizes the anxiolytic action of ethanol in a nonshock conflict task, *Neuropharmacology*, 29, 837, 1990.
10. Becker, H.C. and Hale, R.L., RO15-4513 antagonizes the anxiolytic effects of ethanol in a nonshock conflict task at doses devoid of anxiogenic activity, *Pharmacol. Biochem. Behav.*, 39, 803, 1991.
11. Glowa, J.R., Crawley, J., Suzdak, P.D., and Paul, S.M., Ethanol and the GABA receptor complex: studies with the partial inverse benzodiazepine receptor agonist Ro 15-4513, *Pharmacol. Biochem. Behav.*, 31, 767, 1988.
12. Goldstein, D.B., Alcohol withdrawal reactions in mice: effects of drugs that modify neurotransmission, *J. Pharmacol. Exp. Ther.*, 186, 1, 1973.
13. Morrow, A.L., Suzdak, P.D., and Paul, S.M., Benzodiazepine, barbiturate, ethanol and hypnotic steroid hormone modulation of GABA-mediated chloride ion transport in rat brain synaptoneurosomes, in *Chloride Channels and Their Modulation by Neurotransmitters and Drugs*, Biggio, G. and Costa, E., Eds., Raven Press, New York, 1988, p. 247.
14. Suzdak, P.D., Schwartz, R.D., Skolnick, P., and Paul, S.M., Ethanol stimulates γ-aminobutyric acid receptor-mediated chloride transport in rat brain synaptoneurosomes, *Proc. Natl. Acad. Sci. U.S.A.*, 83, 4071, 1986.
15. Allan, A.M. and Harris, R.A., Acute and chronic ethanol treatments alter GABA receptor-operated chloride channels, *Pharmacol. Biochem. Behav.*, 27, 665, 1987.
16. Ticku, M.K. and Burch, T., Alterations in γ-aminobutyric acid receptor sensitivity following acute and chronic ethanol treatments, *J. Neurochem.*, 34, 417, 1980.
17. Mehta, A.K. and Ticku, M.K., Ethanol potentiation of GABAergic transmission in cultured spinal cord neurons involves γ-aminobutyric acid-gated chloride channels, *J. Pharmacol. Exp. Ther.*, 246, 558, 1988.
18. Suzdak, P.D., Paul, S.M., and Crawley, J.N., Effects of Ro15-4513 and other benzodiazepine inverse agonists on alcohol induced intoxication in the rat, *J. Pharmacol. Exp. Ther.*, 245, 880, 1988.
19. Aguayo, L.G., Pancetti, F.C., Klein, R.L., and Harris, R.A., Differential effects of GABAergic ligands in mouse and rat hippocampal neurons, *Brain Res.*, 647, 97, 1994.
20. Celentano, J.J., Gibbs, T.T., and Farb, D.H., Ethanol potentiates GABA- and glycine-induced chloride currents in chick spinal cord neurons, *Brain Res.*, 455, 377, 1988.
21. Frye, G.D., Fincher, A.S., Grover, C.A., and Griffith, W.H., Interaction of ethanol and allosteric modulators with $GABA_A$-activated currents in adult medial septum/diagonal band neurons, *Brain Res.*, 635, 283, 1994.
22. Givens, B.S. and Breese, G.R., Site-specific enhancement of γ-aminobutyric acid-mediated inhibition of neural activity by ethanol in the rat medial septum, *J. Pharmacol. Exp. Ther.*, 254, 528, 1990.
23. Proctor, W.R., Soldo, B.L., Allan, A.M., and Dunwiddie, T.V., Ethanol enhances synaptically evoked $GABA_A$ receptor-mediated responses in cerebral cortical neurons in rat brain slices, *Brain Res.*, 595, 220, 1992.
24. Reynolds, J.N., Prasad, A., and MacDonald, J.F., Ethanol modulation of GABA receptor-activated Cl^- currents in neurons of the chick, rat and mouse central nervous system, *Eur. J. Pharmacol.*, 224, 173, 1992.
25. Sapp, D.W. and Yeh, H.H., Ethanol-$GABA_A$ receptor interactions: a comparison between cell lines and cerebellar Purkinje cells, *J. Pharmacol. Exp. Ther.*, 284, 768, 1998.

26. Soldo, B.L., Proctor, W.R., and Dunwiddie, T.V., Ethanol differentially modulates $GABA_A$ receptor-mediated chloride currents in hippocampal, cortical, and septal neurons in rat brain slices, *Synapse*, 18, 94, 1994.
27. Criswell, H.E., Simson, P.E., Duncan, G.E., McCown, T.J., Herbert, J.S., Morrow, A.L., and Breese, G.R., Molecular basis for regionally specific action of ethanol on gamma-aminobutyric $acid_A$ receptors: generalization to other ligand-gated ion channels, *J. Pharmacol. Exp. Ther.*, 267, 522, 1993.
28. Givens, B.S. and McMahon, K., Effects of ethanol on nonspatial working memory and attention in rats, *Behav. Neurosci.*, 111, 275, 1997.
29. Mancillas, J.R., Siggins, J.R., and Bloom, F.E., Systemic ethanol: enhancement of responses to acetylcholine and somatostatin in hippocampus, *Science*, 231, 161, 1986.
30. Nestores, J.N., Ethanol specifically potentiates GABA-mediated neurotransmission in the feline cerebral cortex, *Science*, 209, 708, 1980.
31. Siggins, G.R., Pittman, Q.J., and French, E.D., Effects of ethanol on CA1 and CA3 pyramidal cells in the hippocampal slice preparation: an intracellular study, *Brain Res.*, 414, 22, 1987.
32. Simson, P.E., Criswell, H.E., and Breese, G.R., Ethanol potentiates γ-aminobutyric acid-mediated inhibition in the inferior colliculus: evidence for local ethanol/γ-aminobutyric acid interactions, *J. Pharmacol. Exp. Ther.*, 259, 1288, 1991.
33. Roberto, M., Madamba, S.G., Moore, S.D., Tallent, M.K., and Siggins, G.R., Ethanol increases GABAergic transmission at both pre- and postsynaptic sites in rat central amygdala neurons, *Proc. Natl. Acad. Sci. U.S.A.*, 100, 2053, 2003.
34. Nakahiro, M., Arakawa, O., and Narahashi, T., Modulation of gamma-aminobutyic acid receptor-channel complex by alcohols, *J. Pharmacol. Exp. Ther.*, 259, 235, 1991.
35. Takebayashi, M., Kagaya, A., Uchitomi, Y., Yokota, N., Horiguchi, J., and Yamawaki, S., Differential regulation by pregnenolone sulfate of intracellular Ca^{2+} increase by amino acids in primary cultured rat cortical neurons, *Neurochem. Int.*, 32, 205, 1998.
36. Crews, F., Morrow, A.L., Criswell, H., and Breese, G., Effects of ethanol on ion channels, in *International Review of Neurobiology*, Vol. 39, Bradley, R.J. and Harris, R.A., Eds., Academic Press, New York, 1996, p. 283.
37. Marszalec, W., Aistrup, G.L., and Narahashi, T., Ethanol modulation of excitatory and inhibitory synaptic interactions in cultured cortical neurons, *Alcohol. Clin. Exp. Res.*, 22, 1516, 1998.
38. Sundstrom-Poromaa, I., Smith, D.H., Gong, Q.H., Sabado, T.N., Li, X., Light, A., Wiedmann, M., Williams, K., and Smith, S., Hormonally regulated $\alpha_4\beta_2\delta$ $GABA_A$ receptors are a target for alcohol, *Nat. Neurosci.*, 5, 721, 2002.
39. Brown, N., Kerby, J., Bonnert, T.P., Whiting, P.J., and Wafford, K.A., Pharmacological characterization of a novel cell line expressing human alpha(4)beta(3)delta GABA(A) receptors, *Br. J. Pharmacol.*, 136, 965, 2002.
40. Fritschy, J.M. and Brünig, I., Formation and plasticity of γ-aminobutyric acidergic synapses: physiological mechanisms and pathophysiological implications, *Pharmacol. Ther.*, 5539, 1, 2003.
41. Nusser, Z., Sieghart, W., and Somogyi, P., Segregation of different GABAA receptors to synaptic and extrasynaptic membranes of cerebellar granule cells, *J. Neurosci.*, 18, 1693, 1998.
42. Nusser, Z. and Mody, I., Selective modulation of tonic and phasic inhibitions in dentate gyrus granule cells, *J. Neurophysiol.*, 87, 2624, 2002.
43. Morrow, A.L., Suzdak, P.D., Karanian, J.W., and Paul, S.M., Chronic ethanol administration alters γ-aminobutyric acid, pentobarbital and ethanol-mediated $^{36}Cl^-$ uptake in cerebral cortical synaptoneurosomes, *J. Pharmacol. Exp. Ther.*, 246, 158, 1988.

44. Suzdak, P.D., Schwartz, R.D., and Paul, S.M., Alcohols stimulate GABA receptor-mediated chloride uptake in brain vesicles: correlation with intoxication potency, *Brain Res.*, 444, 340, 1987.
45. Breese, G.R., Morrow, A.L., Simson, P.E., Criswell, H.E., McCown, T.J., Duncan, G.E., and Keir, W.J., The neuroanatomical specificity of ethanol action with ligand-gated ion channels: a hypothesis, *Alcohol Alcohol.*, Suppl. 2, 309, 1993.
46. Harris, R.A., Mihic, S.J., and Valenzuela, C.F., Alcohol and benzodiazepines: recent mechanistic studies, *Drug Alcohol Depend.*, 51, 155, 1998.
47. Mihic, S.J., Ye, Q., Wick, M.J., Koltchine, V.V., Krasowski, M.A., Finn, S.E., Mascia, M.P., Valenzuela, C.F., Hanson, K.K., Greenblatt, E.P., Harris, R.A., and Harrison, N.L., Sites of alcohol and volatile anaesthetic action on $GABA_A$ and glycine receptors, *Nature*, 389, 385, 1997.
48. Akk, G. and Steinbach, J.H., Low doses of ethanol and a neuroactive steroid positively interact to modulate rat $GABA_A$ receptor function, *J. Physiol.*, 546, 641, 2003.
49. Weiner, J.L., Zhang, L., and Carlen, P.L., Potentiation of $GABA_A$-mediated synaptic current by ethanol in hippocampal CA1 neurons: possible role of protein kinase C, *J. Pharmacol. Exp. Ther.*, 268, 1388, 1994.
50. Weiner, J.L., Gu, C., and Dunwiddie, T.V., Differential ethanol sensitivity of subpopulations of $GABA_A$ synapses onto rat hippocampal CA1 pyramidal neurons, *J. Neurophysiol.*, 77, 1306, 1997.
51. Wan, F.-J., Berton, F., Madamba, S.G., Francesconi, W., and Siggins, G.R., Low ethanol concentrations enhance GABAergic inhibitory postsynaptic potentials in hippocampal pyramidal neurons only after block of $GABA_B$ receptors, *Proc. Natl. Acad. Sci. U.S.A.*, 93, 5049, 1996.
52. Weiner, J.L., Valenzuela, C.F., Watson, P.L., Frazier, C.J., and Dunwiddie, T.V., Elevation of basal protein kinase C activity increases ethanol sensitivity of $GABA_A$ receptors in rat hippocampal CA1 pyramidal neurons, *J. Neurochem.*, 68, 1949, 1997.
53. Freund, R.K. and Palmer, M.R., Beta adrenergic sensitization of gamma-aminobutyric acid receptors to ethanol involves a cyclic AMP/protein kinase A second-messenger mechanism, *J. Pharmacol. Exp. Ther.*, 280, 1192, 1997.
54. Wang, Y., Freund, R.K., and Palmer, M.R., Potentiation of ethanol effects in cerebellum by activation of endogenous noradrenergic inputs, *J. Pharmacol. Exp. Ther.*, 288, 211, 1999.
55. Morrow, A.L., VanDoren, M.J., and Devaud, L.L., Effects of progesterone or neuroactive steroid? *Nature*, 395, 652, 1998.
56. Morrow, A.L., Pace, J.R., Purdy, R.H., and Paul, S.M., Characterization of steroid interactions with gamma-aminobutyric acid receptor-gated chloride ion channels: evidence for multiple steroid recognition sites, *Mol. Pharmacol.*, 37, 263, 1990.
57. Morrow, A.L., Suzdak, P.D., and Paul, S.M., Steroid hormone metabolites potentiate GABA receptor-mediated chloride ion flux with nanomolar potency, *Eur. J. Pharmacol.*, 142, 483, 1987.
58. Crawley, J.N., Glowa, J.R., Majewska, M.D., and Paul, S.M., Anxiolytic activity of an endogenous adrenal steroid, *Brain Res.*, 398, 382, 1986.
59. Devaud, L.L., Purdy, R.H., and Morrow, A.L., The neurosteroid, 3α-hydroxy-5α-pregnan-20-one, protects against bicuculline-induced seizures during ethanol withdrawal in rats, *Alcohol. Clin. Exp. Res.*, 19, 350, 1995.
60. Barbaccia, M.L., Affricano, D., Trabucchi, M., Purdy, R.H., Colombo, G., Agabio, R., and Gessa, G.L., Ethanol markedly increases "GABAergic" neurosteroids in alcohol-preferring rats, *Eur. J. Pharmacol.*, 384, R1, 1999.

61. VanDoren, M.J., Matthews, D.B., Janis, G.C., Grobin, A.C., Devaud, L.L., and Morrow, A.L., Neuroactive steroid 3α-hydroxy-5α-pregnan-20-one modulates electrophysiological and behavioral actions of ethanol, *J. Neurosci.*, 20, 1982, 2000.
62. Morrow, A.L., VanDoren, M.J., Penland, S.N., and Matthews, D.B., The role of GABAergic neuroactive steroids in ethanol action, tolerance and dependence, *Brain Res. Brain Res. Rev.*, 37, 98, 2001.
63. Normington, K. and Russell, D.W., Tissue distribution and kinetic characteristics of rat steroid 5α-reductase isozymes, *J. Biol. Chem.*, 267, 19548, 1992.
64. Van Rijn, C.M., Willems-van Bree, E., Van der Velden, T.J.A.M., and Rodrigues de Miranda, J.F., Binding of the cage convulsant, [^{3}H]TBOB, to sites linked to the $GABA_A$ receptor complex, *Eur. J. Pharmacol.*, 179, 419, 1990.
65. Aldrich, M., Effect of alcohol on sleep, in *Alcohol Problems and Aging*, National Institute on Alcohol Abuse and Alcoholism Research Monograph 33, Gomberg, E.S.L., Hegedus, A.M., and Zucker, R.A., Eds., National Institutes of Health, Bethesda, MD, 1998, p. 281.
66. Ehlers, C.L., Alcohol and sleep, in *Review of NIAAA's Neuroscience and Behavioral Research Portfolio*, National Institute on Alcohol Abuse and Alcoholism Research Monograph 33, Noronha, A., Eckardt, M., and Warren, K., Eds., National Institutes of Health, Bethesda, MD, 2000, p. 417.
67. Roehrs, T. and Roth, T., Sleep, sleepiness, and alcohol use, *Alcohol. Res. Health*, 25, 101, 2001.
68. Martz, A., Dietrich, R.A., and Harris, R.A., Behavioral evidence for the involvement of γ-aminobutyric acid in the actions of ethanol, *Eur. J. Pharmacol.*, 89, 53, 1983.
69. Mok, W.M., Bukusoglu, C., and Krieger, N.R., 3 alpha-Hydroxy-5 alpha-pregnan-20-one is the only active anesthetic steroid in anesthetized mouse brain, *Steroids*, 58, 112, 1993.
70. Korneyev, A. and Costa, E., Allopregnanolone (THP) mediates anesthetic effects of progesterone in rat brain, *Horm. Behav.*, 30, 37, 1996.
71. Guidotti, A., Dong, E., Matsumoto, K., Pinna, G., Rasmusson, A.M., and Costa, E., The socially-isolated mouse: a model to study the putative role of allopregnanolone and 5alpha-dihydroprogesterone in psychiatric disorders, *Brain Res. Brain Res. Rev.*, 37, 110, 2001.
72. Atzori, G., Fratta, W., Gessa, G.L., and Burov, Y., Inhibition of neurosteroid synthesis fails to reduce the anaesthetic effect of alcohol and other anaesthetics in mice, *Society for Neuroscience Abstracts*, 27, 1714, 2001.
73. Gao, C.Q., Dhooge, W.S., Kaufman, J.M., Weyne, J.J., and Eechaute, W.P., Hypothalamic 5 alpha-reductase and 3 alpha-oxidoreductase activity in the male rat, *J. Steroid Biochem. Mol. Biol.*, 80, 91, 2002.
74. Laszlo, F.A., Varga, C., Pavo, I., Gardi, J., Vecsernyes, M., Galfi, M., Morschl, E., Laszlo, F., and Makara, G.B., Vasopressin pressor receptor-mediated activation of HPA axis by acute ethanol stress in rats, *Am. J. Physiol. Regul. Integr. Comp. Physiol.*, 280, R458, 2001.
75. Rivier, C. and Lee, S., Acute alcohol administration stimulates the activity of hypothalamic neurons that express corticotropin-releasing factor and vasopressin, *Brain Res.*, 726, 1, 1996.
76. Hoffman, P.L. and Tabakoff, B., Neurohypophyseal peptides maintain tolerance to the incoordinating effects of ethanol, *Pharmacol. Biochem. Behav.*, 21, 535, 1984.

77. Czlonkowska, A.I., Krzascik, P., Sienkiewicz-Jarosz, H., Siemiatkowski, M., Szyndler, J., Bidzinski, A., and Plaznik, A., The effects of neurosteroids on picrotoxin-, bicuculline- and NMDA-induced seizures, and a hypnotic effect of ethanol, *Pharmacol. Biochem. Behav.*, 67, 345, 2000.
78. Vanover, K.E., Suruki, M., Robledo, S., Huber, M., Wieland, S., Lan, N.C., Gee, K.W., Wood, P.L., and Carter, R.B., Positive allosteric modulators of the $GABA_A$ receptor: differential interaction of benzodiazepines and neuroactive steroids with ethanol, *Psychopharmacology*, 141, 77, 1999.
79. Mok, W.M., Herschkowitz, S., and Krieger, N.R., Evidence that 3 alpha-hydroxy-5 alpha-pregnan-20-one is the metabolite responsible for anesthesia induced by 5 alpha-pregnanedione in the mouse, *Neurosci. Lett.*, 135, 145, 1992.
80. Givens, B.S. and Breese, G.R., Electrophysiological evidence for an involvement of the medial septal area in the acute sedative effects of ethanol, *J. Pharmacol. Exp. Ther.*, 253, 95, 1990.
81. Steffensen, S.C. and Henriksen, S.J., Comparison of the effects of ethanol and chlordiazepoxide on electrophysiological activity in the fascia dentata and hippocampus regio superior, *Hippocampus*, 2, 201, 1992.
82. Tokunaga, S., McDaniel, J.R., Morrow, A.L., and Matthews, D.B., Effect of acute ethanol administration and acute allopregnanolone administration on spontaneous hippocampal pyramidal cell neural activity, *Brain Res.*, 967, 273, 2003.
83. Simson, P.E., Criswell, H.E., Johnson, K.B., Hicks, R.E., and Breese, G.R., Ethanol inhibits NMDA-evoked electrophysiological activity *in vivo*, *J. Pharmacol. Exp. Ther.*, 257, 225, 1991.
84. Cox, T., The effects of caffeine, alcohol, and previous exposure to the test situation on spontaneous alternation, *Psychopharmacologia*, 17, 83, 1970.
85. Berry, R.B. and Matthews, D.B., Acute ethanol administration selectively impairs spatial memory in C57BL/6J mice, *Alcohol. Clin. Exp. Res.*, 26, 153A, 2002.
86. Shimizu, K., Matsubara, K., Uezono, T., Kimura, K., and Shiono, H., Reduced dorsal hippocampal glutamate release significantly correlates with the spatial memory deficits produced by benzodiazepines and ethanol, *Neuroscience*, 83, 701, 1998.
87. Vandergriff, J.L., Matthews, D.B., Best, P.J., and Simson, P.E., Effects of ethanol and diazepam on spatial and non-spatial tasks in rats on an 8-arm maze, *Alcohol. Clin. Exp. Res.*, 64, 19, 1996.
88. Mizumori, S.J., Miya, D.Y., and Ward, K.E., Reversible inactivation of the lateral dorsal thalamus disrupts hippocampal place representation and impairs spatial learning, *Brain Res.*, 644, 168, 1994.
89. Hoffmann, S.E. and Matthews, D.B., Ethanol-induced impairments in spatial working memory are not due to deficits in learning, *Alcohol. Clin. Exp. Res.*, 25, 856, 2001.
90. White, A.M. and Best, P.J., Effects of ethanol on hippocampal place-cell and interneuron activity, *Brain Res.*, 876, 154, 2000.
91. Matthews, D.B., Ilgen, M., White, A.M., and Best, P.J., Acute ethanol administration impairs spatial performance while facilitating nonspatial performance in rats, *Neurobiol. Learn. Mem.*, 72, 169, 1999.
92. Alexandrov, Y.I., Grinchenko, Y.V., Laukka, S., Jarvilehto, R., Maz, V.N., and Korpusova, A.V., Effect of ethanol on hippocampal neurons depends on their behavioural specialization, *Acta Physiol. Scand.*, 149, 105, 1993.
93. Matthews, D.B., Simson, P.E., and Best, P.J., Ethanol alters spatial processing of hippocampal place cells: a mechanism for impaired navigation when intoxicated, *Alcohol. Clin. Exp. Res.*, 20, 404, 1996.

94. Ludvig, N., Fox, S.E., Kubie, J.L., Altura, B.M., and Altura, B.T., Application of the combined single-cell recording/intracerebral microdialysis method to alcohol research in freely behaving animals, *Alcohol. Clin. Exp. Res.*, 22, 41, 1998.
95. McDaniel, J.R., Tokunaga, S., Morrow, A.L., and Matthews, D.B., Chronic intermittent ethanol exposure in adolescent rats produces tolerance to ethanol-induced spatial memory deficits and ethanol-induced allopregnanolone increases, *Alcohol. Clin. Exp. Res.*, 26, 153A, 2002.
96. Harrison, N.L., Majewska, M.D., Harrington, J.W., and Barker, J.L., Structure-activity relationships for steroid interaction with the gamma-aminobutyric acid-A receptor complex, *J. Pharmacol. Exp. Ther.*, 241, 346, 1987.
97. Barbaccia, M.L., Roscetti, G., Trabucchi, M., Purdy, R.H., Mostallino, M., Concas, A., and Biggio, G., The effects of inhibitors of GABAergic transmission and stress on brain and plasma allopregnanolone concentrations, *Br. J. Pharmacol.*, 120, 1582, 1997.
98. Barbaccia, M.L., Concas, A., Serra, M., and Biggio, G., Stress and neurosteroids in adult and aged rats, *Exp. Gerontol.*, 33, 697, 1998.
99. Purdy, R.H., Morrow, A.L., Moore, P.H., Jr., and Paul, S.M., Stress-induced elevations of gamma-aminobutyric acid type A receptor-active steroids in the rat brain, *Proc. Natl. Acad. Sci. U.S.A.*, 88, 4553, 1991.
100. Diamond, D.M., Park, C.R., Heman, K., and Rose, G.M., Exposing rats to a predator impairs spatial working memory in the radial arm maze, *Hippocampus*, 9, 542, 1999.
101. Holscher, C., Stress impairs performance in spatial water maze learning tasks, *Behav. Brain Res.*, 100, 225, 1999.
102. Stillman, M.J., Shukitt-Hale, B., Levy, A., and Lieberman, H.R., Spatial memory under acute cold and restraint stress, *Physiol. Behav.*, 83, 701, 1998.
103. Ladurelle, N., Eychenne, B., Denton, D., Blair-West, J., Schumacher, M., Robel, P., and Baulieu, E., Prolonged intracerebroventricular infusion of neurosteroids affects cognitive performances in the mouse, *Brain Res.*, 858, 371, 2000.
104. Reddy, D.S. and Kulkarni, S.K., Sex and estrous cycle-dependent changes in neurosteroid and benzodiazepine effects on food consumption and plus-maze learning behaviors in rats, *Pharmacol. Biochem. Behav.*, 62, 53, 1999.
105. Matthews, D.B., Morrow, A.L., Tokunaga, and McDaniel, J.R., Acute ethanol administration and acute allopregnanolone administration impair spatial memory in the Morris water task, *Alcohol. Clin. Exp. Res.*, 26, 1747, 2002.
106. Rivier, C., Bruhn, T., and Vale, W., Effect of ethanol on the hypothalamic-pituitary-adrenal axis in the rat: role of corticotropin-releasing factor (CRF), *J. Pharmacol. Exp. Ther.*, 229, 127, 1984.
107. Budec, M., Koko, V., Milovanovic, T., Balint-Peric, L., and Petkovic, A., Acute ethanol treatment increases level of progesterone in ovariectomized rats, *Alcohol*, 26, 173, 2002.
108. Hundt, W., Zimmermann, U., Pottig, M., Spring, K., and Holsboer, F., The combined dexamethasone-suppression/CRH-stimulation test in alcoholics during and after acute withdrawal, *Alcohol. Clin. Exp. Res.*, 25, 687, 2001.
109. Ferreira, V.M., Takahashi, R.N., and Morato, G.S., Dexamethasone reverses the ethanol-induced anxiolytic effect in rats, *Pharmacol. Biochem. Behav.*, 66, 585, 2000.
110. Sze, P.Y., Glucocorticoids antagonize the sedative action of ethanol in mice, *Pharmacol. Biochem. Behav.*, 45, 991, 1993.
111. Carreno, C.F., Ferreira, V.M., and Morato, G.S., Ethanol-induced hypothermia in rats is antagonized by dexamethasone, *Braz. J. Med. Biol. Res.*, 30, 69, 1997.

112. Randin, D., Vollenweider, P., Tappy, L., Jequier, E., Nicod, P., and Scherrer, U., Suppression of alcohol-induced hypertension by dexamethasone, *N. Engl. J. Med.*, 332, 1733, 1995.
113. King, S.R., Manna, P.R., Ishii, T., Syapin, P.J., Ginsberg, S.D., Wilson, K., Walsh, L.P., Parker, K.L., Stocco, D.M., Smith, R.G., and Lamb, D.J., An essential component in steroid synthesis, the steroidogenic acute regulatory protein, is expressed in discrete regions of the brain, *J. Neurosci.*, 22, 10613, 2002.
114. Khisti, R.T., Penland, S.N., VanDoren, M.J., Grobin, A.C., and Morrow, A.L., GABAergic neurosteroid modulation of ethanol actions, *World J. Biol. Psychiatry*, 3, 87, 2002.
115. Boisse, N.N. and Okamoto, M., Ethanol as a sedative-hypnotic: comparison with barbiturate and non-barbiturate sedative-hypnotics, in *Alcohol Tolerance and Dependence*, Rigter, H. and Crabbe, J.C., Eds., Elsevier, Amsterdam, 1980, p. 265.
116. Brown, M.E., Anton, R.F., Malcolm, R., and Ballenger, J.C., Alcohol detoxification and withdrawal seizures: clinical support for a kindling hypothesis, *Biol. Psychol.*, 23, 507, 1988.
117. Le, A.D., Khanna, J.M., Kalant, H., and Grossi, F., Tolerance to and cross-tolerance among ethanol, pentobarbital and chlordiazepoxide, *Pharmacol. Biochem. Behav.*, 24, 93, 1986.
118. Janis, G.C., Devaud, L.L., Mitsuyama, H., and Morrow, A.L., Effects of chronic ethanol consumption and withdrawal on the neuroactive steroid 3α-hydroxy-5α-pregnan-20-one in male and female rats, *Alcohol. Clin. Exp. Res.*, 22, 2055, 1998.
119. Kokka, N., Sapp, D.W., Taylor, A.M., and Olsen, R.W., The kindling model of alcohol dependence: similar persistent reduction in seizure threshold to pentylenetetrazol in animals receiving chronic ethanol or chronic pentylenetetrazol, *Alcohol. Clin. Exp. Res.*, 17, 525, 1993.
120. Silveri, M.M. and Spear, L.P., Ontogeny of rapid tolerance to the hypnotic effects of ethanol, *Alcohol. Clin. Exp. Res.*, 23, 1180, 1999.
121. Melchior, C.L. and Tabakoff, B., Features of environment-dependent tolerance to ethanol, *Psychopharmacology*, 87, 94, 1985.
122. Hoffman, P.L., The influence of neurohypophysial hormones on central nervous system processes of adaptation: functional tolerance to ethanol, *Ann. N.Y. Acad. Sci.*, 689, 300, 1993.
123. Grobin, A.C., Matthews, D.B., Devaud, L.L., and Morrow, A.L., The role of $GABA_A$ receptors in the acute and chronic effects of ethanol, *Psychopharmacology*, 139, 2, 1998.
124. Buck, K.J., Hahner, L., Sikela, J., and Harris, R.A., Chronic ethanol treatment alters brain levels of gamma-aminobutyric $acid_A$ receptor subunit mRNAs: relationship to genetic differences in ethanol withdrawal seizure severity, *J. Neurochem.*, 57, 1452, 1991.
125. Mhatre, M.C., Pena, G., Sieghart, W., and Ticku, M.K., Antibodies specific for $GABA_A$ receptor alpha subunits reveal that chronic alcohol treatment down-regulates alpha-subunit expression in rat brain regions, *J. Neurochem.*, 61, 1620, 1993.
126. Mhatre, M.C. and Ticku, M.K., Chronic ethanol administration alters γ-aminobutyric acid A receptor gene expression, *Mol. Pharmacol.*, 42, 415, 1992.
127. Montpied, P., Morrow, A.L., Karanian, J.W., Ginns, E.I., Martin, B.M., and Paul, S.M., Prolonged ethanol inhalation decreases gamma-aminobutyric $acid_A$ receptor α subunit mRNAs in the rat cerebral cortex, *Mol. Pharmacol.*, 39, 157, 1991.

128. Morrow, A.L., Herbert, J.S., and Montpied, P., Differential effects of chronic ethanol administration on $GABA_A$ receptor α1 and α6 subunit mRNA levels in rat cerebellum, *Mol. Cell. Neurosci.*, 3, 251, 1992.
129. Morrow, A.L., Montpied, P., Lingford-Hughes, A., and Paul, S.M., Chronic ethanol and pentobarbital administration in the rat: effects on $GABA_A$ receptor function and expression in brain, *Alcohol*, 7, 237, 1990.
130. Devaud, L.L., Fritschy, J.-M., and Morrow, A.L., Influence of gender on chronic ethanol-induced alternations of $GABA_A$ receptors in rats, *Brain Res.*, 796, 222, 1998.
131. Devaud, L.L., Purdy, R.H., Finn, D.A., and Morrow, A.L., Sensitization of γ-aminobutyric $acid_A$ receptors to neuroactive steroids in rats during ethanol withdrawal, *J. Pharmacol. Exp. Ther.*, 278, 510, 1996.
132. Puia, G., Santi, M., Vincini, S., Pritchett, D.B., Purdy, R.H., Paul, S.M., Seeburg, P.H., and Costa, E., Neurosteroids act on recombinant human $GABA_A$ receptors, *Neuron*, 4, 759, 1990.
133. Puia, G., Vincini, S., Seeburg, P.H., and Costa, E., Influence of recombinant GABA-A receptor subunit composition on the action of allosteric modulators of GABA-gated Cl^- currents, *Mol. Pharmacol.*, 39, 691, 1991.
134. Finn, D.A., Gallaher, E.J., and Crabbe, J.C., Differential change in neuroactive steroid sensitivity during ethanol withdrawal, *J. Pharmacol. Exp. Ther.*, 292, 394, 2000.
135. Romeo, E., Brancati, A., De Lorenzo, A., Fucci, P., Furnari, C., Pompili, E., Sasso, G.F., Spalletta, G., Troisi, A., and Pasini, A., Marked decrease of plasma neuroactive steroids during alcohol withdrawal, *J. Pharmacol. Exp. Ther.*, 247, 309, 1996.
136. Gallo, M.A. and Smith, S.S., Progesterone withdrawal decreases latency to and increases duration of electrified prod burial: a possible rat model of PMS anxiety, *Pharmacol. Biochem. Behav.*, 46, 897, 1993.
137. Smith, S.S., Gong, Q.H., Hsu, F.-C., Markowitz, R.S., Ffrench-Mullen, J.M.H., and Li, X., $GABA_A$ receptor α4 subunit suppression prevents withdrawal properties of an endogenous steroid, *Nature*, 392, 926, 1998.
138. Moran, M.H. and Smith, S.S., Progesterone withdrawal I: pro-convulsant effects, *Brain Res.*, 807, 84, 1998.
139. Smith, S.S., Gong, Q.H., Li, X., Moran, M.H., Bitran, D., Frye, C.A., and Hsu, F., Withdrawal from 3α-OH-5α-pregnan-20-one using a pseudopregnancy model alters the kinetics of hippocampal $GABA_A$-gated current and increases the $GABA_A$ receptor α4 subunit in association with increased anxiety, *J. Neurosci.*, 18, 5275, 1998.
140. Moran, M.H., Goldberg, N., and Smith, S.S., Progesterone withdrawal II: insensitivity to the sedative effects of a benzodiazepine, *Brain Res.*, 807, 91, 1998.
141. Devaud, L.L., Smith, F.D., Grayson, D.R., and Morrow, A.L., Chronic ethanol consumption differentially alters the expression of γ-aminobutyric $acid_A$ receptor subunit mRNAs in rat cerebral cortex: competitive, quantitative reverse transcriptase-polymerase chain reaction analysis, *Mol. Pharmacol.*, 48, 861, 1995.
142. Devaud, L.L., Fritschy, J.-M., Sieghart, W., and Morrow, A.L., Bidirectional alterations of $GABA_A$ receptor subunit peptide levels in rat cortex during chronic ethanol consumption and withdrawal, *J. Neurochem.*, 69, 126, 1997.

11 Allopregnanolone Modulates the Action of Ethanol and Stress on the Activity of Mesocortical Dopaminergic Neurons and HPA Axis

Giovanni Biggio, Laura Dazzi, and Mariangela Serra

CONTENTS

0-8493-2392-4/04/$0.00+$1.50

ABSTRACT

In this chapter, we summarize some of our most important recent results indicating that in rats, the central amount of this 5α-reduced, 3α-hydroxilated progesterone (PROG) derivative, which is increased and decreased by acute and chronic stress, respectively, is responsible for the physiologic and pharmacologic modulation of basal and stress-induced activation of mesocortical dopamine neurons. Given that chronic stress is known to alter the sensitivity of different monoamines in the rat brain and the activity of the hypothalamic-pituitary-adrenal (HPA) axis using the model of social isolation, we also studied whether such a stressful condition changes the sensitivity of steroidogenesis to the action of novel acute stressful stimuli such as foot shock and ethanol, two treatments known to increase plasma and brain content of 3α-hydroxy-5α-pregnan-20-one (3α,5α-TH PROG) and to alter the function of mesocortical dopamine neurons.

11.1 INTRODUCTION

Several works suggest that mesocortical dopamine systems participate in the control of the emotional processes that are strictly related to stress and anxiety,[1–5] as well as in the modulation of mood alterations.[6] Accordingly, dopamine plays an important role in the regulation of brain function associated with emotion and fear.[7]

The discovery that facilitation and inhibition of $GABA_A$ receptor-mediated neurotransmission results in reduction and enhancement in the activity of mesocortical dopaminergic neurons, respectively, provided the evidence that such a dopaminergic system is tonically modulated by GABAergic neurotransmission.[8–14] Thus, anxiolytic benzodiazepines and anxiogenic β-carbolines, drugs that selectively enhance and inhibit $GABA_A$-mediated neurotransmission, are able to inhibit and enhance mesocortical dopamine output, respectively.

The evidence that steroid derivatives synthesized *de novo* from cholesterol in peripheral organs or in the central nervous system[15] modulate $GABA_A$ receptor function with potencies and efficacies similar to or greater than those of benzodiazepines and barbiturates[16] has indicated that these compounds might be the putative endogenous modulators of mesocortical dopaminergic neurons. Accordingly, as reported in Chapters 2–4 and 10, the progesterone (PROG) metabolite 3α-hydroxy-5α-pregnan-20-one (allopregnanolone, 3α,5α-TH PROG) induces opening at the $GABA_A$ receptor-associated chloride channel at nanomolar concentrations *in vitro* and, when administered systemically, induces pharmacological and behavioral effects similar to those elicited by anxiolytic drugs that, through an action on $GABA_A$ receptor function, modulate the activity of mesocortical dopamine neurons.

Given that the brain concentration of 3α,5α-TH PROG is regulated by the peripheral as well as central synthesis of PROG, the physiological changes in the synaptic concentration of this steroid derivative might therefore contribute to regulation of $GABA_A$ receptor-mediated transmission, and this in turn may result in modulation of the mesocortical dopamine system and its associated behavior. This mechanism might assume great importance in the physiological conditions, such as stress, pregnancy, menstrual cycle, and menopause, in which there are great fluctuations of this PROG metabolite both in plasma and the brain. Moreover, the evidence that brain levels of 3α,5α-TH PROG are increased by different psychoactive drugs such as antidepressants,[17–19] atypical antipsychotics,[20,21] and ethanol,[22,23] together with the finding that a selective reduction in the concentration of 3α,5α-TH PROG has been detected in the plasma and cerebrospinal fluid (CSF) of individuals with major depression,[24,25] strongly suggests that this neuroactive steroid and its modulatory action on brain monoamines might play a role in the neurobiology and pharmacology of emotional and affective disorders.

In this chapter, we summarize some of our most important recent results indicating that in rats, the central amount of this 5α-reduced, 3α-hydroxilated PROG derivative, which is increased and decreased by acute and chronic stress, respectively, is responsible for the physiologic and pharmacologic modulation of basal and stress-induced activation of mesocortical dopamine neurons. Given that chronic stress is known to alter the sensitivity of different monoamines in the rat brain[26–28] and the activity of the hypothalamic-pituitary-adrenal (HPA) axis,[29] using the model of social isolation we also studied whether such stressful condition changes the sensitivity of steroidogenesis to the action of novel acute stressful stimuli such as foot shock and ethanol, two treatments known to increase plasma and brain content of 3α,5α-TH PROG and to alter the function of mesocortical dopamine neurons.

11.2 EFFECT OF STRESS ON CORTICAL DOPAMINERGIC NEURONS

Dopamine plays an important role in the regulation of brain functions associated with emotion and fear as well as with the response to stress. Dopamine release in the prefrontal cortex is preferentially modulated by both acute and chronic stress.[1–5] Mild acute stress thus increases the turnover and release of dopamine to a greater degree in the prefrontal cortex than in other brain regions enriched in dopaminergic terminals. The increase in dopaminergic transmission elicited by acute stress is antagonized by the acute administration of anxiolytic agonists of benzodiazepine receptors[8–12] and mimicked by acute treatment with anxiogenic inverse agonists of these receptors, such as FG 7142.[12–14] Like acute mild stress, FG 7142 elicits a preferential increase in dopaminergic transmission in the prefrontal cortex when compared with nucleus accumbens and striatum.[30–32]

Exposure to stress not only exacerbates or triggers many neuropsychiatric disorders, including depression and schizophrenia, among others, but also impairs working memory in rats or monkeys; this latter effect is blocked by agents that prevent the stress-induced increase in dopamine turnover[33] or that antagonize dopamine receptors,[33,34] indicating a dopaminergic mechanism.[35] The observation that cognitive and behavioral deficits are induced by reductions in cortical dopamine transmission[36–38] also suggests that the dopaminergic innervation of the prefrontal cortex plays a key role in cognition and in the modulation of affective responses.

The finding that pretreatment with benzodiazepine receptor agonists and long-term treatment with antidepressant drugs antagonized the effect of acute stress on mesocortical dopamine neurons[10,39] suggests that an overactivation of mesocortical dopaminergic neurons might play an important role in the pathophysiology of mood disorders and their associated symptoms of anxiety.[31,40] Thus, some homeostatic compensatory mechanisms should develop in order to counteract the overexcitation of dopaminergic neurons elicited by stress or anxiogenic stimuli.

11.3 STRESS INCREASES 3α,5α-TH PROG IN RAT BRAIN

The demonstration that acute stress increases the abundance of 3α,5α-TH PROG in the rat brain was given first by Purdy et al.,[41] who showed that forced swimming induced a time-dependent increase in the amount of this PROG metabolite in the rat brain and plasma. We subsequently showed that the content of 3α,5α-TH PROG was much higher in the brain of naïve rats than in those of handling-habituated animals.[42] Moreover, handling-habituated rats exposed to foot shock or forced to inhale a mixture of 65% oxygen and 35% CO_2 for 1 min showed a time-dependent increase in the cortical concentration of 3α,5α-TH PROG.[42]

The evidence that different kinds of stress increase at the same time the abundance of 3α,5α-TH PROG and the activity of mesocortical dopaminergic neurons, together with the previous evidence that this steroid derivative has a potent facilitatory action at the $GABA_A$ receptor, which plays a major role in modulating the activity of mesocortical dopamine neurons, therefore suggested that 3α,5α-TH PROG may function to reduce the excitatory effect of stress on these neurons.

Given that 3α,5α-TH PROG is one of the most potent and efficacious positive allosteric modulators of $GABA_A$ receptor function, which exerts a tonic inhibitory action on the activity of the HPA axis,[43] an increase in the brain content of this steroid derivative should also modulate the activation of the HPA axis induced by stress. Accordingly, it has been shown that pretreatment of rats with a single dose of 3α,5α-TH PROG significantly attenuated the elevation of plasma adrenocorticotropic hormone (ACTH) and serum corticosterone after emotional stress.[44] These results prompted us to evaluate whether pharmacological doses of PROG antagonized the stress-induced increase in 3α,5α-TH PROG. Administration of PROG (5 mg/kg, i.p., once a day) to group-housed rats for 5 days induced a significant increase in the cerebrocortical concentration of 3α,5α-TH PROG (Figure 11.1), measured 25 to 30 h after the last injection. The treatment induced similar effects on the plasma concentrations of this neurosteroid (data not shown). Moreover, subchronic administration of PROG significantly inhibited the increase in the concentration of 3α,5α-TH PROG elicited by foot shock stress in the cerebral cortex and in plasma (Figure 11.1). This treatment also markedly reduced the shock-induced increase in the plasma concentration of 3α,5α-TH PROG and corticosterone (data not shown). On the basis of these results, we also investigated the capability of pharmacological concentrations of this PROG metabolite to reduce the basal and stress-induced increase in the output of cortical dopamine.

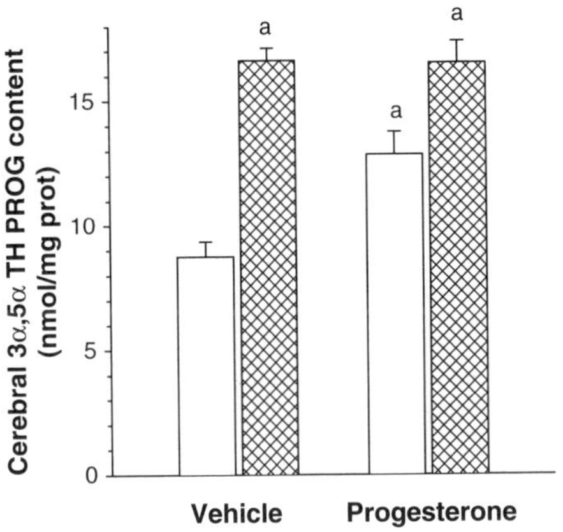

FIGURE 11.1 Effect of pretreatment with PROG on basal and foot shock stress-induced increase in the cerebrocortical levels of 3α,5α-TH PROG. Rats received a subchronic administration of vehicle or progesterone (5 mg/kg, i.p., once a day for 5 days) and, on the day of the experiment, were exposed (hatched bars) or not (open bars) to foot shock stress for 5 min and killed 30 min after. a, $p < .01$ vs. chronic vehicle-treated controls; b, $p < .01$ vs. chronic PROG-treated controls.

11.4 3α,5α-TH PROG REDUCES BASAL AND STRESS-ENHANCED FUNCTION OF DOPAMINERGIC NEURONS

$GABA_A$ receptors play a role in the modulation of mesocortical dopamine neurons. Thus, selective, partial, and full agonists of benzodiazepine receptors affect the firing rate and turnover of dopamine and reduce both the basal and stress-induced increases of extracellular dopamine concentration in the prefrontal cortex.[8–14] Given that $GABA_A$ receptors appear to be the main target of 3α,5α-TH PROG, this PROG metabolite, when administered intracerebroventricularly, shared with benzodiazepines the ability to reduce both the basal extracellular concentration of dopamine in the rat cerebral cortex and the stress-induced increase in this parameter.[45] In contrast to benzodiazepines, 3α,5α-TH PROG failed to reduce dopamine output in the striatum, indicating that this steroid does not contribute to the modulation of motor function related to the extrapyramidal system, whereas its effects on extracellular dopamine concentration in the cerebral cortex are consistent with the putative physiological role of this steroid derivative in the modulation of the emotional state.[46,47] In fact, this PROG metabolite exhibits a pharmacological profile similar to that of benzodiazepines, drugs that by enhancing the activity of $GABA_A$ receptors induce anxiolytic effects[16] and reduce the stress-induced activation of mesocortical dopamine neurons.[45] 3α,5α-TH PROG, like benzodiazepines, decreases the basal extracellular concentration of dopamine in the prefrontal cortex and the increase in the content of this monoamine elicited by foot shock stress.[45] In fact, intracerebroventricular injection of this steroid derivative (10 μg/5 μl) reduced by 40% basal cortical dopamine output and, when administered 20 min before stress exposure, completely prevented the effect of stress on cortical dopamine extracellular concentration. This finding is consistent with the demonstration that 3α,5α-TH PROG inhibits also the decrease in $GABA_A$ receptor function induced by the same stress paradigm in the cerebral cortex[48] and elicits an anxiolytic action following systemic administration.[49] Together, these pharmacological data suggest that endogenous 3α,5α-TH PROG might contribute directly or indirectly to the physiological regulation of basal and stress-induced dopamine release in the rat cerebral cortex.

To better evaluate the role of the endogenous 3α,5α-TH PROG in the tonic modulation of the excitatory response of cortical dopaminergic neurons to stressful and anxiogenic stimuli, the effect of finasteride, a specific 5α-reductase inhibitor able to produce a marked reduction of 3α,5α-TH PROG, was studied in foot shock- and FG 7142-treated rats.

11.5 FINASTERIDE-INDUCED DEPLETION OF 3α,5α-TH PROG POTENTIATES THE EFFECT OF STRESS ON CORTICAL DOPAMINERGIC NEURONS

The role of 3α,5α-TH PROG in the modulation of mesocortical dopamine output neurons was studied in rats pretreated with finasteride (25 mg/kg, s.c., once a day for 2 days). In fact, such a treatment by inhibiting the 5α-reductase activity is able to markedly reduce (75%) the basal content of 3α,5α-TH PROG in the rat brain and to completely prevent the foot shock-induced increase in the cortical content of this PROG metabolite (Figure 11.2). In contrast, the same treatment did not significantly affect the basal and stress-induced increase in the plasmatic and cortical concentrations of PROG (data not shown).

Experiments using the microdialysis technique were performed to evaluate in the cerebral cortex the extracellular concentration of dopamine in control rats and in animals exposed to foot shock stress or subjected to an intraperitoneal injection of FG 7142 (20 mg/kg, i.p.).

As expected,[10] foot shock stress induced a marked and transient increase (+125%) in the extracellular concentration of dopamine output, returning to basal values 20 min after the termination of stress exposure (Figure 11.3A). A second episode of foot shock stress administered 60 min after the end of the first exposure induced a smaller increase (+80%) in dopamine output in control animals, with the extracellular dopamine concentration again returning to basal values within 20 min.

In rats pretreated with finasteride, foot shock increased the extracellular concentration of dopamine by extents (+120 and +70% for the first and second exposures, respectively) similar to those observed in control rats (Figure 11.3A). However, the duration of the effects of stress in the finasteride-pretreated rats was greater than that apparent in the control animals. Thus, the extracellular dopamine concentration remained significantly increased for at least 40 min after the end of stress exposure. Similarly, finasteride pretreatment significantly enhanced the stimulatory effect of the anxiogenic drug FG 7142 (20 mg/kg, i.p.) on cortical dopamine output (Figure 11.3B). Thus, the maximal increase in the cortical extracellular concentration of

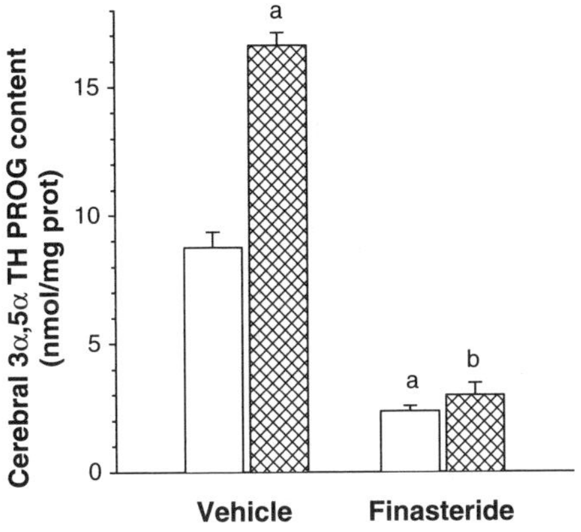

FIGURE 11.2 Effect of pretreatment with finasteride on basal and foot shock stress-induced increase in the cerebrocortical levels of 3α,5α- PROG. Rats received a subchronic administration of vehicle or finasteride (25 mg/kg, s.c., once a day for 6 days) and, on the day of the experiment, were exposed (hatched bars) or not (open bars) to foot shock stress for 5 min and killed 30 min after. a, $p < .01$ vs. chronic vehicle-treated controls; b, $p < .01$ vs. chronic finasteride-treated controls.

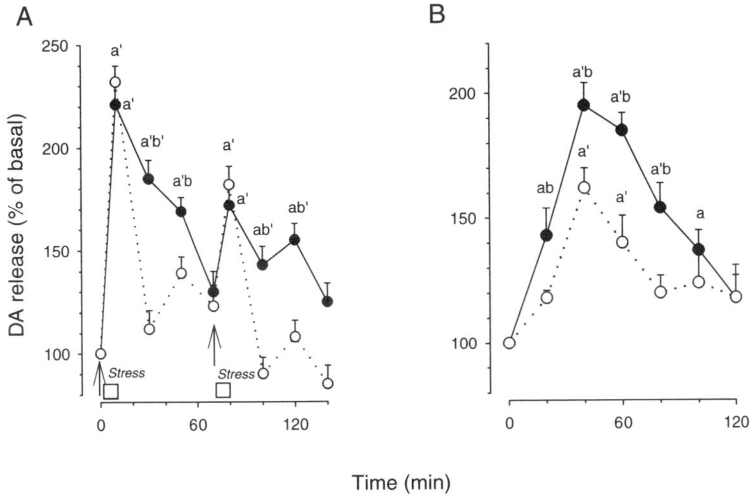

FIGURE 11.3 Effects of finasteride pretreatment on dopamine (DA) release in the rat prefrontal cortex induced by foot shock stress (A) or FG 7142 (B). Rats were pretreated with finasteride (25 mg/kg, s.c.) (closed circles) or vehicle (open circles) once a day for 2 days. They were then subjected either to foot shock for 8 min beginning at time zero (~4 h after the last injection of finasteride or vehicle) and again 60 min after the end of the first episode (A) or to a single injection of FG 7142 (20 mg/kg, i.p.) at time zero (~4 h after the last injection of finasteride or vehicle). Data are means ± SEM of values from at least five rats per group and are expressed as a percentage of basal values. a, $p < .05$; a', $p < .01$ vs. basal values; b, $p < .05$, b'; $p < .01$ vs. the corresponding value for vehicle-pretreated controls. (From Dazzi, L. et al., *Brain Res.*, 932, 135, 2002. With permission.)

dopamine induced by a single injection of FG 7142 was significantly greater in rats pretreated with finasteride (+95%) than in those pretreated with vehicle (+65%). Moreover, the FG 7142-induced increase in dopamine output remained significant for a longer period of time in finasteride-pretreated rats (~100 min after drug injection) than it did in control rats, in which the extracellular dopamine concentration had returned to basal values by ~80 min.

The potentiation by finasteride of the stress- and FG 7142-induced increases in cortical dopamine output is likely related to the marked reduction in the cortical concentration of 3α,5α-TH PROG that resulted from the inhibition of 5α-reductase. Foot shock thus failed to increase the cortical content of this neuroactive steroid in finasteride-treated rats (Figure 11.2).

Given that 3α,5α-TH PROG enhances $GABA_A$ receptor function both *in vitro* and *in vivo*,[48,50,51] reduces basal and stress-induced dopamine output in the rat cerebral cortex,[45] reverses the stress-induced reduction of $GABA_A$ receptor function,[46] and elicits an anxiolytic effect when its brain levels are increased,[47,51] our observation that depletion of cortical 3α,5α-TH PROG is associated with potentiation of dopamine release induced by stress or by FG 7142 thus suggests that the increase in the cortical concentration of this PROG metabolite normally elicited by stress or by anxiogenic drugs might be an important homeostatic response whose function is to increase the excitation threshold of dopaminergic neurons. This conclusion is consistent with previous data showing that the increase in the cortical concentration of 3α,5α-TH PROG during stress contributes to restoration of GABAergic tone and conflict behavior after stress.[46]

Whereas the cortical concentration of PROG increases rapidly and markedly after an acute stressful stimulus, that of 3α,5α-TH PROG undergoes a slower and longer-lasting increase, reaching a peak at 30 min and remaining significantly higher than control values 80 min after stress.[52] Thus, the cortical concentration of 3α,5α-TH PROG would normally peak at around the same time that the foot shock- or FG 7142-induced release of dopamine in the prefrontal cortex is potentiated in finasteride-treated rats. The previous demonstration that, in contrast to the effect of finasteride, the intracerebroventricular administration of 3α,5α-TH PROG antagonized the increase in cortical dopamine output induced by acute stress in freely moving rats[45] further supports our present data.

11.6 SOCIAL ISOLATION-INDUCED DEPLETION OF 3α,5α-TH PROG POTENTIATES THE EFFECT OF STRESS ON HPA AXIS ACTIVITY

Social isolation for 30 days of rats immediately after weaning, without any additional stressor, results in a decrease in the brain and plasma concentrations of neuroactive steroids such as pregnenolone (PREG), PROG, 3α,5α-TH PROG, and 3α,5α-TH DOC.[53] Social isolation of adult mice for 6 to 10 weeks was also shown to reduce the concentration of 3α,5α-TH PROG in the cerebral cortex; however, in this experimental model the effect was selective for this neurosteroid, being cortical PREG and PROG not affected.[54]

The molecular mechanisms that underlie the persistent decrease in the abundance of neuroactive steroids induced by social isolation of rats remain unclear. However, consistent with the notion that a facilitatory trace, characterized by hyperresponsiveness of the HPA axis to new stimuli, may develop during chronic stress,[55] we have previously shown that the functional response of the HPA axis to an acute stressful stimulus is increased in isolated rats.[53] Thus, the increases in the brain and plasma concentration of 3α,5α-TH PROG induced by foot shock,[52,56] used in this instance as a novel acute stressor, were markedly greater on a percentage basis in socially isolated rats than in group-housed animals. This finding suggests that a reduction in the brain content of this PROG metabolite sensitizes the HPA axis to the action of stress, a conclusion consistent with the evidence that PROG administration prevents acute stress-induced increase of 3α,5α-TH PROG (Figure 11.1). These results, together with the evidence that different chronic stressful stimuli induce a similar phenomenon on monoaminergic neurons,[26–28] further suggest that 3α,5α-TH PROG has physiological actions in the modulation of basal and stress-induced activation of specific neuronal populations such as mesocortical dopamine neurons.

11.7 SOCIAL ISOLATION-INDUCED DEPLETION OF 3α,5α-TH PROG INCREASES THE STEROIDOGENIC EFFECT OF ETHANOL

Ethanol, like stress, modifies the function of the HPA axis.[57] The acute administration of ethanol thus increases ACTH release in humans and laboratory animals by inducing the secretion of both corticotropin-releasing factor (CRF) and arginine vasopressin (AVP).[58–60] These actions also result in an increase in the plasma and brain concentrations of various steroids, including corticosterone and PROG.[58,59,61] Thus, like acute stress and pharmacological inhibitors of $GABA_A$ receptor-mediated neurotransmission,[56] ethanol increases the abundance of 3α,5α-TH PROG in the brain and plasma.[23] As shown in Figure 11.4, a single injection of ethanol (1 g/kg, i.p.) markedly increased both the cerebral cortical and plasma concentrations of 3α,5α-TH PROG in both group-housed and isolated rats. However, the percentage increases in the brain concentrations of 3α,5α-TH PROG induced by ethanol in isolated rats were much greater (+308%) than those induced in group-housed rats (+96%). The ability of ethanol to increase the amount of 3α,5α-TH PROG and corticosterone in plasma was also more pronounced in isolated rats than in group-housed animals (data not shown).

Because ethanol injection represents a stressful condition, the greater effects of this drug on the cerebrocortical and plasma concentrations of 3α,5α-TH PROG in isolated rats may reflect the greater sensitivity of the HPA axis in these animals to stress. An acute administration of ethanol able to stimulate the HPA axis in control rats[57,59] was shown to have no effect on the concentrations of neuroactive steroids in the cerebral cortex and plasma of adrenalectomized and orchietomized animals.[62,63] These results are thus consistent with the observations that the large and rapid increases in the brain and plasma concentrations of neuroactive steroids induced by various acute stressful stimuli in rats[41,52,56] are abolished by removal of

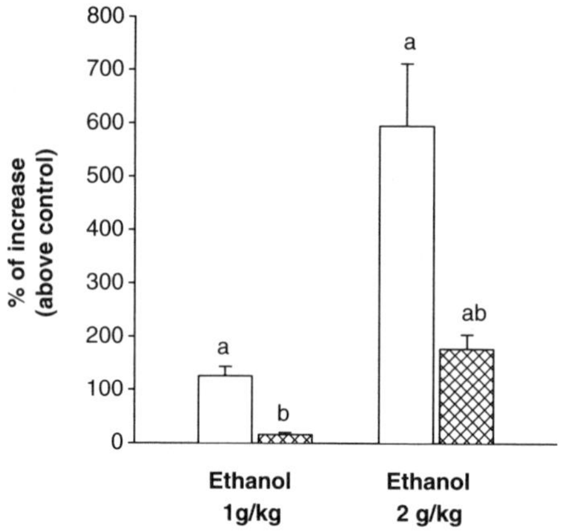

FIGURE 11.4 Effect of a single injection of ethanol on the concentration of 3α,5α-TH PROG in the cerebral cortex of PROG-treated rats. Animals chronically treated with PROG (5 mg/kg, i.p., once a day for 5 days) (hatched bars) or vehicle (open bars) were injected with 1 or 2 g/kg of ethanol (20% w/v) and killed 20 min after. Data are expressed as percentage of increase relative to the respective control values (acute saline-treated animals). a, $p < .01$ vs. respective controls; b, $p < .05$ vs. the same dose of ethanol in chronic vehicle-treated controls.

the adrenals and gonads.[56] Our present data therefore suggest that the facilitatory trace, characterized by hyperresponsiveness of the HPA axis to new stimuli, contributes to the enhanced sensitivity of socially isolated rats to the effects of ethanol on the brain and plasma concentrations of 3α,5α-TH PROG. In agreement with this conclusion is the evidence that subchronic administration of PROG to grouped-housed rats for 5 days, enhanced by 47% basal cerebrocortical levels of 3α,5α-TH PROG (Figure 11.1), completely abolished the capability of 1 g/kg ethanol to stimulate the formation of 3α,5α-TH PROG (Figure 11.4). However, this subchronic treatment does not completely abolish the capability of this drug to activate the HPA axis; in fact, a higher dose of ethanol (2 g/kg, i.p.) significantly enhanced cerebrocortical levels of 3α,5α-TH PROG (Figure 11.4), although this effect is still reduced compared to that induced in vehicle-treated animals.

11.8 3α,5α-TH PROG POTENTIATES THE ETHANOL-INDUCED CHANGES ON CORTICAL DOPAMINERGIC NEURONS: ANTAGONISM BY FINASTERIDE

To further clarify the modulatory role of 3α,5α-TH PROG on the function of cortical dopaminergic neurons, we studied whether increases in the basal brain concentration of this PROG metabolite might affect the action of ethanol on the cortical dopamine extracellular concentration.

The potentiation of GABAergic transmission by 3α,5α-TH PROG has been proposed as the neurochemical mechanism underlying the mediation of the effects of short- and long-term administration of ethanol. This conclusion is strongly suggested by recent results showing that ethanol intake increases the brain content of 3α,5α-TH PROG.[23] This finding, together with the evidence that $GABA_A$ receptors play a major role in the modulation of mesocortical dopaminergic neurons, suggests that 3α,5α-TH PROG might be the compound responsible for some of the central effects of ethanol,[64] including modulation of mesocortical dopamine neurons. Pretreatment of rats with PROG shifted the dose–response relation for the effect of acute intraperitoneal administration of ethanol on dopamine content in the prefrontal cortex to higher ethanol concentrations. In rats pretreated with PROG, which alone had no effect on the basal extracellular concentration of dopamine, a low dose (0.25 g/kg) of ethanol that was ineffective in naïve or vehicle pretreated animals thus induced a marked increase in cortical dopamine extracellular concentration (Figure 11.5). In contrast, a higher dose (0.5 g/kg) of ethanol that significantly increased dopamine content in naïve or vehicle-pretreated rats induced a pronounced decrease in the cortical extracellular concentration of this monoamine in PROG-pretreated animals. This latter effect was similar to that elicited by an even higher dose (1 g/kg) of ethanol in naïve rats. Moreover, similar PROG treatment significantly increased the cortical concentration of the PROG metabolite 3α,5α-TH PROG.

The modulatory action of PROG of the effects of both low and high doses of ethanol on cortical dopamine content was abolished by concomitant blockade of 5α-reductase by finasteride. The same treatment also completely prevented the increase in the brain content of 3α,5α-TH PROG elicited by PROG administration (Figure 11.2). These results further confirm the modulatory role of 3α,5α-TH PROG

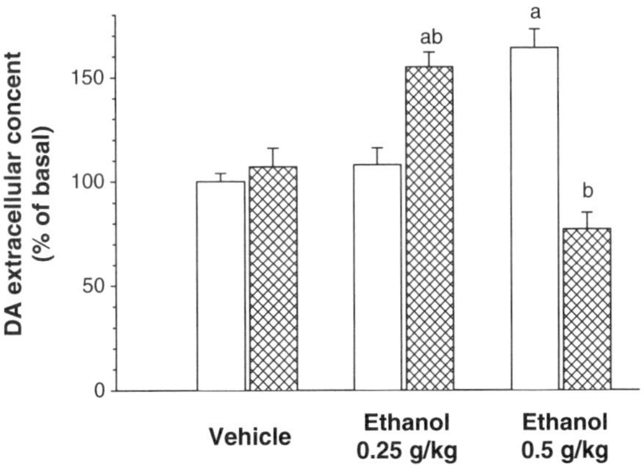

FIGURE 11.5 Effect of pretreatment with PROG on the modulation of dopamine content in the prefrontal cortex by ethanol. Rats were injected intraperitoneally with PROG (5 mg/kg, once a day for 5 days) (hatched bars) or vehicle (open bars). They were subjected to an intraperitoneal challenge with ethanol (0.25 or 0.5 g/kg) or vehicle 28 or 29 h after the last injection of PROG (or vehicle), and the cortical extracellular concentration of dopamine was determined 40 min later. Data are expressed as a percentage of basal values and are means ± SEM of values from at least five rats per group. a, $p < .01$ vs. animals pretreated and challenged with vehicle; b, $p < .01$ vs. animals pretreated with vehicle and challenged with ethanol. (From Dazzi, L. et al., *J. Neurochemistry*, 83, 1103, 2002. With permission.)

on the activity of mesocortical dopaminergic neurons and the hypothesis that this PROG metabolite plays an important role in determining the efficacy with which ethanol affects the function of this neuronal pathway. The abilities of PROG to increase the brain content of 3α,5α-TH PROG and to potentiate the action of ethanol on cortical dopamine content, together with the shared ability of ethanol and 3α,5α-TH PROG to modulate the function of $GABA_A$ receptors,[65,66] suggest that the interaction between 3α,5α-TH PROG and ethanol is important for the modulation by ethanol of mesocortical dopaminergic neurons. Increases in the activity of $GABA_A$ receptors located on the somas of dopaminergic neurons in the ventral tegmental area, which regulate the activity of the mesocortical dopaminergic pathway,[6] induced by the synergistic action of ethanol and 3α,5α-TH PROG, might explain the potentiation of the effects of ethanol on these neurons elicited by subchronic treatment with PROG. A change in the activity of these neurons in response to fluctuations in the brain content of 3α,5α-TH PROG might thus result in behavioral changes associated with various mental disorders, as well as in changes in the responsiveness of these neurons to ethanol.

Given that marked fluctuations in the plasma and CSF concentrations of 3α,5α-TH PROG are associated with physiological or pathological conditions such as menopause,[67] premenstrual syndrome,[68–70] aging,[71] acute and chronic stress,[41,42,52,53,56] and depression,[24,25] it is possible that changes in the brain content of this neurosteroid reflect a neurochemical mechanism responsible for modulation of the activity of the mesocortical dopamine system as well as for the development of anxiety and affective disorders.

These changes may also explain the increase in ethanol intake that has been observed in such pathological situations. Accordingly, several studies suggest that activation of central dopaminergic transmission plays a crucial role in mediating the behavioral modifications induced by drugs of abuse,[72] specifically by moderate to high doses of ethanol in rodents and humans.[73–75]

11.9 CONCLUSION

Our results suggest that changes in the brain content of 3α,5α-TH PROG may reflect a neurochemical mechanism responsible for the modulation of the activity of selective neuronal populations such as the mesocortical dopamine system, as well as those neuronal systems regulating the HPA axis activity, and for some of the behavioral changes associated with dysfunction of these two systems. This conclusion is greatly supported by the evidence that:

1. Finasteride-induced reduction in the cerebrocortical content of 3α,5α-TH PROG is associated with a marked potentiation of the stimulatory effect of stress on both mesocortical dopamine extracellular concentrations[76] and the HPA axis.[53]
2. PROG administration potentiates the effect of ethanol on the function of these dopaminergic neurons while completely antagonizing ethanol-induced activation of the HPA axis.
3. The dramatic reduction in the plasma and brain contents of this PROG metabolite during menopause and in postpartum, as well as the altered

fluctuations of this compound in premenstrual syndrome and after different stress conditions, is often associated with the development of emotional and affective disorders.

4. Administration of antidepressant and atypical antipsychotic drugs or hormone replacement therapy, treatments that relieve some of the symptoms associated with anxiety or affective disorders, normalizes the concentrations of 3α,5α-TH PROG in the plasma and CSF of patients[24,25] and increases its content in the brain of animals.[17–21]

The evidence that isolated animals are more sensitive to an acute effect of ethanol and foot shock stress[53] is consistent with the established relations among stress, glucocorticoids, and vulnerability to drugs of abuse.[77] Indeed, plasma glucocorticoid concentrations appear to be related to the vulnerability to self-administration of ethanol. Thus, adrenalectomy reduced ethanol consumption in alcohol-preferring rats to the level apparent in non-alcohol-preferring animals, whereas administration of corticosterone to the adrenalectomized rats restored their intake of ethanol to preoperative levels.[78] Furthermore, glucocorticoids not only interact with the reinforcing properties of drugs of abuse,[77] but also have intrinsic reinforcing effects, as revealed by the demonstration of corticosterone self-administration in rats.[79] In contrast to the other steroids measured, the plasma concentration of corticosterone was significantly increased in isolated rats compared with that in group-housed animals (+32%). This higher basal level of corticosterone may account for the increase in ethanol intake induced by social isolation.[80–82] Given that social isolation represents a model of mild chronic stress,[53] that 3α,5α-TH PROG is one the most potent endogenous positive modulators of $GABA_A$ receptors,[16] and that the concentration of this steroid is reduced in socially isolated animals,[53] the increased rate of ethanol intake in isolated rats may also result from a rewarding effect of this drug that is attributable to the marked increase in the concentrations of the endogenous anxiolytic steroid induced by its consumption. Thus, the natural preference for ethanol may be related to the effect of this drug on the levels of neuroactive steroids, an effect that is enhanced by social isolation.

In conclusion, the demonstration that isolated rats exhibit increased self-administration of ethanol and are more sensitive to the action of ethanol and stress on the brain and plasma concentrations of 3α,5α-TH PROG, which per se potently modulates the action of ethanol and stress on the activity of the mesocortical dopamine system, suggests that 3α,5α-TH PROG may have a role in modulating the function of specific neuronal populations involved in the development of some mental disorders associated with vulnerability to alcohol.

REFERENCES

1. Thierry, A.M., Tassin, J.P., Blanc, G., et al., Selective activation of the mesocortical DA system by stress, *Nature*, 263, 242, 1976.
2. Abercrombie, E.D., Keefe, K.A., DiFrischia, D.S., et al., Differential effect of stress on *in vivo* dopamine release in the striatum, nucleus accumbens, and medial prefrontal cortex, *J. Neurochem.*, 52, 1655, 1989.

3. Deutch, A.Y. and Roth, R.H., The determinants of stress-induced activation of the prefrontal cortical dopamine system, *Prog. Brain Res.*, 85, 367, 1990.
4. Imperato, A., Puglisi-Allegra, S., Zocchi, A., et al., Changes in brain dopamine and acetylcholine release during and following stress are independent of the pituitary-adrenocortical axis, *Brain Res.*, 538, 111, 1991.
5. Inglis, F.M. and Moghaddam, B., Dopaminergic innervation of amygdala is highly responsive to stress, *J. Neurochem.*, 72, 1088, 1999.
6. Tzschentke, T.M., Pharmacology and behavioral pharmacology of the mesocortical dopamine system, *Prog. Neurobiol.*, 63, 241, 2001.
7. Levita, L., Dalley, J.W., and Robbins, T.W., Nucleus accumbens dopamine and learned fear revisited: a review and some new findings, *Behav. Brain Res.*, 137, 115, 2002.
8. Ida, Y., Tsuda, A., Sueyoshi, K., et al., Blockade by diazepam of conditioned fear-induced activation of rat mesoprefrontal dopamine neurons, *Pharmacol. Biochem. Behav.*, 33, 477, 1989.
9. Kaneyuki, H., Yokoo, H., Tsuda, A., et al., Psychological stress increases dopamine turnover selectively in mesofrontal dopamine neurons of rats: reversal by diazepam, *Brain Res.*, 557, 154, 1991.
10. Dazzi, L., Motzo, C., Imperato, A., et al., Modulation of basal and stress-induced release of acetylcholine and dopamine in rat brain by abecarnil and imidazenil, two anxioselective γ-aminobutyric acid$_A$ receptor modulators, *J. Pharmacol. Exp. Ther.*, 273, 241, 1995.
11. Hegarty, A.A. and Vogel, W.H., The effect of acute and chronic diazepam treatment on stress-induced changes in cortical dopamine in the rat, *Pharmacol. Biochem. Behav.*, 52, 771, 1995.
12. Hutson, P.H. and Barton, C.L., L-701,324, a glycine/NMDA receptor antagonist, blocks the increase of cortical dopamine metabolism by stress and DMCM, *Eur. J. Pharmacol.*, 326, 127, 1997.
13. Tam, S.Y. and Roth, R.H., Selective increases in dopamine metabolism in the prefrontal cortex by the anxiogenic beta-carboline FG7142, *Biochem. Pharmacol.*, 34, 1595, 1985.
14. Tam, S.Y. and Roth, R.H., Modulation of mesoprefrontal dopamine neurons by central benzodiazepine receptors. I. Pharmacological characterization, *J. Pharmacol. Exp. Ther.*, 252, 989, 1990.
15. Baulieu, E.E., Steroid hormones in the brain: several mechanisms? in *Steroid Hormone Regulation of the Brain*, Fuxe, K., Gustafssson, J.A., and Wettemberg, L., Eds., Pergamon Press, Oxford, 1981, p. 3.
16. Majewska, M.D., Harrison, N.L., Schartz, R.D., et al., Steroid hormone metabolites are barbiturate-like modulators of the GABA receptor, *Science*, 232, 1004, 1986.
17. Uzunov, D.P., Cooper, T.B., Costa, E., et al., Fluoxetine-elicited changes in the brain neurosteroid content measured by negative ion mass fragmentography, *Proc. Natl. Acad. Sci. U.S.A.*, 93, 12599, 1996.
18. Serra, M., Pisu, M.G., Muggironi, M., et al., Opposite effects of short- versus long-term administration of fluoxetine on the concentration of neuroactive steroids in rat plasma and brain, *Psychopharmacology*, 158, 48, 2001.
19. Serra, M., Pisu, M.G., Dazzi, L., et al., Prevention of the stress-induced increase in the concentration of neuroactive steroids in rat brain by long-term administration of mirtazapine but not of fluoxetine, *J. Psychopharmacol.*, 16, 133, 2002.
20. Marx, C.E., Duncan, G.E., Gilmore, J.H., et al., Olanzapine increase allopregnanolone in the rat cerebral cortex, *Biol. Psychiatry*, 47, 1000, 2000.

21. Barbaccia, M.L., Affricano, D., Purdy, R.H., et al., Clozapine, but not haloperidol, increases brain concentrations of neuroactive steroids in the rat, *Neuropsychopharmacology*, 25,489, 2001.
22. Morrow, A.L., Janis, G.C., Van Doren, M.J., et al., Neurosteroids mediate pharmacological effects of ethanol: a new mechanism of ethanol action, *Alcohol. Clin. Exp. Res.*, 23, 1933, 1999.
23. VanDoren, M.J., Matthews, D.B., Janis, G.C., et al., Neuroactive steroid 3alpha-hydroxy-5alpha-pregnan-20-one modulates electrophysiological and behavioral actions of ethanol, *J. Neurosci.*, 20, 1982, 2000.
24. Romeo, E., Ströhle, A., Spalletta, G., et al., Effects of antidepressant treatment on neuroactive steroids in major depression, *Am. J. Psychiatry*, 155, 910, 1998.
25. Uzunova, V., Sheline, Y., Davis, J.M., et al., Increase in the cerebrospinal fluid content of neurosteroids in patients with unipolar major depression who are receiving fluoxetine or fluvoxamine, *Proc. Natl. Acad. Sci. U.S.A.*, 95, 3239, 1998.
26. Nisenbaum, L.K., Zigmond, M.J., Sved, A.F., et al., Prior exposure to chronic stress results in enhanced synthesis and release of hippocampal norepinephrine in response to a novel stressor, *J. Neurosci.*, 11, 1478, 1991.
27. Gresch, P.J., Sved, A.F., Zigmond, M.J., et al., Stress-induced sensitization of dopamine and norepinephrine efflux in medial prefrontal cortex of the rat, *J. Neurochem.*, 63, 575, 1994.
28. Gresch, P.J., Sved, A.F., Zigmond, M.J., et al., Local influence of endogenous norepinephrine on extracellular dopamine in rat medial prefrontal cortex, *J. Neurochem.*, 65, 111, 1995.
29. Makino, S., Hasimoto, K., and Gold, P.W., Multiple feedback mechanism activating corticotropin-releasing hormone system in the brain during stress, *Biochem. Behav.*, 73, 147, 2002.
30. Horger, B.A., Elsworth, J.D., and Roth, R.H., Blockade of FG 7142-induced increased dopamine utilization by the glycine/NMDA receptor antagonist (+)HA 966, *J. Neurochem.*, 66, 1959, 1996.
31. Horger, B.A. and Roth, R.H., The role of mesoprefrontal dopamine neurons in stress, *Crit. Rev. Neurobiol.*, 10, 395, 1996.
32. Bradberry, C.W., Lory, J.D., and Roth, R.H., The anxiogenic β-carboline FG 7142 selectively increases dopamine release in the rat prefrontal cortex as measured by microdialysis, *J. Neurochem.*, 56, 748, 1991.
33. Murphy, B.L., Arnsten, A.F.T., Jentsch, J.D., et al., Dopamine and spatial working memory in rats and monkeys: pharmacological reversal of stress-induced impairment, *J. Neurosci.*, 16, 7768, 1996.
34. Arnsten, A.F.T. and Goldman-Rakic, P.S., Noise stress impairs prefrontal cortical cognitive function in monkeys, *Arch. Gen. Psychiatry*, 5, 362, 1998.
35. Mizoguchi, K., Yuzurihara, M., Ishige, A., et al., Chronic stress induces impairment of spatial working memory because of prefrontal dopaminergic dysfunction, *J. Neurosci.*, 20, 1568, 2000.
36. Brozoski, T.J., Brown, R.M., Rosvold, H.E., et al., Cognitive deficits caused by regional depletion of dopamine in prefrontal cortex of rhesus monkey, *Science*, 205, 929, 1979.
37. Bubser, M. and Schmidt, W.J., 6-Hydroxydopamine lesion of the rat prefrontal cortex increases locomotor activity, impairs acquisition of delayed alternation tasks, but does not affect uninterrupted tasks in the radial maze, *Behav. Brain Res.*, 37, 157, 1990.

38. Jentsch, J.D., Redmond, D.E., Jr., Elsworth, J.D., et al., Enduring cognitive deficits and cortical dopamine dysfunction in monkeys after long-term administration of phencyclidine, *Science*, 277, 953, 1997.
39. Dazzi, L., Spiga, F., Pira, L., et al., Inhibition of stress- or anxiogenic-drug-induced increases in dopamine release in the rat prefrontal cortex by long-term treatment with antidepressant drugs, *J. Neurochem.*, 76, 1212, 2001.
40. Jentsch, J.D., Roth, R.H., and Taylor, J.R., Role for dopamine in the behavioral functions of the prefrontal-corticostriatal system: implications for mental disorders and psychotropic drug action, in *Cognition, Emotion and Autonomic Responses: The Integrative Role of the Prefrontal Cortex and Limbic Structures*, Uylings, H.B.M., van Eden, C.G., Feenstra, M.G.P., Pennartz, C.M.A., and de Bruin, J.P.C., Eds., Elsevier, Amsterdam, 2000, p. 35.
41. Purdy, R.H., Morrow, A.L., Moore, P.H., et al., Stress-induced elevations of gamma-aminobutyric acid type A receptor-active steroids in the rat brain, *Proc. Natl. Acad. Sci. U.S.A.*, 88, 553, 1991.
42. Barbaccia, M.L., Roscetti, G., Trabucchi, M., et al., Neurosteroids in the brain of handling-habituated and naive rats: effect of CO_2 inhalation, *Eur. J. Pharmacol.*, 261, 317, 1994.
43. Calogero, A.E., Gallucci, W.T., Chousos, G.P., et al., Interaction between GABAergic neurotransmission and rat hypothalamic corticotropin releasing hormone secretion in vitro, *Brain Res.*, 463, 223, 1988.
44. Patchev, V.K., Hassan, A.H.S., Holsboer, F., et al., The neurosteroid tetrahydroprogesterone attenuates the endocrine response to stress and exerts glucocorticoid-like effects on vasopressin gene transcription in the rat hypothalamus, *Neuropharmacology*, 15, 533, 1996.
45. Motzo, C., Porceddu, M.L., Maira, G., et al., Inhibition of basal and stress-induced dopamine release in the cerebral cortex and nucleus accumbens of freely moving rats by the neurosteroid allopregnanolone, *J. Psychopharmacol.*, 10, 266, 1996.
46. Barbaccia, M.L., Serra, M., Purdy, R.H., et al., Stress and neuroactive steroids, in *International Review of Neurobiology*, Vol. 46, *Neurosteroids and Brain Function*, Biggio, G. and Purdy, R.H., Eds., Academic Press, San Diego, 2001, p. 244.
47. Engel, S.R. and Grant, K.A., Neurosteroids and behavior, in *International Review of Neurobiology*, Vol. 46, *Neurosteroids and Brain Function*, Biggio, G. and Purdy, R.H., Eds., Academic Press, San Diego, 2001, p. 321.
48. Concas, A., Mostallino, M.C., Perra, C., et al., Functional correlation between allopregnanolone and [^{35}S]TBPS binding in the brain of rats exposed to isoniazid, pentylentetrazol or stress, *Br. J. Pharmacol.*, 118, 839, 1996.
49. Bitran, D., Hilvers, R.J., and Kellogg, C.K., Anxiolytic effects of 3α-hydroxy-5α[β]-pregnan-20-one: endogenous metabolites of progesterone that are active at the $GABA_A$ receptor, *Brain Res.*, 561, 157, 1991.
50. Lambert, J.J., Harney, S.C., Belelli, D., et al., Neurosteroid modulation of recombinant and synaptic $GABA_A$ receptors, in *International Review of Neurobiology*, Vol. 46, *Neurosteroids and Brain Function*, Biggio, G. and Purdy, R.H., Eds., Academic Press, San Diego, 2001, p. 178.
51. Bitran, D., Shiekh, M., McLeod, G., et al., Anxiolytic effect of progesterone is mediated by the neurosteroid allopregnanolone at $GABA_A$ receptors, *J. Neuroendocrinol.*, 7, 171, 1995.
52. Barbaccia, M.L., Roscetti, G., Trabucchi, M., et al., Time-dependent changes in rat brain neuroactive steroid concentrations and $GABA_A$ receptor function after acute stress, *Neuroendocrinology*, 63, 166, 1996.

53. Serra, M., Pisu, M.G., Littera, M., et al., Social isolation-induced decreases in both the abundance of neuroactive steroids and $GABA_A$ receptor function in rat brain, *J. Neurochem.*, 75, 732, 2000.
54. Matzumoto, K., Uzunova, V., Pinna, G., et al., Permissive role of brain allopregnanolone content in the regulation of pentobarbital-induced righting reflex loss, *Neuropharmacology*, 38, 955, 1999.
55. Akana, S.F., Scribner, K.A., Bradbury, M.J., et al., Feedback sensitivity of the rat hypothalamo-pituitary-adrenal axis and its capacity to adjust to exogenous corticosterone, *Endocrinology*, 131, 585, 1992.
56. Barbaccia, M.L., Roscetti, G., Trabucchi, M., et al., The effects of inhibitors of GABAergic transmission and stress on brain and plasma allopregnanolone concentrations, *Br. J. Pharmacol.*, 20, 1582, 1997.
57. Rivier, C., Alcohol stimulates ACTH secretion in the rat: mechanism of action and interaction with other stimuli, *Alcohol. Clin. Exp. Res.*, 20, 240, 1996.
58. Ellis, F.W., Effect of ethanol on plasma corticosterone levels, *J. Pharmacol. Exp. Ther.*, 153, 121, 1996.
59. Rivier, C., Bruhn, T., and Vale, W., Effect of ethanol on the hypothalamic-pituitary-adrenal axis in the rat: role of corticotropin-releasing factor (CRF), *J. Pharmacol. Exp. Ther.*, 229, 127, 1984.
60. Ogilvie, K., Lee, S., and Rivier, C., Role of arginine vasopressin and corticotropin-releasing factor in mediating alcohol-induced adrenocorticotropin and vasopressin secretion in male rats bearing lesions of the paraventricular nuclei, *Brain Res.*, 744, 83, 1997.
61. Korneyev, A., Costa, E., and Guidotti, A., During anesthetic-induced activation of hypothalamic pituitary adrenal axis, blood borne steroids fail to contribute to the anesthetic effect, *Neuroendocrinology*, 57, 559, 1993.
62. Khisti, R.T., Kralic, J.E., Van Doren, M.J., et al., Adrenalectomy attenuates increase in cortical allopregnanolone and behavioral effects induced by acute ethanol administration, *Alcohol. Clin. Exp. Res.*, 26, 103A, 2002.
63. Khisti, R.T., Penland, S.N., Van Doren, M.J., et al., Neurosteroid modulation of ethanol action, *World J. Biol. Psychiatry*, 3, 87, 2002.
64. Morrow, A.L., VanDoren, M.J., Penland, S.N., et al., The role of GABAergic neuroactive steroids in ethanol action, tolerance and dependence, *Brain Res. Rev.*, 37, 98, 2001.
65. Lambert, J.J., Belelli, D., Hill-Venning, C., et al., Neurosteroids and $GABA_A$ receptor function, *Trends Pharmacol. Sci.*, 16, 295, 1995.
66. Grobin, A.C., Matthews, D.B., Devaud, L.L., et al., The role of $GABA_A$ receptors in the acute and chronic effects of ethanol, *Psychopharmacology*, 139, 2, 1998.
67. Barbaccia, M.L., Lello, S., Sidiropoulou, T., et al., Plasma 5α-androstane-3α,17βdiol, an endogenous steroid that positively modulates $GABA_A$ receptor function, and anxiety: a study in menopausal women, *Psychoneuroendocrinology*, 25, 659, 2000.
68. Rapkin, A.J., Morgan, M., Goldman, L., et al., Progesterone metabolite allopregnanolone in women with premenstrual syndrome, *Obstetr. Gynecol.*, 90, 709, 1997.
69. Girdler, S.S., Straneva, P.A., Light, K.C., et al., Allopregnanolone levels and reactivity to mental stress in premenstrual dysphoric disorder, *Biol. Psychiatry*, 49, 788, 2001.
70. Rasgon, N., Serra, M., Biggio, G., et al., Neuroactive steroid-serotonergic interaction: responses to an intravenous L-tryptophan challenge in women with premenstrual syndrome, *Eur. J. Endocrinol.*, 145, 25, 2001.

71. Genazzani, A.R., Petraglia, F., Bernardi, F., et al., Circulating levels of allopregnanolone in humans: gender, age, and endocrine influences, *J. Clin. Endocrinol. Metab.*, 83, 2099, 1998.
72. Grace, A.A., The tonic/phasic model of dopamine system regulation and its implications for understanding alcohol and psychostimulant craving, *Addiction*, 95 (Suppl. 2), 119, 2000.
73. Gessa, G.L., Muntoni, F., Collu, M., et al., Low doses of ethanol activate dopaminergic neurons in the ventral tegmental area, *Brain Res.*, 348, 201, 1985.
74. Di Chiara, G. and Imperato, A., Drugs abused by humans preferentially increase synaptic dopamine concentrations in the mesolimbic system of freely moving rats, *Proc. Natl. Acad. Sci. U.S.A.*, 85, 5274, 1988.
75. Weiss, F. and Porrino, L.J., Behavioral neurobiology of alcohol addiction: recent advances and challenges, *J. Neurosci.*, 22, 3332, 2002.
76. Dazzi, L., Serra, M., Vacca, G., et al., Depletion of cortical allopregnanolone potentiates stress-induced increase in cortical dopamine output, *Brain Res.*, 932, 135, 2002.
77. Piazza, P.V. and Le Moal, M., Pathophysiological basis of vulnerability to drug abuse: interaction between stress, glucocorticoids, and dopaminergic neurons, *Annu. Rev. Pharmacol. Toxicol.*, 36, 359, 1996.
78. Fahlke, C., Engel, J.A., Eriksson, P., et al., Involvement of corticosterone in the modulation of ethanol consumption in the rat, *Alcohol*, 11, 195, 1994.
79. Piazza, P.V., Deroche, V., Deminiere, J.M., et al., Reinforcing properties of corticosterone demonstrated by intravenous self-administration: possible biological basis of sensation-seeking, *Proc. Natl. Acad. Sci. U.S.A.*, 90, 11738, 1993.
80. Schenk, S., Gorman, K., and Amit, Z., Age-dependent effects of isolation housing on the self-administration of ethanol in laboratory rats, *Alcohol*, 7, 321, 1990.
81. Wolffgramm, J., Free choice ethanol intake of laboratory rats under different social conditions, *Psychopharmacology*, 101, 233, 1990.
82. Hall, F.S., Huang, S., Fong, G.W., et al., Effect of isolation-rearing on voluntary consumption of ethanol, sucrose and saccharin solutions in Fawn Hooded and Wistar rats, *Psychopharmacology*, 139, 210, 1998.

12 3α,5α-THP's Actions in the Ventral Tegmental Area for Lordosis: A Model System for Defining Mechanisms and Function of Progestins

Cheryl A. Frye and Sandra M. Petralia

CONTENTS

0-8493-2392-4/04/$0.00+$1.50

12.1 INTRODUCTION

Our research has focused on the mechanisms of progestins' actions in the midbrain ventral tegmental area (VTA) that mediate the onset and duration of sexual behavior of female rodents. To date, the majority of our investigations have utilized a model system in which progestins' mechanisms are manipulated in the ventromedial hypothalamus (VMH) or VTA of ovariectomized (OVX), estrogen (E)-primed rodents. The effects on lordosis, the stereotypic posture that female rodents assume to enable mating to occur (given appropriate hormonal and environmental stimulation), are then used as a bioassay to indicate which mechanisms of progestins are required for lordosis. Using this approach, we have found that although progesterone (P) has classic actions through intracellular progestin receptors (PRs) (genomic actions) in the VMH to initiate lordosis, in the VTA, the 5α-reduced, 3-hydroxylated metabolite of P, 3α-hydroxy-5α-pregnan-3α-ol-20-one (3α,5α-THP) has actions independent of PRs (nongenomic actions) to modulate the lordosis response. Using this paradigm, we have identified that 3α,5α-THP's actions in the VTA mediate the incidence and quality of lordosis of rodents via actions at cell surface receptors, such as γ-aminobutyric acid (GABA)–benzodiazepine receptor complexes (GBRs), glutamate type N-methyl-D-asparate receptors (NMDARs), and dopamine (DA) type 1 receptors (D_1).

This review focuses on research from our laboratory and others regarding the role of progestins in the VMH and VTA for lordosis, as well as the source and substrates for progestins' actions in these sites. The putative role of progestins' nongenomic actions on nonreproductive behaviors, including relevance for neuropsychiatric disorders, is also discussed.

12.2 THE VTA AND VMH ARE IMPORTANT FOR PROGESTIN-FACILITATED LORDOSIS

Sequential actions of E and P in the brain mediate the onset and duration of sexual activity in rodents,[1] which is typically defined as the female's display of the stereotypic mating posture, lordosis. Localization, hormone implant, and lesion studies reveal that P's actions in the VMH *and* VTA are important for mediating lordosis of E-primed rodents. Levels of ^{3}H P are greater in the midbrain than the hippocampus, hypothalamus, or cortex of female rodents.[2] P implants to the VTA or VMH of OVX, E-primed rats consistently induce proceptivity and receptivity.[3–5] In hamsters, lordosis is only reliably induced by P to both the VMH *and* VTA.[6–8] For mice, systemic P or P to the VMH and VTA similarly facilitate lordosis.[9–11] Lesions to the VTA or VMH, respectively decrease or abolish P-facilitated lordosis of rats and hamsters.[12–14] Knife cuts between the VTA and VMH attenuate lordosis of rodents.[15] Thus, in these rodent species, P's actions in the VTA enhance lordosis of rats and, in conjunction with P's actions in the VMH, are necessary for lordosis of hamsters and mice. The conservation of progestins' actions in the VTA for facilitating lordosis across these rodent species implies their importance in reproductive behavior.

12.3 P'S EFFECTS MAY BE MEDIATED, IN PART, BY 3α,5α-THP

Circulating concentrations of P and 3α,5α-THP are increased coincident with natural variations in receptivity of rodents. During behavioral estrus, when female rodents typically display lordosis (with appropriate stimulation), plasma P and 3α,5α-THP levels are elevated compared to those seen during the diestrous phase of the estrous cycle, when female rodents are not receptive.[16–20] Rodents that are OVX and hormone-primed with E and P or 3α,5α-THP also have higher levels of circulating P and 3α,5α-THP and increased lordosis responses than OVX rats primed with E alone.[21,22] OVX, E-primed rats or mice intravenously infused with P or 3α,5α-THP demonstrate rapid increases in sexual receptivity with a latency as short as 2.5 min.[11,23–28] Naturally receptive mice show considerable variations in their initial lordosis responses. However, there is a robust linear relationship between mice whole-brain 3α,5α-THP levels and lordosis that is less evident for P and sexual responses.[10] These data suggest P's facilitation of lordosis may involve actions of 3α,5α-THP; however, they do not indicate the brain regions important for 3α,5α-THP's effects on lordosis.

Experiments have directly compared the effectiveness of P and 3α,5α-THP to the VMH on lordosis. In OVX, E-primed rats, 1-μg infusions of P are far more effective at increasing lordosis 2 and 4 h later than are comparable infusions of 3α,5α-THP; however, both significantly increased lordosis compared to vehicle infusions.[29] The enzyme 5α-reductase metabolizes P to 5α-pregnane-3,20-dione (DHP). The 3α-hydroxysteroid dehydrogenase (3α-HSD) enzyme then reduces DHP to 3α,5α-THP and also oxidize 3α,5α-THP back to DHP. Infusions of DHP, the pro-hormone for 3α,5α-THP, to the VMH produce increases in lordosis 2 and 4 h later that are similar to those seen with P infusions.[29] Despite P's and DHP's high affinity, and 3α,5α-THP's weak affinity, for PRs,[30] VMH infusions of the PR and glucocorticoid antagonist, RU38486, block P-, DHP-, or 3α,5α-THP-facilitated lordosis.[31] These findings suggest that P, when applied to the VMH, or its 5α-reduced metabolites DHP and 3α,5α-THP facilitate lordosis via actions at PRs, which further indicates that 3α,5α-THP's effects in the VMH may be due to oxidation back to DHP.

In the VTA, a different picture emerges, with 3α,5α-THP being far more effective at facilitating lordosis than is P itself. When P or 3α,5α-THP is applied to the VTA of rats or hamsters, with prior P priming in the VMH, 3α,5α-THP implants or infusions produce higher levels of lordosis than are observed for P.[5,32,33] Further, 3α,5α-THP to the VTA significantly increases lordosis within 5 min, whereas P requires a longer latency to produce similar effects. These data imply that P's actions in the VTA may require its metabolism to 3α,5α-THP.

12.3.1 Do P's Effects Require Metabolism to 3α,5α-THP?

As previously stated, 3α,5α-THP is formed from P by two metabolizing enzymes, 5α-reductase and 3α-HSD, in various mammalian tissues.[34] 5α-Reductase and 3α-

HSD expression in the adult rodent brain is region specific.[35] These enzymes have been localized to the olfactory bulb, cortex, thalamus, hypothalamus, hippocampus, basal ganglia, midbrain, cerebellum, pons, and medulla.[35] Because these studies have not closely investigated the VTA, we have done so and found that both 5α-reductase and 3α-HSD enzymes are present in the midbrain VTA.[9]

Administration of systemic 5α-reductase or 3α-HSD inhibitors attenuates progestin-facilitated lordosis. Systemic administration of the 5α-reductase inhibitor N,Ndiethyl-4-aza-4-methyl-3-oxo-5 alpha-androstane-17 beta-carboxamide (4-MA) hastens estrous termination of rats in behavioral estrus.[36] Similarly, 4-MA inhibits lordosis of OVX, E + P-primed hamsters, compared to vehicle administration.[37] Systemic administration of the 3α-HSD inhibitor indomethacin also attenuates lordosis of rodents. OVX, E-primed rats administered intravenous (IV) P and systemic indomethacin show significantly less lordosis than vehicle rats.[38] As well, indomethacin significantly reduces lordosis of E + P-primed rats and hamsters compared to vehicle.[9] This effect is also seen in intact rats or hamsters in behavioral estrus.[9] Together, these data indicate that metabolism of P to 3α,5α-THP may be an important component of P's actions for lordosis of rodents. Thus, inhibiting 5α-reductase or 3α-HSD activity throughout the brain attenuates lordosis of rats and hamsters.

In the VMH, is metabolism of P by 5α-reductase necessary for P-facilitated lordosis? Intrahypothalamic implants of DHP or 20α-hydroxyprogesterone (20HP) were significantly less effective than implants of P at facilitating lordosis in OVX, E-primed rats.[39] Although these data suggest that in the VMH, the reduction of P by the 5α-reductase enzyme is not required for P's modulation of lordosis, other studies suggest that P or DHP infusions to the VMH are equally effective at increasing lordosis 2 and 4 h following infusion.[29]

In contrast to the VMH, altering the 5α-reductase or 3α-HSD metabolism enzymes in the VTA has a robust effect on lordosis. Infusions to the VTA of fluoxetine, which shifts 3α-HSD activity to favor reduction over oxidation,[40] facilitate lordosis and increase midbrain 3α,5α-THP levels.[41] Conversely, 5α-reductase or 3α-HSD inhibitors to the VTA decrease sexual behavior and midbrain 3α,5α-THP levels.[9,10,37,42–44] Interestingly, systemic or intra-VTA administration of 5α-reductase or 3α-HSD inhibitors similarly attenuates lordosis and proceptive behaviors of behavioral estrous or hormone-primed rats and hamsters.[9,21,42–44] Therefore, not only is P's metabolism to 3α,5α-THP in the VTA important for lordosis, but the VTA may be the primary site where P is metabolized to 3α,5α-THP for mediating lordosis.

3α,5α-THP can be produced from *de novo* neurosteroidogenesis in glial cells. Mitochondrial benzodiazepine receptors (MBRs) in glial cells facilitate the translocation of cholesterol across the mitochondrial membranes to make it available to the P450 side chain cleavage enzyme (P450scc), which converts it to pregnenolone.[45] Pregnenolone is subsequently metabolized to P by 3β-hydroxysteroid dehydrogenase (3β-HSD), which is metabolized by 5α-reductase to DHP, and DHP is further reduced to 3α,5α-THP by 3α-HSD. All of the enzymes, P450 scc, 3β-HSD, 5α-reductase, and 3α-HSD, that are necessary for *de novo* synthesis of 3α,5α-THP have been localized to the rat midbrain.[9,46] Thus, it may be that *de novo* synthesis of neurosteroids in the VTA is an important component in the mediation of lordosis.

De novo synthesis of $3\alpha,5\alpha$-THP in the VTA may be important for lordosis. Systemic or intra-VTA administration of 5α-reductase or 3α-HSD inhibitors fails to completely abolish lordosis or produce midbrain $3\alpha,5\alpha$-THP levels at nadir. Mating-relevant stimuli increase $3\alpha,5\alpha$-THP in the midbrain of intact rats and those without peripheral steroidogenic organs (ovaries and adrenals).[42] Inhibiting neurosteroidogenesis in the VTA with MBR antagonist infusions decreases lordosis and midbrain $3\alpha,5\alpha$-THP levels of hormone-primed and behavioral estrous rats and hamsters.[47,48] As well, VTA infusions of the P450scc inhibitor, digitoxin, or the 3β-HSD inhibitor, epostane, attenuate lordosis of cycling rats in behavioral estrus.[49] Conversely, MBR agonists, or other neurosteroidogenic agents, such as fluoxetine or olanzapine, to the VTA increase lordosis and midbrain $3\alpha,5\alpha$-THP concentrations of hormone-primed and naturally receptive rodents.[41,47,48,50] Together these data support the notion that induction of female sexual behavior is, in part, dependent upon biosynthesis of neurosteroids in the VTA.

12.4 SUBSTRATES OF PROGESTINS' ACTIONS IN THE VTA

12.4.1 Are Actions at PRs Required for Progestins' Effects to Facilitate Sexual Receptivity?

The classic view of P's mechanisms involves genomic, ligand-dependent actions at E-induced intracellular PRs. In this model, P crosses neuronal membranes, binds to cytostolic or nuclear PRs, and produces a conformational change in PRs. This P–PR complex binds to steroid response elements on the genome to alter gene transcription, and subsequently protein synthesis, and ultimately cellular function.[51,52] These P–PR-initiated changes in protein synthesis may require a minimum of 10 to 15 min to occur.[53]

Progesterone may have classic actions via PRs in the VMH to mediate sexual receptivity. Receptor binding, immunocytochemistry and *in situ* hybridization indicate that there are many PRs in the VMH.[9,42,54,55] A number of studies have demonstrated that E-induced PRs in the hypothalamus are upregulated coincident with lordosis responsiveness.[56] Further, blocking actions of P at PRs, or the consequence of P–PR interactions — protein synthesis — in the hypothalamus inhibits progestin-faciliated lordosis.[31,57–59] Although these data suggest that P's actions at PRs in the VMH may mediate lordosis, P likely also has other actions in the VMH that may enhance sexual receptivity. For example, there is a poor correlation between the efficacy of progestins applied to the VMH of E-primed rodents to facilitate lordosis and their capacity to alter PR binding.[39] Thus, although P may have actions via PRs in the VMH to facilitate lordosis, this does not preclude other mechanisms of P in the VMH.

In contrast, within the VTA, progestins' actions seem to be independent of PRs. Using RU5020 binding and immunocytochemistry for PRs, many E-induced PRs were localized in the VMH, but few non-E-induced PRs were observed in the VTA of rats, hamsters, or mice.[9,11,42,55,60,61] Blocking PRs with the antiprogestin RU38486 or antisense oligonucleotide infusions for PRs to the VMH attenuates lordosis of hamsters or rats, but similar infusions to the VTA had negligible effects on lordosis.[60,62] Progestins' binding affinities for PRs do not correspond to their ability to facilitate lordosis when applied to the VTA of E-primed rats or hamsters.[32] Further, progestins to the VMH and VTA of PR knockout (PRKO) mice facilitate lordosis[11]; although these mice have few classic PRs, they do have higher levels of membrane PRs.[63] Together these data suggest that while P's actions in the VMH can be PR mediated, in the VTA they are not.

12.4.2 Are P's Membrane Actions in the VTA Sufficient to Facilitate Sexual Receptivity?

P's effects in the VTA for lordosis can occur through rapid actions at cell membranes. P increases cell firing in the VTA 60 sec after IV infusion to E-primed rats; within 120 sec lordosis is enhanced and cell firing in the VTA is increased further.[28] Free P, or P conjugated to bovine serum album (P:BSA), which does not cross cell membranes, can facilitate lordosis of E-primed rats or hamsters within 5 to 10 min of application to the VTA. Only free P (not P:BSA) to the VMH is effective at facilitating lordosis,[5,64] which suggests that P:BSA does not become unbound to produce its effects and that classic PR effects in the VMH may precede membrane PR actions in the VMH.[63] P:BSA can facilitate lordosis of PRKO mice,[11] indicating that P:BSA does not have actions at classic PRs to produce its behavioral effects. Whether membrane PRs that have been characterized[63,65] are localized to the VTA and mediate these effects is not known.

12.4.3 Can P Have Actions via GBRs to Mediate Lordosis?

One identified membrane substrate that may mediate some of progestins' actions for lordosis are GBRs. Progestins are positive, allosteric modulators of GBRs, which increase binding of GABA to GBRs.[66,67] Progestins vary in their ability to enhance the function of GBRs. Structural requirements associated with progestins that are the most effective positive modulators of GBRs include A-ring reduction at the 5-carbon.[67] This structure is characteristic of 3α,5α-THP, the most potent naturally occurring ligand for GBRs, which acts as a positive modulator of GBRs.[67]

Anatomical and biochemical studies indicate that progestins modify GBRs. Number,[68] density,[69] and affinity[70] of GBRs are enhanced during behavioral estrus, when progestin levels are high, compared to other phases of the estrous cycle, when progestin levels are low.

GABA, acting at GBRs in the VMH, facilitates lordosis behavior of hormone-primed rodents. Within the VMH, there are very high levels of muscimol binding that are regulated by E.[71] Progesterone enhances muscimol binding[72] and GABA-induced chloride currents when administered in the micromolar range.[73] In female rats, endogenous levels of GABA are decreased in the VMH at the time of sexual receptivity and increased prior to receptivity (on the afternoon of proestrous) and 24 h after sexual responsiveness.[74] The elevation in GABA in the VMH on the afternoon of proestrous may be due to increased activity of glutamic acid decarboxylase (GAD) enzymes, which synthesize GABA from glutamate. GAD 67, which may be the GAD enzyme isoform that is important for the synthesis of tonic, synaptically released GABA, is increased on the afternoon of proestrous within the VMH.[75]

In the VTA, 3α,5α-THP may facilitate lordosis by enhancing function of GABAergic neurons. GABA-containing neurons have been localized to the VTA.[76] Robust changes in midbrain 3α,5α-THP over the estrous cycle and with mating occur concomitant with endogenous variations in midbrain GABA and GBR sensitivity.[9,42] Midbrain GABA levels are greater during behavioral estrus, than diestrus, and are highest after mating.[9,42] As well, during behavioral estrus, less GABA is needed to displace muscimol, a GBR agonist, from its binding sites in midbrain tissues of rats and hamsters.[9,42] These findings suggest that coincident with sexual receptivity, GABA and GBR function in the midbrain VTA is greater than that observed during diestrus.

Manipulating GABA or GBR function in the VTA alters progestin-facilitated lordosis. GAD antisense infusions to the VTA disrupt lordosis and reduce GAD immunoreactivity of hormone-primed or naturally receptive rats or hamsters, compared to that seen following vehicle or sense infusions.[60,62] Preventing GABA degradation by infusing a GABA-transaminase inhibitor to the VTA of hamsters facilitates lordosis.[77] Infusions of GBR agonists and antagonists to the VTA respectively increase and decrease P-mediated lordosis of rats or hamsters.[9,42,43,62,77–79] Similarly, progestins' rapid facilitation of lordosis when applied to the VTA positively correlates with their activation of GBRs in the midbrain VTA. Progestins' facilitation of lordosis when applied to the VTA of rodents concurrent with P to the VMH corresponds to their enhancement of GABA-stimulated chloride influx in cortical tissue and SR95531 receptor binding in the VTA.[32] Thus, altering GABA levels or GBR function in the VTA alters progestin-facilitated sexual receptivity of rodents. These data support the notion that progestins facilitate lordosis, in part, by enhancing GABAergic function via actions at GBRs in the VTA.

12.4.4 Can P Have Actions via NMDARs to Mediate Lordosis?

Progesterone may also have actions to modulate sexual receptivity by interacting with NMDARs. First, NMDARs have been localized to the VMH and VTA,[80] and pharmacological manipulations of NMDARs in these regions may elicit electrophysiological and behavioral changes in rats.[81] Second, progestins can modulate NMDARs in a rapid fashion by increasing or decreasing NMDA-activated calcium influx, and blocking P's metabolism to 3α,5α-THP attenuates this effect.[82,83] Third, NMDA binding, measured by autoradiography, is decreased in the frontal cortex of OVX rats following administration of P or E and P.[84] Fourth, NMDA agonists produce both excitatory responses in cerebellar Purkinje cells and persistent inflammatory hyperalgesia that is attenuated by P.[83] Fifth, in the hypothalamus and preoptic area, progestins can alter NMDAR function by modulating the release of glutamate.[85] Sixth, altering NMDARs produces changes in sexual receptivity of rodents. Blocking NMDAR activity by systemic administration of (+)-MK-801 hydrogen maleate (MK-801) or dextrorphan — selective noncompetitive antagonists at NMDARs —

attenuates lordosis and motor behavior of OVX, hormone-primed rats[86] and hamsters.[87] Together these findings suggest that progestins are capable of mediating changes in NMDAR function that produce alterations in behavior and leave open the possibility that in the VMH or VTA, progestins may have effects through NMDARs to produce changes in lordosis of rodents.

Pharmacological manipulations of hypothalamic NMDARs alter lordosis of rodents. MK-801 infusions to the VMH modestly decrease lordosis of E- but not E + P-primed rats.[9] Infusions of MK-801 to the VMH also decrease lordosis of E + P-primed hamsters.[87] Infusions to the VMH of the competitive NMDAR-selective antagonist, D,L2-amino-5-phosphonopentoic acid (AP-5), did not alter lordosis of fully receptive female rats.[88] However, more anterior infusions of AP-5 of sexually receptive female rats did produce significant reductions in lordosis.[88] Decreases in lordosis of rats have also been observed with intrahypothalamic infusions of nonspecific glutamate agonists, NMDA, kainic acid, and potassium chloride.[89] While these findings may seem inconsistent, one would not expect that the effects of exciting all glutamatergic neurons in the hypothalamus would be similar to a manipulation that expressly blocks the NMDA subtype of glutamate receptors. In sum, blockage of NMDA receptors in the hypothalamus may have a modest inhibitory effect on lordosis of rodents.

NMDARs in the VTA may be an important substrate through which progestins regulate sexual behavior of female rodents. Infusions of MK-801 to the VTA increase lordosis of E-primed rats and E + P-primed rats or hamsters above that seen with vehicle infusions.[9,87] Notably, blocking NMDARs in the VTA of rats or hamsters produces increases in lordosis that are similar to those elicited by intra-VTA infusions of the $GABA_A$ agonist muscimol.[9,78,79] In the VTA, progestins' actions at NMDARs and GBRs may modulate female sexual behavior, in part, through NMDAR's and GBR's subsequent effects at dopaminergic neurons. Manipulations that increase GBR function or inhibit NMDARs in the VTA respectively elevate and reduce extracellular DA concentrations in the prefrontal cortex.[90] Notably, changes in DA levels have been associated with alterations in lordosis.[9,91] Thus, it may be that in the VTA, NMDA neurons excite and GABAergic neurons inhibit dopaminergic cell bodies, and these opposing influences regulate DA concentrations, which in turn can have a modulatory effect on lordosis. All of these may be mediated by actions at D_1 receptors.

12.4.5 Can P Have Actions via D_1 to Mediate Lordosis?

Lordosis, which has appetitive, motivation, and motor components, may be mediated in part by release of DA in the mesolimbic and nigrostriatal systems. First, ovarian steroids enhance DA concentrations. Mesolimbic DA function is increased in estrous compared to diestrous rats.[92] Amphetamine-stimulated DA release is increased during behavioral estrus, attenuated by OVX, and reinstated by E + P, but not E or P alone.[93] Hormone-priming increases the reuptake of DA of OVX rats.[94] Second, DA levels are altered with mating. DA concentrations in the VTA are greatest among mated compared to receptive, nonmated rats that have higher DA levels than do nonreceptive rats.[9] During paced mating, which is associated with higher DA levels

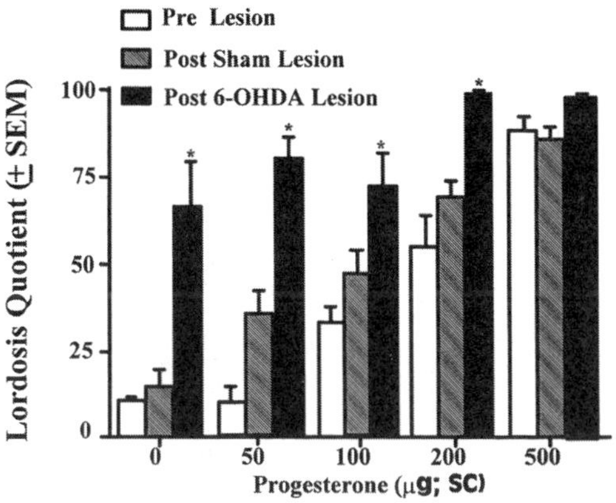

FIGURE 12.1 6-OHDA lesions of the ventral tegmental area increase P-facilitated lordosis of rats. Lordosis quotients (LQs) of OVX E + P (0, 50, 100, 200 or 500 μg)-primed rats before (open) and after (black) bilateral 6-OHDA (4 μg) or sham (striped) lesions to the VTA. Asterisks indicate a significant difference ($p < .05$) from prelesion scores.

than is nonpaced mating, DA levels are increased preceding sexual contacts.[95] Third, DA release rather than absolute concentrations of DA may be important for facilitation of lordosis. Although 6-hydroxydopamine (6-OHDA) lesions lower DA concentrations,[91,96] the functional capacity of the remaining DA neurons is increased.[97,98] 6-OHDA to the lateral ventricle[91] or VTA (Figure 12.1) increases lordosis of hormone-primed rats and hamsters compared to sham lesions. These results suggest that DA release in the VTA may have phasic effects on lordosis.

In addition to progestins' actions at PRs, in the VMH, D_1 ligands may also facilitate lordosis. Intracerebroventricular (ICV) infusions of the D_1 agonist SKF38393 or P are equally effective at producing lordosis in OVX, E-primed rats.[99,100] Blocking D_1 receptors with antagonists or antisense oligonucleotides inhibits lordosis of OVX, E- or E + P-primed rats.[99,100] In the hypothalamus, the facilitative effects of D_1 activation are dependent upon functioning PRs. In support, the ability of SKF38393 to facilitate lordosis of OVX, E-primed rats is blocked by co-administration of the PR antagonist RU38486 or antisense, but not sense, oligonucleotides against PR mRNA.[99] Furthermore, infusions of SKF38393 are ineffective at facilitating lordosis in PR knockout, but not wild-type mice.[99] Together, these data suggest that in the VMH, the effects of P and D_1 ligands may be through PR-mediated actions.

D_1 receptors may also be important for progestins' actions in the VTA; however, the few E-induced PRs in the VTA suggest such effects are unlikely PR mediated in this region. Antagonizing D_1 receptors in the VTA attenuates lordosis of rodents. VTA infusions of antisense oligonucleotides against D_1 mRNA or of the D_1 antagonist SCH23390 significantly reduce lordosis of cycling rats and hamsters in behavioral estrus, compared to appropriate vehicle-infused animals.[60] Congruous with D_1 activity in the VTA being necessary for lordosis of rodents, VTA infusions of D_1 agonists and antagonists can respectively augment and inhibit P's ability to facilitate lordosis. OVX, E-primed rats that received VTA infusions of the D_1 antagonist SCH23390 to the VTA and P 30 min later showed significantly less lordosis than rats that

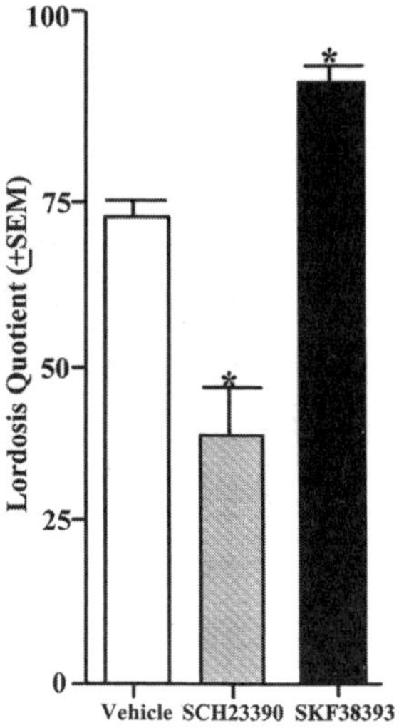

FIGURE 12.2 Inhibiting and activating D_1 alters P's ability to facilitate lordosis of rats. LQs of OVX, E-primed rats pretreated with infusions of SCH23390 (100 ng; stippled), SKF38393 (100 ng; black), or vehicle (open) to the VTA. Rats received systemic P (100 μg), 30 min after infusion and were tested for lordosis 2.5 h after P administration. Asterisks indicate a significant difference ($p < .05$) from vehicle infusion.

received vehicle infusions (Figure 12.2). As well, pretreatment with VTA infusions of the D_1 agonist SKF38393, followed by P 30 min later, produced significantly greater lordosis responses in OVX, E-primed rats compared to vehicle pretreatment (Figure 12.2). Together these data suggest progestins may have actions at membrane-bound D_1 receptors to facilitate lordosis.

12.4.6 Is cAMP a Substrate for Progestin-Facilitated Lordosis?

Progestins' actions for lordosis may involve activation of adenosine 3′,5′-monophosphate (cAMP). *In vitro*, P augments forskolin-induced cAMP within 15 min of exposure to neocortical slices from OVX rats (Figure 12.3). Interestingly, the presence of bicuculline or picrotoxin attenuates P's ability to enhance forskolin-induced increases in cAMP (Figure 12.4), suggesting that this is a GBR-dependent process. As well, concentrations of 3α,5α-THP, which is a potent positive modulator of GBRs, were positively correlated with levels of P, a poor modulator of GBRs, and increases in cAMP levels. Therefore, in the neocortex P-induced increases in cAMP may be mediated by 3α,5α-THP's actions at GBRs.[101]

Systemic manipulations that increase cAMP accumulation also augment lordosis. Systemic administration of P, but not vehicle, to OVX, E-primed rats significantly increases central cAMP concentrations and lordosis.[102] OVX, E-primed rats given subcutaneous injections of the cAMP agonist dibutyryl cAMP showed increases in lordosis that were not evident in vehicle-injected rats.[103] Additionally, systemic injections of dibutyryl cAMP, but not vehicle, augment lordosis of OVX,

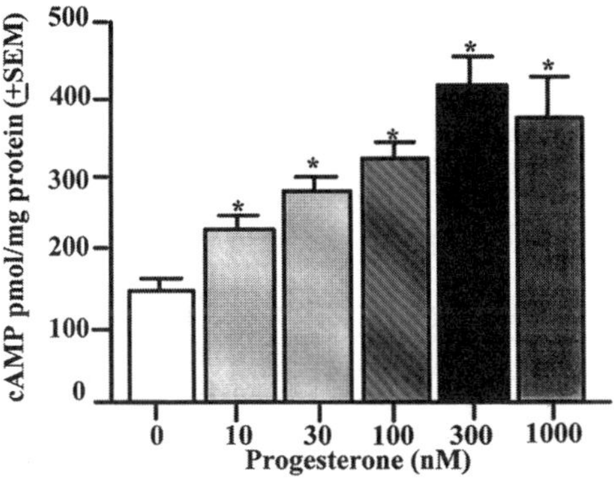

FIGURE 12.3 P enhances forskolin-induced increases in cAMP. Forskolin-induced cAMP levels of neocortical slices after application of P. Asterisks indicate a significant difference ($p < .05$) between P dosage and vehicle.

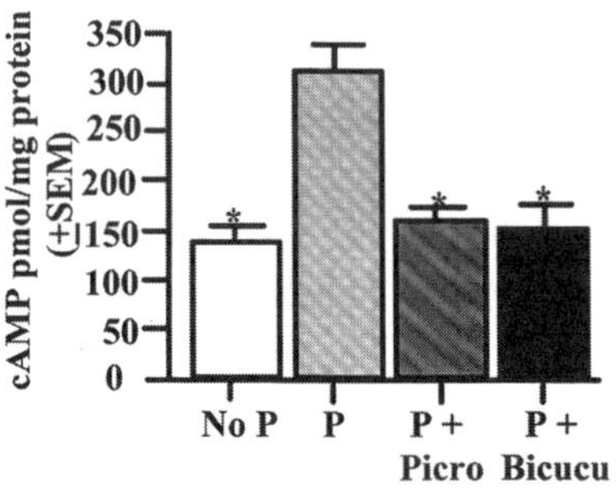

FIGURE 12.4 P (100 n*M*; stippled), but not vehicle (open), enhances forskolin-induced cAMP in neocortical slices. Picrotoxin (50 μ*M*; striped) or bicuculline (100 μ*M*; black) attenuates P's enhancing effect on forskolin-induced cAMP. Asterisks indicate condition is significantly ($p < .05$) lower than P condition.

E-primed rats.[104] Notably, the facilitative effects of dibutyryl cAMP were attenuated by co-administration of the antiprogestin RU38486.[104] Thus, cAMP may be a substrate through which progestins facilitate lordosis, and this may be dependent, in part, on PRs in the VMH.

In the VMH, cAMP may be an important substrate for facilitating lordosis. OVX, E-primed rats given systemic P had significantly higher levels of cAMP in the VMH following mating than vehicle controls.[102] Central manipulations that augment cAMP in the VMH also produce increases in lordosis. ICV administration of the cAMP analogue 8-bromo-cAMP facilitates lordosis of E-primed rats.[105] P or D_1 activation with SKF38393 increases cAMP and cAMP-dependent protein kinase A (PKA) levels in the hypothalamus.[105] Inhibiting PKA with ICV infusions of the Rp isomer of 8-bromo-cAMP (Rp-cAMP) attenuates lordosis of E + P-primed rats and prevents the facilitation of lordosis by SKF38393 or 8-bromo-cAMP.[105] Together, these data implicate cAMP as an important substrate through which P and D_1 activation facilitates lordosis.

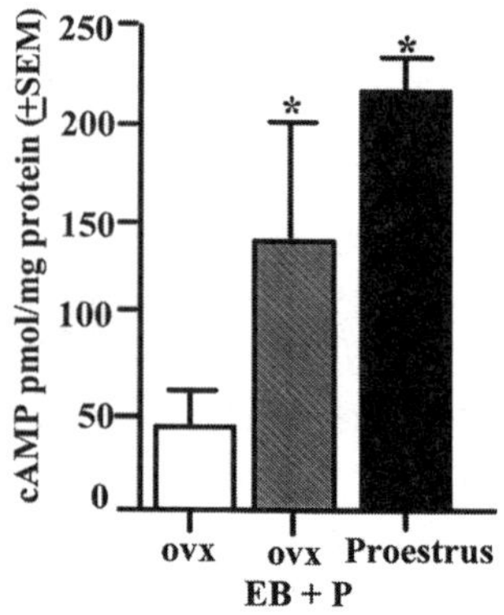

FIGURE 12.5 Midbrain cAMP is greater in receptive OVX, E + P-primed (striped) or behavioral estrous (black) rats than in nonreceptive OVX (open) rats. Asterisks indicate a significant increase ($p < .05$) compared to OVX rats.

In the VTA, progestins' ability to facilitate lordosis may be dependent upon cAMP. Midbrain cAMP levels are higher in hormone-primed or proestrous vs. OVX rats (Figure 12.5). Furthermore, direct pharmacological manipulations of cAMP in the VTA produce changes in lordosis. VTA infusions of the cAMP analogue 8-bromo-cAMP, but not vehicle, increase lordosis of rats late in behavioral estrus, when sexual behavior is waning.[106] Similarly, VTA infusions of 8-bromo-cAMP significantly increase lordosis of OVX, E + P-primed rats.[106] Together these findings suggest that in the VTA, P-facilitated lordosis may be dependent upon the intracellular actions of cAMP.

In the VMH, progestin-facilitated lordosis may occur through direct actions at membrane DA receptors, which initiate signal transduction and phosphorylation-dependent mechanisms to mediate lordosis.[107] In the VMH, PRs are integral for facilitation of lordosis. PRs can be activated through traditional ligand-dependent activation by P or through ligand-independent actions. Briefly, D_1 agonists can increase cAMP, PKA, and DA- and cAMP-regulated phosphoprotein of 32-kDa (DARPP-32) phosphorylation.[108] This inhibits protein phosphatase 1 (PP-1) and results in increased phosphorylation of many substrate proteins,[108] including PRs, steroid receptor coactivator proteins (SRCs), or other proteins that may increase lordosis of rats.[105] Additionally, ICV infusions of P or SKF38393, but not vehicle, similarly and significantly increase concentrations of cAMP, PKA, and the phosphorylated form of DARPP-32 in the hypothalamus of OVX, E-primed rats.[105] Finally, SKF38393's facilitation of lordosis is blocked by co-administration of the PR antagonist RU38486,[99] suggesting that in the VMH, DA may require PRs to facilitate lordosis.

Given that progestins' actions in the VTA may involve D_1, we have begun to consider whether P's actions in the VTA include second messengers. It may be that in the VTA, progestins facilitate lordosis by initiating the classical D_1 signaling pathway. D_1 activation initiates a signaling cascade that results in increases in cAMP and PKA, which facilitate the phosphorylation of DARPP-32.[105,108] DARPP-32, in its phosphorylated state, reduces protein dephosphorylation by inhibiting PP-1,[105,108]

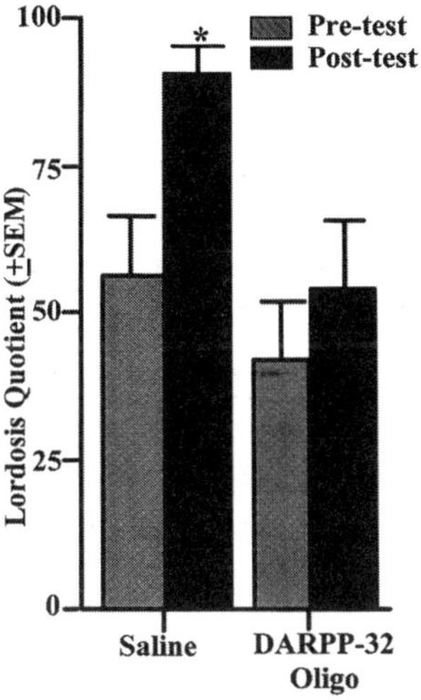

FIGURE 12.6 Antisense oligonucleotides for DARPP-32 attenuate P-facilitated lordosis. Pre (striped)- and post (solid)-test LQs of OVX, E-primed rats administered VTA infusions of antisense oligonucleotides for DARPP-32 (right) or vehicle (left) 48 and 24 h prior to VTA infusions of P. Asterisks indicate a significant difference ($p < .05$) between pre- and posttest.

which may result in a prolonged phosphorylation of proteins that are important for mediating lordosis. In support, activation and inhibition of D_1 receptors in the VTA respectively increase and decrease P's ability to facilitate lordosis of rats (Figure 12.2). Furthermore, increasing cAMP in the VTA enhances lordosis of OVX, E- or E + P-primed rats.[106] Finally, inhibition of DARPP-32 in the VTA with infusions of antisense oligonucleotides reduces P-facilitated lordosis of OVX, E-primed rats (Figure 12.6). In sum, progestins in the VTA may facilitate lordosis through a D_1-initiated intracellular pathway that includes cAMP, PKA, or DARPP-32. Future experiments in our laboratory are designed to further elucidate the precise mechanisms that are involved.

12.5 SUMMARY OF RESEARCH ON PROGESTINS' MECHANISMS AND LORDOSIS

This reductionist approach of using lordosis as a bioassay for progestins' actions has been very fruitful and has revealed that progestins can have multiple actions for mediating this behavioral response. Progestins' mechanisms include, in the VMH, genomic effects through interaction with DNA and induction of gene transcription. However, in the VTA, P has nongenomic or nonclassical membrane-mediated effects that involve rapid actions through changes in neurosteroids, which are synthesized in the brain and act locally at the membrane, ion flux, and second messengers. The true power of this reductionist, bioassay approach to understanding progestins' actions involves placing discovered mechanisms in a broader functional context. The remainder of this chapter will consider the functional role of progestins' nonclassical actions for reproductive and nonreproductive behaviors, as well as putative relevance for the etiology and treatment of neuropsychiatric disorders.

12.6 PROGESTINS' NONCLASSICAL MECHANISMS AND INHIBITION OF LORDOSIS

Progestins' actions in the VTA may be involved in progestins' facilitation *and* sequential inhibition of lordosis. It is well known that E-primed female rodents can be behaviorally unresponsive to a second injection of P administered 24 to 48 h after P dosing that facilitates lordosis.[109] The VTA seems to be an important site for P's effects on sequential inhibition.[110] Notably, P or P:BSA, but not BSA, implants to the VTA at hour 68 (after E priming) inhibit lordosis at hour 72 if P implants were previously administered at hour 48 to facilitate lordosis (Figure 12.7). Further, 3α,5α-THP was the most effective progestin when applied to the VTA at producing sequential inhibition.[9] These data suggest that P can have membrane-mediated actions in the VTA, not only for facilitation of lordosis, but also for sequential inhibition of lordosis.

12.7 3α,5α-THP'S ROLE IN SOLICITATION AND SOCIAL BEHAVIORS

Mating is characterized by exploration and affiliations, social behaviors that bring individuals together, as well as reproductive and aggressive behaviors. Reproduction

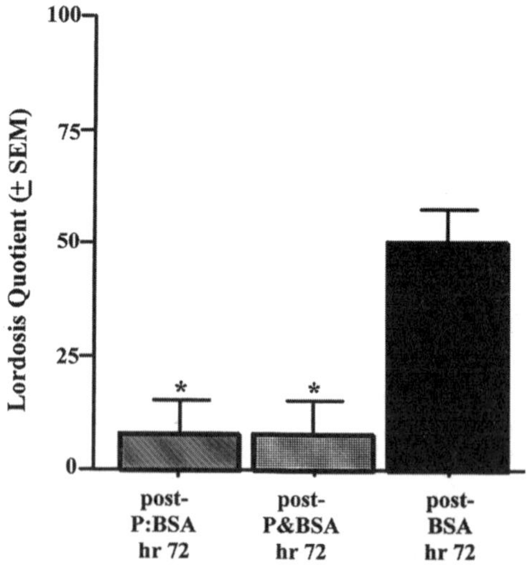

FIGURE 12.7 P has membrane actions to inhibit lordosis of rats. LQs of OVX, E-primed rats following a second administration of P (striped) or P:BSA (checked), but not BSA (black), 24 h later (hour 72) inhibits lordosis. Asterisks indicate a significant difference ($p < .05$) from BSA (control condition).

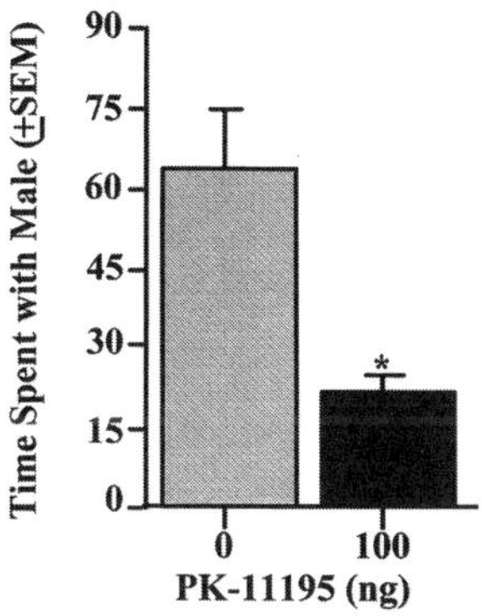

FIGURE 12.8 Inhibiting neurosteroidogenesis in the VTA decreases the amount of time spent by the female in close proximity to a male. Duration of time spent (seconds) by OVX, E + P-primed rats following VTA infusions of PK-11195 (100 ng; black) or vehicle (stippled). Asterisks indicate a significant difference ($p < .05$) from vehicle.

requires that exploration occur to find mates, that fearful responses to potential mates are suppressed, and that approaches are made to stimuli that previously elicited aggressive responses.[111] In natural or seminatural laboratory mating situations, female rats spontaneously exhibit approach and avoidance behaviors (pacing behaviors) and other solicitation behaviors (e.g., hopping, darting, and ear wiggling) toward males, which enables them to control the temporal pattern of mating and optimizes their fertility and fecundity.[112]

Solicitation behaviors, such as pacing, are predominantly progestin-mediated.[113] Preliminary data from our laboratory suggest some solicitation behaviors may be mediated, in part, by 3α,5α-THP. In support, systemic administration of P or 3α,5α-THP similarly increases solicitation behaviors. Co-administration of inhibitors of 3α,5α-THP biosynthesis or P's central metabolism to 3α,5α-THP decrease solicitation behaviors and completely eliminate pacing behavior.[21] Moreover, solicitation behaviors such as pacing may enhance 3α,5α-THP biosynthesis. Female rats that pace their sexual contacts have significantly higher whole-brain 3α,5α-THP levels than do females mated in standard arenas that cannot pace their sexual contacts.[21] Further, both 3α,5α-THP and P have been demonstrated to be rewarding.[114,115]

Other models of approach–avoidance behaviors suggest a role of 3α,5α-THP. Administration of P or 3α,5α-THP to OVX, E-primed rats significantly increases the time female rats spend in proximity to a male in a T-maze, compared to vehicle administration. Co-administration of systemic inhibitors of 3α,5α-THP to hormone-primed rats attenuates female rats' preference for proximity to a male in the T-maze.[21] Further, decreasing neurosteroidogenesis in the VTA with infusions of MBR antagonist PK-11195 significantly reduces the amount of time that female rats spend in close proximity to a male rat in a T-maze (Figure 12.8).

3α,5α-THP can influence other social behaviors that mitigate reproduction, such as aggressive behaviors. Systemic administration of 3α,5α-THP to mice dose-dependently reduces alcohol-induced aggression.[116] Preliminary data from our laboratory suggest that P administration decreases the number of aggressive acts (2.1 ± 2.1) exhibited by resident, wild-type female mice in response to an intruder, compared to that seen with vehicle administration (5.7 ± 4.3). However, 5α-reductase knockout mice that have a deficiency in their ability to metabolize P to 3α,5α-THP fail to show P-induced decreases in aggression (5.3 ± 3.5). 3α,5α-THP can have actions in the VTA to influence agonistic behavior. For example, implants of 3α,5α-THP to the VTA reduce fighting of aggressive male hamsters (Meisel, unpublished data).

12.8 EVIDENCE FOR A ROLE OF 3α,5α-THP ON AFFECTIVE BEHAVIORS

3α,5α-THP may influence affective or anxiety behavior concomitant with effects on lordosis, solicitation, approach–avoidance, and aggressive behaviors. Plasma and central 3α,5α-THP levels are increased during proestrus coincident with increases in anxiolytic behavior, as evidenced by increased open arm time on the elevated plus maze, compared to that seen in other phases of the estrous cycle or by male rats.[117] Administration of P decreases anxiety behavior of OVX, E-primed rats, but co-administration of finasteride or indomethacin, which inhibits metabolism of P to 3α,5α-THP, increases anxiety behaviors. Additionally, biosynthesis of 3α,5α-THP can be anxiolytic. Mating-induced increases in 3α,5α-THP enhance time spent on the open arms of the plus maze over and above that seen in proestrous rats.[9]

It is important to reveal 3α,5α-THP's effects and mechanisms due to its relevance for stress, neuropsychiatric disorders, and women's health. Sex differences in depression and schizophrenia may be due, in part, to sex differences in hormones. Some neuropsychiatric symptoms are alleviated when 3α,5α-THP levels are increased relative to lower concentrations. Women are at greater risk for some neuropsychiatric disorders and are more sensitive to fluctuations in hormones, such as 3α,5α-THP, than are men.

12.8.1 Sex Differences in Stress and 3α,5α-THP

There are sex differences in 3α,5α-THP. From birth until menopause, females have higher and more labile progestin levels than do males. During the follicular phase, progestin levels for women and men are similarly low; however, during the luteal phase and pregnancy, progestin levels among women are much higher than among men.[118] This same pattern of progestin levels is seen in female and male rats.[17,20] These sex differences in concentrations of circulating progestins are due to gonadal and adrenal sources; however, central levels of 3α,5α-THP also change.

In addition to sex differences in peripheral 3α,5α-THP levels, female and male rats differ in their 3α,5α-THP response to stress. Cold-water swimming and other extreme stressors increase brain 3α,5α-THP of gonadectomized and ADX rats.[119,120] Notably, females have increased 3α,5α-THP production in response to lower levels of stress stimuli than do males.[121] 3α,5α-THP can attenuate hypothalamic-pituitary-adrenal (HPA) activation in response to stress[122]; as such, it may be an important neuroendocrine mediator or adaptive mechanism that buffers an individual's response to behavioral/environmental stimuli.

Research showing 3α,5α-THP increases in response to fight-or-flight stimuli suggests that a functional role of 3α,5α-THP may be to serve as a rapid neuroendocrine mediator of an organism's response to environmental stressors. Concentrations of 3α,5α-THP that are sufficient to enhance GBR and HPA function and produce anxiolysis are reported for a number of avian, amphibian, and mammalian species.[9,123,124] In mammals, 3α,5α-THP has been detected as early in development as embryonic day 17.[125] By postnatal day 6, brain 3α,5α-THP concentrations increase in response to the psychosocial stress of brief removal from the nest, dam, and siblings.[126,127]

Stress-induced changes in 3α,5α-THP may be particularly important for females. Females are more sensitive/responsive to stress and have more readily increased 3α,5α-THP in response to stress than do males. Social isolation for 1 h on postnatal days 2 to 9 produces greater biosynthesis of 3α,5α-THP of female than male rat pups.[127] As adults, more females that were neonatally handled failed to adapt, as measured by weight loss, to chronic restraint stress.[121] Thus, female rats may be more sensitive to stress and 3α,5α-THP than male rats, which may reflect an adaptive mechanism.

12.8.2 Sex Differences in Depression, Schizophrenia, and 3α,5α-THP

There are stark sex differences in the prevalence, onset, and symptomology of depression and schizophrenia. More women than men suffer from depression,[128] and women are uniquely at risk for depressive disorders that are linked to hormonal states, such as premenstrual syndrome, postpartum depression, and associated psychoses. In contrast, the sex differences in schizophrenia favor women. Women with schizophrenia experience later age of onset, less debilitating psychiatric symptoms, fewer psychiatric hospitalizations, better pre- and postfunctioning, and more rapid and greater response to antipsychotics than do men.[129] Together, these findings demonstrate clear sex differences in neuropsychiatric disorders.

Stress reactivity is an important component in the incidence of psychiatric episodes related to depression and schizophrenia.[128] Notably, people with schizophrenia and depression have altered stress responsiveness compared to those without these psychopathologies.[128,130] Genetic deficits, such as a novel polymorphism and genetic mutations in the sequence encoding the gene for MBRs, which is necessary for 3α,5α-THP biosynthesis in glial cells, have been demonstrated among schizophrenics and may create a predisposition to oversensitivity to stress.[131] 3α,5α-THP attenuates hypereactivity of the HPA axis.[122] For example, administration of 3α,5α-THP to female rats attenuates the elevation of plasma adrenocorticotropin hormone (ACTH) and serum corticosterone following emotional stress.[122] Therefore, depression and schizophrenia may be characterized by a decreased capacity to synthesize 3α,5α-THP in the brain, which would heighten sensitivity to stress. Furthermore, given the sex differences in response to stress (discussed above), it may be that sex differences in sensitivity to changes in 3α,5α-THP concentrations account for some of the sex differences in depression and schizophrenia.

12.9 CONCLUSIONS

Our model investigating progestins' actions in the VTA for lordosis has enabled us to investigate progestins' source, mechanisms, and function. This approach has revealed that neurosteroids such as 3α,5α-THP are synthesized in the brain and act locally at the membrane to produce their effects. Some membrane effects include rapid actions through changes in nonclassical steroid substrates such as GBRs, NMDARS, and D_1, as well as second messengers. These actions of 3α,5α-THP on sociosexual behaviors are also relevant for neuropsychiatric disorders. Indeed, we have found that therapeutic drugs for depression and schizophrenia demonstrate classic neurosteroidogenic effects and that these neurosteroidogenic effects improve social functioning.

REFERENCES

1. Feder, H.H., Hormones and sexual behavior, *Annu. Rev. Psychol.*, 35, 165, 1984.
2. Whalen, R.E. and Luttge, W.G., Differential localization of progesterone uptake in brain: role of sex, estrogen pretreatment and adrenalectomy, *Brain Res.*, 33, 147, 1971.
3. Rubin, B.S. and Barfield, R.J., Progesterone in the ventromedial hypothalamus facilitates estrous behavior in ovariectomized, estrogen-primed rats, *Endocrinology*, 113, 797, 1983.
4. Pleim, E.T., Baumann, J., and Barfield, R.J., A contributory role for midbrain progesterone in the facilitation of female sexual behavior in rats, *Horm. Behav.*, 25, 19, 1991.
5. Frye, C.A. and Gardiner, S.G., Progestins can have a membrane-mediated action in rat midbrain for facilitation of sexual receptivity, *Horm. Behav.*, 30, 682, 1996.
6. Takahashi, L.K., and Lisk, R.D., Dual progesterone action in diencephalon facilitates the induction of sexual receptivity in estrogen-primed golden hamsters, *Physiol. Behav.*, 44, 741, 1988.
7. DeBold, J.F. and Malsbury, C.W., Facilitation of sexual receptivity by hypothalamic and midbrain implants of progesterone in female hamsters, *Physiol. Behav.*, 46, 655, 1989.
8. Pleim, E.T., Lisciotto, C.A., and DeBold, J.F., Facilitation of sexual receptivity in hamsters by simultaneous progesterone implants into the VMH and ventral mesencephalon, *Horm. Behav.*, 24, 139, 1990.
9. Frye, C.A., The role of neurosteroids and non-genomic effects of progestins and androgens in mediating sexual receptivity of rodents, *Brain Res. Brain Res. Rev.*, 37, 201, 2001.
10. Frye, C.A. and Vongher, J.M., Progesterone and 3α,5α-THP enhance sexual receptivity in mice, *Behav. Neurosci.*, 115, 1118, 2001.
11. Frye, C.A. and Vongher, J.M., Progesterone has rapid and membrane effects in the facilitation of female mouse sexual behavior, *Brain Res.*, 815, 259, 1999.
12. Malsbury, C.W., Kow, L.M., and Pfaff, D.W., Effects of medial hypothalamic lesions on the lordosis response and other behaviors in female golden hamsters, *Physiol. Behav.*, 19, 223, 1977.
13. Mathews, D. and Edwards, D.A., Involvement of the ventromedial and anterior hypothalamic nuclei in the hormonal induction of receptivity in the female rat, *Physiol. Behav.*, 19, 319, 1977.

14. Muntz, J.A., Rose, J.D., and Shults, R.C., Disruption of lordosis by dorsal midbrain lesions in the golden hamster, *Brain Res. Bull.*, 5, 359, 1980.
15. Rose, J.D., Brainstem influences on sexual behavior, in *Brainstem Mechanisms of Behavior*, Klemm, W.R. and Vertes, R.P., Eds., John Wiley & Sons, New York, 1990, p. 407.
16. Freeman, M.C., Dupke, K.C., and Croteau, C.M., Extinction of the estrogen-induced daily signal for LH release in the rat: a role for the proestrous surge of progesterone, *Endocrinology*, 99, 223, 1976.
17. Holzbauer, M., Physiological variations in the ovarian production of 5α-pregnane derivatives with sedative properties in the rat, *J. Steroid Biochem.*, 6, 1307, 1975.
18. Purdy, R.H. et al., Radioimmunoassay of 3α-hydroxy-5α-pregnan-20-one in rat and human plasma, *Steroids*, 55, 290, 1990.
19. Corpechot, C. et al., Neurosteroids: 3α-hydroxy-5α-pregnan-20-one and its precursors in the brain, plasma, and steroidogenic glands of male and female rats, *Endocrinology*, 133, 1003, 1993.
20. Frye, C.A. and Bayon, L.E., Mating stimuli influence endogenous variations in the neurosteroids 3α,5α-THP and 3α-diol, *J. Neuroendocrinol.*, 11, 839, 1999.
21. Frye, C.A. et al., The neurosteroids, progesterone and 3α,5α-THP, enhance sexual motivation, receptivity, and proceptivity in female rats, *Brain Res.*, 808, 72, 1998.
22. Frye, C.A. and Vongher, J.M., 3α,5α-THP in the ventral tegmental areas of rats and hamsters is increased in exogenous hormonal states associated with estrous cyclicity and sexual receptivity, *J. Endocrinol. Invest.*, 22, 455, 1999.
23. Lisk, R.D., A comparison of the effectiveness of intravenous, as opposed to subcutaneous, injections of progesterone for the induction of estrous behavior in the rat, *Can. J. Biochem. Psychol.*, 38, 1381, 1960.
24. Meyerson, B.J., Relationship between the anesthetic and gestagenic action and estrous behavior-inducing activity of different progestins, *Endocrinology*, 81, 369, 1967.
25. Meyerson, B., Latency between intravenous injection of progestins and the appearance of estrous behavior in estrogen-treated ovariectomized rats, *Horm. Behav.*, 3, 1, 1972.
26. Kubli-Garfias, C. and Whalen, R.E., Induction of lordosis behavior in female rats by intravenous administration of progestins, *Horm. Behav.*, 9, 380, 1977.
27. Beyer, C., Gonzalez-Flores, O., and Gonzalez-Mariscal, G., Ring A reduced progestins potently stimulate estrous behavior in rats: paradoxical effect through the progesterone receptor, *Physiol. Behav.*, 58, 985, 1995.
28. Frye, C.A., Bayon, L.E., and Vongher, J.M., Intravenous progesterone elicits a more rapid induction of lordosis in rats than does SKF38393, *Psychobiology*, 28, 99, 2000.
29. Beyer, C. et al., Lordosis facilitation in estrogen primed rats by intra-brain injection of pregnanes, *Pharmacol. Biochem. Behav.*, 22, 919, 1988.
30. Moguilewsky, M. and Raynaud, J.P., Progestin binding sites in the rat hypothalamus-pituitary and uterus, *Steroids*, 30, 99, 1977.
31. Gonzalez-Mariscal, G., Gonzalez-Flores, O., and Beyer, C., Intrahypothalamic injection of RU486 antagonizes the lordosis induced by ring A-reduced progestins, *Physiol. Behav.*, 46, 435, 1989.
32. Frye, C.A. and Vongher, J.M., Progestins' rapid facilitation of lordosis when applied to the ventral tegmentum corresponds to efficacy at enhancing $GABA_A$ receptor activity, *J. Neuroendocrinol.*, 11, 829, 1999.

33. Frye, C.A. and DeBold, J.F., 3α-OH-DHP and 5α-THDOC implants to the ventral tegmental area facilitate sexual receptivity in hamsters after progesterone priming to the ventral medial hypothalamus, *Brain Res.*, 612, 130, 1993.
34. Robel, P. et al., *Pregnenolone, Dehydroepiandrosterone, and Their Sulfate Esters in Rat Brain*, Raven Press, New York, 1984, p. 185.
35. Mellon, S.H., Griffin, L.D., and Compagnone, N.A., Biosynthesis and action of neurosteroids, *Brain Res. Brain Res. Rev.*, 37, 3, 2001.
36. Erskine, M.S., Effects of an anti-androgen and 5α-reductase inhibitors on estrus duration in the cycling female rat, *Physiol. Behav.*, 30, 519, 1983.
37. Frye, C.A. and Leadbetter, E.A., 5α-Reduced progesterone metabolites are essential in hamster VTA for sexual receptivity, *Life Sci.*, 54, 653, 1994.
38. Beyer, C. et al., Indomethacin inhibits lordosis induced by ring A-reduced progestins: possible role of 3α-oxioreduction in progestin-facilitated lordosis, *Horm. Behav.*, 35, 1, 1999.
39. Glaser, J.H., Etgen, A.M., and Barfield, R.J., Intrahypothalamic effects of progestin agonists on estrous behavior and progestin receptor binding, *Physiol. Behav.*, 34, 871, 1985.
40. Griffin, L.D. and Mellon, S.H., Selective serotonin reuptake inhibitors directly alter activity of neurosteroidogenic enzymes, *Proc. Natl. Acad. Sci. U.S.A.*, 96, 13512, 1999.
41. Frye, C.A. et al., Fluoxetine to the ventral tegmental area increases lordosis and midbrain 3α,5α-THP concentrations of rats, *Ann. N.Y. Acad. Sci.*, 2003, in press.
42. Frye, C.A., The role of neurosteroids and nongenomic effects of progestins in the ventral tegmental area in mediating sexual receptivity of rodents, *Horm. Behav.*, 40, 226, 2001.
43. Frye, C.A., Inhibition of 5α-reductase enzyme or $GABA_A$ receptors in the VMH and the VTA attenuates progesterone-induced sexual behavior in rats and hamsters, *J. Endocrinol. Invest.*, 24, 399, 2001.
44. Frye, C.A. and Vongher, J.M., Ventral tegmental area infusions of inhibitors of the biosynthesis and metabolism of 3α,5α-THP attenuate lordosis of hormone-primed and behavioural oestrous rats and hamsters, *J. Neuroendocrinol.*, 13, 1076, 2001.
45. Papadopoulos, V. et al., Peripheral benzodiazepine receptor in cholesterol transport and steroidogenesis, *Steroids*, 62, 21, 1997.
46. Stoffel-Wagner, B., Neurosteroid metabolism in the human brain, *Eur. J. Endocrinol*, 145, 669, 2001.
47. Frye, C.A. and Petralia, S.M., Lordosis of rats is modified by neurosteroidogenic effects of membrane benzodiazepine receptors in the ventral tegmental area, *Neuroendocrinology*, 77, 71, 2003.
48. Frye, C.A. and Petralia, S.M., Mitochondrial benzodiazepine receptors in the ventral tegmental area modulate sexual behaviour of hormone-primed or cycling hamsters, *J. Neuroendocrinol.*, 15, 677, 2003.
49. Jahagirdar, V., Petralia, S.M., and Frye, C.A., Infusions of P450 side chain cleavage or 5α-reductase enzyme inhibitors to the ventral tegmental area decrease lordosis of rats in behavioral estrus, *Horm. Behav.*, 44, 56, 2003.
50. Frye, C.A. and Seliga, A., Olanzapine infusions to the ventral tegmental area increase lordosis and midbrain 3α,5α-THP concentrations, *Psychopharmacology*, 2003, in press.
51. Etgen, A.M., Progestin receptors and the activation of female reproductive behavior: a critical review, *Horm. Behav.*, 18, 411, 1984.

52. Power, R.F., Conneely, O.M., and O'Malley, B.W., New insights into activation of the steroid hormone receptor superfamily, *Trends Pharmacol. Sci.*, 13, 318, 1992.
53. Pfaff, D.W. and McEwen, B.S., Action of estrogens and progestins on nerve cells, *Science*, 219, 808, 1983.
54. Lauber, A.H., Romano, G.J., and Pfaff, D.W., Sex difference in estradiol regulation of progestin receptor mRNA in rat mediobasal hypothalamus as demonstrated by *in situ* hybridization, *Neuroendocrinology*, 53, 608, 1991.
55. MacLusky, N.J. and McEwen, B.S., Progestin receptors in the rat brain: distribution and properties of cytoplasmic progestin binding sites, *Endocrinology*, 106, 192, 1980.
56. Parsons, B. et al., The temporal relationship between estrogen-inducible progestin receptors in the female rat brain and the time course of estrogen activation of mating behavior, *Endocrinology*, 107, 774, 1980.
57. Vathy, I.U., Etgen, A.M., and Barfield, R.J., Actions of progestins on estrous behaviour in female rats, *Physiol. Behav.*, 40, 591, 1987.
58. Ogawa, S. and Pfaff, D.W., Application of antisense DNA method for the study of molecular bases of brain function and behavior, *Behav. Genet.*, 26, 279, 1996.
59. Rainbow, T.C. et al., Application of anisomycin to the lateral ventromedial nucleus of the hypothalamus inhibits the activation of sexual behavior by estradiol and progesterone, *Brain Res.*, 233, 417, 1982.
60. Frye, C.A. and Vongher, J.M., $GABA_A$, D_1, and D_5, but not progestin receptor, antagonist and anti-sense oligonucleotide infusions to the ventral tegmental area of cycling rats and hamsters attenuate lordosis, *Behav. Brain Res.*, 103, 23, 1999.
61. Munn, A.R., Sar, M., and Stumpf, W.E., Topographic distribution of progestin target cells in hamster brain and pituitary after injection of $[^3H]$R5020, *Brain Res.*, 274, 1, 1983.
62. Frye, C.A., Murphy, R.E., and Platek, S.M., Anti-sense oligonucleotides, for progestin receptors in the VMH and glutamic acid decarboxylase in the VTA, attenuate progesterone-induced lordosis in hamsters and rats, *Behav. Brain Res.*, 115, 55, 2000.
63. Krebs, C.J. et al., A membrane-associated progesterone-binding protein, 25-Dx, is regulated by progesterone in brain regions involved in female reproductive behaviors, *Proc. Natl. Acad. Sci. U.S.A.*, 97, 12816, 2000.
64. Frye, C.A., Mermelstein, P.G., and DeBold, J.F., Evidence for a non-genomic action of progestins on sexual receptivity in hamster ventral tegmental area but not hypothalamus, *Brain Res.*, 578, 87, 1992.
65. Zhu, Y. et al., Cloning, expression, and characterization of a membrane progestin receptor and evidence it is an intermediary in meiotic maturation of fish oocytes, *Proc. Natl. Acad. Sci. U.S.A.*, 100, 2231, 2003.
66. Harrison, N.L. et al., Structure-activity relationships for steroid interaction with the γ-aminobutyric acid A receptor complex, *J. Pharmacol. Exp. Ther.*, 241, 346, 1987.
67. Majewska, M.D. et al., Steroid hormone metabolites are barbiturate-like modulators of the GABA receptor, *Science*, 232, 1004, 1986.
68. Masco, A., Weiger, R., and Carrer, H.F., γ-Aminobutyric acid mediates ventromedial hypothalamic mechanisms controlling the execution of lordotic responses in the female rat, *Behav. Brain Res.*, 19, 153, 1986.
69. Westerling, P., Lindgren, S., and Meyerson, B., Functional changes in $GABA_A$ receptor stimulation during the estrous cycle of the rat, *Br. J. Pharmacol.*, 103, 1580, 1991.

70. Wilson, M.A., Influences of gender, gonadectomy, and estrous cycle on GABA/BZ receptors and benzodiazepine responses in rats, *Brain Res. Bull.*, 29, 165, 1992.
71. O'Connor, L.H., Nock, B., and McEwen, B.S., Regional specificity of γ-aminobutyric acid receptor regulation by estradiol, *Neuroendocrinology*, 47, 473, 1988.
72. Lopez-Colome, A.M., McCarthy, M., and Beyer, C., Enhancement of [^{3}H]muscimol binding to brain synaptic membranes by progesterone and related pregnanes, *Eur. J. Pharmacol.*, 176, 297, 1990.
73. Wu, F.S., Gibbs, T.T., and Farb, D.H., Inverse modulation of γ-aminobutyric acid- and glycine-induced currents by progesterone, *Mol. Pharmacol.*, 37, 597, 1990.
74. McCarthy, M.M. et al., GABAergic control of receptivity in the female rat, *Neuroendocrinology*, 53, 473, 1991.
75. McCarthy, M.M. et al., Estrogen modulation of mRNA levels for the two forms of glutamic acid decarboxylase (GAD) in female rat brain, *J. Comp. Neurol.*, 360, 685, 1995.
76. Bayer, V.E. and Pickel, V.M., GABA-labeled terminals form proportionally more synapses with dopaminergic neurons containing low densities of tyrosine hydroxylase-immunoreactivity in rat ventral tegmental area, *Brain Res.*, 559, 44, 1991.
77. Frye, C.A., Mermelstein, P.G., and DeBold, J.F., Bicuculline infused into the hamster ventral tegmentum inhibits, while sodium valproate facilitates, sexual receptivity, *Pharmacol. Biochem. Behav.*, 46, 1, 1993.
78. Frye, C.A. and Gardiner, S.G., Progestins can have a membrane-mediated action in rat midbrain for facilitation of sexual receptivity, *Int. J. Anat. Embryol.*, 100, 162, 1996.
79. Frye, C.A. and DeBold, J.F., Muscimol facilitates sexual receptivity in hamsters when infused into the ventral tegmentum, *Pharmacol. Biochem. Behav.*, 42, 879, 1992.
80. Kole, M.H. et al., Changes in 5-HT_{1A} and NMDA binding sites by a single rapid transcranial magnetic stimulation procedure in rats, *Brain Res.*, 826, 309, 1999.
81. Zhang, J., Chiodo, L.A., and Freeman, A.S., Electrophysiological effects of MK-801 on rat nigrostriatal and mesoaccumbal dopaminergic neurons, *Brain Res.*, 590, 153, 1992.
82. Gibbs, T.T. et al., Modulation of ionotropic glutamate receptors by neuroactive steroids, in *Contemporary Endocrinology: Neurosteroids: A New Regulatory Function in the Nervous System*, Baulieu, E.E., Robel, P., and Schumacher, M., Eds., Humana Press Inc., Totowa, NJ, 1999, 167.
83. Smith, S.S., The effect of estrogen and progesterone on GABA and glutamate responses at extrahypothalamic sites, in *FIDA Research Foundation Symposium Series on Neurosteroids and Brain Function*, Vol. 8., Costa, E.E.E. and Paul, A.M., Eds., Theime Medical Publishers., New York, 1991, p. 87.
84. Cyr, M., Ghribi, O., and DiPaolo, T., Regional and selective effects of oestradiol and progesterone on NMDA and AMPA receptors in the rat brain, *J. Neuroendocrinol.*, 12 445, 2000.
85. Carbone, S. et al., Effect of ovarian hormones on the hypothalamic excitatory amino acids system during sexual maturation in female rats, *Neuroendocrinology*, 61, 235, 1995.
86. Fleischmann, A., Vincent, P.A., and Etgen, A.M., Effects of non-competitive NMDA receptor antagonists on reproductive and motor behaviors in female rats, *Brain Res.*, 568, 138, 1991.

87. DeBold, J.F., Parow, A.M., and Fish, E.W., MK-801 Infused into the VTA of Female Hamsters Enhances Progesterone Effects on Sexual Receptivity but Inhibits in the VMH, paper presented at Society for Neuroscience 30th Annual Meeting, New Orleans, 2000.
88. McCarthy, M.M., Curran, G.H., and Feder, H.H., Excitatory amino acid modulation of lordosis in the rat, *Neurosci. Lett.*, 126, 94, 1991.
89. Kow, L.M. et al., Inhibition of the lordosis reflex in rats by intrahypothalamic infusion of neural excitatory agents: evidence that the hypothalamus contains separate inhibitory and facilitatory elements, *Brain Res.*, 341, 26, 1985.
90. Westerink, B.H. et al., The pharmacology of mesocortical dopamine neurons: a dual-probe microdialysis study in the ventral tegmental area and prefrontal cortex of the rat brain, *J. Pharmacol. Exp. Ther.*, 285, 143, 1998.
91. Caggiula, A.R. et al., Dissociation of active from immobility components of sexual behavior in female rats by central 6-hydroxydopamine: implications for CA involvement in sexual behavior and sensorimotor responsiveness, *Brain Res.*, 31, 505, 1979.
92. Fernandez-Ruiz, J.J. et al., Nigrostriatal and mesolimbic dopaminergic activities were modified throughout the ovarian cycle of female rats, *J. Neural Transm. Gen. Sect.*, 85, 223, 1991.
93. Becker, J.B. and Ramirez, V.D., Sex differences in the amphetamine stimulated release of catecholamines from rat striatal tissue in vitro, *Brain Res.*, 204, 361, 1981.
94. Morissette, M. and Di Paolo, T., Sex and estrous cycle variations of rat striatal dopamine uptake sites, *Neuroendocrinology*, 58, 16, 1993.
95. Becker, J.B., Rudick, C.N., and Jenkins, W.J., The role of dopamine in the nucleus accumbens and striatum during sexual behavior in the female rat, *J. Neurosci.*, 21, 3236, 2001.
96. Herndon, J.G., Jr. et al., Selective enhancement of the lordotic component of female sexual behavior in rats following destruction of central catecholamine-containing systems, *Brain Res.*, 141, 137, 1978.
97. Harden, D.G. et al., Depletion of dopamine in the prefrontal cortex decreases the basal electrophysiological activity of mesolimbic dopamine neurons, *Brain Res.*, 794, 96, 1998.
98. Thomas, J. et al., A 6-hydroxydopamine-induced selective Parkinsonian rat model: further biochemical and behavioral characterization, *Exp. Neurol.*, 126, 159, 1994.
99. Mani, S.K. et al., Convergent pathways for steroid hormone- and neurotransmitter-induced rat sexual behavior, *Science*, 265, 1246, 1994.
100. Apostolakis, E.M. et al., Dopaminergic regulation of progesterone receptors: brain D_5 dopamine receptors mediate induction of lordosis by D_1-like agonists in rats, *J. Neurosci.*, 16, 4823, 1996.
101. Thalmann, R.H. et al., Progesterone Regulation of Cyclic AMP in Female Rat Neocortex in the Presence of Tetrodotoxin and $GABA_A$ Receptor Blockers, paper presented at Society for Neuroscience 28th Annual Meeting, Los Angeles, 1998.
102. Collado, M.L., Rodriguez-Manzo, G., and Cruz, M.L., Effect of progesterone upon adenylate cyclase activity and cAMP levels on brain areas, *Pharmacol. Biochem. Behav.*, 23, 501, 1985.
103. Beyer, C., Canchola, E., and Larsson, K., Facilitation of lordosis behavior in the ovariectomized estrogen primed rat by dibutyryl cAMP, *Physiol. Behav.*, 26, 249, 1981.
104. Beyer, C., Gonzalez-Flores, O., and Gonzalez-Mariscal, G., Progesterone receptor participates in the stimulatory effect of LHRH, prostaglandin, E_2, and cyclic AMP on lordosis and proceptive behaviours in rats, *J. Neuroendocrinol.*, 9, 609, 1997.

105. Mani, S.K. et al., Requirement for DARPP-32 in progesterone-facilitated sexual receptivity in female rats and mice, *Science*, 287, 1053, 2000.
106. Petralia, S.M. and Frye, C.A., Progestins in the ventral tegmental area may mediate lordosis in part by actions at dopamine type 1 receptors, cAMP, and DARPP-32, *Trabajos del Instituto Cajal*, 79, 139, 2003.
107. Mani, S., Ligand-independent activation of progestin receptors in sexual receptivity, *Horm. Behav.*, 40, 183, 2001.
108. Greengard, P. et al., The DARPP-32/protein phosphatase-1 cascade: a model for signal integration, *Brain Res. Brain Res. Rev.*, 26, 274, 1998.
109. Feder, H.H. and Marrone, B.L., Progesterone: its role in the central nervous system as a facilitator and inhibitor of sexual behavior and gonadotropin release, *Ann. N.Y. Acad. Sci.*, 286, 331, 1977.
110. DeBold, J.F., Martin, J.V., and Whalen, R.E., The excitation and inhibition of sexual receptivity in female hamsters by progesterone: time and dose relationships, neural localization and mechanisms of action, *Endocrinology*, 99, 1519, 1976.
111. Carter, C., Lederhendler, I., and Kirkpatrick, B., Eds., *The Integrative Neurobiology of Affiliation*, MIT Press, Cambridge, 1999, p. ix.
112. Frye, C.A. and Erskine, M.S., Influence of time of mating and paced copulation on induction of pseudopregnancy in cyclic female rats, *J. Reprod. Fertil.*, 90, 375, 1990.
113. Erskine, M.S., Solicitation behavior in the estrous female rat: a review, *Horm. Behav.*, 23, 473, 1989.
114. Finn, D.A. et al., Rewarding effect of the neuroactive steroid 3α-hydroxy-5α-pregnan-20-one in mice, *Pharmacol. Biochem. Behav.*, 56, 261, 1997.
115. Paredes, R.G. and Vazquez, B., What do female rats like about sex? Paced mating, *Behav. Brain Res.*, 105, 117, 1999.
116. Fish, E.W., DeBold, J.F., and Miczek, K.A., Repeated alcohol: behavioral sensitization and alcohol-heightened aggression in mice, *Psychopharmacology*, 160, 39, 2002.
117. Frye, C.A., Petralia, S.M., and Rhodes, M.E., Estrous cycle and sex differences in performance on anxiety tasks coincide with increases in hippocampal progesterone and 3α,5α-THP, *Pharmacol. Biochem. Behav.*, 67, 1, 2000.
118. Pearson-Murphy, B.E. and Allison, C.M., Determination of progesterone and some of its neuroactive ring A-reduced metabolites in human serum, *J. Steroid Biochem. Mol. Biol.*, 74, 137, 2000.
119. Drugan, R.C. et al., Etiology of the sexual dimorphism in renal peripheral benzodiazepine receptor response to stress in rats, *Horm. Behav.*, 27, 348, 1993.
120. Purdy, R. et al., Stress-induced elevations of γ-aminobutyric acid type A receptor-active steroids in the rat brain, *Proc. Natl. Acad. Sci. U.S.A.*, 88, 4553, 1991.
121. Papaioannou, A. et al., Sex differences in the effects of neonatal handling on the animal's response to stress and the vulnerability for depressive behaviour, *Behav. Brain Res.*, 129, 131, 2002.
122. Patchev, V.K. et al., The neurosteroid tetrahydroprogesterone attenuates the endocrine response to stress and exerts glucocorticoid-like effects on vasopressin gene transcription in the rat hypothalamus, *Neuropsychopharmacology*, 15, 533, 1996.
123. Mensah-Nyagan, A.G. et al., Regulation of neurosteroid biosynthesis in the frog diencephalon by GABA and endozepines, *Horm. Behav.*, 40, 218, 2001.
124. Schlinger, B.A., Soma, K.K., and London, S.E., Neurosteroids and brain sexual differentiation, *Trends Neurosci.*, 24, 429, 2001.

125. Kellogg, C.K. and Frye, C.A., Endogenous levels of 5α-reduced progestins and androgens in fetal vs. adult rat brains, *Brain Res. Dev. Brain Res.*, 115, 17, 1999.
126. Kehoe, P. et al., Central allopregnanolone is increased in rat pups in response to repeated, short episodes of neonatal isolation, *Brain Res. Dev. Brain Res.*, 124, 133, 2000.
127. McCormick, C.M. et al., Neonatal isolation alters stress hormone and mesolimbic dopamine release in juvenile rats, *Pharmacol. Biochem. Behav.*, 73, 77, 2002.
128. Young, E.A., Sex differences and the HPA axis: implications for psychiatric disease, *J. Gender Specif. Med.*, 1, 21, 1998.
129. Leung, A. and Chue, P., Sex differences in schizophrenia, a review of the literature, *Acta Psychiatr. Scand. Suppl.*, 401, 3, 2000.
130. Norman R.M. and Malla, A.K., Stressful life events and schizophrenia. I. A review of the research, *Br. J. Psychiatry*, 162, 161, 1993.
131. Kurumaji, A. et al., An association study between two missense variations of the benzodiazepine receptor (peripheral) gene and schizophrenia in a Japanese sample, *J. Neural Transm.*, 107, 491, 2000.

13 Premenstrual Dysphoric Disorder

Inger Sundström-Poromaa

CONTENTS

13.1 EPIDEMIOLOGY AND DIAGNOSIS

Premenstrual dysphoric disorder (PMDD) is a psychoneuroendocrine disorder where neurosteroids are likely to play a major role in the pathophysiology and surfacing of symptoms. During the luteal phase of the menstrual cycle, fluctuations in gonadal hormones, psychiatric vulnerability traits, and psychosocial factors act in concert to provoke symptoms in affected women, and the purpose of this chapter is to highlight the roles of the GABA system and neurosteroids for the syndrome.

Premenstrual dysphoric disorder is commonly encountered in the Western world and has a substantial impact on several aspects of the daily lives of women, and although most women experience mild mood and somatic symptoms premenstrually, a small but significant number are severely disabled by the disorder.[1] The disorder is defined by the cyclical recurrence of a cluster of negative mood symptoms and physical symptoms. Symptoms develop in the luteal phase of the menstrual cycle and remit within a few days after the onset of menstrual bleeding. The American Psychiatric Association's Diagnostic and Statistical Manual of Mental Disorders, 4th edition[2] (DSM-IV) defines diagnostic criteria for premenstrual dysphoric disorder (Table 13.1). To fulfill the criteria for PMDD, a patient needs to present with at least *five* of the listed symptoms during the premenstrual week. At least one of these symptoms must be a mood symptom such as depressed mood, anxiety, affective lability, or irritability. Other symptoms described by the DSM-IV criteria are decreased interest in usual activities, difficulty in concentrating, marked lack of energy, hypersomnia, or insomnia. Physical symptoms are also included among the listed symptoms in the diagnostic criteria, and common physical symptoms encountered among PMDD patients are

0-8493-2392-4/04/$0.00+$1.50

TABLE 13.1
Summary of PMS Criteria

A. In most menstrual cycles during the past year, five or more of the following symptoms were present for most of the time during the last week of the luteal phase, began to remit within a few days after the onset of the follicular phase, and were absent in the week postmenses with at least one of the symptoms being either (1), (2), (3), or (4).

(1) Markedly depressed mood, feelings of hopelessness, or self-deprecating thoughts
(2) Marked anxiety, tension, feelings of being "keyed up" or "on edge"
(3) Marked affective lability
(4) Persistent and marked anger or irritability or increased interpersonal conflicts
(5) Decreased interest in usual activities
(6) Difficulty in concentrating
(7) Easy fatigability or marked lack of energy
(8) Marked change in appetite, overeating, or specific food cravings
(9) Hypersomnia or insomnia
(10) Sense of being overwhelmed or out of control
(11) Other physical symptoms, e.g., breast tenderness or swelling, headaches, joint or muscle pain, a sensation of bloating, weight gain

B. Symptoms must interfere with work, school, usual activities, or relationships with others.

C. Symptoms must not merely be an exacerbation of another disorder, such as major depressive disorder, panic disorder, dysthymic disorder or a personality disorder.

D. Criteria (A), (B), and (C) must be confirmed by prospective daily ratings for at least two cycles.

breast tenderness, swelling, joint pain, and headaches. Besides the presence of a number of typical symptoms, the DSM-IV criteria also state that symptoms must interfere with usual activities (school, work performance, or interpersonal relationships) to provide some measure on the severity of symptoms. Furthermore, patients must be devoid of symptoms in the follicular phase to ensure that the premenstrual complaint is not merely an exacerbation of an underlying mood disorder. Finally, unlike other DSM-IV criteria, PMDD diagnosis must be confirmed by prospective ratings for at least 2 months. As none of the symptoms of PMDD are unique to the syndrome, patients need to keep a daily diary of symptoms to establish the temporal relationship between onset of symptoms and the premenstrual period. At least 2 months of symptom ratings are needed to confirm the consistency of premenstrual symptoms over time.[3] The discrepancy between retrospective reports of premenstrual complaints and reports based on daily prospective evaluation has been thoroughly demonstrated.[4] Retrospective assessment favors recall of premenstrual symptoms but not symptoms at other times across the menstrual cycle, and approximately 14 to 50% of women complaining of premenstrual symptoms do not show a true relation between the menstrual cycle and cyclicity of symptoms.[4,5] The likelihood of a prospective confirmation of reported premenstrual depression is higher in women reporting severe symptoms than in women reporting mild or moderate symptoms.[5]

As diagnosis of PMDD depends on prospective ratings rather than retrospective reports, valid estimations of a population-based prevalence of PMDD are difficult to achieve. The point prevalence of PMDD has been estimated to approximately 2 to 6% among fertile women in a population-based study conforming to the DSM-IV criteria,[1] and this figure is similar to the estimated 12-month prevalence of 5.8%.[6]

Only 10% of women between 18 to 45 years experience *no* cyclical changes or fluctuations in mood or physical symptoms throughout the menstrual cycle.[1] Two thirds of fertile women report mental symptoms and feelings of body swelling in the premenstrual week; of these, 10.8% state that they wish to consult a physician because of their premenstrual symptoms.[7] Severe and discomforting symptoms are reported by 2 to 3% of fertile women.[7] These figures give an understanding of the continuum within the female fertile population, from those with only minute changes to those who are severely handicapped by their symptoms 14 days per month. The regularly occurring fluctuations in sex hormone concentrations and their effects on body and soul are obviously felt by a majority of women. Most of them seem to accept, tolerate, and integrate these effects into their lives as being normal physiological variations. Thus, even though a large part of the female population experiences cyclical mood changes, only a small fraction can be considered sufferers of PMDD.

13.2 CLINICAL PICTURE

Premenstrual symptoms usually develop 7 to 14 days before onset of menses, i.e., in parallel with the early development of corpus luteum.[8] Symptom severity gradually increases with time as the onset of menstrual bleeding approaches (Figure 13.1). Within 2 to 3 days after the onset of menstrual bleeding, when estradiol and progesterone levels have reached follicular phase levels, symptoms remit.[8] Patients seeking help for premenstrual symptoms form a heterogeneous group. In clinical practice about 30% of women seeking treatment for premenstrual complaints are completely devoid of symptoms in the follicular phase.[5] Approximately 40 to 50% of attending women are not entirely symptom-free in the follicular phase; instead, their premenstrual complaints can be regarded as a premenstrual exacerbation or aggravation of always existing minor mood symptoms.[5,9] Finally, approximately 15% of women seeking help for PMS will not display any cyclical pattern at all on prospective ratings.[4,5] These findings emphasize the value of determining the presence of other psychiatric disorders. Patients with an underlying affective disorder can experience a premenstrual aggravation of their general psychiatric condition whereby criteria for PMDD are not met.

The prevalence and intensity of PMDD have been reported to increase with age, whereas marital status and parity do not affect the severity of symptoms.[7] Many women report their premenstrual symptoms beginning after childbirth, after a postpartum depression, or after a major depression. Most women state that symptoms interfere with their family life.[5,10] The severity, however, varies between patients. Some patients are severely disabled and meet cross-sectional criteria for major depressive disorder during the premenstrual phase, and occasional patients report suicidal thoughts in the luteal phase.[11–13] If one is not aware of the cyclical nature of the symptoms, the premenstrual state is sometimes impossible to distinguish from an ordinary anxiety or depressive state. PMDD patients, in particular those with severe mental symptoms, report recurrent occupational impairment,

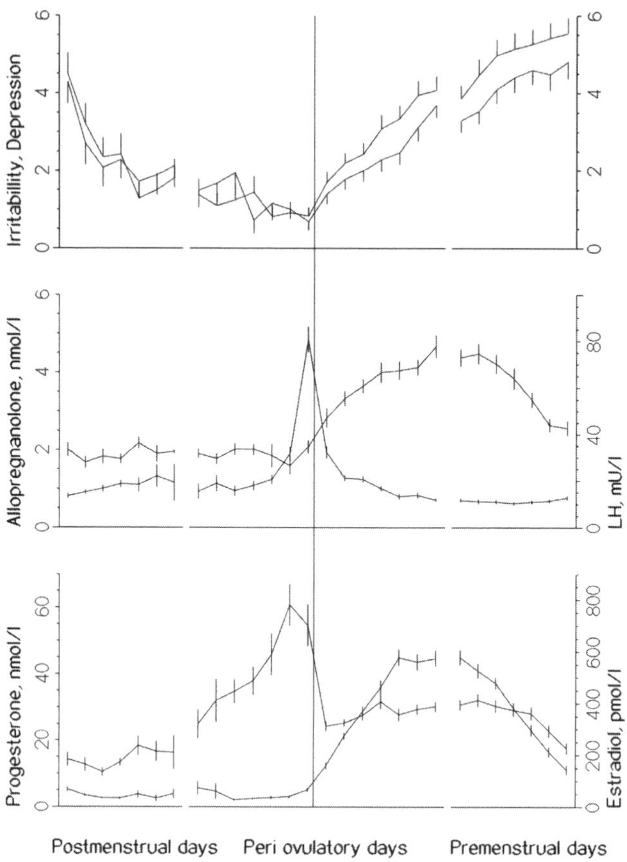

FIGURE 13.1 Allopregnanolone, progesterone, and symptom relationship during the menstrual cycle in women with PMDD. The relationship between symptom development and hormonal variations during the menstrual cycle was studied by asking women with PMDD to rate their symptoms prospectively during one menstrual cycle and to give blood samples for plasma allopregnanolone, estradiol, progesterone, and LH determinations during the same cycle. The rating scales used were visual analogue scales (VASs) for adverse mood symptoms. Premenstrual symptom severity was most intense during the last five premenstrual days with a delay of 5 to 6 days from the allopregnanolone and progesterone maximum.

and approximately 3% of fertile women will on more than two occasions within a 6-month period be unable to work during the premenstrual period.[7]

The association between major depressive disorder and premenstrual syndrome has gained special interest, and there are several lines of evidence indicating an interaction between the two disorders. First, some women with no evidence of other current mental disorder have premenstrual depressive symptoms that meet cross-sectional criteria for major depressive disorder.[11,12,14] Second, although recently questioned, the lifetime prevalence of depressive disorder appears to be increased in women with PMDD compared with control subjects;[15] however, see Breaux et al.[16] Among patients with PMDD, 46% on interview had a prior experience of major depression and 27 to 29% of the parous subjects had experienced a postpartum depression of more than 1 month's duration.[15] Moreover, patients with a prior but not ongoing mental disorder have an increased relative risk of developing premenstrual dysphoric disorder.[4] Also, it has been reported that premenstrual depression significantly correlates with the occurrence of major depressive disorder within a 4-year follow-up.[17] The predictive value of a retrospective report of PMS was above two known risk factors for major depressive disorder, namely, family history of depression and prior personal history of depression.[17] Hartlage and co-workers[18] found that 88% of women diagnosed with PMDD developed an episode of major depression within 2 years compared to 33% of women who were not diagnosed with PMDD. However, more longitudinal, prospective studies of women with confirmed PMDD are needed to clarify whether PMDD is to be regarded as a risk factor for future major depressive disorder. Finally, perhaps the most obvious association between PMDD and major depression is that premenstrual dysphoric disorder can successfully be treated with antidepressant drugs such as selective serotonin reuptake inhibitors (SSRIs).[19] SSRI treatment is effective in approximately 60% of patients, and compared to treatment of major depression, lower doses and intermittent dosing can be used in PMDD patients.[20]

13.3 ENDOCRINE ASPECTS

Ever since the premenstrual syndrome was recognized in the 1930s, the disorder has been considered a hormonal disease. Over the years many attempts have been made to explain premenstrual symptoms in terms of fluctuations of gonadal hormones. However, the only consistent endocrine finding in PMS patients is that symptom development in the luteal phase requires an ovulation and the consequent formation of a corpus luteum.[21–23] During spontaneous anovulatory cycles the cyclicity in symptoms disappears.[22] Anovulation can also be induced by GnRH agonists, and these compounds have in several studies been proven beneficial for PMDD.[21,23–27]

The symptom development in PMDD coincides with the luteal phase of the menstrual cycle, in which levels of estradiol and progesterone are high. The intensity gradually increases during the luteal phase, and the maximum symptom severity occurs 3 to 5 days after the progesterone peak in the luteal phase. The remittance of symptoms does not occur until a few days after the onset of menstrual bleeding, when gonadal hormone levels have declined.[28] Although the temporal relationship with the progesterone peak in the luteal phase is obvious, PMDD patients do not have higher or lower peripheral levels of progesterone than healthy women.[28,29] Furthermore, there is a general opinion on the absence of peripheral markers of hypothalamus-pituitary-gonadal (HPG) axis dysfunction in PMDD. Hence, plasma levels of progesterone and estradiol do not differ between PMDD patients and control subjects in the luteal phase, nor do luteinizing hormone (LH), follicle-stimulating hormone (FSH), dehydroepiandrostendione sulfate, dihydrotestosterone, SHBG, prolactin, or cortisol.[28–32]

Although no differences in progesterone and estradiol plasma levels have been demonstrated between PMDD paticnts and control subjects, particular attention has been given to progesterone as a symptom-provoking factor in PMDD. Evidence to substantiate this assumption is found from sequential hormone replacement therapy (HRT) in postmenopausal women. During the progestin addition to estrogen treatment, postmenopausal women respond with a significant cyclicity in mood and physical symptoms, whereas women treated with unopposed estrogen do not report any mood deterioration.[33–35] Add-back HRT during GnRH agonist treatment in PMDD (to relieve the hypoestrogenic symptoms) induced progestagenic side effects similar to the underlying condition, although the efficacy of the GnRH treatment on mood symptoms was still evident.[23,25]

However, not only progesterone is symptom provoking. Within PMDD patients, menstrual cycles with increased estradiol and progesterone plasma levels during the luteal phase were associated with more severe symptoms than cycles with lower luteal phase levels of these hormones,[36] and the symptom severity was particularly related to the luteal estradiol levels.[36] Moreover, estradiol treatment during the luteal phase induced more negative symptoms than placebo in a group of PMDD patients.[37] However, estradiol alone cannot provoke symptoms. During the follicular phase of the menstrual cycle, when estrogen predominates and also reaches it highest levels, PMDD patients feel at their best.[28] Obviously, estradiol and progesterone acting together seem to induce a different mood response in the central nervous system (CNS) than when they act separately.

13.4 PATHOPHYSIOLOGY OF PMDD

As peripheral levels of ovarian steroids do not differ between PMDD patients and healthy controls, focus has shifted to the interaction of ovarian steroids with the central nervous system. Several neurotransmitter receptor families have been examined as potential regulators of premenstrual dysphoric disorder. These putative causative agents include (but are by no means limited to) GABA receptors, sigma receptors (σ),[38] serotonin (5HT) receptors,[39–41] and noradrenaline receptors.[42] Although all of these neurotransmitter systems are likely to be involved in symptom provocation in PMDD patients, this chapter will focus on research that has illuminated the interactions between the GABA system and PMDD. Recent studies have also examined the potential role of endogenous modulators of neurotransmitter receptors, such as the neurosteroid metabolites of steroid hormones.

Altered $GABA_A$ receptor function has been demonstrated in clinical studies of PMDD. By use of proton magnetic resonance spectroscopy it was recently established that the GABAergic system is substantially modulated in PMDD patients, particularly during the follicular phase, where cortical GABA levels were significantly lower among PMDD patients.[43] Furthermore, whereas cortical GABA levels decreased across the menstrual cycle in healthy controls, PMDD patients displayed an increase in cortical GABA levels from the follicular phase to the luteal phase.[43] This finding is partly at odds with previous studies of peripheral GABA levels in PMDD patients.[44] Peripheral GABA levels were reported to be lower in both phases of the menstrual cycle in PMDD patients with a history of prior major depression, whereas plasma levels of GABA decreased from the follicular phase to the late luteal phase in PMDD patients devoid of previous mood disorders.[44] However, the discrepancies between these two studies mainly point toward the suggestion that peripheral measures of GABA function may not accurately reflect central function. Another way of estimating the functional $GABA_A$ receptor sensitivity is by measuring specific behavioral responses to drugs that act at the $GABA_A$ receptor. The saccadic eye velocity (SEV) is one such measure, since the SEV reliably decreases in response to sedative drugs, such as benzodiazepines, operating via the $GABA_A$ receptor. In this pharmacodynamic model patients with PMDD have a reduced sensitivity to benzodiazepines.[45,46] This subsensitivity to benzodiazepines was present in both phases of the menstrual cycle and was further accentuated in patients with severe premenstrual symptoms, compared to those with moderate symptoms. Decreased sensitivity to benzodiazepines is consistent with the mixed response of treatment with benzodiazepines in patients with PMDD.[47–50] Also consistent with the subsensitivity to benzodiazepines in PMDD patients is the finding of greater panic response to the benzodiazepine antagonist flumazenil during the luteal phase of the menstrual cycle among PMDD patients.[51] Increased panic response to flumazenil has also been reported in patients with panic disorder[52] and has been interpreted as a shift in benzodiazepine sensitivity toward inverse agonism and, consequently, for flumazenil toward partial inverse agonism.[51] Whether these findings of altered GABA system function in PMDD patients represent psychiatric vulnerability traits or state variables affected by the ovarian hormones across the menstrual cycle is thus far not fully established.

13.5 NEUROSTEROIDS AND PREMENSTRUAL SYNDROME

Though there is evidence of underlying changes in $GABA_A$ receptor function in PMDD, the question of what causes these phenomena remains. One plausible mechanism is that $GABA_A$ receptors may be regulated by endogenous modulators, such as the neuroactive metabolites of steroid hormones. Generally, steroid hormones, such as corticosterone and progesterone, exert their effects over a relatively long time by acting as transcription factors that regulate protein expression. These compounds can, however, also be further processed into neuroactive metabolites, which are known as neurosteroids since they have rapid actions at nerve cell membrane receptors and are synthesized in the brain.[53] Neuroactive steroids include tetrahydrodeoxycorticosterone (THDOC), a derivative of corticosterone, and allopregnanolone and pregnanolone (3α-hydroxy-5α-pregnane-20-one and 3α-hydroxy-5β-pregnane-20-one, respectively), which are neuroactive metabolites of progesterone.

Although there is evidence for *de novo* synthesis of neurosteroids in the human brain,[54] it is conceivable that progesterone, produced by the corpus luteum, is the major source for neurosteroids in fertile women.[55,56] In fertile women, allopregnanolone serum levels increase from 2 n*M* in the follicular phase to approximately 4 to 6 n*M* in the luteal phase, and allopregnanolone levels are temporally correlated with progesterone levels.[57,58] Not only is neurosteroid synthesis increased during the luteal phase of the menstrual cycle, but neurosteroids are also accumulated in the human brain.[59] CNS levels of allopregnanolone are roughly 14 to 21 ng/g, depending on the examined brain region.[59]

Over the past years a number of studies have tried to elucidate the role of neurosteroids in the pathophysiology of PMDD. An indication that ovarian steroids may regulate $GABA_A$ receptors via the interaction of their GABA-modulatory metabolites with this receptor, and not due to their genomic actions as transcription factors, is the fact that treatment of PMDD patients with the intracellular progesterone receptor blocker RU486 does not ameliorate the premenstrual symptoms.[60,61] Hence, neurosteroids are likely to play a role in the luteal phase symptom surfacing in PMDD, although a number of issues remain to be determined before their specific actions are fully understood.

First, there appears to be no simple relationship between peripheral levels of neurosteroids and PMDD. Studies that have examined neurosteroids in prospectively diagnosed PMDD patients have yielded mixed results, indicting no difference in allopregnanolone levels,[58,62,63] significantly lower levels in PMDD patients,[64–66] or significantly higher levels of allopregnanolone among PMDD patients.[67] Also, no differences in pregnanolone (a stereoisomer of allopregnanolone), pregnenolone, pregnanolone sulfate, and 5α-DHP levels have been demonstrated between PMDD patients and controls.[58,62] In light of these findings and the absence of differences in peripheral levels of ovarian steroids, cortisol, prolactin, or other hormones between PMDD patients and controls, it is unlikely that PMDD can be explained as a simple deficiency or excess of neurosteroids, at least not as can be measured by crude peripheral serum concentrations. Furthermore, the density of peripheral benzodiazepine receptors, which are located on outer mitochondrial and plasma membranes and have been suggested to be the primary regulator of the rate of neurosteroid biosynthesis,[68] does not differ between PMDD patients and controls, indicating that neurosteroid biosynthesis in PMDD patients is not altered.[69]

Nevertheless, within subjects two studies have indicated that plasma levels of allopregnanolone during the luteal phase are associated with the severity of symptom in PMDD patients.[58,67] Wang and colleagues[58] reported that menstrual cycles with high luteal phase allopregnanolone levels were related to symptom improvement, and likewise, Girdler et al.[67] found that patients with greatest levels of premenstrual anxiety and irritability had significantly lower luteal phase allopregnanolone levels.

Although the neurosteroids are probably involved in the pathophysiology of PMDD, their role as potentially intrinsically rewarding or aversive in PMDD patients is thus far unclear. One study has suggested that women with PMDD have an abnormal response to ovarian steroids.[70] During the early luteal phase, symptom severity increases in parallel with increasing progesterone and allopregnanolone levels, whereas during the late luteal phase the final increase in symptomatology before onset of menses is accompanied by decreasing allopregnanolone levels (Figure 13.1).[71] In animal models, acute injections of progesterone and allopregnanolone are reliably anxiolytic within hours.[72–74] However, a few days of progesterone treatment, producing high neurosteroid levels, can increase anxiety in laboratory animals.[75] In humans, administration of micronized progesterone (which, as opposed to vaginal treatment, results in high levels of allopregnanolone[76]) is unable to relieve symptoms of anxiety and depression in PMDD patients.[19,77] Furthermore, a single high dose of progesterone, resulting in high levels of allopregnanolone, to healthy women produced increased fatigue, confusion, and poor concentration, symptoms similar to those encountered among women with PMDD.[78] Likewise, whereas successful treatment with serotonin reuptake inhibitors raises allopregnanolone levels in patients with major depression,[79,80] findings in PMDD patients are less conclusive.[81] In PMDD patients treated with SSRIs, allopregnanolone levels were not affected by treatment.[81] However, improvement as such, whether induced by SSRI treatment or placebo, was associated with lower levels of allopregnanolone, rather than increased levels.[81]

Finally, an issue that remains to be addressed in humans concerning neurosteroids and PMDD is the potentially adverse mood effects that can be induced by fluctuations of allopregnanolone throughout the menstrual cycle. The homeostatic response of the GABAergic system to prolonged exposure and withdrawal from progesterone may be similar to its response to conventional GABA-modulatory drugs, such as the benzodiazepines and barbiturates. Animal models of progesterone withdrawal have indicated that upon chronic exposure or withdrawal from this steroid, symptoms of increased anxiety may be precipitated.[82–84] The luteal phase of the menstrual cycle is characterized by the abrupt decline of progesterone (and neurosteroids) after chronic exposure to high levels of hormones, and the late luteal phase can therefore be regarded as a state of progesterone withdrawal. Female rats, besides displaying anxiety-like behavior during progesterone withdrawal, are also comparatively insensitive to the sedative and anxiolytic effects of benzodiazepines and neurosteroids.[84,85] In PMDD patients, the progesterone withdrawal hypothesis is supported by the accentuation of adverse symptomatology during the last 5 days of the menstrual cycle, a time point during which allopregnanolone levels rapidly decline to follicular phase levels (Figure 13.1).[71] Furthermore, changes in sensitivity to neurosteroids and other GABA-modulatory agents across the menstrual cycle have been reported in PMDD patients. For instance, using saccadic eye velocity and subjective scores of sedation as independent measures, patients with PMDD have been reported to have a decreased responsiveness to pregnanolone (a stereoisomer to allopregnanolone) and benzodiazepines.[45,46,86] Moreover, neurosteroid and benzodiazepine insensitivity was also related to symptom severity within the PMDD group. High-severity PMDD patients were even less responsive to GABA-modulatory neurosteroids and benzodiazepines than low-severity patients.[86] Although progesterone withdrawal cannot readily explain why symptoms are exacerbated during the early and mid-lutel phase, as is the case in the majority of PMDD patients, the pharmacological similarities between the progesterone withdrawal model and PMDD certainly point toward the need of more thorough studies on the potentially adverse effects induced by homeostatic response to fluctuations in progesterone levels across the menstrual cycle.

Also, in other aspects PMDD patients have also been shown to have an altered neurosteroid response. For instance, PMDD patients have a different allopregnanolone reactivity to mental stress. Whereas 83% of healthy controls responded with the expected stress-induced increase in allopregnanolone, only 42% of PMDD patients responded with an increase in allopregnanolone levels after stress.[67]

Another interesting area of research in PMDD patients is the interaction between neurosteroids and the serotonin system. It has been suggested that some of the effects mediated by SSRIs on mood could result from increased allopregnanolone production. Administration of SSRIs increases allopregnanolone concentrations, most likely through a direct effect of the enzyme 3α-hydroxysteroid oxidoreductase, one of the two enzymes involved in the conversion of progesterone to allopregnanolone.[87,88] In PMDD patients, SSRIs, as first-line therapy for treatment of premenstrual dysphoria, have been shown to normalize the impaired pregnanolone sensitivity.[89] Furthermore, the allopregnanolone response to repeated challenges of the serotonin precursor L-tryptophan across the menstrual cycle was enhanced in women with PMDD, particularly during the luteal phase.[90]

Clearly, the only way to fully elucidate the topic of neurosteroids in premenstrual dysphoric disorder would be to conduct a clinical trial to evaluate their possible symptom-relieving or symptom-provoking actions. Thus far, due to poor oral bioavailability, no such studies have been conducted.

In conclusion, premenstrual dysphoric disorder is defined by DSM-IV criteria. By distinguishing this severe form of premenstrual syndrome from other milder-symptom constellations or naturally occurring fluctuations in somatic symptoms over the menstrual cycle, we form a better understanding of the disorder. Given the accumulating evidence of interactions between ovarian steroids, neurosteroids, and the GABA system, it appears plausible that estradiol and progesterone may affect this particular neurotransmitter system and thereby surface the mood symptoms among PMDD patients in the luteal phase. However, PMDD is unlikely to be the result of a single deficiency or excess of a particular substance. It is not possible that only one neurotransmitter system is involved in the symptom provocation throughout the entire luteal phase. Multiple neurotransmitter systems, of which we have elucidated one, the GABA system, are undoubtedly involved.

REFERENCES

1. Sveindottir, H. and Bäckström, T., Prevalence of menstrual cycle symptom cyclicity and premenstrual dysphoric disorder in a random sample of women using and not using oral contraceptives, *Acta Obstet. Gynecol. Scand.*, 79, 405, 2000.
2. American Psychiatric Association, *Diagnostic and Statistical Manual of Mental Disorders*, 4th ed., U.S. Department of Health and Human Services, Washington D.C., 1994, p. 714.
3. Bloch, M., Schmidt, P.J., and Rubinow, D.R., Premenstrual syndrome: evidence for symptom stability across cycles, *Am. J. Psychiatry*, 154, 1741, 1997.
4. Hurt, S.W. et al., Late luteal phase dysphoric disorder in 670 women evaluated for premenstrual complaints, *Am. J. Psychiatry*, 149, 525, 1992.
5. Hammarbäck, S. and Bäckström, T., A demographic study in subgroups of women seeking help for premenstrual syndrome, *Acta Obstet. Gynecol. Scand.*, 68, 247, 1989.
6. Wittchen, H.U. et al., Prevalence, incidence and stability of premenstrual dysphoric disorder in the community, *Psychol. Med.*, 32, 119, 2002.
7. Andersch, B. et al., Premenstrual complaints. I. Prevalence of premenstrual symptoms in a Swedish urban population, *J. Psycosom. Obstet. Gynaecol.*, 5, 39, 1986.
8. Hammarbäck, S., Damber, J.E., and Bäckström, T., Relationship between symptom severity and hormone changes in women with premenstrual syndrome, *J. Clin. Endocrinol. Metab.*, 68, 125, 1989.

9. Bailey, J.W. and Cohen, L.S., Prevalence of mood and anxiety disorders in women who seek treatment for premenstrual syndrome, *J. Womens Health Gend. Based Med.*, 8, 1181, 1999.
10. Ekholm, U.B. and Bäckström, T., Influence of premenstrual syndrome on family, social life, and work performance, *Int. J. Health Serv.*, 24, 629, 1994.
11. Fava, M., Comorbid anxiety and depression among patients with late luteal phase dysphoric disorder, *J. Anxiety Disord.*, 6, 325, 1992.
12. Steege, J.F. et al., Reduced platelet tritium-labeled imipramine binding sites in women with premenstrual syndrome, *Am. J. Obstet. Gynecol.*, 167, 168, 1992.
13. Baca-Garcia, E. et al., Menstrual cycle and profiles of suicidal behaviour, *Acta Psychiatr. Scand.*, 97, 32, 1998.
14. Severino, S.K., Late luteal phase dysphoric disorder: a scientific puzzle, *Med. Hypotheses*, 41, 229, 1993.
15. Pearlstein, T.B. et al., Prevalence of axis I and axis II disorders in women with late luteal phase dysphoric disorder, *J. Affect. Disord.*, 20, 129, 1990.
16. Breaux, C., Hartlage, S., and Gehlert, S., Relationships of premenstrual dysphoric disorder to major depression and anxiety disorders: a re-examination, *J. Psychosom. Obstet. Gynaecol.*, 21, 17, 2000.
17. Graze, K.K., Nee, J., and Endicott, J., Premenstrual depression predicts future major depressive disorder, *Acta Psychiatr. Scand.*, 81, 201, 1990.
18. Hartlage, S.A., Arduino, K.E., and Gehlert, S., Premenstrual dysphoric disorder and risk for major depressive disorder: a preliminary study, *J. Clin. Psychol.*, 57, 1571, 2001.
19. Wyatt, K.M., Dimmock, P.W., and O'Brien, P.M., Selective serotonin reuptake inhibitors for premenstrual syndrome, *Cochrane Database Syst. Rev.*, CD001396, 2002 (Cochrane review).
20. Endicott, J. et al., Is premenstrual dysphoric disorder a distinct clinical entity? *J. Womens Health Gend. Based Med.*, 8, 663, 1999.
21. Hammarbäck, S. and Bäckström, T., Induced anovulation as treatment of premenstrual tension syndrome: a double-blind cross-over study with GnRH-agonist versus placebo, *Acta Obstet. Gynecol. Scand.*, 67, 159, 1988.
22. Hammarbäck, S., Ekholm, U.B., and Bäckström, T., Spontaneous anovulation causing disappearance of cyclical symptoms in women with the premenstrual syndrome, *Acta Endocrinol. (Copenhagen)*, 125, 132, 1991.
23. Mortola, J.F., Girton, L., and Fischer, U., Successful treatment of severe premenstrual syndrome by combined use of gonadotropin-releasing hormone agonist and estrogen/progestin, *J. Clin. Endocrinol. Metab.*, 72, 252A, 1991.
24. Brown, C.S. et al., Efficacy of depot leuprolide in premenstrual syndrome: effect of symptom severity and type in a controlled trial, *Obstet. Gynecol.*, 84, 779, 1994.
25. Leather, A.T. et al., The treatment of severe premenstrual syndrome with goserelin with and without 'add-back' estrogen therapy: a placebo-controlled study, *Gynecol. Endocrinol.*, 13, 48, 1999.
26. Mezrow, G. et al., Depot leuprolide acetate with estrogen and progestin add-back for long-term treatment of premenstrual syndrome, *Fertil. Steril.*, 62, 932, 1994.
27. Sundström, I. et al., Treatment of premenstrual syndrome with gonadotropin-releasing hormone agonist in a low dose regimen, *Acta Obstet. Gynecol. Scand.*, 78, 891, 1999.
28. Bäckström, T. et al., Mood, sexuality, hormones, and the menstrual cycle. II. Hormone levels and their relationship to the premenstrual syndrome, *Psychosom. Med.*, 45, 503, 1983.

29. Rubinow, D.R. et al., Changes in plasma hormones across the menstrual cycle in patients with menstrually related mood disorder and in control subjects, *Am. J. Obstet. Gynecol.*, 158, 5, 1988.
30. Rabin, D.S. et al., Hypothalamic-pituitary-adrenal function in patients with the premenstrual syndrome, *J. Clin. Endocrinol. Metab.*, 71, 1158, 1990.
31. Redei, E. and Freeman, E.W., Daily plasma estradiol and progesterone levels over the menstrual cycle and their relation to premenstrual symptoms, *Psychoneuroendocrinology*, 20, 259, 1995.
32. Reame, N.E., Marshall, J.C., and Kelch, R.P., Pulsatile LH secretion in women with premenstrual syndrome (PMS): evidence for normal neuroregulation of the menstrual cycle, *Psychoneuroendocrinology*, 17, 205, 1992.
33. Björn, I. et al., The impact of different doses of medroxyprogesterone acetate on mood symptoms in sequential hormonal therapy, *Gynecol. Endocrinol.*, 16, 1, 2002.
34. Björn, I. et al., Negative mood changes during hormone replacement therapy: a comparison between two progestogens, *Am. J. Obstet. Gynecol.*, 183, 1419, 2000.
35. Hammarbäck, S. et al., Cyclical mood changes as in the premenstrual tension syndrome during sequential estrogen-progestagen postmenopausal replacement therapy, *Acta Obstet. Gynecol. Scand.*, 64, 393, 1985.
36. Seippel, L. and Bäckström, T., Luteal-phase estradiol relates to symptom severity in patients with premenstrual syndrome, *J. Clin. Endocrinol. Metab.*, 83, 1988, 1998.
37. Dhar, V. and Murphy, B.E., The premenstrual syndrome and its treatment, *J. Steroid Biochem. Mol. Biol.*, 39, 275, 1991.
38. Straneva, P.A. et al., Menstrual cycle, beta-endorphins, and pain sensitivity in premenstrual dysphoric disorder, *Health Psychol.*, 21, 358, 2002.
39. Eriksson, E. et al., The serotonin reuptake inhibitor paroxetin is superior to the noradrenaline reuptake inhibitor maprotiline in the treatment of premenstrual syndrome, *Neuropsychopharmacology*, 12, 167, 1995.
40. Halbreich, U. and Tworek, H., Altered serotonergic activity in women with dysphoric premenstrual syndromes, *Int. J. Psychiatry Med.*, 23, 1, 1993.
41. Parry, B.L., The role of central serotonergic dysfunction in the aetiology of premenstrual dysphoric disorder: therapeutic implications, *CNS Drugs*, 15, 277, 2001.
42. Halbreich, U. et al., Increased imidazoline and alpha 2 adrenergic binding in platelets of women with dysphoric premenstrual syndromes, *Biol. Psychiatry*, 34, 676, 1993.
43. Epperson, C.N. et al., Cortical gamma-aminobutyric acid levels across the menstrual cycle in healthy women and those with premenstrual dysphoric disorder: a proton magnetic resonance spectroscopy study, *Arch. Gen. Psychiatry*, 59, 851, 2002.
44. Halbreich, U. et al., Low plasma gamma-aminobutyric acid levels during the late luteal phase of women with premenstrual dysphoric disorder, *Am. J. Psychiatry*, 153, 718, 1996.
45. Sundström, I., Ashbrook, D., and Bäckström, T., Reduced benzodiazepine sensitivity in patients with premenstrual syndrome: a pilot study, *Psychoneuroendocrinology*, 22, 25, 1997.
46. Sundström, I., Nyberg, S., and Bäckström, T., Patients with premenstrual syndrome have reduced sensitivity to midazolam compared to control subjects, *Neuropsychopharmacology*, 17, 370, 1997.
47. Evans, S.M. et al., Mood and performance changes in women with premenstrual dysphoric disorder: acute effects of alprazolam, *Neuropsychopharmacology*, 19, 499, 1998.
48. Schmidt, P.J., Grover, G.N., and Rubinow, D.R. Alprazolam in the treatment of premenstrual syndrome: a double-blind, placebo-controlled trial, *Arch. Gen. Psychiatry*, 50, 467, 1993.

49. Smith, S. et al., Treatment of premenstrual syndrome with alprazolam: results of a double-blind, placebo-controlled, randomized crossover clinical trial, *Obstet. Gynecol.*, 70, 37, 1987.
50. Freeman, E.W. et al., A double-blind trial of oral progesterone, alprazolam, and placebo in treatment of severe premenstrual syndrome, *JAMA*, 274, 51, 1995.
51. Le Melledo, J.M. et al., Response to flumazenil in women with premenstrual dysphoric disorder, *Am. J. Psychiatry*, 157, 821, 2000.
52. Nutt, D.J. et al., Flumazenil provocation of panic attacks: evidence for altered benzodiazepine receptor sensitivity in panic disorder, *Arch. Gen. Psychiatry*, 47, 917, 1990.
53. Baulieu, E.E., Neurosteroids: a new function in the brain, *Biol. Cell*, 71, 3, 1991.
54. Stoffel-Wagner, B., Neurosteroid metabolism in the human brain, *Eur. J. Endocrinol.*, 145, 669, 2001.
55. Genazzani, A.R. et al., Circulating levels of allopregnanolone in humans: gender, age, and endocrine influences, *J. Clin. Endocrinol. Metab.*, 83, 2099, 1998.
56. Bäckström, T. et al., The human corpus luteum secretes 5 α-pregnane-3,20-dione, *Acta Endocrinol. (Copenhagen)*, 111, 116, 1986.
57. Paul, S.M. and Purdy, R.H., Neuroactive steroids, *FASEB J.*, 6, 2311, 1992.
58. Wang, M. et al., Relationship between symptom severity and steroid variation in women with premenstrual syndrome: study on serum pregnenolone, pregnenolone sulfate, 5α-pregnane-3,20-dione and 3α-hydroxy-5α-pregnan-20-one, *J. Clin. Endocrinol. Metab.*, 81, 1076, 1996.
59. Bixo, M. et al., Progesterone, 5α-pregnane-3,20-dione and 3α-hydroxy-5α-pregnane-20-one in specific regions of the human female brain in different endocrine states, *Brain Res.*, 764, 173, 1997.
60. Chan, A.F. et al., Persistence of premenstrual syndrome during low-dose administration of the progesterone antagonist RU 486, *Obstet. Gynecol.*, 84, 1001, 1994.
61. Schmidt, P.J. et al., Lack of effect of induced menses on symptoms in women with premenstrual syndrome, *N. Engl. J. Med.*, 324, 1174, 1991.
62. Sundström, I. and Bäckström, T., Patients with premenstrual syndrome have decreased saccadic eye velocity compared to control subjects, *Biol. Psychiatry*, 44, 755, 1998.
63. Schmidt, P.J. et al., Circulating levels of anxiolytic steroids in the luteal phase in women with premenstrual syndrome and in control subjects, *J. Clin. Endocrinol. Metab.*, 79, 1256, 1994.
64. Bicikova, M. et al., Allopregnanolone in women with premenstrual syndrome, *Horm. Metab. Res.*, 30, 227, 1998.
65. Rapkin, A.J. et al., Progesterone metabolite allopregnanolone in women with premenstrual syndrome, *Obstet. Gynecol.*, 90, 709, 1997.
66. Monteleone, P. et al., Allopregnanolone concentrations and premenstrual syndrome, *Eur. J. Endocrinol.*, 142, 269, 2000.
67. Girdler, S.S. et al., Allopregnanolone levels and reactivity to mental stress in premenstrual dysphoric disorder, *Biol. Psychiatry*, 49, 788, 2001.
68. Bitran, D. et al., Activation of peripheral mitochondrial benzodiazepine receptors in the hippocampus stimulates allopregnanolone synthesis and produces anxiolytic-like effects in the rat, *Psychopharmacology*, 151, 64, 2000.
69. Daly, R.C. et al., Effects of gonadal steroids on peripheral benzodiazepine receptor density in women with PMS and controls, *Psychoneuroendocrinology*, 26, 539, 2001.
70. Schmidt, P.J. et al., Differential behavioral effects of gonadal steroids in women with and in those without premenstrual syndrome, *N. Engl. J. Med.*, 338, 209, 1998.

71. Bäckström, T. et al., The role of hormones and hormonal treatments in premenstrual syndrome, *Drugs, CNS*, 17, 325, 2003.
72. Bitran, D., Purdy, R.H., and Kellogg, C.K., Anxiolytic effect of progesterone is associated with increases in cortical allopregnanolone and GABAA receptor function, *Pharmacol. Biochem. Behav.*, 45, 423, 1993.
73. Akwa, Y. et al., The amygdala mediates the anxiolytic-like effect of the neurosteroid allopregnanolone in rat, *Behav. Brain. Res.*, 106, 119, 1999.
74. Wieland, S. et al., Anxiolytic activity of the progesterone metabolite 5 alpha-pregnan-3 alpha-o1-20-one, *Brain Res.*, 565, 263, 1991.
75. Gulinello, M. et al., Short-term exposure to a neuroactive steroid increases α4 GABA(A) receptor subunit levels in association with increased anxiety in the female rat, *Brain Res.*, 910, 55, 2001.
76. de Lignieres, B., Dennerstein, L. and Bäckström, T., Influence of route of administration on progesterone metabolism, *Maturitas*, 21, 251, 1995.
77. Vanselow, W. et al., Effect of progesterone and its 5 alpha and 5 beta metabolites on symptoms of premenstrual syndrome according to route of administration, *J. Psychosom. Obstet. Gynaecol.*, 17, 29, 1996.
78. Freeman, E.W. et al., Anxiolytic metabolites of progesterone: correlation with mood and performance measures following oral progesterone administration to healthy female volunteers, *Neuroendocrinology*, 58, 478, 1993.
79. Uzunova, V. et al., Increase in the cerebrospinal fluid content of neurosteroids in patients with unipolar major depression who are receiving fluoxetine or fluvoxamine, *Proc. Natl. Acad. Sci. U.S.A.*, 95, 3239, 1998.
80. Romeo, E. et al., Effects of antidepressant treatment on neuroactive steroids in major depression, *Am. J. Psychiatry*, 155, 910, 1998.
81. Freeman, E.W. et al., Allopregnanolone levels and symptom improvement in severe premenstrual syndrome, *J. Clin. Psychopharmacol.*, 22, 516, 2002.
82. Gallo, M.A. and Smith, S.S., Progesterone withdrawal decreases latency to and increases duration of electrified prod burial: a possible rat model of PMS anxiety, *Pharmacol. Biochem. Behav.*, 46, 897, 1993.
83. Smith, S.S. et al., Withdrawal from 3α-OH-5α-pregnan-20-one using a pseudopregnancy model alters the kinetics of hippocampal $GABA_A$-gated current and increases the $GABA_A$ receptor α4 subunit in association with increased anxiety, *J. Neurosci.*, 18, 5275, 1998.
84. Smith, S.S. et al., GABA(A) receptor alpha4 subunit suppression prevents withdrawal properties of an endogenous steroid, *Nature*, 392, 926, 1998.
85. Moran, M.H., Goldberg, M., and Smith, S.S., Progesterone withdrawal. II. Insensitivity to the sedative effects of a benzodiazepine, *Brain Res.*, 807, 91, 1998.
86. Sundström, I. et al., Patients with premenstrual syndrome have a different sensitivity to a neuroactive steroid during the menstrual cycle compared to control subjects, *Neuroendocrinology*, 67, 126, 1998.
87. Uzunov, D.P. et al., Fluoxetine-elicited changes in brain neurosteroid content measured by negative ion mass fragmentography, *Proc. Natl. Acad. Sci. U.S.A.*, 93, 12599, 1996.
88. Griffin, L.D. and Mellon, S.H., Selective serotonin reuptake inhibitors directly alter activity of neurosteroidogenic enzymes, *Proc. Natl. Acad. Sci. U.S.A.*, 96, 13512, 1999.
89. Sundström, I. and Bäckström, T., Citalopram increases pregnanolone sensitivity in patients with premenstrual syndrome: an open trial, *Psychoneuroendocrinology*, 23, 73, 1998.
90. Rasgon, N. et al., Neuroactive steroid-serotonergic interaction: responses to an intravenous L-tryptophan challenge in women with premenstrual syndrome, *Eur. J. Endocrinol.*, 145, 25, 2001.

14 Role of Neurosteroids in Epilepsy

Zakaria Mtchedlishvili and Jaideep Kapur

CONTENTS

14.1 INTRODUCTION AND DEFINITIONS

Epilepsies are a group of disorders characterized by recurrent, unprovoked seizures. Approximately 0.5 to 2% of the population suffers from epilepsy. An epileptic seizure is a behavioral manifestation of a paroxysmal, excessive, and synchronous discharge of neurons. Partial seizures are generated by a paroxysmal discharge that begins in a group of cortical neurons confined to one cerebral hemisphere. Generalized seizures begin with the simultaneous activation of both cerebral hemispheres, typically involving the thalamus.

The role of neurosteroids in epilepsy is best understood in terms of their ability to modulate seizure frequency and mediate the effects of endogenous biological rhythms. One of the best-characterized endogenous rhythms that can modulate seizure frequency is the menstrual cycle. There is growing evidence that neurosteroids and γ-aminobutyric acid type A ($GABA_A$) receptors play an important role in mediating the effects of changing sex hormone levels during the menstrual cycle. In addition to modulating seizures, natural and synthetic neurosteroids can be used to treat epileptic seizures via their anticonvulsant actions on $GABA_A$ receptors. Some

0-8493-2392-4/04/$0.00+$1.50

studies suggest that unlike other $GABA_A$ receptor agonists, tolerance to the anticonvulsant effect of neurosteroids does not develop.

14.2 CYCLICAL EXACERBATION OF SEIZURES, THE MENSTRUAL CYCLE, AND $GABA_A$ RECEPTORS: A HISTORICAL PERSPECTIVE

Cyclical exacerbation of seizures in patients with epilepsy has been recognized since antiquity. Phases of the moon were believed to cause exacerbation of seizures, and hence patients were referred to as lunatic. This theory was dominant in the nineteenth century, and the noted neurologist Romberg believed that the phases of new and full moons were especially influential in epilepsy.

Sir Charles Locock, an obstetrician to Queen Victoria, suggested a very different view of the cyclical pattern of epileptic seizures in short remarks in response to a study presented at the Royal Medical and Chirurgical Society on May 11, 1857 (Locock, 1857). His remarks were reported as follows:

> There was also a form of hysterical epilepsy connected with the menstrual cycle and as periodic as that function. This form of disease was very difficult to treat. The attacks only occurred during the catamenial period, except under otherwise strong exciting causes. He had been baffled in every way in the treatment of this affliction. Some years since, however, he had read in the British and Foreign Review, an account of some experiments performed by a German on himself with bromide of potassium. Dr. Locock determined to try this remedy in cases of hysteria in young women unaccompanied by epilepsy. He found it of greatest service. In a case of hysterical epilepsy that had occurred every month for nine years, and had resisted every kind of treatment he had administered the bromide of potassium. This patient had no epilepsy since the commencement of the use of bromide of potassium. Out of fourteen or fifteen cases treated by this medicine, only one had remained uncured.

With a few words, Locock laid the foundation of the modern treatment era of epilepsy. He related the periodicity of seizures to the menstrual cycle rather than the phases of the moon. He was the first to suggest that bromides were an effective treatment of seizures. Following his lead, in 1861, Hutchinson and Hughlings Jackson reported the successful treatment of 12 cases of epilepsy with bromides. By the mid-1870s, over 2.5 tons of bromides were consumed at the National Hospital, London (Max, 1990). Bromides remained the only effective treatment of seizures until 1912, when phenobarbital became available. Bromides are rarely used now and largely remain in the English lexicon as referring to a sedative, for example, "take a bromide" or a soothing statement, "The old bromide that poetry never sells is once again proved to be wrong" (*Oxford English Dictionary*).

In retrospect, Locock's remarks provide initial clues that periodicity of seizures in epilepsy and the menstrual cycle may be linked via $GABA_A$ receptors, because it appears that bromides exert their anticonvulsant effects by permeating these receptors. Similar to many drugs known to act on $GABA_A$ receptors, bromides are sedatives. Bromides block seizures induced by the $GABA_A$ receptor antagonists pentylenetetrazol and picrotoxin. Bromides enhance the binding affinity of flunitrazepam and diazepam to $GABA_A$ receptors (Max, 1990). The bromide ion, like the chloride ion, is a monovalent anion that belongs to the halogen family. Bromide ions permeate the $GABA_A$ receptors more efficiently than chloride ions (Bormann et al., 1987). A recent detailed study demonstrated that bromides block seizures in the combined entorhinal cortex–hippocampal slice preparation and enhance paired pulse inhibition and GABA-mediated inhibitory postsynaptic currents (Meierkord et al., 2000).

14.3 CATAMENIAL EPILEPSY: ROLE OF PROGESTERONE WITHDRAWAL

Since the original report, several investigators over the past one and a half centuries have noted that seizure frequency increases with menstruation (Newmark and Penry, 1980). Early studies were somewhat limited by incomplete descriptions of seizure types, seizure frequency, lack of a definition of the catamenial phase, and lack of quantitative criteria for defining exacerbation. Recently, a series of reports from Herzog et al. (1997, 1999) have accounted for many of these confounders and have proposed that there exist three distinct patterns of catamenial epilepsy in women with partial seizures. Exacerbation was defined as a twofold increase in seizure frequency over baseline. Most commonly, the exacerbation occurs during the 3 days prior to menstruation and the first 3 days of menstruation (Klein and Herzog, 1998). The second pattern of exacerbation occurs in the preovulatory phase, and the third pattern consists of a luteal phase exacerbation.

During the menstrual cycle, progesterone levels peak in the midluteal phase, remain high for 3 days, and then decline to low levels prior to menstruation. Progesterone has a strong anticonvulsant action in humans and experimental animals, and there is growing evidence that withdrawal from progesterone during the premenstrual phase leads to exacerbation of seizures. Progesterone therapy attenuates catamenial exacerbation of seizures (Herzog, 1986, 1995). In women with partial epilepsy, intravenously administered progesterone reduces the frequency of epileptiform discharges recorded on EEG (Backstrom et al., 1984). In experimental animals, progesterone has been recognized as a powerful anticonvulsant since 1942 when Selye (1942) demonstrated its effects on metrazol-induced seizures.

A series of studies suggest that the anticonvulsant effect of progesterone is mediated largely by its conversion to its tertahydro-derivative (3α-hydroxy-5α-pregnane-20-one (3α,5α-THP), or allopregnanolone) in the brain. When injected peripherally, progesterone and allopregnanolone have similar efficacies as anticonvulsants. However, when the conversion of progesterone to allopregnanolone is blocked by inhibiting the enzyme 5α-reductase, the anticonvulsant activity of progesterone is lost (Frye et al., 1998; Kokate et al., 1999a). Furthermore, pentylenetetrazol (PTZ)-induced seizures are blocked by direct infusion of allopregnanolone into the pontine reticular formation, but not by progesterone infusion (Frye et al., 2000). Withdrawal from neurosteroids increases seizure susceptibility (Reddy and Rogawski, 2000b; Reddy et al., 2001). In naïve rats, allopregnanolone blocks seizures induced by cocaine, N-methyl D-aspartate (NMDA), pilocarpine, and kainic acid in a dose-dependent fashion (Kokate et al., 1996; Gasior et al., 1997).

Anticonvulsant action of allopregnanolone is likely mediated by its action on $GABA_A$ receptors. The 5α-reduced neurosteroids like allopregnanolone act as positive allosteric modulators of $GABA_A$ receptors. In physiological concentrations (10 to 30 n*M*), allopregnanolone enhances $GABA_A$ receptors (Majewska, 1992) by allosteric modulation. Higher concentrations of allopregnanolone (0.1 to 1 μ*M*) directly open $GABA_A$ receptor-operated chloride channels. Allopregnanolone reduces the binding of [^{35}S] t-butylbicyclophosphorothionate (TBPS), a ligand for the picrotoxin binding site of $GABA_A$ receptors, to fresh unwashed cortical membrane preparations in a noncompetitive fashion.

$GABA_A$ receptors are formed by five subunits (pentamers) (Macdonald and Olsen, 1994). $GABA_A$ receptor subunits are derived from seven different gene families: α, β, γ, δ, ε, π, and θ that have multiple members, α1–6, β1–4, γ1–3, and ρ1–2. A majority of $GABA_A$ receptors are proposed to be composed of two α, two β, and one γ or δ, ε, or π subunit. The precise location of allopregnanolone binding sites for allosteric modulation, direct activation, and open-channel block of the $GABA_A$ receptor are discussed in other chapters. The mechanisms of allopregnanolone modulation of $GABA_A$ receptors are also discussed in other chapters.

14.4 ANIMAL MODELS OF PROGESTERONE WITHDRAWAL

When animals are treated with progesterone for a prolonged period of time and are subjected to abrupt withdrawal, they demonstrate increased sensitivity to convulsants. Current animal models achieve persistent elevations of progesterone levels in female rats by administering gonadotrophin. In one model, withdrawal of progesterone is obtained by ovariectomy (Moran and Smith, 1998). In this model, there was increased seizure susceptibility observed 24 h after progesterone withdrawal, increased sensitivity to the convulsant effects of β carbolines and picrotoxin, and diminished sensitivity to the anxiolytic properties of benzodiazepines. In another model, after making rats pseudopregnant with gonadotrophins, the conversion of progesterone to allopregnanolone was blocked by finasteride (Reddy et al., 2001). This resulted in decreasing the latency, frequency, and intensity of seizures induced by PTZ in experimental animals (Reddy et al., 2001). In these models of progesterone withdrawal, the seizure threshold was tested by administration of high doses of convulsants like PTZ. While these drugs are commonly used to evaluate seizure threshold and the efficacy of anticonvulsants, their relevance to chronic epilepsy is uncertain. It is also unclear whether these acute drugs model partial or generalized seizures.

In contrast to the PTZ model, stimulation of the perforant pathway leads to focal limbic seizures resembling complex partial seizures (Mazarati et al., 1998; Michalakis et al., 1998). Similar to perforant path stimulation, administration of kainic acid also leads to limbic seizures resembling complex partial seizures (Lothman and Collins, 1981). These models of partial seizures are particularly relevant to progesterone withdrawal studies because cyclical exacerbation of seizure frequency is best documented in patients with complex partial seizures. It is particularly significant that in these models, progesterone and allopregnanolone had antiseizure effects. Withdrawal from progesterone and allopregnanolone lowered the threshold for kainic acid-induced seizures (Frye and Bayon, 1999). Increased sensitivity to seizures following progesterone and allopregnanolone withdrawal also occurred in a mouse alcohol withdrawal model of seizure sensitivity (Reilly et al., 2000).

14.5 MECHANISMS OF PROGESTERONE WITHDRAWAL-INDUCED HYPEREXCITABILITY

There are many similarities between progesterone withdrawal syndromes and withdrawal from GABA modulators such as benzodiazepines, alcohol, and barbiturates (Moran and Smith, 1998). In addition to increased sensitivity to seizures, this results in increased anxiety (Smith et al., 1998a) and diminished sensitivity to benzodiazepines (Smith et al., 1998b).

The precise mechanisms of neurosteroid withdrawal are discussed in detail by Smith in this book; however, for the purposes of this chapter, it is important to note that progesterone withdrawal is associated with increased expression of the α4 subunit of $GABA_A$ receptors in hippocampal neurons (Smith et al., 1998a, 1998b). It is also important to note that the α4 subunit has some unique properties: it renders $GABA_A$ receptors diazepam insensitive; it commonly co-assembles with the δ subunit (Sur et al., 1999); and it is expressed in the extrasynaptic membrane in hippocampal dentate granule cells (Sun et al., 2002). Interestingly, expression of the α4 subunit or diazepam-insensitive receptors is prominent in many other seizure-prone states. For example, there is increased expression of α4 subunit mRNA in dentate granule cells of rats rendered chronically epileptic by the administration of pilocarpine (Brooks-Kayal et al., 1998). Prolonged seizures of status epilepticus render the $GABA_A$ receptors on dentate granule cells diazepam insensitive (Kapur and Macdonald, 1997). The immature brain is much more susceptible to seizures than the adult brain, and in neonatal rats, dentate granule cells express diazepam-insensitive $GABA_A$ receptors (Kapur and Macdonald, 1999). Chronic intermittent alcohol treatment is also associated with increased expression of diazepam-insensitive receptors and the α4 subunit (Mhatre et al., 1988; Mhatre and Ticku, 1989; Mahmoudi et al., 1997). While there is clear association of the α4 (and δ) subunits with hyperexcitable states, the mechanism by which these subunits increase seizure susceptibility remains unclear.

Interestingly, there is no tolerance to the anticonvulsant effects of allopregnanolone in animal models of catamenial epilepsy (Reddy and Rogawski, 2000a). Studies indicate that the anticonvulsant activity of ganaxolone, a synthetic neuro-steroid, is enhanced after neurosteroid withdrawal (Reddy and Rogawski, 2000b). While the authors did not determine the cause of enhanced neurosteroid sensitivity in these animals, Smith's (Smith et al., 1998a, 1998b) studies do provide a potential mechanism for this phenomenon. It is suggested that progesterone withdrawal leads to increased expression of the α4 subunit, which commonly co-assembles with the δ subunit. Recent studies indicate that the δ subunit increases the neurosteroid sensitivity of the $GABA_A$ receptor (Mihalek et al., 1999; Wohlfarth et al., 2002). Thus, if progesterone withdrawal were to increase the expression of the δ subunit along with the α4 subunit, increased sensitivity to neurosteroids would result.

14.6 TEMPORAL LOBE EPILEPSY

The majority of patients with medial temporal lobe epilepsy have hippocampal sclerosis as the predominant pathology in their surgical specimens; removal of sclerotic hippocampus cures epilepsy in over 70% of the patients, and during intracranial recordings, seizures originate in the sclerotic hippocampus or closely associated structures (Sloviter, 1994b). For these reasons, mechanisms of epileptogenesis are extensively studied in the hippocampus. The hippocampus is a three-layered cortex consisting of the cornu ammonis region 1 (CA1), cornu ammonis region 3 (CA3), and dentate gyrus. Pyramidal neurons are the principal neurons in the CA1 and CA3 regions, while granule cells are the primary neurons of the dentate gyrus. A trisynaptic circuit links the three major structures of the hippocampus. The dentate granule cells receive their input from the entorhinal cortex in form of the perforant path and in turn project to CA3 pyramidal neurons via mossy fibers. The CA3 pyramidal neurons project through the fimbria and fornix but also send a large collateral projection (Schaffer collaterals) to the CA1 pyramidal neurons. Each region has a local circuit in which inhibitory and excitatory interneurons participate and mediate feed-forward and feed-back inhibition of the principal neuron.

Hippocampal sclerosis refers to pathological changes in the brains of patients with medial temporal lobe epilepsy. There is extensive loss of CA1 and CA3 pyramidal neurons and loss of hilar mossy cells. There is also loss of CA2 pyramidal neurons and dentate granule cells, but substantial numbers of the latter are preserved (Sloviter, 1994a).

14.7 ALTERED $GABA_A$ RECEPTOR FUNCTION IN TEMPORAL LOBE EPILEPSY

Recently, it has been hypothesized that altered inhibition in temporal lobe epilepsy is due to alteration in the molecular structure of $GABA_A$ receptors in the hippocampus. As discussed in the previous chapter, the $GABA_A$ receptor has been cloned and appears to be composed of five subunits belonging to seven families of polypeptides designated α, β, γ, δ, ρ, σ, and ε. The properties of $GABA_A$ receptors depend upon their subunit composition (Macdonald and Olsen, 1994). Early studies on animal models of temporal lobe epilepsy demonstrated increased benzodiazepine binding in the dentate gyrus of kindled rats (Valdes et al., 1982; Shin et al., 1985). More recent studies showed that there was an increased number of $GABA_A$ receptors in dentate granule cells in kindled rats (Czeizel et al., 1990) and that the amplitude of $GABA_A$ receptor-mediated miniature inhibitory postsynaptic currents was increased (Otis et al., 1994).

Zinc, a divalent cation, is released by the mossy fibers, the axon terminals of granule cells, and has long been known to inhibit $GABA_A$ receptors (Smart et al., 1991; Legendre and Westbrook, 1991; Smart, 1992). As discussed in the section above, in epileptic rats and humans, these mossy fibers may aberrantly innervate dentate granule cells and release Zn^{2+}. While $GABA_A$ receptors present on dentate granule cells of normal adult rats are not sensitive to Zn^{2+}, $GABA_A$ receptors present on dentate granule cells of epileptic rats and humans are Zn^{2+} sensitive (Buhl et al., 1996; Gibbs et al., 1997). This change in the properties of dentate granule cell $GABA_A$ receptors is associated with changes in the pattern of $GABA_A$ receptor subunit mRNA expressed in dentate granule cells. Thus, epileptogenesis may alter the subunit structure of $GABA_A$ receptors present in the hippocampus (Brooks-Kayal et al., 1998).

In addition to Zn^{2+}, the $GABA_A$ receptor is modulated by neuroactive steroids, such as allopregnanolone and pregnenolone sulfate (PS) (Macdonald and Olsen, 1994). PS is a natural sulfated derivative of pregnenolone. In contrast to allopregnanolone, PS potently inhibits $GABA_A$ receptors (Baulieu, 1998). The potential physiological significance of PS was demonstrated in studies that assessed the spatial memory performance of cognitively impaired, aged rats. In these studies, cognitive impairment was correlated by bilateral intrahippocampal injection of PS (El-Etr et al., 1998).

14.8 LOSS OF ALLOPREGNANOLONE SENSITIVITY MAY INCREASE SUSCEPTIBILITY TO SEIZURES

Allopregnanolone enhancement of $GABA_A$ receptor currents in control and epileptic dentate granule cells has been investigated recently (Mtchedlishvili et al., 2001). Allopregnanolone (10 n*M*) enhancement of 10 μ*M* GABA-evoked currents in epileptic dentate granule cells was significantly diminished compared to that in control dentate granule cells. In these studies, 10 μ*M* GABA was 0.5 log units less than the EC_{50} in control and epileptic DGCs, thereby allowing clear discrimination of peak current enhancement. In epileptic DGCs, there was a significant right shift of the allopregnanolone concentration response curve. The potency (EC_{50}) of allopregnanolone enhancement of GABA-evoked currents was markedly diminished in epileptic DGCs (92.7 ± 13 n*M*), compared to that in control DGCs (12.9 ± 2 n*M*). Allopregnanolone efficacy was similar in control and epileptic DGCs; the maximal augmentation of GABA current by allopregnanolone was 96% ± 13.4 in epileptic dentate granule cells and 106.5% ± 10.8 in control dentate granule cells.

14.9 PREGNENOLONE SULFATE INHIBITION OF $GABA_A$ RECEPTOR CURRENTS

Pregnenolone sulfate is a physiologically active sulfated derivative of pregnenolone. Endogenous pregnenolone sulfate in the hippocampus is believed to enhance memory, and age-related decline in memory function of rats correlates well with decline in hippocampal pregnenolone sulfate levels. A number of studies have shown that the physiological action of pregnenolone sulfate is reciprocal to that of allopregnanolone. Pregnenolone sulfate has convulsant properties (Kokate et al., 1999b), inhibits $GABA_A$ receptor currents (Baulieu, 1998), and enhances NMDA currents (Wu et al., 1991).

Pregnenolone sulfate modulation of $GABA_A$ receptor currents was compared in epileptic and control dentate granule cells. In epileptic dentate granule cells, pregnenolone sulfate (30 μ*M*) inhibited 30 μ*M* GABA-evoked currents to 24.4% ± 5 of that evoked by 30 μ*M* GABA alone. In control dentate granule cells, peak current amplitude evoked by co-application of 30 μ*M* pregnenolone sulfate and 30 μ*M* GABA was 27.4% ± 4 of that evoked by 30 μ*M* GABA alone. There was no difference in inhibition of GABA current by 30 μ*M* pregnenolone sulfate in epileptic and control dentate granule cells ($p = 0.64$, unpaired *t*-test). The pregnenolone sulfate concentration (300 n*M* to 300 μ*M*)–$GABA_A$ receptor current inhibition relationship was studied further in epileptic and control dentate granule cells. The potency and efficacy of pregnenolone sulfate inhibition of $GABA_A$ receptor currents were also similar in epileptic and control dentate granule cells. In epileptic dentate granule cells, the IC_{50} for pregnenolone sulfate was 17.4 ± 2 μ*M*, and in control dentate granule cells it was 13.1 ± 1 μ*M*. The maximal inhibition of GABA currents by pregnenolone sulfate was similar in epileptic DGCs (7.2% ± 0.7) and control dentate granule cells (7.6% ± 1.9). Therefore, neither the potency nor efficacy of pregnenolone sulfate inhibition of $GABA_A$ receptor-mediated currents was altered in dentate granule cells in chronic temporal lobe epilepsy.

This study demonstrated that $GABA_A$ receptor currents were less sensitive to allopregnanolone enhancement in epileptic DGCs than control DGCs. Thus, allopregnanolone, an endogenous anticonvulsant, is likely to be less effective in epileptic hippocampus. This study also found that the potency and efficacy of pregnenolone sulfate remained unchanged. Sustained inhibition of GABA currents by pregnenolone sulfate, in parallel with its ability to enhance NMDA-mediated currents, could shift an excitatory–inhibitory balance toward increased excitation. When augmentation of $GABA_A$ receptor-mediated currents by allopregnanolone is diminished in epileptic DGCs, this increased excitation can trigger a seizure. Neurosteroid modulation of $GABA_A$ receptors may be of particular relevance in the catamenial exacerbation of seizures. It was suggested that the catamenial exacerbation of seizures was due to rapid withdrawal from allopregnanolone, which was the result of the rapid decline of serum levels of progesterone prior to menses (see above). The findings of this study suggest that in addition to changes in sex hormone concentrations during the menstrual cycle, loss of allopregnanolone enhancement of $GABA_A$ receptor function may contribute to the catamenial exacerbation of seizures.

REFERENCES

Backstrom, T., Zetterlund, B., Blom, S., and Romano, M., (1984), Effects of intravenous progesterone infusions on the epileptic discharge frequency in women with partial epilepsy, *Acta Neurol. Scand.*, 69, 240–248.

Baulieu, E.E., (1998), Neurosteroids: a novel function of the brain, *Psychoneuroendocrinology*, 23, 963–987.

Bormann, J., Hamill, O.P., and Sakmann, B., (1987), Mechanism of anion permeation through channels gated by glycine and gamma-aminobutyric acid in mouse cultured spinal neurones, *J. Physiol.*, 385, 243–286.

Brooks-Kayal, A.R., Shumate, M.D., Jin, H., Rikhter, T.Y., and Coulter, D.A., (1998), Selective changes in single cell GABA(A) receptor subunit expression and function in temporal lobe epilepsy, *Nat. Med.*, 4, 1166–1172.

Buhl, E.H., Otis, T.S., and Mody, I., (1996), Zinc-induced collapse of augmented inhibition by GABA in a temporal lobe epilepsy model, *Science*, 271, 369–373.

Czeizel, A., Sankaranarayanan, K., and Szondy, M., (1990), The load of genetic and partially genetic diseases in man. III. Mental retardation, *Mutat. Res.*, 232, 291–303.

El-Etr, M., Akwa, Y., Robel, P., and Baulieu, E.E., (1998), Opposing effects of different steroid sulfates on GABAA receptor-mediated chloride uptake, *Brain Res.*, 790, 334–338 (published erratum appears in *Brain Res.*, 798, 337–338, 1998).

Frye, C.A., Scalise, T.J., and Bayon, L.E., (1998), Finasteride blocks the reduction in ictal activity produced by exogenous estrous cyclicity, *J. Neuroendocrinol.*, 10, 291–296.

Frye, C.A. and Bayon, L.E., (1999), Cyclic withdrawal from endogenous and exogenous progesterone increases kainic acid and perforant pathway induced seizures, *Pharmacol. Biochem. Behav.*, 62, 315–321.

Frye, C.A., Manjarrez, J., and Camacho-Arroyo, I., (2000), Infusion of 3alpha,5alpha-THP to the pontine reticular formation attenuates PTZ-induced seizures, *Brain Res.*, 881, 98–102.

Gasior, M., Carter, R.B., Goldberg, S.R., and Witkin, J.M., (1997), Anticonvulsant and behavioral effects of neuroactive steroids alone and in conjunction with diazepam, *J. Pharmacol. Exp. Ther.*, 282, 543–553.

Gibbs, J.W., Shumate, M.D., and Coulter, D.A., (1997), Differential epilepsy-associated alterations in postsynaptic $GABA_A$ receptor function in dentate granule and CA1 neurons, *J. Neurophysiol.*, 77, 1924–1938.

Herzog, A.G., (1986), Intermittent progesterone therapy and frequency of complex partial seizures in women with menstrual disorders, *Neurology*, 36, 1607–1610.

Herzog, A.G., (1995), Progesterone therapy in women with complex partial and secondary generalized seizures, *Neurology*, 45, 1660–1662.

Herzog, A.G., Klein, P., and Ransil, B.J., (1997), Three patterns of catamenial epilepsy, *Epilepsia*, 38, 1082–1088.

Herzog, A.G., (1999), Psychoneuroendocrine aspects of temporolimbic epilepsy. Part I. Brain, reproductive steroids, and emotions, *Psychosomatics*, 40, 95–101.

Kapur, J., Lothman, E.W., and DeLorenzo, R.J., (1994), Loss of $GABA_A$ receptors during partial status epilepticus, *Neurology*, 44, 2407–2408.

Kapur, J. and Macdonald, R.L., (1997), Rapid seizure-induced reduction of benzodiazepine and Zn2+ sensitivity of hippocampal dentate granule cell GABAA receptors, *J. Neurosci.*, 17, 7532–7540.

Kapur, J. and Macdonald, R.L., (1999), Postnatal development of hippocampal dentate granule cell gamma-aminobutyric acidA receptor pharmacological properties, *Mol. Pharmacol.*, 55, 444–452.

Klein, P. and Herzog, A.G., (1998), Hormonal effects on epilepsy in women, *Epilepsia*, 39 (Suppl. 8), S9–16.

Kokate, T.G., Cohen, A.L., Karp, E., and Rogawski, M.A., (1996), Neuroactive steroids protect against pilocarpine- and Kainic acid-induced limbic seizures and status epilepticus in mice, *Neuropharmacology*, 35, 1049–1056.

Kokate, T.G., Banks, M.K., Magee, T., Yamaguchi, S., and Rogawski, M.A., (1999a), Finasteride, a 5alpha-reductase inhibitor, blocks the anticonvulsant activity of progesterone in mice, *J. Pharmacol. Exp. Ther.*, 288, 679–684.

Kokate, T.G., Juhng, K.N., Kirkby, R.D., Llamas, J., Yamaguchi, S., and Rogawski, M.A., (1999b), Convulsant actions of the neurosteroid pregnenolone sulfate in mice, *Brain Res.*, 831, 119–124.

Legendre, P. and Westbrook, G.L., (1991), Noncompetitive inhibition of γ-aminobutyric acidA channels by Zn, *Mol. Pharmacol.*, 39, 267–274.

Locock, C., (1857), Discussion of paper by Edward H. Sieveking: analysis of fifty-two cases of epilepsy observed by the author, *Lancet*, 1, 527–528.

Lothman, E.W. and Collins, R.C., (1981), Kainic acid induced limbic seizures: metabolic, behavioral, electroencephalographic and neuropathological correlates, *Brain Res.*, 218, 299–318.

Macdonald, R.L. and Olsen, R.W., (1994), GABAA receptor channels, *Annu. Rev. Neurosci.*, 17, 569–602.

Mahmoudi, M., Kang, M.H., Tillakaratne, N., Tobin, A.J., and Olsen, R.W., (1997), Chronic intermittent ethanol treatment in rats increases GABA(A) receptor alpha4-subunit expression: possible relevance to alcohol dependence, *J. Neurochem.*, 68, 2485–2492.

Majewska, M.D., (1992), Neurosteroids: endogenous bimodal modulators of the GABAA receptor. Mechanism of action and physiological significance, *Prog. Neurobiol.*, 38, 379–395.

Max, B., (1990), This and that: on artificial fats and the first effective treatment for fits, *Trends Pharmacol. Sci.*, 11, 225–229.

Mazarati, A.M., Wasterlain, C.G., Sankar, R., and Shin, D., (1998), Self-sustaining status epilepticus after brief electrical stimulation of the perforant path, *Brain Res.*, 801, 251–253.

Meierkord, H., Grunig, F., Gutschmidt, U., Gutierrez, R., Pfeiffer, M., Draguhn, A., Bruckner, C., and Heinemann, U., (2000), Sodium bromide: effects on different patterns of epileptiform activity, extracellular pH changes and GABAergic inhibition, *Naunyn Schmiedebergs Arch. Pharmacol.*, 361, 25–32.

Mhatre, M., Mehta, A.K., and Ticku, M.K., (1988), Chronic ethanol administration increases the binding of the benzodiazepine inverse agonist and alcohol antagonist [3H]RO15-4513 in rat brain, *Eur. J. Pharmacol.*, 153, 141–145.

Mhatre, M. and Ticku, M.K., (1989), Chronic ethanol treatment selectively increases the binding of inverse agonists for benzodiazepine binding sites in cultured spinal cord neurons, *J. Pharmacol. Exp. Ther.*, 251, 164–168.

Michalakis, M., Holsinger, D., Ikeda-Douglas, C., Cammisuli, S., Ferbinteanu, J., DeSouza, C., DeSouza, S., Fecteau, J., Racine, R.J., and Milgram, N.W., (1998), Development of spontaneous seizures over extended electrical kindling. I. Electrographic, behavioral, and transfer kindling correlates, *Brain Res.*, 793, 197–211.

Mihalek, R.M., Banerjee, P.K., Korpi, E.R., Quinlan, J.J., Firestone, L.L., Mi, Z.P., Lagenaur, C., Tretter, V., Sieghart, W., Anagnostaras, S.G., Sage, J.R., Fanselow, M.S., Guidotti, A., Spigelman, I., Li, Z., DeLorey, T.M., Olsen, R.W., and Homanics, G.E., (1999), Attenuated sensitivity to neuroactive steroids in gamma-aminobutyrate type A receptor delta subunit knockout mice, *Proc. Natl. Acad. Sci. U.S.A.*, 96, 12905–12910.

Moran, M.H. and Smith, S.S., (1998), Progesterone withdrawal I: pro-convulsant effects, *Brain Res.*, 807, 84–90.

Mtchedlishvili, Z., Bertram, E.H., and Kapur, J., (2001), Diminished allopregnanolone enhancement of GABA(A) receptor currents in a rat model of chronic temporal lobe epilepsy, *J. Physiol*, 537 (Pt. 2), 453–465.

Newmark, M.E. and Penry, J.K., (1980), Catamenial epilepsy: a review, *Epilepsia*, 21, 281–300.

Otis, T.S., De Koninck, Y., and Mody, I., (1994), Lasting potentiation of inhibition is associated with an increased number of γ-aminobutyric acid type A receptors activated during miniature inhibitory postsynaptic currents, *Proc. Natl. Acad. Sci. U.S.A.*, 91, 7698–7702.

Reddy, D.S. and Rogawski, M.A., (2000a), Chronic treatment with the neuroactive steroid ganaxolone in the rat induces anticonvulsant tolerance to diazepam but not to itself, *J. Pharmacol. Exp. Ther.*, 295, 1241–1248.

Reddy, D.S. and Rogawski, M.A., (2000b), Enhanced anticonvulsant activity of ganaxolone after neurosteroid withdrawal in a rat model of catamenial epilepsy, *J. Pharmacol. Exp. Ther.*, 294, 909–915.

Reddy, D.S., Kim, H.Y., and Rogawski, M.A., (2001), Neurosteroid withdrawal model of perimenstrual catamenial epilepsy, *Epilepsia*, 42, 328–336.

Reilly, M.T., Crabbe, J.C., Rustay, N.R., and Finn, D.A., (2000), Acute neuroactive steroid withdrawal in withdrawal seizure-prone and withdrawal seizure-resistant mice, *Pharmacol. Biochem. Behav.*, 67, 709–717.

Selye, H., (1942), The antagonism between anesthetic steroid hormones and pentamethylenetetrazol (metrazol), *J. Lab. Clin. Med.*, 27, 1051–1053.

Shin, C., Pedersen, H.B., and McNamara, J.O., (1985), γ-Aminobutyric acid and benzodiazepine receptors in the kindling model of epilepsy: a quantitative radiohistochemical study, *J. Neurosci.*, 5, 2696–2701.

Sloviter, R.S., (1994a), On the relationship between neuropathology and pathophysiology in the epileptic hippocampus of humans and experimental animals, *Hippocampus*, 4, 250–253.

Sloviter, R.S., (1994b), The functional organization of the hippocampal dentate gyrus and its relevance to the pathogenesis of temporal lobe epilepsy, *Ann. Neurol.*, 35, 640–654.

Smart, T.G., Moss, S.J., Xie, X., and Huganir, R.L., (1991), GABAA receptors are differentially sensitive to zinc: dependence on subunit composition, *Br. J. Pharmacol.*, 103, 1837–1839.

Smart, T.G., (1992), A novel modulatory binding site for zinc on the $GABA_A$ receptor complex in cultured rat neurones, *J. Physiol.*, 447, 587–625.

Smith, S.S., Gong, Q.H., Hsu, F.C., Markowitz, R.S., Ffrench-Mullen, J.M., and Li, X., (1998a), GABA(A) receptor alpha4 subunit suppression prevents withdrawal properties of an endogenous steroid, *Nature*, 392, 926–930.

Smith, S.S., Gong, Q.H., Li, X., Moran, M.H., Bitran, D., Frye, C.A., and Hsu, F.C., (1998b), Withdrawal from 3alpha-OH-5alpha-pregnan-20-one using a pseudopregnancy model alters the kinetics of hippocampal GABAA-gated current and increases the GABAA receptor alpha4 subunit in association with increased anxiety, *J. Neurosci.*, 18, 5275–5284.

Sun, C., Sieghart, W., and Kapur, J., (2002), Synaptic and extrasynaptic distribution of γ-aminobutyric acid type A receptor subunits in rat hippocampal dentate granule cells, *Epilepsia*, 43 (Suppl. 7), 145.

Sur, C., Farrar, S.J., Kerby, J., Whiting, P.J., Atack, J.R., and McKernan, R.M., (1999), Preferential coassembly of alpha4 and delta subunits of the gamma-aminobutyric acidA receptor in rat thalamus, *Mol. Pharmacol.*, 56, 110–115.

Valdes, F., Dasheiff, R.M., Birmingham, F., Crutcher, K.A., and McNamara, J.O., (1982), Benzodiazepine receptor increases after repeated seizures: evidence for localization to dentate granule cells, *Proc. Natl. Acad. Sci. U.S.A.*, 79, 193–197.

Wohlfarth, K.M., Bianchi, M.T., and Macdonald, R.L., (2002), Enhanced neurosteroid potentiation of ternary GABA(A) receptors containing the delta subunit, *J. Neurosci.*, 22, 1541–1549.

Wu, F.S., Gibbs, T.T., and Farb, D.H., (1991), Pregnenolone sulfate: a positive allosteric modulator at the N-methyl-D-aspartate receptor, *Mol. Pharmacol.*, 40, 333–336.

15 Animal Models of Anxiety and Stress-Induced Behavior: Effects of Neuroactive Steroids

Deborah A. Finn, Robert H. Purdy, and George F. Koob

CONTENTS

0-8493-2392-4/04/$0.00+$1.50

15.1 INTRODUCTION

Behaviorally, γ-aminobutyric acid (GABA)–agonist neuroactive steroids now are recognized to possess especially potent anesthetic, hypnotic, anxiolytic, and anticonvulsant properties. It has been over 60 years since Hans Selye (1942)[1] reported the sedative-anesthetic activity of the hormones progesterone and deoxycorticosterone, where of 75 steroids tested by i.p. administration in rodents, 5β-pregnanedione was the most active. This led to the introduction of hydroxydione sodium (21-hydroxy-5β-pregnane-3,20-dione sodium succinate) as the first steroidal anesthetic in 1955.[2] However, it was Margarethe Holzbauer and her colleagues, from 1969 to 1985, who isolated and identified pregnenolone, progesterone, allopregnanolone (3α,5α-THP), epiallopregnanolone (3β,5α-THP), allopregnanedione (5α-DHP), 20α-dihydroprogesterone, and allopregnanediol (5α-pregnane-3α,20α-diol) from ovarian venous blood of the rat, and measured the ovarian content and secretion rates of these steroids during proestrus (reviewed in Holzbauer[3]). Their work was seminal, but often is forgotten. Holzbauer and colleagues[4] further demonstrated the *in vivo* secretion of pregnenolone, progesterone, and 3α,5α-THP by the adrenal gland of the rat in quantities similar to those secreted by the ovary in estrus.

It was only within the last two decades that the reported rapid (seconds) onset of action of steroids was found to be mediated via an interaction with $GABA_A$ receptor complexes (GRCs). The initial *in vitro* evidence was provided by electrophysiological studies demonstrating enhancement of GABA-stimulated chloride conductance by the synthetic steroid anesthetic alfaxalone (3α-hydroxy-5α-pregnane-11,20-dione) in rat brain.[5] These findings soon were extended to both the 5α- and 5β-reduced, 3α-hydroxy metabolites of progesterone and deoxycorticosterone. Subsequent studies identified these neuroactive steroids as the most potent GABA–agonist steroid modulators at the GRC.

An understanding of the physiological role of neuroactive steroids is essential for their acceptance as modulators of stress and anxiety. It is apparent that these endogenously occurring steroids can reach levels in the rat brain that are capable of modulating the GRC. Allopregnanolone and allotetrahydrodeoxycorticosterone (alloTHDOC) have been measured in the brain and plasma, where their levels have been shown to fluctuate in response to stress in rats (reviewed in Barbaccia et al.[6]) and during the estrous and menstrual cycles of rats and humans, respectively (reviewed in Paul and Purdy[7]). Brain and plasma levels of 3α,5α-THP temporally follow those of progesterone in the female rat (i.e., peak brain levels are observed in estrus in the rat and dramatically increase during pregnancy). Both 3α,5α-THP and alloTHDOC are detected in the male rat brain, but the levels are low under most circumstances. Exposure of male rats to swim stress resulted in a rapid 4- to 20-fold increase in 3α,5α-THP and alloTHDOC.[8] The brain levels of these 3α-hydroxysteroids measured after swim stress in males (approximately 3 to 10 ng/g or 10 to 30 n*M*) and during the estrous cycle (approximately 4 to 12 ng/g) and pregnancy (≥30 ng/ml or 100 n*M*) in females are therefore within the range of concentrations previously shown to potentiate the *in vitro* action of GABA at the GRC.

15.2 BEHAVIORAL ANIMAL MODELS OF ANXIETY

Rodent models of human generalized anxiety symptoms take advantage of a variety of behaviors within the natural behavioral repertoire of rodents that have been interpreted to be anxiety-like. While these tasks are based on putative parallels between human and rodent symptoms of anxiety, it is not known whether a rodent experiences anxiety per se in the same way as humans. With this in mind, different behavioral principles seem to underlie various animal models of anxiety, such that paradigms can be based on spontaneous exploration, a learned response, conflict behavior, or a combination of these ideas. Most of these tasks utilize approach–avoidance behaviors in rodents that can provide a good indication of a rodent's response to conflict in the natural environment. A number of detailed reviews are available on this topic.[9–14] Here, we will briefly describe tasks that have been well validated in terms of their specificity for anxiety-like behaviors, ability to demonstrate genotypic differences in anxiety-like behaviors, and ability to demonstrate anxiolytic effects with GABAergic neuroactive steroids. The paradigms described below are good choices for testing hypotheses related to the modulatory effect of neuroactive steroids on anxiety-related behavior. However, these tests also can be influenced by factors unrelated to anxiety. It is therefore recommended that any animal model be tested on more than one anxiety test so that different sensory and motor modalities can be examined. Alternatively, antianxiety effects in multiple tasks that measure both increases and decreases in responding can control for motor effects.

We will focus our discussion on animal models relevant to generalized anxiety, rather than those that encompass classical fear conditioning. While the terms *fear* and *anxiety* have been used interchangeably at times to describe similar symptoms, fear has been regarded as being more stimulus specific than anxiety in the clinical setting.

15.2.1 Geller–Seifter and Vogel Conflict Tests

Both of these paradigms are conflict tests that utilize operant responding for a food reward[15] or water reward,[16] with electric shock as the aversive component in both tasks. The number of lever presses for either the food or water reward, in conjunction with a mild shock that is delivered on a fixed-ratio schedule, is the variable of interest. Anxiolytic drugs typically increase the number of lever presses, or the number of shocks that are acceptable, for the food or water reward. Although weeks of training on the lever press task are required, these paradigms have the advantage that once stable performance baselines are established, the same animal can be used repeatedly to examine drug effects by comparing these data with its baseline (i.e., it is its own control). These tasks also have the advantage of an additional internal control, unpunished responding, that gives an indication of nonspecific drug effects. However, both the Geller–Seifter and Vogel paradigms have been used primarily in rats.

15.2.2 Elevated Plus Maze

One of the most well-validated animal models of anxiety is the elevated plus maze task,[17–20] as it is able to detect both anxiolytic and anxiogenic agents in rodents (e.g.,

Lister[18]). This task is based on a rodent's natural avoidance of open, elevated alleys that was described almost 50 years ago.[21] Thus, the test sets up a conflict for the rodent between its tendency to explore a novel environment and its tendency to avoid the aversive properties of an open, elevated environment. While mice and rats perform similarly on the elevated plus maze, strain distribution analyses indicate that there is a genetic contribution to basal levels of anxiety measured on this task (e.g., Trullas and Skolnick[19] and Rodgers and Cole[22]), which can be influenced by light level.[23]

The elevated plus maze apparatus consists of two open and two enclosed horizontal perpendicular arms of the same size and shape that are at right angles to each other. These four arms extend from a central platform (approximately 5 × 5 cm), forming the shape of a plus, and are elevated approximately 50 cm above the floor. The two open arms are simple runways with a minimal lip on the edges that allows the animal to see the edge without falling off the arm. The enclosed arms are closed runways with high walls that can be opaque. Typically, an animal is placed on the central platform (i.e., at the junction of the four arms) and allowed to explore freely for 5 min. During the 5-min test, the number of entries into the open and closed arms, as well as the amount of time spent in the open and closed arms, is measured. Ideally, for an arm entry to be measured, all four paws should be within the arm. If the task is videotaped, additional anxiety-related behaviors can be quantified (see below and Rodgers et al.[11,24] for details). However, the most common reported index of anxiety with this task is the percentage of open-arm entries or open-arm time, which is a proportion of the total number of entries or total time of the test, multiplied by 100. Normally, mice prefer the closed arms of the plus maze. Anxiolytic drugs typically increase the proportion of open-arm entries and the time spent on the open arms,[17,24–30] whereas anxiogenic drugs or treatments decrease the percentage of open-arm entries or time.[18,27]

Comprehensive profiling of behavior also can be achieved by performing detailed videotaped analysis of untreated rodents on the elevated plus maze, yielding a catalog of readily identifiable behaviors that have been referred to as *risk assessment*.[11] Examples include rearing, head-dipping, stretched attend postures (SAPs), closed-arm returns (i.e., doubling back, rather than leaving a closed arm), and several nonexploratory behaviors[24]; protected vs. unprotected forms of these behaviors can be differentiated based on wall contact. This distinction is based on recent evidence that the most important feature producing open-arm avoidance is the absence of thigmotactic cues (i.e., walls), rather than the height of the maze.[27] Thus, the ethological approach provides a multidimensional approach to the analysis of plus maze behavior.[11]

An important consideration for interpretation of elevated plus maze data is that any treatment that affects motor behavior may influence the percentage of open-arm entries or time in the open arms. This is due to the fact that the paradigm requires the animal to locomote for 5 min. However, the total number of entries as well as the number of entries into the closed arms will give an indication of whether the treatment is influencing activity per se, and closed-arm entries are the most validated measure of activity.[10,31] In this case, independent paradigms that specifically analyze locomotor behavior are recommended.[29]

15.2.3 Elevated Zero Maze

This task is a variant of the elevated plus maze and was designed to eliminate the time spent in the center area, which was a potential confound inherent with the plus maze configuration.[32] Briefly, the zero maze contains open and closed arms, but they are configured around the perimeter of an elevated circular runway. Similar to the elevated plus maze, anxiolytic drugs increase open-arm entries and time.

15.2.4 Light–Dark Box

Conceptually similar to the elevated plus maze, the light–dark exploration task is based on the conflict between a rodent's tendency to explore a novel environment and the aversive properties of a brightly lit open field.[33,34] The apparatus is an acrylic box divided by a panel into a large open area (28 cm long × 28 cm wide) and a slightly smaller enclosed area that is painted black (28 cm long × 19 cm wide). The open area is illuminated to 355 lux, whereas the light level in the dark chamber is low (2 lux). Typically, the animal is placed in the center of the light area, facing away from the entrance to the dark chamber, and behavior is recorded for 5 to 10 min. The number of transitions between the light and dark chambers, the time spent in both chambers, as well as the latency to enter the dark chamber can be quantified by photocell array or manually. A chamber crossing is defined as all four paws of the mouse inside the chamber. The number of transitions and the time spent in the light chamber are the most commonly used variables. Anxiolytic drugs increase the number of transitions as well as the amount of time spent in the light chamber, relative to the dark chamber, which are believed to be indicators of the animal's willingness to explore the brightly lit open area.[34–36]

Interpretation of the light–dark box data also can be confounded by effects of treatments on locomotor activity, particularly when the number of transitions is the only variable that is utilized. This appears to be more of a problem with anxiogenic drugs or treatments, as low levels of general activity have been shown to produce false positives.[37] Thus, it is recommended that independent testing for locomotor function be conducted in new classes of drugs in order to detect stimulant or sedative properties that could confound interpretation of the light–dark transition measure.

15.2.5 Canopy Test

This task was developed in order to create a test situation based on the open vs. closed design of the zero maze, producing an approach–avoidance conflict situation in rodents that would elicit a high level of SAPs.[38] SAP is a characteristic behavior in rodents that originally was classified as reflecting an approach–avoidance conflict.[40] More recent studies have suggested that SAP is an integral component of the *risk assessment* repertoire of defensive behaviors in rodents, occurring in response to potential, rather than discrete, threat stimuli (e.g., Blanchard et al.[40]). Taken in conjunction with studies suggesting that risk assessment is an important behavioral indicator of anxiety in rodents,[11,24,40] SAP may be a useful index of anxiety in rodent models.

The mouse canopy SAP apparatus consists of an opaque black Plexiglas circular platform that is 70 cm in diameter and elevated 60 cm above the ground. Centered directly above this black platform at a distance of 4 cm is a transparent red circular platform that is 38 cm in diameter.[38,41] Thus, these concentric circular platforms of different diameters create a covered area and an exposed area, with illumination on the covered area of approximately 13 to 15 lux and on the exposed area of 400 to 450 lux.[38] The rat test apparatus is an enlarged version of the mouse model. A video camera suspended directly above the apparatus allows for behavioral observations to be made from a monitor on the opposite side of the room. Typically, the rodent is placed in the boundary between the open and enclosed areas, facing the covered area, and behavior is scored for the first 5 min of the test. Behavioral measurements include the number of occurrences of SAPs, time spent in the exposed area of the apparatus, and number of line crossings (an index of overall locomotor activity). Anxiolytic drugs decrease SAPs and increase time spent in the exposed area of the apparatus.[38,41]

15.2.6 Mirrored Chamber

The chamber-of-mirrors procedure was first described in 1990 as a measure of anxiety in rodents.[42] This test generates approach–avoidance conflict behavior, which is based on early work[43] indicating that several vertebrate species demonstrate behavioral responses to mirrors such as aggressive threats and approach–withdrawal behavior. The chamber consists of a mirrored cube, 30 cm on a side, which is open on one side and placed inside the center of an opaque Plexiglas box (40 × 40 × 30.5 cm). Thus, a 5-cm corridor completely surrounds the five-sided mirrored chamber. A sixth mirror is placed against the container wall opposite the single open side of the mirrored chamber. With the exception of the initial report in which mice were tested for 30 min,[42] mice are placed in the corridor outside the mirrored chamber. Latency to enter the mirrored chamber, number of entries, time per entry, and total time in the mirrored chamber are recorded for 5 to 10 min.[44–47] In this assay, a mouse avoids entering the mirrored chamber, whereas it will enter readily if the walls of the chamber are not mirrored.[42] However, a recent study in three inbred strains of mice in which the mirrors in the chamber were replaced with white or gray tiles suggested that a brightness or position effect in the chamber also could explain the avoidance behavior in certain genotypes.[47] Nonetheless, the mirrored-chamber procedure has been validated in terms of being responsive to a wide spectrum of anxiolytic agents at doses that do not affect locomotor activity.[42,43,45,46] Recently, a modified version of the mirrored chamber was validated in the mouse to detect both anxiolytic and anxiogenic agents.[48]

15.2.7 Holeboard Activity

The holeboard apparatus provides independent measures of motor activity and exploration in rodents[17,18,31,49] and comprises a square Plexiglas or wooden box measuring 60 × 60 cm for rats or 40 × 40 cm or 26 × 26 cm for mice. Holes are spaced equally in the floor and vary in number from 4 to 16 and in diameter from 2.2 to 6.5 cm. Photocells below the surface of the holes provide measures of the number of head-dips and time spent head-dipping, while photocells within the sides of the box provide measures of locomotor activity and rearing. Typically, behavior within the holeboard apparatus is measured for 5 min. While anxiolytic drug effects on holeboard behavior (i.e., head-dips) did not correspond with measures of anxiety on the elevated plus maze (percent open-arm entries), exploration and locomotor activity in the holeboard were correlated with overall activity in the maze.[17,18,42] Factor analysis for behavioral variables from the elevated plus maze and the holeboard found that the two indices of anxiety (i.e., percent open-arm entries and time) were highly correlated with each other and loaded on one factor, whereas the two measures of directed exploration (i.e., number of head-dips and duration of head-dipping) loaded on a separate factor.[18,31] Nonetheless, anxiolytic drugs can affect holeboard behavior.[17,18,42] Recent work in mice demonstrated that restraint stress and anxiolytic drugs produce opposite effects on head-dipping that are independent of locomotor activity[50] and that strain differences in exploratory behavior differed from those found for locomotor activity.[51]

15.2.8 Open-Field Activity

Open-field activity is the oldest and simplest measure of rodent emotional behavior.[52,53] Briefly, spontaneous exploratory locomotion, proximity to the walls and central arena, and number of fecal boli deposited are quantified in a brightly lit novel open field for 5 to 10 min. Whereas fully automated systems are widely used, the scoring also can be done manually. Thigmotaxis is measured by comparing activity in the center vs. activity in the perimeter of the open field. An animal exhibiting high-perimeter and low-center activity would be interpreted as possessing high levels of anxiety. However, an important distinction is that open-field activity is not a specific measure of anxiety or as specific a measure of anxiolytic drug effects as other paradigms described in this section. Nonetheless, open-field activity does provide a useful measure of normal vs. abnormal exploratory behavior.

15.2.9 Social Interaction

This task is similarly based on the aversion that rodents exhibit to a brightly lit open area.[54,55] Although one factor in this task is whether the environment is familiar, rats typically will spend less time engaged in social behaviors when in an illuminated open field than when they are in a dimly lit environment. Previously isolated rats are placed into a neutral environment, and behaviors such as sniffing, following, and grooming each other are scored for 5 to 10 min. Low levels of illumination, familiar environment, and anxiolytic drugs increase the number of social interactions.[54–57]

An advantage of the social interaction task is that it shares the ethological relevance to the natural conflicts experienced by rodents, as do the elevated plus maze, elevated zero maze, and light–dark box tasks. However, the social interaction results also can be confounded by treatment effects on locomotor activity. While social interaction has the advantage of providing a new set of behavioral substrates related to interactions with a conspecific, this paradigm has been reported to be less useful for mice,[14] perhaps due to the higher levels of novel environment exploration in mice vs. rats.[58,59]

15.2.10 Defensive Burying

This task measures the amount of time spent burying an aversive object, such as a small metal object that had previously delivered an electric shock.[60] Typically, rodents will cover the shock source with cage litter. Anxiolytic drugs suppress conditioned defensive burying as measured by a reduction in the mean duration of burying behavior.[61,62] While this paradigm has the advantage of requiring much less ambulatory locomotor activity than other tasks, it does have the potential confound that treatments producing behavioral sedation may produce false positives on this task (i.e., decreased defensive burying due to sedation). This task also has been well characterized for rats, but not for mice.

15.2.11 Mouse Defense Test Battery

Whereas studies using a visible burrow system have been used to characterize the overall pattern of defensive behaviors of rodents to a potential threat, the creation of test batteries allows for the examination of individually elicited defense behaviors. Thus, the mouse defense test battery was designed to investigate the defensive responses of mice confronted with a natural predator such as a rat (e.g., Griebel et al.[63,64]). Behaviors assessed include flight, avoidance, defensive threat/attack responses, and risk assessment activities. Use of a wide variety of defensive behaviors allowed for the detection of a differential pattern of drug response, with anxiolytic drug action being associated with an alteration in defensive threat/attack responses and risk assessment measures and with anxiogenic drug action being associated with an alteration in flight/avoidance responses (e.g., Griebel et al.[63]).

The test is conducted in an oval runway (0.4 m wide, 0.3 m high, 6.0 m total length), consisting of 2-m straight segments joined by two 0.4-m curved segments and separated by a median wall ($2.0 \times 0.3 \times 0.06$ m). The apparatus is made of black Plexiglas and has floor markings every 20 cm to facilitate distance measurements. Behavioral and activity assessments are made from videotaped recordings of the procedure (see Griebel et al.[63] for details). Briefly, after a 3-min familiarization period in the runway, the mouse is exposed to a handheld conscious or unconscious rat, and reactions to the predator are recorded. Recent work indicates that genetic factors contribute to defensive behaviors in this animal model of anxiety,[64] with some genotypes using flight as the dominant defensive strategy, whereas the defensive responses of other genotypes consisted of flight reactions and risk assessment activities.

15.3 EFFECTS OF NEUROSTEROIDS ON ANXIETY-RELATED BEHAVIORS

The effect of neuroactive steroid administration on rodent behavior in several of the animal models of anxiety described above is summarized in Table 15.1. Some of this information also has been recently reviewed.[80] In general, neuroactive steroids that potentiate the action of GABA are very potent anxiolytic agents in a variety of behavioral animal models of anxiety. For example, $3\alpha,5\alpha$-THP, $3\alpha,5\beta$-THP, alloTHDOC, alfaxalone, and 5α-pregnane-$3\alpha,20\alpha$-diol consistently exhibit an anxiolytic

TABLE 15.1
Examples of Animal Models Used for Evaluation of Anxiolytic Effects of Neuroactive Steroids in Rodents

Animal Model	Species	Neuroactive Steroid	Anxiolytic	References
Geller–Seifter and Vogel conflict test	Male rats	3α,5α-THP, alloTHDOC	+, +	36, 65
	Male rats	Alfaxalone	+	66
	Male rats	3α,5α-THP, 3α,5β-THP	+, +	67
	Male rats	3α,5α-THP, 5α-pregnane-3α,20α-diol	+, +	68
	Ovariectomized female rats	3α,5α-THP	+	69
Elevated plus maze	Female rats	3α,5α-THP, alloTHDOC, 3β,5α-THP	+, +, –	70
	Male rats	3α,5β-THP	+	71
	Male rats	Testosterone propionate	+	72
	Male mice	DHEA, DHEA sulfate	+, +	73
	Male mice	Pregnenolone, pregnenolone sulfate	Anxiogenic, biphasic	74
	Male mice	3α,5α-THP	+	30, 67, 75
	Male mice	3α,5β-THP	+	67, 75
	Male mice	5α-Pregnane-3α,20α-diol	+	68
	Male mice	AlloTHDOC, 5β-pregnane-3,20-dione	+, +	75
	Male mice	5α-Pregnane-3,20-dione, 3β,5α-THP	–, –	75
Light–dark box	Male mice	3α,5α-THP, alloTHDOC	+, +	36, 65, 67
	Male mice	3β,5α-THP	–	65
Mirrored chamber	Male mice	3α,5α-THP, pregnenolone sulfate	+, +	46
	Male mice	DHEA sulfate	Anxiogenic	46
Open-field activity	Male mice	3α,5α-THP, alloTHDOC, 3β,5α-THP	+, +, –	65
Burying behavior	Male rats	3α,5β-THP	+	71
Staircase test	Male mice	AlloTHDOC, alfaxalone	+, +	76, 77
	Male mice	DHEA, DHEA sulfate	+, +	78
Separation-induced ultrasonic vocalization	7-day old male and female mouse pups	3α,5α-THP	+	79

Note: The effect of neuroactive steroid administration on behavior in several animal models of anxiety is summarized. An anxiolytic effect is depicted by the plus sign (+), whereas a lack of effect is depicted by the minus sign (–). The term *anxiogenic* refers to a drug that increased the behavioral indices of anxiety. The designation of *biphasic* for pregnenolone sulfate reflects an anxiolytic effect at low doses and an anxiogenic effect at higher doses.

profile in several anxiety tests. The 3β-stereoisomer of 3α,5α-THP (i.e., 3β,5α-THP) was inactive in the elevated plus maze, light–dark box, and open-field tests, indicating that it had no effect on anxiety-related behavior. This lack of effect of 3β,5α-THP on animal models of anxiety is consistent with reports of its inability to potentiate GABA-gated currents. In contrast, the neuroactive steroids that are noncompetitive antagonists of $GABA_A$ receptors have mixed effects on anxiety-related behavior. Depending on the animal model utilized, both pregnenolone sulfate and dehydroepiandrosterone (DHEA) sulfate were found to possess anxiolytic or anxiogenic properties. In one study, pregnenolone sulfate was found to have a biphasic effect on elevated plus maze behavior, exhibiting anxiolytic effects at low doses and anxiogenic effects at increasing doses. Although the data for DHEA sulfate and pregnenolone sulfate are limited, it is possible that the mixed results reflect confounding effects of these steroids on activity.

Overall, it is apparent that GABA–agonist neurosteroids have anxiolytic properties, yet their role as endogenous modulators of anxiety is less clear. For example, systemic administration of progesterone has anxiolytic effects via its metabolism to 3α,5α-THP because inhibiting the enzymatic conversion of progesterone to 3α,5α-THP blocked the anxiolytic effect of progesterone.[81] While the putative modulatory role of brain- vs. peripheral-derived 3α,5α-THP on anxiety-related behavior is not clear, recent work has examined the effect of ligands that stimulate biosynthesis of 3α,5α-THP in adrenalectomized-gonadectomized (ADX-GDX) and intact animals.[82] These ligands, which activate peripheral benzodiazepine receptors, significantly increased brain levels of 3α,5α-THP to a greater extent in the intact rats than the ADX-GDX rats, yet the increase in brain 3α,5α-THP also was associated with a marked anticonflict effect in the Vogel test. Thus, it is possible that peripheral sources of 3α,5α-THP may have a greater impact in modulating anxiety-related behavior than brain-derived 3α,5α-THP.

15.4 NEUROBIOLOGY OF ANXIETY-RELATED BEHAVIOR AND EFFECT OF NEUROSTEROID MICROINJECTION

Evidence about the neural circuits of anxiety comes from research in animals (i.e., pharmacological, stimulation, and lesion studies) and humans (neuroimaging studies), and it is beyond the scope of the present review to describe all these data. Excellent reviews on this topic are available (e.g., see references 83–85). Two important concepts are worth mentioning, however. First, the various types of anxiety can be broadly divided into different systems, such as defense, behavioral inhibition, and avoidance systems, that work in concert. Second, the neural circuits underlying anxiety are organized at different levels, with each system having the ability to feedback onto those below it (see Figure 15.1). The different levels of organization reflect the different demands for cognitive processing and will be described briefly.

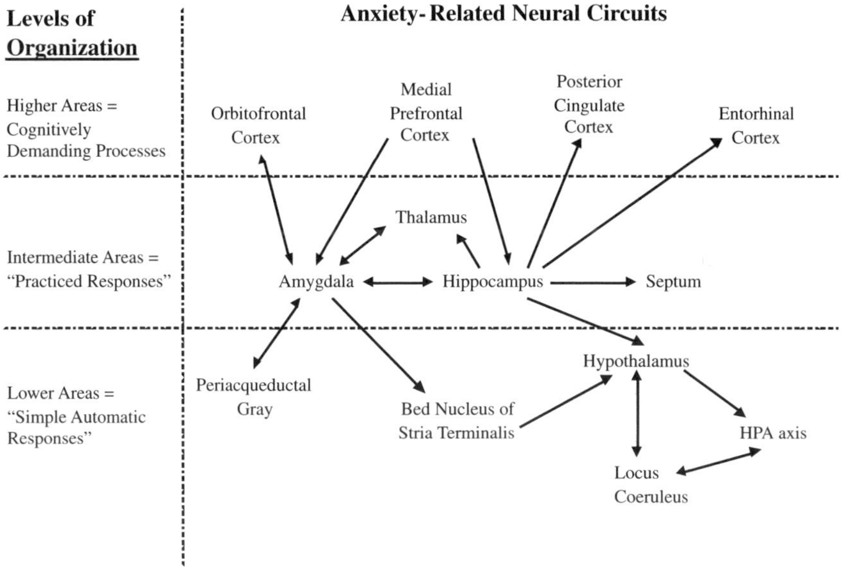

FIGURE 15.1 A simplified organization of the neural circuits underlying anxiety is shown, with each level having the ability to feedback onto those below it. Simple automatic responses are mediated by the periaqueductal gray and locus coeruleus. The periaqueductal gray has been hypothesized to coordinate specific response patterns, based on the nature of the threatening stimuli, via amygdaloid pathways. The locus coeruleus, a critical component of the efferent response systems, is activated by stress and aversive stimuli and has been implicated in the orchestration of the fundamental alarm reaction. The hypothalamus is a critical site for the integration of the autonomic and neuroendocrine responses to threat or anxiety, namely, the activation of the HPA axis and sympathoadrenal system, through the release of peptides and releasing hormones. Intermediate levels that mediate more practiced responses include the amygdala and septohippocampal systems. The amygdala has extensive connections to the cortex and locus coeruleus and projections to the striatum, midbrain, and brainstem, and thus plays a pivotal role in the assessment of, and response to, danger. The role of the hippocampus has been described as a comparator system that can detect whether a threat is familiar or novel, requiring either a conditioned automatic response or higher-order processing, respectively. Higher cortical regions, such as the paralimbic cortex (e.g., orbitofrontal, insular, anterior temporal, and anterior cingulate) and prefrontal cortex manage the more cognitively demanding stimuli. The orbitofrontal cortex is involved in determining the significance of anxiety-producing sensory events, the choice and implementation of behaviors important for survival, and the extinction of conditioned fear responses via its reciprocal interactions with the amygdala. The medial prefrontal cortex also has projections to the amygdala that are involved in the suppression of responses to anxiety-related stimuli, helping to regulate the peripheral responses to stress. (From Finn, D.A. et al., *Neurogenetics*, 4, 109, 2003. With permission.)

Simple automatic responses are mediated by midbrain/pons areas, such as the periaqueductal gray (PAG) and locus coeruleus. The PAG has been hypothesized to coordinate specific response patterns based upon the nature of the threatening stimuli, via amygdaloid pathways. See Graeff et al.[87] for further discussion of the role of the PAG in different anxiety states. The locus coeruleus, a critical component of the efferent response systems, is activated by stress and aversive stimuli[88] and has been implicated in the orchestration of the fundamental alarm reaction.[89,90] The hypothalamus is a critical site for the integration of the autonomic and neuroendocrine responses to threat or anxiety, namely, the activation of the hypothalamic-pituitary-adrenal (HPA) axis and sympathoadrenal system, through the release of peptides and releasing hormones. This is an important common pathway in all stress and anxiety reactions. Thus, the brainstem areas are responsible for the coordination of the physiological response patterns to anxiety, fear, and panic.

Intermediate levels that mediate more behaviorally sophisticated responses involve the amygdala and septohippocampal systems. As depicted in Figure 15.1, the amygdala plays a pivotal role in the assessment of, and response to, danger. The amygdala has extensive connections to the cortex and locus coeruleus and projections to the striatum, hypothalamus midbrain, and brainstem. Thus, it can exert control over locomotor, neuroendocrine, autonomic, and respiratory responses. In addition, the amygdala also is seen as the common pathway and processor of fear (e.g., Davis[91]). Recent work demonstrated that infusion of $3\alpha,5\alpha$-THP (8 μg per side) into the central nucleus of the amygdala of male and ovariectomized female rats produced a significant anxiolytic effect, measured by a significant increase in punished responding in the modified Geller–Seifter conflict test.[92]

The role of the hippocampus has been described as a comparator system that can detect whether a threat is familiar or novel, thus requiring either a conditioned automatic response or higher-order processing, respectively.[93] The septum, posterior cingulate, and thalamic nuclei also have been implicated in this role.[94] Overall, the hippocampus is important in traumatic memory consolidation and, with the entorhinal cortex, in contextual fear-conditioned behaviors. Projections from the hippocampus to the bed nucleus of the stria terminalis (BNST) and from the BNST to hypothalamic and brainstem sites may be involved in the expression of contextual fear conditioning.

Recent findings indicate that microinjection of GABA–agonist neuroactive steroids or agents that increase neurosteroid levels in the hippocampus can produce anxiolytic effects. Specifically, microinjection of 2.5- and 5-μg doses of pregnanolone ($3\alpha,5\beta$-THP) into the dorsal hippocampus of male rats significantly increased the proportion of time spent in the open arms of the elevated plus maze.[95] Similar behavioral effects were seen when the hippocampus was microinjected with agonists of peripheral mitochondrial benzodiazepine receptors, which led to an increase of brain $3\alpha,5\alpha$-THP.[96]

Higher cortical regions, such as the paralimbic cortex (e.g., orbitofrontal, insular, anterior temporal, and anterior cingulate) and prefrontal cortex, manage more cognitively demanding stimuli. The orbitofrontal cortex is involved in determining the significance of anxiety- or fear-producing sensory events, the choice and implementation of behaviors important for survival, and the extinction of conditioned fear responses via its reciprocal interactions with the amygdala. The medial prefrontal cortex also has projections to the amygdala that are involved in the suppression of responses to anxiety-related or fearful stimuli. Overall, the prefrontal cortex, in conjunction with the adjacent orbitofrontal cortex and anterior cingulate, helps regulate the peripheral responses to stress. Thus, these higher cortical regions provide a link between the sensory association, premotor and executive areas, and the limbic system.

15.5 NEUROSTEROID MODULATORS OF STRESS-INDUCED BEHAVIOR

We will briefly review GABAergic-related stress paradigms involving neuroactive steroids, as well as the HPA axis involving both glucocorticoid and peptide hormones. There now is intensive research on anxiolytic drug therapy to develop respective agonists/antagonists for these systems for the treatment and prevention of anxiety.

Acute stress of many types results in the release of corticotropin-releasing factor (CRF), adrenocorticotropic hormone (ACTH), and cortisol in humans, and corticosterone in rodents (reviewed in Korte[97]). Corticosteroids bind to mineralocorticoid (MR) and glucocorticoid (GR) receptors, which are colocalized in brain regions important in the regulation of anxiety, such as the hippocampus, septum, and amygdala. Increased corticosterone levels in rodents recently was found to be highly correlated with measures of risk assessment.[98] Several animal models of anxiety have demonstrated that both MR and GR can modulate specific aspects of anxiety, as antagonists for both receptor subtypes have been shown to have anxiolytic effects (reviewed in Korte[97]).

GRs are found in higher concentrations in regions involved in feedback regulation of the hormonal stress response. GRs have a 10-fold lower affinity for corticosterone than MRs, becoming occupied only during stress and at the circadian peak, whereas MR is almost completely occupied under basal conditions. Because it is believed that GR action is mainly regulated by hormone level, with MR action being influenced by receptor density, endogenous corticosteroids can play an important role in the maintenance of homeostatic equilibrium via its two receptor types.[99,100] High corticosteroid levels, such as those elicited by exposure to an acute stressor, decrease the number of GRs, which in turn results in increased corticosterone secretion and feedback resistance.

Alterations in HPA axis function are suggested in posttraumatic stress disorder, as demonstrated by a decrease in cortisol levels and increased suppression of cortisol levels with the GR agonist dexamethasone, suggesting supersuppression to cortisol feedback.[101,102] Patients with posttraumatic stress disorder also have a blunted ACTH response to CRF vs. control subjects, suggesting hypersecretion of CRF.[103] Consistent with this, patients with combat-related posttraumatic stress disorder had elevated levels of CRF in the cerebrospinal fluid (CSF).[104] CSF levels of CRF also have been shown to be elevated in patients with obsessive–compulsive disorder,[105] but not panic disorder.[106] An additional study demonstrating no difference in baseline CSF levels of CRF between control subjects and patients with generalized anxiety disorder, panic disorder, or obsessive–compulsive disorder[107] suggests that these patients do not exhibit tonic hypersecretion of CRF. Therefore, evidence for dysfunction of CRF or HPA systems in panic disorder has been inconsistent.[84]

A recent study by Strohle and colleagues[108] demonstrated that panic attacks induced by sodium lactate and cholecystokinin tetrapeptide in patients with panic disorder were accompanied by marked decreases in circulating levels of 3α,5α-THP and 3α,5β-THP, with concomitant increases in the neuroactive steroid antagonist 3β,5α-THP. Similar changes were not observed in control subjects. Therefore, neuroactive steroids may play a role in the pathophysiology of panic disorders.

A number of studies in which various animal models of anxiety were utilized demonstrated that injection of CRF into the locus coeruleus, paraventricular nucleus of the hypothalamus, BNST, and the central nucleus of the amygdala produced signs and symptoms that were identical to those in animals in response to stress as well as those observed in patients with anxiety disorders.[109–113] In male rats, alfaxalone, the 11-keto derivative of 3α,5α-THP, was shown to attenuate the anxiogenic behavioral effects of CRF and swim stress, without reducing the CRF increases in plasma ACTH.[66] Alfaxalone also inhibits the startle-enhancing effects of CRF in male rats without disrupting CRF-stimulated locomotor activity.[114] At doses of 5 and 10 μg i.c.v., 3α,5α-THP counteracted the anxiogenic action of 5 μg CRF in male rats, as well as prevented the adrenalectomy-induced up-regulation of CRF gene expression.[115] Thus, CRF appears to play a significant role in anxiety-related and stress-related states that can be attenuated by GABAergic neuroactive steroids. Collectively, the data also suggest that agents acting at CRF receptors may have therapeutic value in anxiety disorders. Indeed, the development of CRF receptor antagonists as novel anxiolytics is being actively pursued.[112]

15.6 KNOCKOUTS, TRANSGENICS, AND OTHER SINGLE-GENE MANIPULATIONS

A powerful approach that is complementary to genetic studies of anxiety disorders involves the study of genes and proteins implicated in the brain neurocircuitry of anxiety.[116] In this regard, genes and molecules that affect specific neural systems can be targeted as candidates; these genes also may correspond to those identified for further study based on therapeutic efficacy. The mouse has emerged as a premier tool for modeling complex human diseases, such as anxiety, and the reader is referred to detailed reviews of these investigations (e.g., see references 86 and 117–120). These mouse knockout and transgenic models support some tentative conclusions. With regard to GABAergic gene manipulations, deletion of the γ2 subunit of the $GABA_A$ receptor, as well as deletion of one isoform of the biosynthetic enzyme glutamic acid decarboxylase (GAD-65), produced an increase in basal anxiety. However, manipulation of other components of the GRC did not have a significant influence on basal levels of anxiety. In contrast, manipulation of nearly any gene in the HPA axis and CRF system affected anxiety. A fairly reliable pattern of results is consistent with the notion that higher CRF levels lead to increased anxiety via interactions with the CRF_1 receptor. Comparisons of findings across laboratories have yielded some inconsistencies that could be due to procedural differences or to the influences of other genes in the genetic construct or the genetic background of the animals,[121] but it is beyond the scope of this chapter to describe all the results in genetic mouse models. These potential discrepancies underscore the difficulty of modeling complex phenotypes using simple tasks.

15.7 CONCLUSIONS

It now is apparent that the neuroactive steroids, whether via peripheral sources or *de novo* synthesis in the brain, can alter brain excitability, and ultimately behavior, through an interaction with $GABA_A$ receptors. In addition, selective enhancement of GABA-gated inhibition in particular brain regions may occur, based on evidence demonstrating brain regional differences in neuroactive steroid potency and in the coupling of steroid recognition sites to the GRC. This suggests that subtypes of the steroid recognition site may exist and provides the potential for additional specificity of neuroactive steroid action.

Within the last decade, considerable attention has been focused on understanding the physiological significance of fluctuations in neuroactive steroids as modulators of brain excitability. Numerous studies have demonstrated that 3α,5α-THP and 5α-THDOC can increase in response to various stressors in male rodents (reviewed by Barbaccia et al.[6]), can fluctuate during the estrous/menstrual cycle in rats/humans (reviewed by Paul and Purdy[7]), and can dramatically increase during pregnancy to levels that are presumed to be physiologically relevant (i.e., within the range of concentrations previously shown to potentiate the action of GABA). The pharmacological profile of these GABA–agonist neuroactive steroids (e.g., anxiolytic, anticonvulsant, sedative) provides some insight into the potential physiological consequence resulting from a marked fluctuation in endogenous neuroactive steroid levels. It is noteworthy that the stress-induced increase in neuroactive steroids in rodents recently was found to occur primarily via secretion from adrenal and gonadal tissues.[122,123] This contrasts with the adrenal glands of humans and nonhuman primates where the compounds are not secreted in significant amounts in response to acute stress and should serve as a caveat to those investigators who are tempted to make exaggerated claims for their rat or mouse model systems involving paradigms of acute stress. And with regard to anxiety-related behavior, recent findings suggest that peripheral sources of 3α,5α-THP may have a greater impact in modulating anxiety-related behavior than brain-derived 3α,5α-THP,[82] yet brain sources of 3α,5α-THP also were significant in this study.

Despite this proviso, the human literature does provide some examples of the potential physiological consequences of fluctuations in endogenous neuroactive steroid levels. Recent studies have demonstrated an inverse relationship between plasma or CSF levels of neuroactive steroids and symptoms of anxiety during ethanol withdrawal and symptoms of depression in patients with unipolar depression (i.e., decreased levels of 3α,5α-THP and 5α-THDOC were associated with increased severity of symptoms[124,125]). And, as discussed above, neuroactive steroids may play a role in the pathophysiology of panic disorders.[108] However, it is important to point out that fluctuations in endogenous neuroactive steroids do not produce psychopathology in all humans, which could explain the difficulty in correlating the emotional symptoms of premenstrual syndrome (PMS) with endogenous levels of 3α,5α-THP. Further support for this notion is provided by the report that up to 75% of all women of fertile age notice cyclical mood changes, but only 5 to 10% of these women actually suffer from PMS.[126] Because there are individual differences in women in their metabolism of progesterone, where these differences were correlated with symptoms of fatigue and anxiety (i.e., levels of 3α,5α-THP were positively correlated with ratings of fatigue and negatively correlated with ratings of tension and anxiety[127]), it is possible that behavioral changes such as anxiety are observed in susceptible individuals when neuroactive steroid metabolism or the affinity of neuroactive steroids for $GABA_A$ receptors is altered, upsetting normal brain homeostasis. Overall, the modulation of brain excitability by steroids via an interaction with $GABA_A$ receptors provides insights into another mechanism by which the endocrine system may influence brain function and behavior.

ACKNOWLEDGMENTS

This is publication number 15805-NP from The Scripps Research Institute. The authors gratefully acknowledge the editorial assistance of Mike Arends.

REFERENCES

1. Selye, H., Correlations between the chemical structure and the pharmacological actions of the steroids, *Endocrinology*, 30, 437, 1942.
2. Laubach, G.D. et al., Steroid anesthetic agent, *Science*, 122, 78, 1955.
3. Holzbauer, M., Physiological aspects of steroids with anaesthetic properties, *Med. Biol.*, 54, 227, 1976.
4. Holzbauer, M. et al., *In vivo* secretion of 3α-hydroxy-5α-pregnan-20-one, a potent anaesthetic steroid, by the adrenal gland of the rat, *J. Steroid Biochem.*, 22, 97, 1985.
5. Harrison, N.L. and Simmonds, M.A., Modulation of the GABA receptor complex by a steroid anesthetic, *Brain Res.*, 323, 287, 1984.
6. Barbaccia, M.L. et al., Stress and neuroactive steroids, in *International Review of Neurobiology*, Vol. 46, *Neurosteroids and Brain Function*, Biggio, G. and Purdy, R.H., Eds., Academic Press, San Diego, 2001, p. 244.
7. Paul, S.M. and Purdy, R.H., Neuroactive steroids, *FASEB J.*, 6, 2311, 1992.
8. Purdy, R.H. et al., Stress-induced elevations of gamma-aminobutyric acid type A receptor-active steroids in the rat brain, *Proc. Natl. Acad. Sci. U.S.A.*, 88, 4553, 1991.
9. Crawley, J.N. and Paylor, R., A proposed test battery and constellations of specific behavioral paradigms to investigate the behavioral phenotypes of transgenic and knockout mice, *Horm. Behav.*, 31, 197, 1997.
10. File, S.E., Animal tests of anxiety, in *Current Protocols in Neuroscience*, Crawley, J.N., Gerfen, C.R., McKay, R., Rogawski, M.A., Sibley, D.R., and Skolnick, P., Eds., Wiley, New York, 1997, p. 8.3.1.
11. Rodgers, R.J. et al., Animal models of anxiety: an ethological perspective, *Braz. J. Med. Biol. Res.*, 30, 289, 1997.
12. Martin, P., Animal models sensitive to anti-anxiety agents, *Acta Psychiatr. Scand. Suppl.*, 393, 74, 1998.
13. Koob, G.F., Henrichs, S.C., and Britton, K., Animal models of anxiety disorders, in *Textbook of Pharmacology*, 2nd ed., Schatzberg, A.F. and Nemeroff, C.B., Eds., American Psychiatric Press, Washington, D.C., 1998, p. 133.

14. Crawley, J.N., Evaluating anxiety in rodents, in *Techniques in the Behavioral and Neural Sciences*, Vol. 13, *Handbook of Molecular-Genetic Techniques for Brain and Behavior Research*, Crusio, W.E. and Gerlai, R.T., Eds., Elsevier Science BV, Amsterdam, 1999, p. 667.
15. Geller, I. and Seifter, J., The effects of meprobamate, barbiturates, d-amphetamine, and promazine on experimentally induced conflict in the rat, *Psychopharmacologia*, 1, 482, 1960.
16. Vogel, J.R. et al., A simple and reliable conflict procedure for testing antianxiety agents, *Psychopharmacologia*, 21, 1, 1971.
17. Pellow, S. et al., Validation of open:closed arm entries in an elevated plus-maze as a measure of anxiety in the rat, *J. Neurosci. Methods*, 14, 149, 1985.
18. Lister, R.G., The use of a plus-maze to measure anxiety in the mouse, *Psychopharmacology*, 92, 180, 1987.
19. Trullas, R. and Skolnick, P., Differences in fear motivated behaviors among inbred mouse strains, *Psychopharmacology*, 111, 323, 1993.
20. Cole, J.C. and Rodgers, R.J., Ethological evaluation of the effects of acute and chronic buspirone treatment in the murine elevated plus-maze test: comparison with haloperidol, *Psychopharmacology*, 114, 288, 1994.
21. Montgomery, K.C., The relation between fear induced by novel stimulation and exploratory drive, *J. Comp. Physiol. Psychol.*, 48, 254, 1955.
22. Rodgers, R.J. and Cole, J.C., Influence of social isolation, gender, strain, and prior novelty on plus-maze behaviour in mice, *Physiol. Behav.*, 54, 729, 1993.
23. Lamberty, Y. and Gower, A.J., Arm width and brightness modulation of spontaneous behaviour of two strains of mice tested in the elevated plus-maze, *Physiol. Behav.*, 59, 439, 1996.
24. Rodgers, R.J. et al., Anxiogenic-like effects of fluprazine and eltoprazine in the mouse elevated plus maze: profile comparisons with 8-OH-DPAT, CGS 12066B, TFMPP and mCPP, *Behav. Pharmacol.*, 3, 621, 1992.
25. Handley, S.L. and Mithani, S., Effects of alpha-adrenoceptor agonists and antagonists in a maze exploration model of "fear-motivated" behaviour, *Naunyn Schmiedebergs Arch. Pharmacol.*, 327, 1, 1984.
26. Handley, S.L. and McBlane, J.W., An assessment of the elevated X-maze for studying anxiety and anxiety-modulating drugs, *J. Pharmacol. Toxicol. Methods*, 29, 129, 1993.
27. Treit, D. et al., Anxiogenic stimuli in the elevated plus maze, *Pharmacol. Biochem. Behav.*, 44, 463, 1993.
28. Dawson, G.R. and Tricklebank, M.D., Use of the elevated plus maze in the search for novel anxiolytic agents, *Trends Pharmacol. Sci.*, 16, 33, 1995.
29. Dawson, G.R. et al., Evidence that the anxiolytic-like effects of chlordiazepoxide on the elevated plus maze are confounded by increases in locomotor activity, *Psychopharmacology*, 118, 316, 1995.
30. Finn, D.A. et al., Genetic differences in behavioral sensitivity to a neuroactive steroid, *J. Pharmacol. Exp. Ther.*, 280, 820, 1997.
31. Fernandes, C. and File, S.E., The influence of open arm ledges and maze experience in the elevated plus-maze, *Pharmacol. Biochem. Behav.*, 54, 31, 1996.
32. Shepherd, J.K. et al., Behavioral and pharmacological characterisation of the elevated "zero-maze" as an animal model of anxiety, *Psychopharmacology*, 116, 56, 1994.
33. Crawley, J. and Goodwin, F.K., Preliminary report of a simple animal behavior model for the anxiolytic effects of benzodiazepines, *Pharmacol. Biochem. Behav.*, 13, 167, 1980.

34. Crawley, J.N., Neuropharmacologic specificity of a simple animal model for the behavioral actions of benzodiazepines, *Pharmacol. Biochem. Behav.*, 15, 695, 1981.
35. Crawley, J.N. et al., Interaction between purine and benzodiazepine: inosine reverses diazepam-induced stimulation of mouse exploratory behavior, *Science*, 211, 725, 1981.
36. Crawley, J.N. et al., Anxiolytic activity of an endogenous adrenal steroid, *Brain Res.*, 398, 382, 1986.
37. Crawley, J.N. et al., Absence of intrinsic actions of benzodiazepine antagonists on a mouse exploratory model of anxiety in the mouse, *Neuropharmacology*, 23, 531, 1984.
38. Grewal, S.S. et al., Behavioural and pharmacological characterisation of the canopy stretched attend posture test as a model of anxiety in mice and rats, *Psychopharmacology*, 133, 29, 1997.
39. Grant, E.W. and Mackintosh, J.H., A comparison of the social postures of some common laboratory rodents, *Behaviour*, 21, 246, 1963.
40. Blanchard, D.C., Blanchard, R.J., and Rodgers, R.J., Risk assessment and animal models of anxiety, in *Animal Models in Psychopharmacology*, Olivier, B., Mos, J., and Slangen, J.L., Eds., Birkhauser, Basel, 1991, p. 117.
41. Atkins, A.L. et al., Anxiety and sensitivity to ethanol and pentobarbital in alcohol Withdrawal Seizure-Prone and Withdrawal Seizure-Resistant mice, *Alcohol. Clin. Exp. Res.*, 24, 1743, 2000.
42. Toubas, P.L. et al., Latency to enter a mirrored chamber: a novel behavioral assay for anxiolytic agents, *Pharmacol. Biochem. Behav.*, 35, 121, 1990.
43. Cao, W. et al., A genetic comparison of behavioral actions of ethanol and nicotine in the mirrored chamber, *Pharmacol. Biochem. Behav.*, 45, 803, 1993.
44. Gallup, G.G., Jr., Mirror-image stimulation, *Psychol. Bull.*, 70, 782, 1968.
45. Seale, T.W. et al., Anxiolysis by ethanol, diazepam and buspirone in a novel murine behavioral assay, *Neuroreport*, 7, 1803, 1996.
46. Reddy, D.S. and Kulkarni, S.K., Differential anxiolytic effects of neurosteroids in the mirrored chamber behavior test in mice, *Brain Res.*, 752, 61, 1997.
47. Lamberty, Y., The mirror chamber test for testing anxiolytics: is there a mirror-induced stimulation? *Physiol. Behav.*, 64, 703, 1998.
48. Kleithermes, C.L. et al., Validation of a modified mirrored chamber sensitive to anxiolytics and anxiogenics in mice, *Psychopharmacology*, 2003, in press.
49. File, S.E. and Wardill, A.G., Validity of head-dipping as a measure of exploration in a modified holeboard, *Psychopharmacologia*, 44, 53, 1975.
50. Takeda, H., Tsuji, M., and Matsumiya, T., Changes in head-dipping behavior in the hole-board test reflect the anxiogenic and/or anxiolytic state in mice, *Eur. J. Pharmacol.*, 305, 21, 1998.
51. van Gaalen, M.M. and Steckler, T., Behavioural analysis of four mouse strains in an anxiety test battery, *Behav. Brain Res.*, 115, 95, 2000.
52. Hall, C.S., Emotional behavior in the rat. III. The relationship between emotionality and ambulatory activity, *J. Comp. Psychol.*, 22, 345, 1936.
53. Henderson, N.D., Prior treatment effects on open field behaviour of mice: a genetic analysis, *Anim. Behav.*, 15, 364, 1967.
54. File, S.E. and Hyde, J.R.G., A test of anxiety that distinguishes between the actions of benzodiazepines and those of other minor tranquilisers and of stimulants, *Pharmacol. Biochem. Behav.*, 11, 65, 1979.
55. File, S.E., The use of social interaction as a method for detecting anxiolytic activity of chlordiazepoxide-like drugs, *J. Neurosci. Methods*, 2, 219, 1980.

56. Dunn, A.J. and File, S.E., Corticotropin-releasing factor has an anxiogenic action in the social interaction test, *Horm. Behav.*, 21, 193, 1987.
57. Higgins, G.A. et al., Behavioral and biochemical consequences following activation of 5-HT1-like and GABA receptors in the dorsal raphe nucleus of the rat, *Neuropharmacology*, 27, 993, 1988.
58. De Angelis, L. and File, S.E., Acute and chronic effects of three benzodiazepines in the social interaction anxiety test in mice, *Psychopharmacology*, 64, 127, 1979.
59. Lister, R.G. and Hilakivi, L.A., The effects of novelty, isolation, light and ethanol on the social behavior of mice, *Psychopharmacology*, 96, 181, 1988.
60. Pinel, J.P. and Treit, D., Burying as a defensive response in rats, *J. Comp. Physiol. Psychol.*, 92, 708, 1978.
61. Treit, D. et al., Conditioned defensive burying: a new paradigm for the study of anxiolytic agents, *Pharmacol. Biochem. Behav.*, 15, 619, 1981.
62. Treit, D., The inhibitory effect of diazepam on defensive burying: anxiolytic vs. analgesic effects, *Pharmacol. Biochem. Behav.*, 22, 47, 1985.
63. Griebel, G. et al., Further evidence that the mouse defense test battery is useful for screening anxiolytic and panicolytic drugs: effects of acute and chronic treatment with alprazolam, *Neuropharmacology*, 34, 1625, 1995.
64. Griebel, G., Sanger, D.J., and Perrault, G., Genetic differences in the mouse defense test battery, *Aggr. Behav.*, 23, 19, 1997.
65. Wieland, S. et al., Anxiolytic activity of the progesterone metabolite 5α-pregnan-3α-ol-20-one, *Brain Res.*, 565, 263, 1991.
66. Britton, K.T. et al., Pregnane steroid alphaxalone attenuates anxiogenic behavioral effects of corticotropin releasing factor and stress, *Pharmacol. Biochem. Behav.*, 41, 399, 1992.
67. Wieland, S. et al., Comparative behavioral characterization of the neuroactive steroids 3α-OH,5α-pregnan-20-one and 3α-OH,5β-pregnan-20-one in rodents, *Psychopharmacology*, 118, 65, 1995.
68. Carboni, E. et al., Anxiolytic properties of endogenously occurring pregnanediols in two rodent models of anxiety, *Psychopharmacology*, 126, 173, 1996.
69. Brot, M.D. et al., The anxiolytic-like effects of the neurosteroid allopregnanolone: interactions with $GABA_A$ receptors, *Eur. J. Pharmacol.*, 325, 1, 1997.
70. Bitran, D. et al., Anxiolytic effects of 3α-hydroxy-5α[β]pregnan-20-one: endogenous metabolites of progesterone that are active at the $GABA_A$ receptor, *Brain Res.* 561, 157, 1991.
71. Gomez, C. et al., Rapid anxiolytic activity of progesterone and pregnanolone in male rats, *Pharmacol. Biochem. Behav.*, 72, 543, 2002.
72. Bitran, D. et al., Treatment with an anabolic-androgenic steroid affects anxiety-related behavior and alters the sensitivity of cortical $GABA_A$ receptors in the rat, *Horm. Behav.*, 27, 568, 1993.
73. Melchior, C.L. and Ritzmann, R.F., Dehydroepiandrosterone is an anxiolytic in mice on the plus maze, *Pharmacol. Biochem. Behav.*, 47, 437, 1994.
74. Melchior, C.L. and Ritzmann, R.F., Pregnenolone and pregnenolone sulfate, alone and with ethanol, in mice on the plus-maze, *Pharmacol. Biochem. Behav.*, 48, 893, 1994.
75. Rodgers, R.J. and Johnson, N.J.T., Behaviorally selective effects of neuroactive steroids on plus-maze anxiety in mice, *Pharmacol. Biochem. Behav.*, 59, 221, 1998.
76. Pick, C.G. et al., Effect of the neuroactive steroid 5α-THDOC on staircase test bchavior in mice, *Psychopharmacology*, 128, 61, 1996.

77. Pick, C.G. et al., Effect of the pregnane-related GABA-active steroid alphaxalone on mice performance in the staircase test, *Brain Res.*, 765, 129, 1997.
78. Weizman, R. et al., Behavioral effects of agents active at the γ-aminobutyric acid receptor complex in the staircase paradigm, *Brain Res.*, 901, 137, 2001.
79. Zimmerberg, B. et al., Reduction of rat pup ultrasonic vocalizations by the neuroactive steroid allopregnanolone, *Pharmacol. Biochem. Behav.*, 47, 735, 1994.
80. Gasior, M. et al., Neuroactive steroids: potential therapeutic use in neurological and psychiatric disorders, *Trends Pharmacol. Sci.*, 20, 107, 1999.
81. Bitran, D. et al., Anxiolytic effect of progesterone is mediated by the neurosteroid allopregnanolone at brain $GABA_A$ receptors, *J. Neuroendocrinol.*, 7, 171, 1995.
82. Serra, M. et al., 2-Phenyl-imidazo[1,2-a]pyridine derivatives as ligands for peripheral benzodiazepine receptors: stimulation of neurosteroid synthesis and anticonflict action in rats, *Br. J. Pharmacol.*, 127, 177, 1999.
83. Hyman, S.E., Brain neurocircuitry of anxiety and fear: implications for clinical research and practice, *Biol. Psychiatry*, 44, 1201, 1998.
84. Charney, D.S. and Bremner, J.D., The neurobiology of anxiety disorders, in *Neurobiology of Mental Illness*, Charney, D.S., Nestler, E.J., and Bunney, B.S., Eds., Oxford University Press, New York, 1999, p. 494.
85. Sandford, J.J. et al., The psychobiology of anxiolytic drugs. Part 1. Basic neurobiology, *Pharmacol. Ther.*, 88, 197, 2000.
86. Finn, D.A. et al., Genetic animal models of anxiety, *Neurogenetics*, 4, 109, 2003.
87. Graeff, F.G. et al., Role of the amygdala and periaqueductal gray in anxiety and panic, *Behav. Brain Res.*, 58, 123, 1993.
88. Stanford, S.C., Central adrenoceptors in response and adaptation to stress, in *The Pharmacology of Noradrenaline in the Central Nervous System*, Heal, D.J. and Marsden, C.A., Eds., Oxford University Press, Oxford, 1990, p. 379.
89. Charney, D.S. and Heninger, G.R., Abnormal regulation of noradrenergic function in panic disorders: effects of clonidine in healthy subjects and patients with agoraphobia and panic disorder, *Am. J. Psychiatry*, 43, 1042, 1986.
90. Redmond, D.E., Jr., The possible role of locus coeruleus noradrenergic activity in anxiety-panic, *Clin. Neuropharmacol.*, 9 (Suppl. 4), 40, 1986.
91. Davis, M., The role of the amygdala in fear-potentiated startle: implications for animal models of anxiety, *Trends Pharmacol. Sci.*, 13, 35, 1992.
92. Akwa, Y. et al., The amygdala mediates the anxiolytic-like effect of the neurosteroid allopregnanolone in rat, *Behav. Brain Res.*, 106, 119, 1999.
93. Gray, J.A., The neuropsychology of anxiety: an enquiry into the functions of the septo-hippocampal system, *Behav. Brain Sci.*, 5, 469, 1982.
94. Gabriel, M., Discriminative avoidance learning: a model system, in *Neurobiology of Cingulate Cortex and Limbic Thalamus: A Comprehensive Handbook*, Vogt, B.A. and Gabriel, M., Eds., Birkhauser, Boston, 1993, p. 478.
95. Bitran, D. et al., Anxiolytic effects of the neuroactive steroid pregnanolone (3α-OH-5β-pregnan-20-one) after microinjection in the dorsal hippocampus and lateral septum, *Brain Res.*, 850, 217, 1999.
96. Bitran, D. et al., Activation of peripheral mitochondrial benzodiazepine receptors in the hippocampus stimulates allopregnanolone synthesis and produces anxiolytic-like effects in the rat, *Psychopharmacology*, 151, 64, 2000.
97. Korte, S.M., Corticosteroids in relation to fear, anxiety and psychopathology, *Neurosci. Biobehav. Rev.*, 25, 117, 2001.
98. Rodgers, R.J. et al., Corticosterone response to the plus-maze: high correlation with risk assessment in rats and mice, *Physiol. Behav.*, 68, 47, 1999.

99. De Kloet, E.R., Brain corticosteroid receptor balance and homeostatic control, *Front. Neuroendocrinol.*, 12, 95, 1991.
100. Sapolsky, R.M., Stress, glucocorticoids, and damage to the nervous system: the current state of confusion, *Stress*, 1, 1, 1996.
101. Yehuda, R. et al., Enhanced suppression of cortisol following dexamethasone administration in posttraumatic stress disorder, *Am. J. Psychiatry*, 150, 83, 1993.
102. Yehuda, R. et al., Increased pituitary activation following metyrapone administration in post-traumatic stress disorder, *Psychoneuroendocrinology*, 21, 1, 1996.
103. Smith, M.A. et al., The corticotropin-releasing hormone test in patients with post-traumatic stress disorder, *Biol. Psychiatry*, 26, 349, 1989.
104. Bremner, J.D. et al., Chronic PTSD in Vietnam combat veterans: course of illness and substance abuse, *Am. J. Psychiatry*, 153, 369, 1996.
105. Altemus, M. et al., Changes in cerebrospinal fluid neurochemistry during treatment of obsessive-compulsive disorder with clomipramine, *Arch. Gen. Psychiatry*, 51, 794, 1994.
106. Jolkkonen, J. et al., CSF corticotropin-releasing factor is not affected in panic disorder, *Biol. Psychiatry*, 33, 136, 1993.
107. Fossey, M.D. et al., Cerebrospinal fluid corticotropin-releasing factor concentrations in patients with anxiety disorders and normal comparison subjects, *Biol. Psychiatry*, 39, 703, 1996.
108. Strohle, A. et al., Induced panic attacks shift γ-aminobutyric acid type A receptor modulatory neuroactive steroid composition in patients with panic disorder: preliminary results, *Arch. Gen. Psychiatry*, 60, 161, 2003.
109. Sutton, R.E. et al., Corticotropin releasing factor produces behavioural activation in rats, *Nature*, 297, 331, 1982.
110. Butler, P.D. et al., Corticotropin-releasing factor produces fear-enhancing and behavioral activating effects following infusion into the locus coeruleus, *J. Neurosci.*, 10, 176, 1990.
111. Dunn, A.J. and Berridge, C.W., Physiological and behavioral responses to corticotropin-releasing factor administration: is CRF a mediator of anxiety or stress responses? *Brain Res. Brain Res. Rev.*, 15, 71, 1990.
112. Griebel, G., Is there a future for neuropeptide receptor ligands in the treatment of anxiety disorders? *Pharmacol. Ther.*, 82, 1, 1999.
113. Koob, G.F. and Heinrichs, S.C., A role for corticotropin releasing factor and urocortin in behavioral responses to stressors, *Brain Res.*, 848, 141, 1999.
114. Swerdlow, N. and Britton, K.T., Alphaxalone, a steroid anesthetic, inhibits the startle-enhancing effects of corticotropin releasing factor, but not strychnine, *Psychopharmacology*, 115, 141, 1994.
115. Patchev, V.K. et al., The neurosteroid tetrahydroprogesterone counteracts corticotropin-releasing hormone-induced anxiety and alters the release and gene expression of corticotropin-releasing hormone in the rat hypothalamus, *Neuroscience*, 62, 265, 1994.
116. Pickar, D. and Rubinow, K., Pharmacogenomics of psychiatric disorders, *Trends Pharmacol. Sci.*, 22, 75, 2001.
117. Picciotto, M.R., Knock-out mouse models used to study neurobiological systems, *Crit. Rev. Neurobiol.*, 13, 103, 1999.
118. Bolivar, V. et al., List of transgenic and knockout mice: behavioral profiles, *Mamm. Genome*, 11, 260, 2000.
119. Tarantino, L.M. and Bucan, M., Dissection of behavior and psychiatric disorders using the mouse as a model, *Hum. Mol. Genet.*, 9, 953, 2000.

120. Anagnostopoulos, A.V. et al., Transgenic and knockout databases: behavioral profiles of mouse mutants, *Physiol. Behav.*, 73, 675, 2001.
121. Crabbe, J.C., Use of genetic analyses to refine phenotypes related to alcohol tolerance and dependence, *Alcohol. Clin. Exp. Res.*, 25, 288, 2001.
122. Purdy, R.H. et al., Ethanol-stimulated increases in neuro-active steroids in brain are derived from endocrine glands, *Int. Behav. Neurosci. Soc. Abstr.*, 12, 59, 2003.
123. Budec, M. et al., Acute ethanol treatment increases level of progesterone in ovariectomized rats, *Alcohol*, 26, 173, 2002.
124. Romeo, E. et al., Marked decrease of plasma neuroactive steroids during alcohol withdrawal, *Clin. Neuropharmacol.*, 19, 366, 1996.
125. Uzunova, V. et al., Increase in the cerebrospinal fluid content of neurosteroids in patients with unipolar major depression who are receiving fluoxetine or fluvoxamine, *Proc. Natl. Acad. Sci. U.S.A.*, 95, 3239, 1998.
126. Andersch, B. et al., Premenstrual complaints. I. Prevalence of premenstrual symptoms in a Swedish urban population, *J. Psychosom. Obstet. Gynaecol.*, 5, 39, 1986.
127. Freeman, E. et al., Anxiolytic metabolites of progesterone: correlation with mood and performance measures following oral progesterone administration to healthy female volunteers, *Clin. Neuroendocrinol.*, 58, 478, 1993.

16 Direct Modulation of Amino Acid Receptors by Neuroactive Steroids: Physiological and Pharmacological Implications

Terrell T. Gibbs and David H. Farb

CONTENTS

0-8493-2392-4/04/$0.00+$1.50

ABSTRACT

Neuroactive steroids can produce rapid effects on central nervous system (CNS) function by acting directly on a variety of neurotransmitter receptors, including $GABA_A$, glycine, and glutamate receptors. Although $GABA_A$ and NMDA receptors are structurally distinct, both exhibit bidirectional modulation by neuroactive steroids, with some steroids enhancing and others inhibiting receptor function. NMDA receptors resemble $GABA_A$ receptors in that they exhibit at least two distinct sites for steroid modulation, one of which mediates the effects of positive modulators, while the other mediates the effects of negative modulators. Pregnenolone sulfate (PS), a relatively abundant neurosteroid, inhibits $GABA_A$, glycine, and non-NMDA glutamate receptors, while its effects on NMDA receptors depend upon subunit composition, with the NR2A and NR2B subunits supporting a potentiating effect, while NR2C and NR2D subunits support inhibition. In contrast, pregnanolone sulfate inhibits all tested NMDA and non-NMDA receptor subtypes. Pregnanolone hemisuccinate, a synthetic analog of pregnanolone sulfate, is active *in vivo* after intraperitoneal or intravenous administration, exhibiting sedative, analgesic, anticonvulsant, and neuroprotective properties, as well as inhibiting NMDA-induced dopamine release in the striatum. Pregnenolone sulfate, when infused directly into rat striatum at physiological concentrations, directly enhances dopamine release, consistent with the hypothesis that it acts as an endogenous modulator of neurotransmission.

16.1 INTRODUCTION

The observation that neuroactive steroids can exert rapid effects upon neurotransmitter receptor function suggests the possibility that endogenous neurosteroids may play a regulatory role in synaptic transmission. γ-Aminobutyric acid A ($GABA_A$) receptors are subject to modulation by a variety of neurosteroids. Progesterone enhances the GABA response,[1] and its 5α-reduced metabolites, such as pregnanolone (3α5α) and allopregnanolone (3α5β), are among the most potent and efficacious positive modulators of $GABA_A$ receptors.[2] Progesterone and its metabolites have been shown to produce anxiolytic, anticonvulsive, sedative, and antinociceptive effects *in vivo* similar to those of other $GABA_A$ receptor positive modulators, such as benzodiazepines and barbiturates.[3] Synthetic steroids such as alphaxalone that enhance $GABA_A$ receptor activity have provided a useful basis for the development of novel anesthetics.[2,4,5] In contrast, sulfated steroids such as PS and dehydroepiandrosterone sulfate (DHEAS) have been shown to act as negative modulators of $GABA_A$ receptors[6–8] and thus decrease $GABA_A$ receptor activity. We wondered whether sensitivity to steroids was a specific feature of the $GABA_A$ receptor or a characteristic of ligand-gated ion channels in general and whether other endogenous steroids exhibited similar activity.

16.2 SELECTIVE MODULATION OF $GABA_A$ AND GLYCINE RECEPTORS BY NEUROSTEROIDS

Initial voltage-clamp studies of primary chick spinal cord neurons in culture demonstrated that progesterone potentiated the GABA response to about the same extent as observed with benzodiazepines.[9] In contrast, progesterone exerted an inhibitory effect on glycine receptors (Figure 16.1). A number of other steroids, including corticosterone and deoxycorticosterone, exhibited differential modulation of GABA and glycine responses (Table 16.1). The results also confirmed that pregnenolone sulfate (100 μ*M*) virtually abolished the response to a low concentration of GABA (3 μ*M*), whereas 3α5α enhanced the GABA-induced current by more than an order of magnitude, in agreement with previous reports. Whereas pregnenolone sulfate inhibited both GABA and glycine responses, 3α5α was without effect on the glycine response. Studies of other steroids revealed additional differences. Of particular interest was progesterone, which potentiated the GABA response by 142%. While this was considerably less than the potentiation induced by 3α5α, it is nevertheless comparable to the maximal potentiation seen with many benzodiazepines. Thus, despite the structural and functional similarity of glycine-gated chloride channels to $GABA_A$ receptors, studies of steroid modulation of glycine receptors revealed a different pharmacology. The observation that steroids can exhibit selectivity with respect to similar receptors suggested that steroid modulation of amino acid receptors was not likely to be due to a nonspecific interaction with the receptor or its phospholipid environment, and that neurosteroids might play a role as functional modulators *in vivo*.

16.3 DISTINCT SITES FOR STEROID POSITIVE AND NEGATIVE MODULATORS OF $GABA_A$ RECEPTORS

To investigate the role of the sulfate group in controlling the polarity of modulation of $GABA_A$ receptors, we performed a more detailed investigation of the effects of sulfated and unsulfated natural and synthetic receptors on $GABA_A$ response. Androsterone produced a strong potentiation comparable to that of 3α5α; androsterone sulfate inhibited the GABA response. Similarly, 3α5α and 3β5α potentiated the GABA response, whereas allopregnanolone sulfate (3α5αS) and pregnanolone sulfate (3α5βS) were inhibitory. Inhibitory activity of 3α5βS was retained when the sulfate group was replaced with a hemisuccinate group, suggesting that other negatively charged groups could substitute for sulfate and that the steroid binding site could accommodate a bulky substitution at C3. Not all inhibitory compounds were sulfated; DHEA also produced an inhibitory effect, but even in this case inhibition was increased with the C3 sulfated derivative, DHEAS (Table 16.1).

11-Keto pregnenolone sulfate (11-ketoPS) produced only a slight potentiation (26 ± 8%) of the peak GABA response of chick spinal cord neurons, but substantially reduced potentiation by 3α5α. However, 11-ketoPS inhibition of 3α5α potentiation could not be surmounted by increasing the concentration of 3α5α, inconsistent with a simple competitive interaction (Figure 16.2A). Moreover, when we examined spontaneous GABAergic inhibitory activity of hippocampal neurons in culture, we found that GABA-mediated synaptic responses were blocked by 11-ketoPS alone. These results suggested that 11-ketoPS might have an inhibitory activity that was partially masked by a potentiating effect mediated by a different site, implying the existence of multiple steroid modulatory sites linked to the $GABA_A$ receptor. This result suggested that other sulfated steroids were probably acting at a different site from potentiating steroids such as pregnanolone and 3α5α.

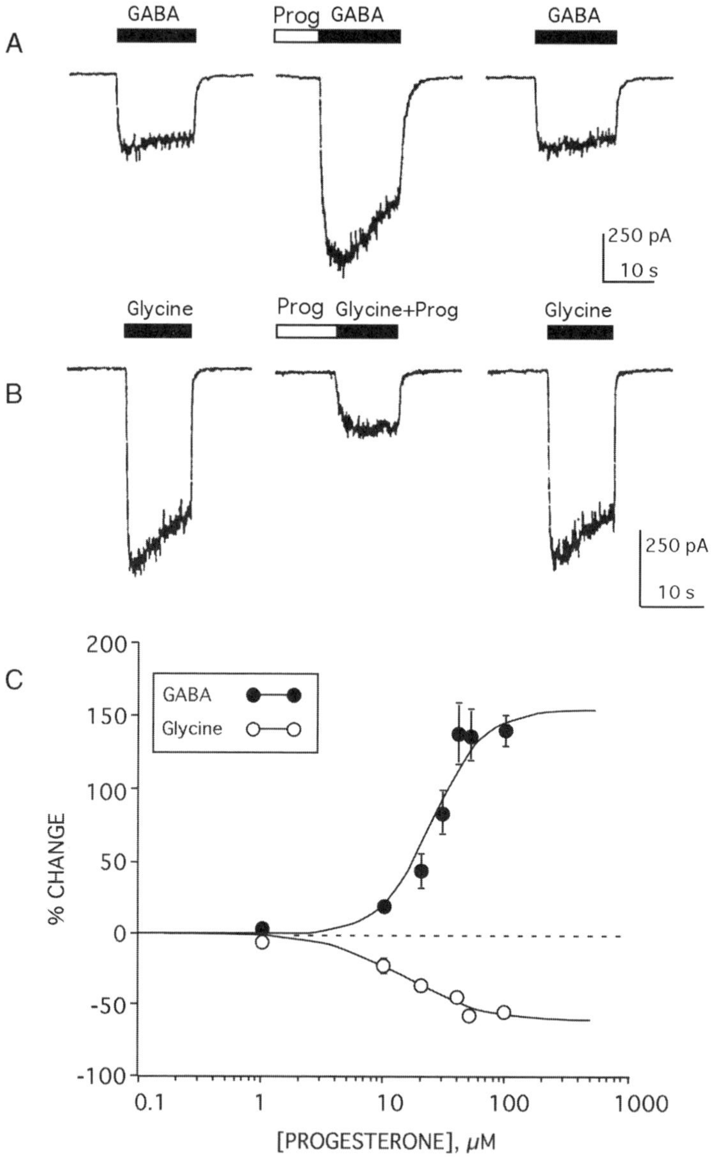

FIGURE 16.1 Progesterone produces opposite effects on whole-cell current responses to GABA and glycine. Primary cultures of spinal cord neurons were prepared from 7-day chicken embryos. Holding potential was –70 mV. (A) The GABA (3 μ*M*)-induced current is reversibly potentiated by 100 μ*M* progesterone (Prog). (B) In another cell, the glycine (50 μ*M*)-induced current is reversibly inhibited by 100 μ*M* progesterone. (C) Concentration–response curves for progesterone. Data points, percentage of change in peak GABA- or glycine-induced current, mean of 4 to 18 experiments. Error bars, SEM. Smooth curves are fits to the logistic equation. For GABA, EC_{50} = 26 μ*M*, n = 2.07, maximum potentiation = 156%. For glycine, EC_{50} = 16 μ*M*, n = 1.42, maximum inhibition = 60%. (From Wu, F.-S. et al., *Mol. Pharmacol.*, 37, 597–602, 1990. With permission from ASPET.)

TABLE 16.1
Effects of Steroids on Responses of Voltage-Clamped Primary Spinal Cord Neurons to GABA, Glycine, or NMDA

Steroid	Percent Change in Response		
	GABA	Glycine	NMDA
11β-Hydroxy-pregnenolone sulfate	–93 ± 2 (5)	—	–31 ± 2 (3)
11-Keto-pregnenolone sulfate	+26 ± 8 (7)	—	+4 ± 2 (7)
17α-Estradiol sulfate	–14 ± 13 (3)	—	—
17α-Hydroxy-progesterone	+67 ± 6 (6)	–32 ± 3 (6)	—
17α-Hydroxy-pregnenolone sulfate	—	—	+178 ± 15 (4)
17β-Estradiol	—	—	–40 ± 5 (3)
17β-Estradiol benzoate (20 μ*M*)	–30 ± 7 (4)	—	+24 ± 18 (3)
17β-Estradiol-17-hemisuccinate (200 μ*M*)	+13 ± 11 (4)	—	–67 ± 10 (3)
17β-Estradiol-3,17-hemisuccinate	—	—	–56 ± 3 (3)
17β-Estradiol-3-hemisuccinate	—	—	–65 ± 3 (4)
17β-Estradiol sulfate	–21± 8 (5)	—	+8 ± 4 (3)
20β-Dihydropregnenolone sulfate	–75 ± 7 (5)	—	+92 ± 16 (5)
21-Acetoxy-pregnenolone sulfate	–85 ± 2 (4)	—	+85 ± 16 (7)
3α-Hydroxy-5α-pregnan-20-one (allopregnanolone; 1 μM)	+1373 ± 225 (7)	+2 ± 2 (5)	—
3α-Hydroxy-5α-pregnan-20-one sulfate (allopregnanolone sulfate)	—	—	–25 ±3 (8)
3α-Hydroxy-5β-pregnan-20-one (pregnanolone)	+1257 ± 245 (4) EC_{50} = 110 n*M*	-	–3 ± 4 (4)
3α-Hydroxy-5β-pregnan-20-one hemisuccinate (pregnanolone hemisuccinate)	–70 ± 5 (4)	—	–31 ± 9 (6)
3α-Hydroxy-5β-pregnan-20-one hemisuccinate methyl ester	—	—	–0.5 ± 1.9 (4)
3α-Hydroxy-5β-pregnan-20-one sulfate (pregnanolone sulfate)	–68 ± 7 (4)	—	–66 ± 3 (5) IC_{50} = 62.1 μ*M*
3β-Hydroxy-5α-pregnan-20-one	+4 ± 3 (5)	0 ± 2 (6)	—
3β-Hydroxy-5α-pregnan-20-one sulfate	–97 ± 3 (3)	—	+39 ± 6 (7)
3β-Hydroxy-5β-pregnan-20-one sulfate	–90 ± 4 (2)	-	–50 ± 4 (12) IC_{50} = 143 μ*M*
5α-Androstan-3α,17β-diol	+341 ± 73 (4)	—	—
5α-Pregnan-3α,20-diol (20 μ*M*)	+166 ± 25 (5) EC_{50} = 120 n*M*	—	—
7-Keto-pregnenolone sulfate	–57 ± 3 (3)		–5 ± 5 (4)
Androstenedione (200 μ*M*)	+118 ± 26 (4)	—	—
Androsterone	+1533 ± 234 (6) EC_{50} = 5.9 μ*M*	—	—
Androsterone sulfate (300 μ*M*)	–58 ± 5 (5) IC_{50} = 57 μ*M*	—	–18 ± 3 (3)
Corticosterone	+60 ± 12 (7)	–23 ± 3 (6)	—
Dehydroepiandrosterone (DHEA)	–40 ± 6 (3)	—	+1 ± 3 (3)

(continued)

TABLE 16.1 (CONTINUED)
Effects of Steroids on Responses of Voltage-Clamped Primary Spinal Cord Neurons to GABA, Glycine, or NMDA

	Percent Change in Response		
Steroid	GABA	Glycine	NMDA
Dehydroepiandrosterone sulfate (DHEAS)	–89 ± 0.04 (4) IC_{50} = 10 μ*M*	—	+29 ± 9 (4)
Deoxycorticosterone	+131 ± 22 (6)	—	–51 ± 3 (5)
Epiandrosterone sulfate (300 μ*M*)	–91 ± 2 (4)	—	—
Hydrocortisone	0 ± 3 (5)	+1 ± 4 (5)	—
Pregnenolone	Inactive	—	–8 ± 4 (3)
Pregnenolone hemisuccinate	–39 ± 4 (4)	—	+204 ± 18 (4)
Pregnenolone methyl phosphonothioate	–74 ± 2 (3)	—	+17 ± 7 (3)
Pregnenolone sulfamate	—	—	+15 ± 7 (3)
Pregnenolone sulfate	–91 ± 2 (6) IC_{50} = 7.2 μ*M*	–91 ± 2 (6)	+150 ± 14 (14) EC_{50} = 57 μ*M*
Progesterone	+142 ± 11 (16) EC_{50} = 26 μ*M*	–53 ± 2.5 (18)	–33 ± 13 (4)
Testosterone	+82 ± 9 (4) EC_{50} = 78 μ*M*	—	—

Note: Effects of steroids on responses of voltage-clamped primary spinal cord neurons to 3 μ*M* GABA, 50 μ*M* glycine, or 30 μ*M* NMDA. Steroid was tested at 100 μ*M* unless otherwise noted. Results are mean percentage change (+ for increase, – for decrease) in response to 3 μ*M* GABA, 50 μ*M* glycine, or 30 μ*M* NMDA. Number of cells tested is given in parentheses.

Data are from Wu, F.-S. et al., *Mol. Pharmacol.*, 37, 597–602, 1990; Park-Chung, M. et al., *Brain Res.*, 830, 72–87, 1999; Wu, F.-S. et al., *Mol. Pharmacol.*, 40, 333–336, 1991; Park-Chung, M. et al., *Mol. Pharmacol.*, 52, 1113–1123, 1997; Park-Chung, M. et al., , *Mol. Pharmacol.*, 46, 146–150, 1994; Weaver, C.E., Jr. et al., *Proc. Natl. Acad. Sci. U.S.A.*, 94, 10450–10454, 1997.

To test this hypothesis, the concentration–response curve for inhibition by DHEAS was determined in the presence and absence of pregnanolone (3α5β), using α1β2γ2 receptors expressed in *Xenopus* oocytes. There was no detectable shift in the DHEAS IC_{50} in the presence of 10 μ*M* 3α5β (EC_{50} = 330 n*M*; Figure 16.3B), demonstrating that the interaction between DHEAS and 3α5β is not competitive (Figure 16.3C)[8] and that potentiation and inhibition must be mediated through distinct sites.

16.4 REGULATION OF STEROID MODULATION OF THE $GABA_A$ RECEPTOR

When recording from chick spinal cord neurons in culture using the whole-cell clamp method, we observed a progressive rundown (decline in amplitude) of the GABA-induced current, combined with an acceleration of desensitization.[10] The

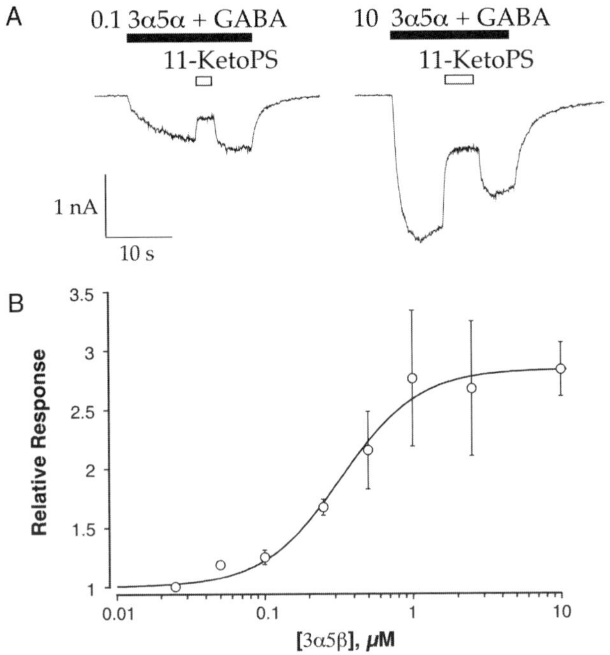

FIGURE 16.2 Two sites for steroid modulation of the $GABA_A$ receptor. (A) Inhibition of the allopregnanolone-potentiated GABA response by 11-ketoPS is insurmountable. Primary cultures of spinal cord neurons were prepared from 7-day chicken embryos. Application of 100 μ*M* 11-ketoPS inhibits (64 ± 7%, n = 4) the response to 3 μ*M* GABA + 100 n*M* allopregnanolone (3α5α). The percentage inhibition by 11-ketoPS is not reduced (64 ± 8%, n=4) when the concentration of allopregnanolone is increased to 10 μ*M*. Solid bar indicates period of application of GABA + allopregnanolone. Open bar indicates period of application of 11-ketoPS. (B) 3α5β enhances the activation of α1β2γ2S receptors. *Xenopus laevis* oocytes were coinjected with cRNA for the rat α1, β2, and γ2S subunits of the $GABA_A$ receptor. Injected oocytes were exposed to GABA in the presence of DHEAS or 3α5β. Oocytes were exposed to 1 μ*M* GABA in the presence of the indicated concentration of 3α5β. The GABA-induced current in the presence of 3α5β is expressed relative to the average of flanking control responses to GABA alone. Points are means of four oocytes. Error bars indicate S.E.M. Smooth curve is a fit of the data to the logistic equation (fitted parameters: maximum current = 2.83 _ control; K = 330 n*M*, n = 1.66). *(continued)*

decreased current was due to a decrease in the maximal GABA-induced current, which was also associated with a decrease in the GABA EC_{50}. No rundown was observed when recordings were made using the nystatin-perforated patch method, suggesting that rundown is related to washout of intracellular factors. Rundown was increased when neurons were repeatedly exposed to a high (30 μ*M*) concentration of GABA, but was virtually absent when neurons were exposed to a low (3 μ*M*) concentration of GABA. However, exposure to 3 μ*M* GABA did produce rundown when the GABA response was potentiated by 500 n*M* 3α5α, arguing that rundown is a use-dependent phenomenon.

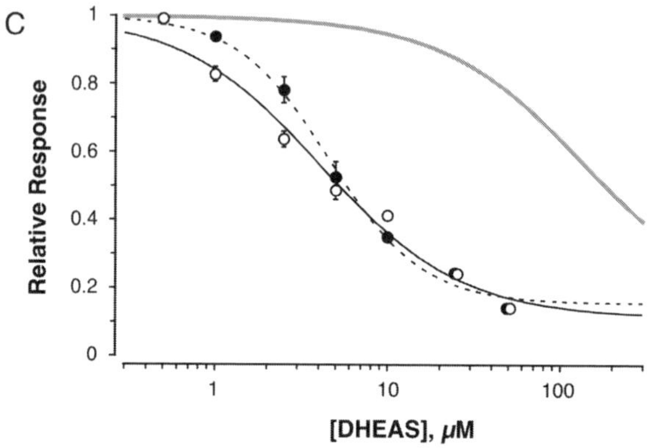

FIGURE 16.2 (CONTINUED) Two sites for steroid modulation of the $GABA_A$ receptor. (C) 3α5β does not shift the DHEAS dose–response curve. Oocytes expressing α1β2γ2S receptors were exposed to 1 μ*M* GABA plus the indicated concentration of DHEAS in the absence (—o—) or presence (- -●- -) of 10 μ*M* 3α5β. The GABA-induced current in the presence of DHEAS is expressed relative to the GABA-induced current in the absence of DHEAS, which was increased 2.8-fold in the presence of 10 μ*M* 3α5β. Points are means of four oocytes. Error bars indicate SEM. Smooth curves are fits to the logistic equation. Fitted parameters: (without 3α5β) maximum inhibition = 88%, K = 4.12 μ*M*, n_H = 1.07; (with 3α5β) maximum inhibition = 84%, K = 4.58 μ*M*, n_H = 1.6. Gray curve indicates predicted shift of the DHEAS concentration–response curve if 3α5β were to compete for the DHEAS recognition site. (From Park-Chung, M. et al., *Brain Res.*, 830, 72–87, 1999, with permission from Elsevier.)

Rundown was antagonized by the inclusion of ATPγS in the intracellular buffer and accelerated by the inclusion of alkaline phosphatase, indicating that rundown of the GABA response is a dephosphorylation-mediated event. Rundown was also associated with decreased potentiation of the GABA response by various modulators, including progesterone and 3α5α, but this loss of potentiation was at least in part a consequence of the decrease in the GABA EC_{50}. These results suggest that GABA receptor activation and its sensitivity to modulators are subject to use-dependent regulation.[11]

We had previously observed that enhancement of ^{3}H-flunitrazepam binding by GABA was diminished after exposure of primary spinal cord cultures to GABA or flurazepam for 48 h, suggesting that chronic GABA or benzodiazepine exposure induces uncoupling of allosteric interactions between benzodiazepine and GABA recognition sites.[12,13] Binding of ^{3}H-flunitrazepam is also enhanced in the presence of the steroid positive modulators 3α5α and 3α5β, so we examined the effect of chronic treatment of cultures with steroids. Cultures exposed for 48 h to 3α5β exhibited complete loss of 3α5β potentiation of ^{3}H-flunitrazepam binding, with no change in the number of binding sites, as well as reduced potentiation of ^{3}H-flunitrazepam binding by GABA. 3α5β-induced uncoupling was specific for the highly efficacious steroid positive modulators 3α5β and 3α5α; steroids with weaker activity at the $GABA_A$ receptor produced at most a small degree of uncoupling (Figure 16.3). No uncoupling was produced when culture membrane homogenates

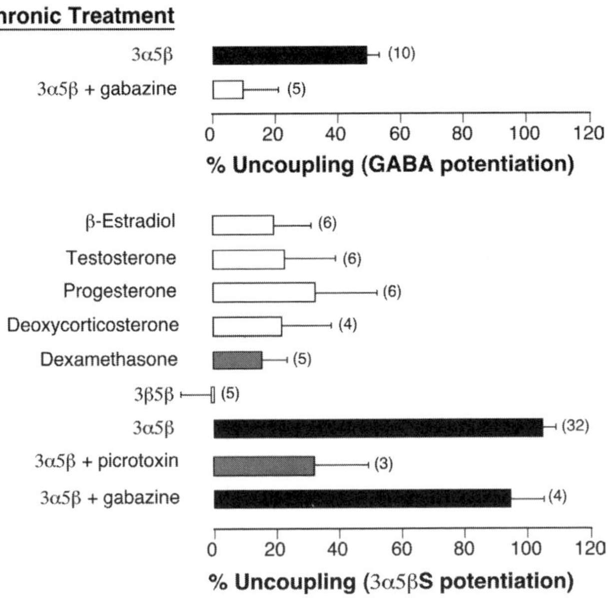

FIGURE 16.3 Pharmacological specificity of steroid-induced $GABA_A$ receptor uncoupling. Cultures were treated for 48 h with the steroid indicated (progesterone, 5 μ*M*; others, 10 μ*M*). Cultures were scraped, homogenized, dialyzed for 24 h against 4 l of phosphate buffer with four changes of buffer, and tested for potentiation of 1 n*M* ^{3}H-FNZ binding by 10 μ*M* 3α5β or 1 m*M* GABA. Percentage of uncoupling was calculated relative to homogenate from untreated sister cultures according to the equation: percent uncoupling = 100% × ($\text{potentiation}_{\text{control}}$ − $\text{potentiation}_{\text{treated}}$)/$\text{potentiation}_{\text{control}}$. Picrotoxin (100 μ*M*) or gabazine (100 μ*M*) was included during the chronic treatment period where indicated. Black bars indicate statistical significance at the $p < .01$ level. Gray bars indicate statistical significance at the $p < .05$ level (unpaired *t*-test). Uncoupling was of borderline significance for 17β-estradiol ($p \leq .067$), testosterone ($p \leq .1$), progesterone ($p \leq .07$), and deoxycorticosterone ($p \leq 0.09$). The decrease in uncoupling of 3α5β potentiation by the presence of picrotoxin is also statistically significant ($p < .0001$), as is the decrease in uncoupling of GABA potentiation in the presence of gabazine ($p < .007$). No uncoupling was observed with cultures treated with the inactive isomer 3β-hydroxy-5β-pregnan-20-one. (Redrawn from Friedman, L.K. et al., *Brain Res.*, 707, 100–109, 1996, with permission from Elsevier; Friedman, L. et al., *Mol. Pharmacol.*, 44, 191–197, 1993, with permission from ASPET.)

were chronically exposed to 3α5β, suggesting that uncoupling is an active process that requires intact cells. Co-treatment of cultures with 3α5β and gabazine did not prevent the loss of 3α5β potentiation of ^{3}H-flunitrazepam binding, but did block the decrease in GABA potentiation of ^{3}H-flunitrazepam binding,[14] demonstrating that 3α5β–induced steroid–benzodiazepine site uncoupling and GABA–benzodiazepine site uncoupling are distinct. This result suggests that the GABA–benzodiazepine site uncoupling is likely an indirect consequence of potentiation of GABA transmission by 3α5β, whereas steroid–GABA site uncoupling is triggered by a different mechanism. Although resistant to gabazine, steroid–benzodiazepine site uncoupling was significantly decreased by co-treatment with the noncompetitive $GABA_A$ receptor antagonist picrotoxin, indicating that this form of uncoupling is mediated through an action of 3α5β at the $GABA_A$ receptor.[15]

16.5 POSITIVE AND NEGATIVE MODULATION OF NMDA RECEPTORS BY NEUROACTIVE STEROIDS

The observation that pregnenolone sulfate, a relatively abundant neurosteroid, inhibits both inhibitory $GABA_A$ and glycine receptors suggested that pregnenolone sulfate might be an endogenous pro-excitatory modulator. We hypothesized that such a compound might be expected to enhance the function of ionotropic glutamate receptors, which are the predominant receptors mediating fast excitatory transmission in the CNS. This was a particularly interesting question because while ionotropic glutamate receptors are ligand-gated ion channels like the $GABA_A$ receptor, they are structurally quite different and evolutionarily unrelated.

$GABA_A$ receptors belong to the Cys-loop superfamily of receptors, which also includes glycine, nicotinic acetylcholine, and 5-HT_3 receptors. These ligand-gated channels are comprised of five subunits, each of which has four transmembrane domains. In contrast, glutamate receptors bear no sequence similarity to Cys-loop receptors and are believed to be comprised of four subunits, each of which has three transmembrane domains, as well as a hydrophobic reentrant loop that dips into the membrane but does not cross.[16] The NMDA receptor is of particular interest in that it is thought to play a critical role in learning, as well as in a number of pathological processes, including stroke and neurodegenerative diseases.[17,18]

To test this prediction, we examined the effect of pregnenolone sulfate on NMDA receptors of chick spinal cord neurons. As shown in Figure 16.4, pregnenolone sulfate (100 μ*M*) dramatically enhanced the NMDA-induced current. The potentiating effect was independent of the effects of other known NMDA receptor modulators, including glycine,[19] arachidonic acid, and spermine,[20] arguing that pregnenolone sulfate acts through a distinct binding site. To determine whether pregnenolone sulfate was diffusing across the membrane to act at an intracellular site, pregnenolone sulfate was introduced into the interior of the cell via the patch electrode. No increase in the NMDA response was observed, whereas subsequent application of pregnenolone sulfate to the exterior of the neuron still enhanced the NMDA-induced current, demonstrating that the site of pregnenolone sulfate action is associated with the extracellular membrane surface.

Not all glutamate receptors were enhanced by pregnenolone sulfate, however; AMPA and kainate receptors (and as we later discovered, some NMDA receptor subtypes) were inhibited.[19–22] Follow-up experiments yielded similar results for modulation of NMDA-induced calcium accumulation in primary rat hippocampal cultures, indicating that modulation by neurosteroids was a general characteristic of NMDA receptors.[23]

16.6 DISTINCT SITES FOR STEROID POSITIVE AND NEGATIVE MODULATORS OF NMDA RECEPTORS

Pregnanolone is inactive at the NMDA receptor, but its sulfated derivative, 3α-hydroxy-5β-pregnan-20-one sulfate (3α5βS), is inhibitory, as is 3β-hydroxy-5β-pregnan-20-

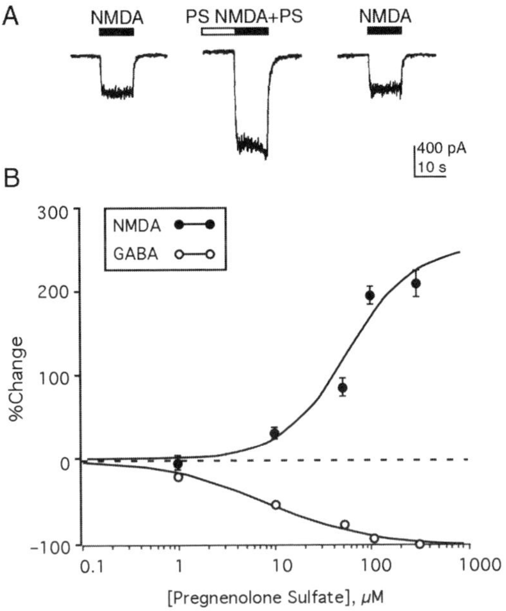

FIGURE 16.4 Pregnenolone sulfate potentiates the NMDA response. Primary cultures of spinal cord neurons were prepared from 7-day chicken embryos. (A) Pregnenolone sulfate greatly potentiates the NMDA-induced current. Horizontal bar above trace indicates period of drug application. (B) Concentration–response curves for potentiation of the NMDA response and inhibition of the GABA response by pregnenolone sulfate. Results were fitted with the logistic equation: for NMDA, $EC_{50} = 57\ \mu M$, n = 1.3, maximum potentiation = 256%; for GABA, $EC_{50} = 7.2\ \mu M$, n = 0.85, maximum inhibition = 100%. (From Wu, F.-S. et al., *Mol. Pharmacol.*, 40, 333–336, 1991. With permission from ASPET.)

one sulfate (3β5βS).[24] To determine whether the potentiating and inhibitory effects of these sulfated steroids on the NMDA receptor were mediated by a common site, the inhibitory dose–response curve for 3β5βS was determined in the presence and absence of 200 μ*M* pregnenolone sulfate, using NR1g/NR2A receptors expressed in *Xenopus* oocytes. No change in the 3β5βS IC_{50} was observed in the presence of PS, indicating that PS and 3β5βS do not compete for a common site (Figure 16.5). Therefore, there are at least two distinct steroid modulatory sites associated with the NMDA receptor.[20] The observation of two steroid sites on the NMDA receptor parallels the similar observation with the $GABA_A$ receptor, suggesting that this could be a general feature of amino acid receptors.

16.7 GEOMETRY AND CHARGE AS DETERMINANTS OF MODULATORY ACTIVITY AT NMDA RECEPTORS

Pregnenolone and pregnenolone formate were without potentiating activity at NMDA receptors, demonstrating that the presence of the negatively charged C3 sulfate group

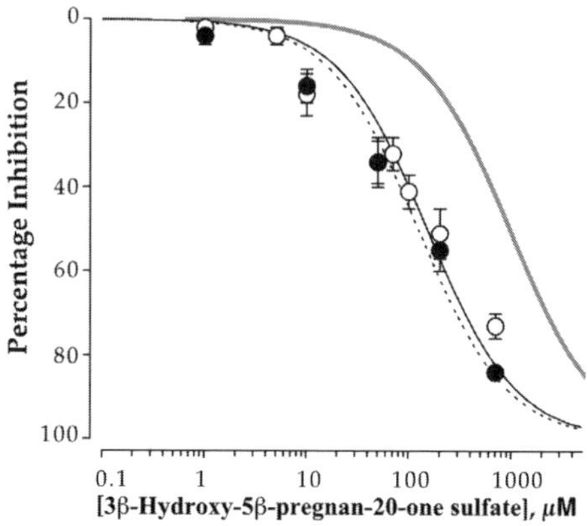

FIGURE 16.5 PS and 3β5βS act at separate sites on recombinant NMDA receptors. Concentration–response curve for inhibition of the NMDA-induced current (peak responses) by 3β5βS in the absence(—o—) and presence (- -●- -) of 200 μ*M* PS (EC_{50} = 40 μ*M*) in oocytes expressing recombinant NR1g/NR2A NMDA receptors. Results are means of pooled data from five to eight oocytes. Error bars indicate standard errors. No shift of the 3β5βS IC_{50} is observed in the presence of PS. 3β5βS IC_{50} values, estimated by nonlinear regression using the Michaelis–Menten equation, are 151 μ*M* in the absence and 124 μ*M* in the presence of 100 μ*M* PS. Gray curve shows the expected competitive shift in the 3β5β concentration–response curve in the presence of PS, taking the PS EC_{50} of 40 μ*M* as an estimate of its K_d. (From Park-Chung, M. et al., *Mol. Pharmacol.*, 52, 1113–1123, 1997. With permission from ASPET.)

is a critical determinant of positive modulatory activity at the NMDA receptor. However, other negatively charged groups could also confer activity. In particular, pregnenolone hemioxylate, hemisuccinate, and hemiglutarate all exhibited positive modulation. Similarly, pregnanolone, pregnanolone formate, and pregnanolone hemisuccinate methyl ester were inactive, whereas pregnanolone hemioxylate, hemisuccinate, and hemiglutarate all exhibited negative modulation, indicating that a negative charge at C3 is also a critical determinant of activity of negative modulators.

To investigate the stereochemistry of steroid modulation, the activity of stereoisomers of pregnanolone sulfate with different C3 or C5 stereochemistry was compared. 3β-Hydroxy-5α-pregnan-20-one sulfate (3β5αS), which most resembles pregnenolone sulfate in having a relatively planar structure, enhanced NMDA currents, whereas the three kinked pregnanolone isomers were inhibitory (Figure 16.6), suggesting that the geometry of the steroid nucleus determines selectivity for the positive and negative modulatory sites of the NMDA receptor.

16.8 PREGNENOLONE SULFATE IS A SUBTYPE SELECTIVE MODULATOR OF THE NMDA RECEPTOR

To explore the subunit dependence of steroid potentiation of the NMDA receptor, the NR1g subunit was coexpressed in *Xenopus* oocytes with the NR2A, NR2B, NR2C, or NR2D subunit. With NR1/NR2A and NR1/NR2B receptors, pregnenolone

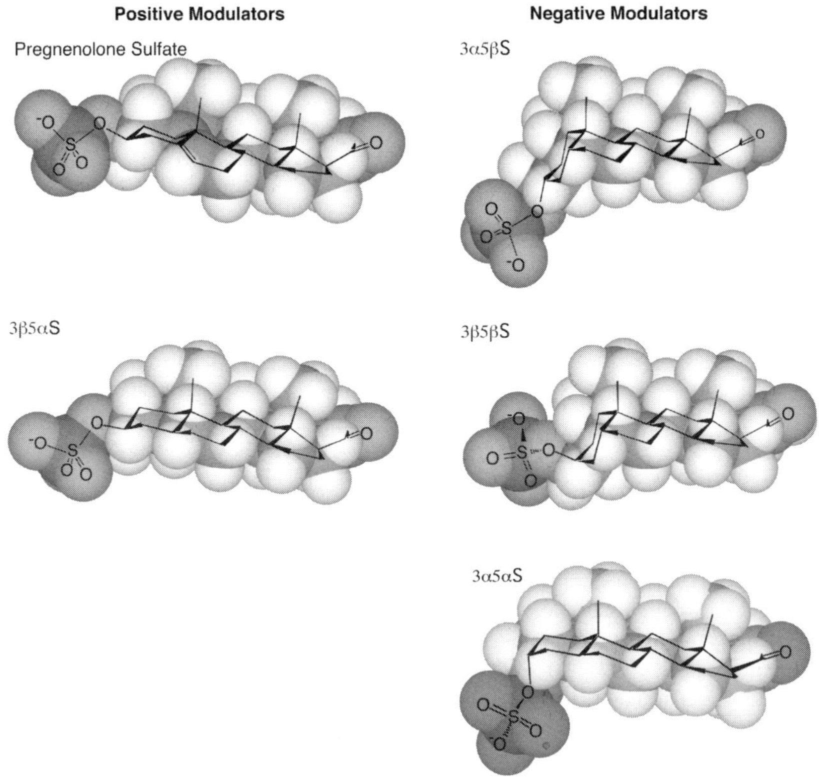

FIGURE 16.6 Geometry of positive and negative modulators of the NMDA response. Steroids are classified as positive or negative modulators based upon enhancement/inhibition of the NMDA response of primary chick neurons and enhancement/inhibition of NMDA-induced cell death of hippocampal neurons in culture.[28] (From Weaver et al., *J. Pharmacol. Exp. Ther.*, 293, 747–754, 2000. With permission.)

sulfate enhanced the NMDA response, similar to the results seen in chick spinal cord neurons, but with NR1/NR2C and NR1/NR2D receptors, the effect of NMDA was inhibitory (Figure 16.7). In contrast, 3α5βS was inhibitory at all four NMDA receptor subunit combinations,[21] consistent with our previous conclusion that sulfated pregnenolone and pregnanolone derivatives act at distinct sites,[20] as well as at non-NMDA GluR1, GluR3, and GluR5 receptors.[25]

16.9 NEUROPROTECTIVE AND PRO-EXCITOTOXIC EFFECTS OF NEUROACTIVE STEROIDS

Excessive accumulation of intracellular calcium mediated by the NMDA receptor is believed to be a critical factor leading to excitotoxic neuronal death in stroke and neurodegenerative diseases such as Parkinson's disease, Huntington's disease, and Alzheimer's disease. The ability of neuroactive steroids to modulate NMDA-induced neuronal calcium accumulation[23,26] suggested that neurosteroids could

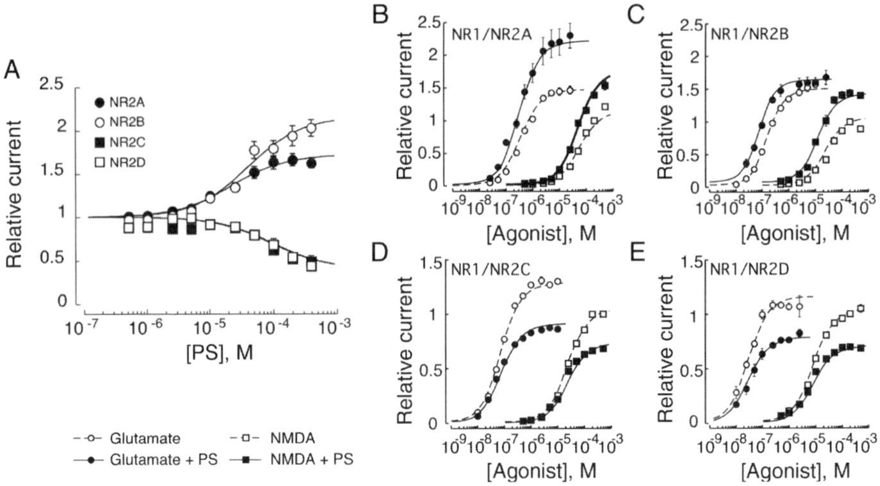

FIGURE 16.7 The choice of NR2 subunit determines the direction of PS modulation of the glutamate and NMDA concentration–response curves. Data points are averaged normalized peak NMDA-induced current responses obtained from oocytes injected with NR1/NR2A, NR1/NR2B, NR1/NR2C, or NR1/NR2D mRNAs. (A) Concentration–response curves for PS's effect on NR1/NR2 receptors. Responses were normalized to the control response obtained by application of 10 μ*M* glycine plus 80 μ*M* NMDA (NR2A), 25 μ*M* NMDA (NR2B, NR2C), or 10 μ*M* NMDA (NR2D). (B–E) Concentration–response data for NMDA and L-glutamate were obtained in the presence of 10 μ*M* glycine in oocytes expressing NR1/NR2A, NR1/NR2B, NR1/NR2C, or NR1/NR2D. The data were normalized relative to the current response from the same oocyte induced by co-application of 200 μ*M* NMDA and 10 μ*M* glycine. Error bars represent SEM. (From Malayev, A.A. et al., *Br. J. Pharmacol.*, 135, 901–909, 2002. With permission.)

be expected to influence NMDA-induced excitotoxicity. In studies of primary hippocampal cultures, potentiating steroids, such as pregnenolone sulfate and 3β-hydroxy-5α-pregnan-20-one sulfate, enhanced NMDA toxicity, whereas inhibitory steroids such as 3α5βS were protective. Effects of steroids on NMDA-induced toxicity were strongly correlated with their effects on NMDA-induced calcium accumulation, suggesting that modulation of neurotoxicity was a consequence of modulation of NMDA receptor-mediated calcium accumulation.[27] Notably, activity was preserved with dicarboxylic acid derivatives in which the sulfate group of pregnenolone sulfate or 3α5βS was replaced with a hemioxylate, hemisuccinate, or hemiglutarate group, but absent with the uncharged pregnanolone formate and pregnanolone hemisuccinate methyl ester.

16.10 PROTECTION AGAINST NMDA-INDUCED EXCITOTOXICITY BY 17B-ESTRADIOL

A notable exception to the general rule that a negatively charged group at C3 is required for NMDA receptor modulation was 17β-estradiol, which inhibited the NMDA-induced current despite having only a hydroxyl group at C3. Consistent

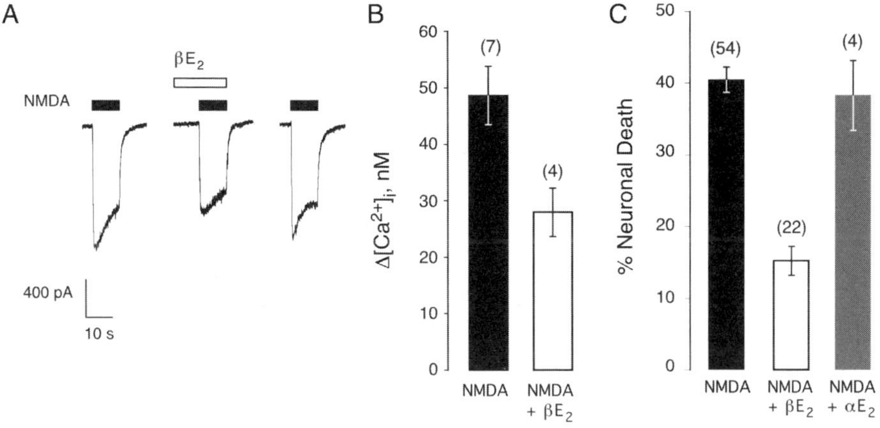

FIGURE 16.8 17β-Estradiol is a negative modulator of the NMDA receptor of rat hippocampal neurons in culture and protects neurons against NMDA excitotoxicity. (A) 17β-Estradiol (βE_2; 50 μ*M*) rapidly and reversibly inhibits the 30 μ*M* NMDA-induced current. (B) 17β-Estradiol (50 μ*M*) significantly ($p < .05$) reduces the 5 μ*M* NMDA-induced rise in intracellular free Ca^{2+}. Error bars depict standard errors. Number of experiments is given in parentheses. (C) βE_2 (10 or 50 μ*M*) significantly ($p < .05$) reduces 30 μ*M* NMDA-induced death. αE_2 has no protective effect. Error bars depict standard errors. Number of experiments is given in parentheses. (From Weaver, C.E., Jr. et al., *Brain Res.*, 761, 338–341, 1997. With permission from Elsevier.)

with its activity as an inhibitor of the NMDA-induced current, 17β-estradiol antagonized NMDA-induced death of hippocampal neurons in culture. The neuroprotective effect of 17β-estradiol was not blocked by tamoxifen, arguing against an action mediated by intracellular estrogen receptors. 17α-Estradiol was not neuroprotective, arguing that the protective effect of 17β-estradiol in this assay was probably due to NMDA receptor antagonism, and not to nonspecific scavenging of free radicals (Figure 16.8).[28]

16.11 A STEROID-BASED THERAPEUTIC STRATEGY

Sulfated steroids are not particularly good candidates as pharmacological agents for *in vivo* use. They are highly charged, and probably diffuse across slowly, if at all, the blood–brain barrier as intact steroid sulfates. Moreover, they are likely to be rapidly degraded by steroid sulfatases. Hemisuccinated derivatives such as pregnanolone hemisuccinate (3α5βHS) are likely to have more favorable pharmacokinetics, as they are less highly charged and resistant to sulfatases. To test this, mice were injected with either 3α5βS or 3α5βHS (25 mg/kg, i.v.). 3α5βS was without effect, whereas 3α5βHS produced rapid, reversible sedation. In other studies, 3α5βHS at a subsedative dose increased latency to seizure after NMDA injection and decreased licking in the late, but not early, phase of formalin-induced pain. Both results suggested that 3α5βHS was most likely acting to inhibit NMDA receptors.

To explore the neuroprotective efficacy of 3α5βHS in a model of stroke, a nonsedative dose of 3α5βHS (6.9 mg/kg, i.v. bolus, followed by i.v. infusion of 6.9

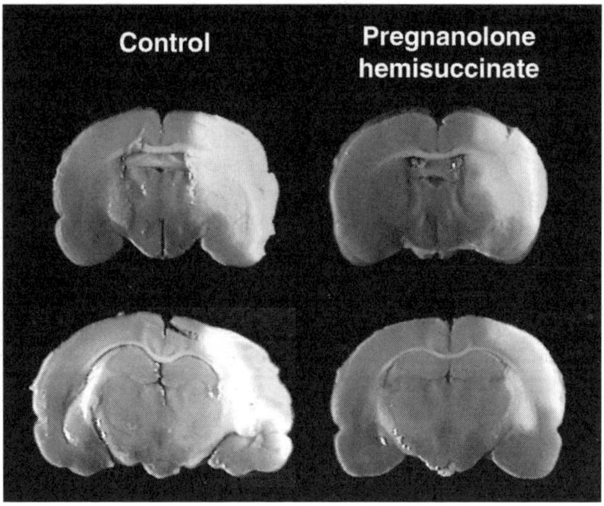

FIGURE 16.9 3α5βHS is neuroprotective in an *in vivo* model of stroke. Rats were infused with vehicle (0.1 *M* phosphate buffer, pH 7.4/5% DMSO/10% cyclodextrin) or 6.9 mg/kg/h (6.9 mg/kg loading dose) of 3α5βHS, beginning immediately after initiation of ischemia by middle cerebral artery occlusion. Infusion of 3α5βHS was continued for an additional 22 h, at which time the rats were killed and their brains were stained with 2,3,5-triphenyl tetrazolium Cl. Representative brain sections are shown. Infarct area appears pale. (From Weaver, C.E., Jr. et al., *Proc. Natl. Acad. Sci. U.S.A.*, 94, 10450–10454, 1997. With permission.)

mg/kg/h) was administered to rats immediately or 30 min after middle cerebral artery occlusion. Rats that received 3α5βHS immediately following occlusion exhibited a reduction in the volume of cortical and subcortical infarct by 47 ± 10% and 26 ± 6%, respectively (n = 10). Rats that received 3α5βHS 30 min after occlusion exhibited a 39 ± 7% (n = 13) reduction in cortical infarct volume, with no significant reduction in subcortical infarct (Figure 16.9). These results suggest that steroid negative modulators may have promise as neuroprotective agents *in vivo*.[29]

16.12 MODULATION OF DOPAMINE RELEASE BY NEUROSTEROIDS

As a further test of the ability of 3α5βHS to inhibit NMDA receptors *in vivo*, the effect of systemically administered 3α5βHS on NMDA-induced striatal dopamine release was examined using *in vivo* microdialysis. Administration of NMDA (1 m*M*) to rats via a striatal microdialysis probe resulted in increased levels of striatal dopamine. NMDA-induced dopamine release was significantly decreased by prior injection of 3α5βHS at doses of 5 mg/kg (i.p.) or greater (Figure 16.10). In addition, 3α5βHS (1 mg/kg) blocked the increase in behavioral activity resulting from striatal microinjection of NMDA (5 μg).[30]

To determine whether steroid positive modulators of NMDA receptors also influence dopamine release, we infused PS into striatum via the microdialysis probe.

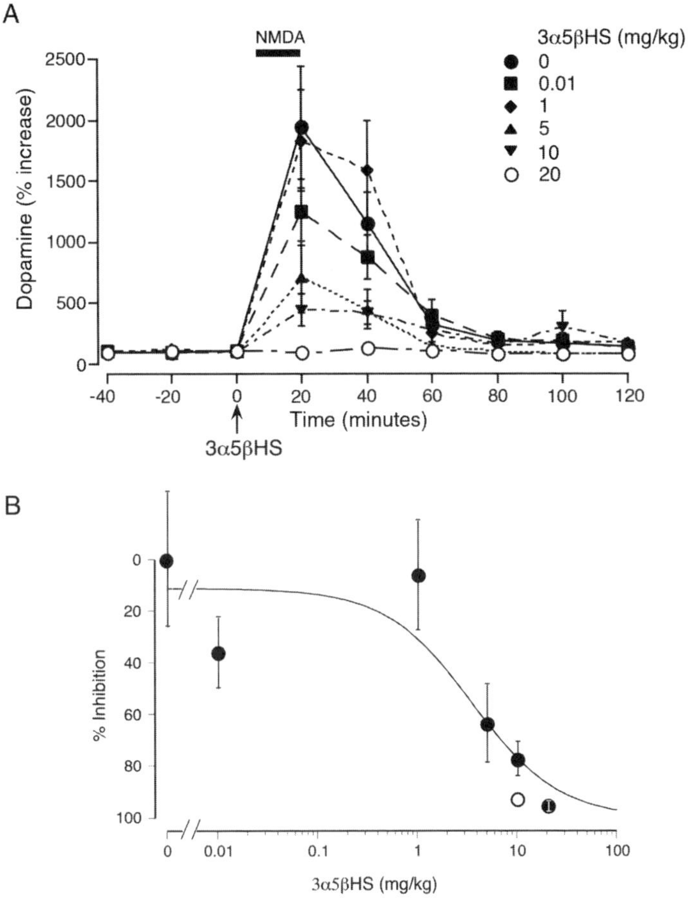

FIGURE 16.10 3α5βHS attenuates NMDA-evoked release of extracellular dopamine in the striatum. (A) Rats were implanted with a striatal dialysis guide cannula a minimum of 2 days prior to the experiment. A microdialysis probe was inserted either 2 or 12 h prior to the initiation of the experiment. Artificial cerebrospinal fluid (aCSF) was infused at 2 μl/min. Samples were analyzed by HPLC with electrochemical detection. Dopamine concentrations were normalized to an external standard and plotted as percentage change from baseline. Rats were injected with either vehicle or 3α5βHS (0 to 20 mg/kg, i.p.) immediately after the collection of the sample at time 0. After 10 min, the perfusion medium was switched from aCSF to a 1 m*M* NMDA solution in aCSF. After collection of the next 20-min sample, the perfusion medium was switched back to aCSF and five more samples were collected. Data are mean ± SEM for extracellular concentrations of dopamine, presented as percentages of baseline values. The infusion of NMDA started at the 10-min time point and lasted for 20 min (indicated by horizontal bar). (B) Dose–response curve for 3α5βHS inhibition of NMDA-induced dopamine release at 20 min. Open circle is from a group of rats that had the microdialysis probe inserted 12 h before the start of the experiment. Data points indicate percentage change in NMDA-induced striatal dopamine release. Error bars indicate SEM. Four to six rats per group were used. (From Sodri-Vakili, G. et al., 86, 92–101, 2003. With permission from Blackwell Publishing Ltd.)

Surprisingly, PS alone induced a large spike in dopamine release when infused at submicromolar concentrations.[31] As this is below the concentration range at which PS has been observed to modulate NMDA response's spinal and hippocampal neurons in culture or *Xenopus* oocytes expressing NMDA receptor subunits, we wondered whether PS stimulation of dopamine release might be mediated by some other neurosteroid target, such as σ receptors.[32] A variety of neurosteroids, including PS, bind to σ1 receptors,[33] and modulation of NMDA-induced norepinephrine release in the rat hippocampus by nanomolar concentrations of neurosteroids has been reported to be mediated by σ1 receptors.[34]

We examined whether the effect of PS on striatal dopamine could be blocked or mimicked by σ receptor ligands. However, a σ receptor antagonist did not block stimulation of dopamine release by PS. Consistent with this observation, a selective σ agonist failed to induce dopamine release. PS stimulation of dopamine release was blocked by the NMDA receptor antagonist D-AP5 and was mimicked by pregnenolone hemisuccinate, suggesting an action mediated by NMDA receptors. However, the lowest reported EC_{50} for PS modulation of NMDA receptors as determined by enhancement of spontaneous EPSCS of hippocampal neurons in culture[20] is 12 μ*M*. It may be that the affinity of PS for adult NMDA receptors *in vivo* is greater than is evident from studies of expressed receptors or primary embryonic neurons in culture, perhaps reflecting the involvement of novel NMDA receptor subunits or its association with other cellular components. Alternatively, it is possible that PS stimulation of dopamine release is mediated through a novel binding site that interacts directly or indirectly with NMDA receptors. While the details of the mechanism remain to be elucidated, this result clearly strengthens the case for a modulatory role for endogenous PS *in vivo*.

REFERENCES

1. Wu, F.-S., Gibbs, T.T., and Farb, D.H., Inverse modulation of γ-aminobutyric acid- and glycine-induced currents by progesterone, *Mol. Pharmacol.*, 37, 597–602, 1990.
2. Harrison, N.L., Majewska, M.D., Harrington, J.W., and Barker, J.L., Structure-activity relationships for steroid interaction with the γ-aminobutyric acidA receptor complex, *J. Pharmacol. Exp. Ther.*, 241, 346–353, 1987.
3. Rupprecht, R., The neuropsychopharmacological potential of neuroactive steroids, *J. Psychiatr. Res.*, 31, 297–314, 1997.
4. Harrison, N.L. and Simmonds, M.A., Modulation of the GABA receptor complex by a steroid anaesthetic, *Brain Res.*, 323, 287–292, 1984.
5. Cottrell, G.A., Lambert, J.J., and Peters, J.A., Modulation of GABAA receptor activity by alphaxalone, *Br. J. Pharmacol.*, 90, 491–500, 1987.
6. Majewska, M.D. and Schwartz, R.D., Pregnenolone-sulfate: an endogenous antagonist of the γ-aminobutyric acid receptor complex in brain? *Brain Res.*, 404, 355–360, 1987.
7. Majewska, M.D., Demirgören, S., Spivak, C.E., and London, E.D., The neurosteroid dehydroepiandrosterone sulfate is an allosteric antagonist of the GABAA receptor, *Brain Res.*, 526, 143–146, 1990.

8. Park-Chung, M., Malayev, A., Purdy, R.H., Gibbs, T.T., and Farb, D.H., Sulfated and unsulfated steroids modulate γ-aminobutyric acidA receptor function through distinct sites, *Brain Res.*, 830, 72–87, 1999.
9. Chan, C.Y. and Farb, D.H., Modulation of neurotransmitter action: control of the γ-aminobutyric acid response through the benzodiazepine receptor, *J. Neurosci.*, 5, 2365–2373, 1985.
10. Gyenes, M., Farrant, M., and Farb, D.H., "Run-down" of γ-aminobutyric acidA receptor function during whole-cell recording: a possible role for phosphorylation, *Mol. Pharmacol.*, 34, 719–723, 1988.
11. Gyenes, M., Wang, Q., Gibbs, T.T., and Farb, D.H., Phosphorylation factors control neurotransmitter and neuromodulator action at the γ-aminobutyric acid type A receptor, *Mol. Pharmacol.*, 46, 542–549, 1994.
12. Roca, D.J., Schiller, G.D., Rozenberg, I., Gibbs, T.T., and Farb, D.H., γ-Aminobutyric acid receptor regulation in culture: altered allosteric interactions following prolonged exposure to benzodiazepines, barbiturates, and methylxanthines, *Mol. Pharmacol.*, 37, 710–719, 1990.
13. Roca, D.J., Rozenberg, I., Farrant, M., and Farb, D.H., Chronic agonist exposure induces down regulation and allosteric uncoupling of the γ-aminobutyric acid/benzodiazepine receptor complex, *Mol. Pharmacol.*, 37, 37–43, 1990.
14. Friedman, L.K., Gibbs, T.T., and Farb, D.H., γ-Aminobutyric acidA receptor regulation: heterologous uncoupling of modulatory site interactions induced by chronic steroid, barbiturate, benzodiazepine, or GABA treatment in culture, *Brain Res.*, 707, 100–109, 1996.
15. Friedman, L., Gibbs, T.T., and Farb, D.H., γ-Aminobutyric acidA receptor regulation: chronic treatment with pregnanolone uncouples allosteric interactions between steroid and benzodiazepine recognition sites, *Mol. Pharmacol.*, 44, 191–197, 1993.
16. Wo, Z. and Oswald, R., Unraveling the modular design of glutamate-gated ion channels, *Trends Neurosci.*, 18, 161–168, 1995.
17. Choi, D.W., Excitotoxic cell death, *J. Neurobiol.*, 23, 1261–1276, 1992.
18. Coyle, J.T. and Puttfarcken, P., Oxidative stress, glutamate, and neurodegenerative disorders, *Science*, 262, 689–695, 1993.
19. Wu, F.-S., Gibbs, T.T., and Farb, D.H., Pregnenolone sulfate: a positive allosteric modulator at the NMDA receptor, *Mol. Pharmacol.*, 40, 333–336, 1991.
20. Park-Chung, M., Wu, F.-S., Purdy, R.H., Malayev, A.A., Gibbs, T.T., and Farb, D.H., Distinct sites for inverse modulation of NMDA receptors by sulfated steroids, *Mol. Pharmacol.*, 52, 1113–1123, 1997.
21. Malayev, A.A., Gibbs, T.T., and Farb, D.H., Inhibition of the NMDA response by pregnenolone sulfate reveals subtype selective modulation of NMDA receptors by sulfated steroids, *Br. J. Pharmacol.*, 135, 901–909, 2002.
22. Yaghoubi, N., Malayev, A., Russek, S.J., Gibbs, T.T., and Farb, D.H., Neurosteroid modulation of recombinant ionotropic glutamate receptors, *Brain Res.*, 803, 153–160, 1998.
23. Irwin, R.P., Maragakis, N.J., Rogawski, M.A., Purdy, R.H., Farb, D.H., and Paul, S.M., Pregnenolone sulfate augments NMDA receptor mediated increases in intracellular Ca2+ in cultured rat hippocampal neurons, *Neurosci. Lett.*, 141, 30–34, 1992.
24. Park-Chung, M., Wu, F.-S., and Farb, D.H., 3α-Hydroxy-5β-pregnan-20-one sulfate: a negative modulator of the NMDA-induced current in cultured neurons, *Mol. Pharmacol.*, 46, 146–150, 1994.

25. Yaghoubi, N., Gibbs, T.T., and Farb, D.H., Evaluation of neurosteroid modulation of kainate receptors using an automated system for oocyte electrophysiology, *Soc. Neurosci. Abstr.*, 20, 1109, 1995.
26. Irwin, R.P., Lin, S.-Z., Rogawski, M.A., Purdy, R.H., and Paul, S.M., Steroid potentiation and inhibition of N-methyl-D-aspartate receptor-mediated intracellular Ca++ responses: structure activity studies, *J. Pharmacol. Exp. Ther.*, 271, 677–682, 1994.
27. Weaver, C.E., Jr., Land, M.B., Purdy, R.H., Richards, K.G., Gibbs, T.T., and Farb, D.H., Geometry and charge determine pharmacological effects of steroids on N-methyl-D-aspartate receptor induced Ca2+ accumulation and cell death, *J. Pharmacol. Exp. Ther.*, 293, 747–754, 2000.
28. Weaver, C.E., Jr., Park-Chung, M., Gibbs, T.T., and Farb, D.H., 17β-Estradiol protects against NMDA-induced excitotoxicity by direct inhibition of NMDA receptors, *Brain Res.*, 761, 338–341, 1997.
29. Weaver, C.E., Jr., Marek, P., Park-Chung, M., Tam, S.W., and Farb, D.H., Neuroprotective activity of a new class of steroidal inhibitors of the N-methyl-D-aspartate receptor, *Proc. Natl. Acad. Sci. U.S.A.*, 94, 10450–10454, 1997.
30. Sadri-Vakili, G., Johnson, D.W., Janis, G.C., Gibbs, T.T., Pierce, R.C., and Farb, D.H., Inhibition of NMDA-induced striatal dopamine release and behavioral activation by the neuroactive steroid 3α-hydroxy-5β-pregnan-20-one hemisuccinate, *J. Neurochem.*, 86, 92–101, 2003.
31. Sadri-Vakili, G., Janis, G.C., Gibbs, T.T., Pierce, R.C., and Farb, D.H., Pregnenolone sulfate increases dopamine release in the rat striatum, *Soc. Neurosci. Abstr.*, 28, 640.8, 2002.
32. Gibbs, T.T. and Farb, D.H., Dueling enigmas: neurosteroids and sigma receptors in the limelight, *Sci. STKE*, 2000, PE1, 2000.
33. Su, T.P., London, E.D., and Jaffe, J.H., Steroid binding at sigma receptors suggests a link between endocrine, nervous, and immune systems, *Science*, 240, 219–221, 1988.
34. Monnet, F.P., Mahe, V., Robel, P., and Baulieu, E.E., Neurosteroids, via sigma receptors, modulate the [3H]norepinephrine release evoked by N-methyl-D-aspartate in the rat hippocampus, *Proc. Natl. Acad. Sci. U.S.A.*, 92, 3774–3778, 1995.

17 On the Neurosteroid Mechanisms of Action and Their Clinical Potentials

Yvette Akwa and Etienne-Emile Baulieu

CONTENTS

17.1 INTRODUCTION

Neurosteroids, by definition, are those steroids that are synthesized in the central (CNS) and peripheral (PNS) nervous systems, *independently of the steroidogenic activity of endocrine glands* (gonads and adrenals). They pertain to the same chemical group as classical steroid hormones, their precursors and metabolites, and they derive from cholesterol.

Pregnenolone (PREG) appears to be among the most quantitatively important neurosteroids, and it is a major precursor of other neurosteroids, including progesterone (PROG). So far, there is no direct evidence permitting the strict classification of estradiol, testosterone, and corticosteroids as neurosteroids according to the above-cited definition.

0-8493-2392-4/04/$0.00+$1.50

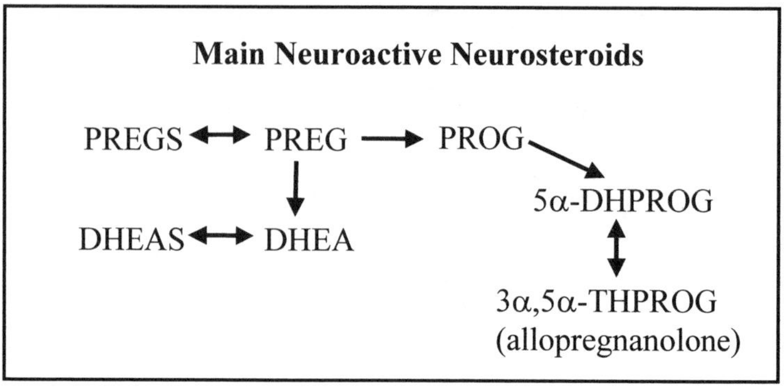

PREG, pregnenolone;PREGS, pregnenolone sulfate
DHEA, dehydroepiandrosterone; DHEAS, dehydroepiandrosterone sulfate
PROG, progesterone
5α-DHPROG, 5α-dihydroprogesterone
3α,5α-THPROG, 3α,5α-tetrahydroprogesterone

FIGURE 17.1

Neurosteroid biosynthesis and metabolism in the CNS and PNS are found in *different cell types*, such as oligodendrocytes, Schwann cells, astrocytes, and neurons. However, not all involved enzymes are found in the same cell type, implicating that if some neurosteroids may be produced in a single cell type (e.g., oligodendrocytes for PREG and Schwann cells for PROG), others require a cooperation between cells of a different nature (Zwain and Yen, 1999; also see Chapter 1 of this handbook).

Regulatory events involved in neurosteroid synthesis and metabolism are far from being completely explored. Significant results may only be obtained if gender, age, and other environmental factors (enriched medium, food, circadian variation, etc.) are taken into account. There are already some indications that the following modulate neurosteroid synthetizing and metabolizing enzymes:

- Peptidic factors — octadecaneuropeptide, a peptide derived from the diazepam binding inhibitor (Mensah-Nyagan et al., 1999), and neuropeptide Y (Beaujean et al., 2002; Ferrara et al., 2001)
- Neurotransmitters — GABA (Do-Rego et al., 2000) and glutamate/N-methyl-D-aspartate (NMDA) (Kimoto et al., 2001)
- Steroid hormones — estradiol (Akwa et al., 1992; Coirini et al., 2003)

There are also indications for a role for cholesterol transporters such as steroidogenic acute regulatory protein (StAR) (Furukawa et al., 1998; King et al., 2002) and peripheral benzodiazepine receptor (PBR) (Lacor et al., 1999; Serra et al., 1999) in neurosteroidogenesis. Effects may differ according to the stage of *cell differentiation* (e.g., oligodendroglial lineage (Gago et al., 2001) and aging cells (Weill-Engerer et al., 2003)). As a result, the metabolic turnover of neurosteroids is far from being understood, and actually no relevant data have been obtained, the global contribution of neurosteroids made in the brain to the blood concentration of steroids being very small.

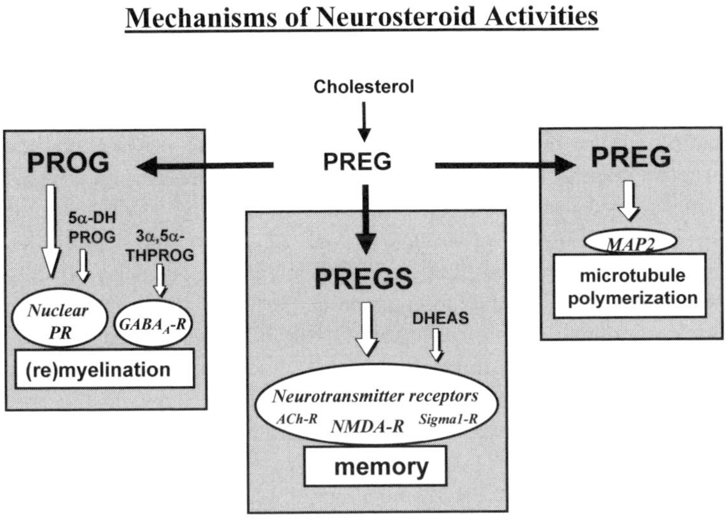

FIGURE 17.2

Further complexity is found at the level of mechanisms of action, whose modulation may differ according to the distinct *topology of steroid distribution*. It is thus particularly difficult to extrapolate from *in vitro* experiments to *in vivo* situations. Neurosteroids act within the nervous system, according to mostly *paracrine* and possibly *autocrine* mechanisms: there is local diffusion across brain cells with efficient concentrations in the micromolar range, differently from hormonal steroids reaching the nervous system via blood circulation and acting only in the nanomolar range concentrations. So far, except for PROG (neuroprogesterone), there is no evidence for the existence of nuclear receptors for neurosteroids, including PREG, pregnanolone sulfate (PREGS), dehydroepiandrosterone (DHEA), dehydroepiandrosterone sulfate (DHEAS), allopregnanolone, etc. No specific membrane receptor has yet been cloned, but membrane mechanisms of neurosteroid activities have been demonstrated (e.g., at the level of neurotransmitter receptors; see Section 17.2.2 and related chapters in this book). New modes of action are still possible (e.g., microtubule receptor; see Section 17.2.3). Finally, it is difficult to distinguish neurosteroid actions from those of hormonal steroids accessing the brain targets.

The clinical potentials of neurosteroids are also not easy to classify because of the above-mentioned complexity, the same neurosteroid being susceptible to multiple modes of action and undergoing transformation into differently active metabolites.

17.2 MECHANISMS OF NEUROSTEROID ACTIVITIES

17.2.1 Nuclear Receptors

So far, the only qualified neurosteroid by which main activity appears to be mediated by a *classical nuclear progesterone receptor* (PR) is PROG. Even though a class of non-estrogen-inducible PR has been detected in certain regions of the brain

(MacLusky and McEwen, 1978), the expression of no other genes than those encoding the classical PRA and PRB has been reported.

We summarized here findings demonstrating that neuroprogesterone is a remyelinating agent in the PNS. *In vivo*, after cryolesion of the mouse sciatic nerve, remyelination of the reelongating axon is dependent on PROG synthesized by Schwann cells (Koenig, 1995). Indeed, PROG also stimulates myelination of neurites *in vitro*, in cultured dorsal root ganglia. This myelinating process implicates the expression of several myelin proteins (Po, PMP-22) and is enhanced by the PROG-PR system (Schumacher et al., 2001). PROG metabolites such as 5α-dihydroprogesterone (5α-DHPROG) and allopregnanolone (3α,5α-THPROG) can also increase the synthesis of some myelin proteins, 5α-DHPROG being able to interact with PR and 3α,5α-THPROG mainly through the $GABA_A$ receptor (Melcangi et al., 1998). Interestingly, sciatic nerves in aging rats display myelination defects that are restored by PROG and 5α-reduced metabolites (but not by testosterone and its 5α-reduced metabolites) (Melcangi et al., 2000).

In the CNS, the role of PROG in *in vivo* myelination remains to be firmly established, but *in vitro* experiments with organotypic cultures of cerebellum have clearly demonstrated a stimulatory PR-mediated effect on myelin synthesis (Ghoumari et al., 2003b).

Other nuclear PR-mediated PROG effects in the nervous system are still to be worked out in details (e.g., in the LHRH controlling system (Sterling et al., 1984)).

17.2.2 Membrane Receptors

The existence of proper steroid receptors suggested by the well-known reports of *rapid steroid action* (Blackmore et al., 1996; Finidori-Lepicard et al., 1981; McEwen et al., 1990; Selye, 1942) has not been substantiated by the definitive isolation of the corresponding receptor structure or its cloning. There is an important, still unique observation for a role of a corticosteroid receptor in the amphibian neuronal membrane (Orchinik et al., 1991). It is not yet known for sure if for a given steroid there are functional membrane receptors that are structurally identical to the corresponding classical nuclear receptors.

What is currently definitively accepted is neurosteroid effects on the function of *neurotransmitter receptors*. The most documented are the $GABA_A$ receptors (justifying this book), but even with it, we are far from understanding the molecular basis of the steroid action (binding site structure, different receptor isoforms containing different subunit subtypes that result in multiple steroid recognition sites on the receptors, possibly distinct modulatory interactions between neuroactive neurosteroids, etc.).

Among the glutamate receptors, only the ionotropic ones and among them the NMDA receptors have been functionally studied in terms of neurosteroid effects. The NMDA receptors are allosterically modulated positively by PREGS (Bowlby, 1993; Wu et al., 1991) and negatively by allopregnanolone sulfate (Park-Chung et al., 1997). Similarly and reciprocally, it is known that these compounds have opposite effects on the $GABA_A$ receptor function, negatively and positively, respectively (El-Etr et al., 1998): this is an additional indication that a specific steroid structure is involved in the regulation of neurotransmitter receptor activity. PREGS is also a negative modulator of the AMPA /kainate receptors.

The function of the recently cloned sigma 1 receptor (Hanner et al., 1996) is also regulated by neurosteroids. DHEAS behaves as an agonist (Bergeron et al., 1996; Monnet et al., 1995), while PREGS has been found to display either stimulatory (Ueda et al., 2001) or inhibitory (Monnet et al., 1995) actions. Moreover, PROG acts as sigma 1 receptor antagonist (Bergeron et al., 1996; Monnet et al., 1995).

Interactions of neurosteroids with other neurotransmitter receptors have also been studied. This is the case of the glycine receptors, the modulation of which by PREGS, DHEAS, and PROG is negative and subunit dependent (Maksay et al., 2001), and of the nicotinic acetylcholine receptor that is mainly antagonized by PROG (Bullock et al., 1997; Valera et al., 1992).

17.2.3 Microtubular Receptor

A new mode of action of steroids has recently been observed in the brain. A specific PREG binding protein was demonstrated and identified, after purification and immunological studies, as microtubule-associated protein 2 (MAP2). This protein is known to interact with tubulin and to stimulate its polymerization *in vitro*. It was found that PREG enhances the stimulatory effect of MAP2 on microtubule polymerization (Murakami et al., 2000). Steroids that do not bind to MAP2 do not have this effect. PROG and PREGS, which bind to MAP2, do not stimulate tubulin polymerization but, in contrast, competitively inhibit the polymerizing activity of PREG. MAP2 is found as several isoforms of large (≈250,000 Da) and small (≈60,000 Da) molecular weight (MW) and thus may be considered as a steroid receptor of a new category, neither nuclear nor membrane active. *In cello* experiments with cultured fetal rat neurons and PC-12 cells confirm the polymerizing effect of PREG leading to the outgrowth of neurites. Whether these expanding structures are functional, particularly as dendrites (where MAP2 is mostly found), remains to be investigated *in vitro* and *in vivo*.

17.2.4 Complementary Remarks on Neurosteroid Function

Five series of remarks may be considered in the near future:

1. The analytical *classification* of neurosteroid mechanisms of action summarized under Sections 17.2.1 through 17.2.3 is clearly too schematic: it is well known that membrane mechanisms often modify gene expression, and reciprocally, nuclear mechanisms frequently lead to changes of channel and receptor structure/function at the membrane level.
2. Much remains to be determined in molecular terms since almost no steroid receptor and enzyme of steroid synthesis and metabolism present and operating in the different regions of the nervous system have yet been examined to determine if specific *isoforms* are involved and may explain specific neuroactivities. Decisive help will now be found thanks to the recent advances with the available complete sequencing of genomes.

3. Possibly related to the complexity that remarks 1 and 2 entertain, studies of the relationships between neurosteroids and *ion channels*, the function of which is so important to the brain function, should become a priority in neurobiology.
4. More attention should be given to the modulation of *neurotransmitters and neuropeptide synthesis* and release by neurosteroids, an important field currently neglected.
5. The chemical nature of neurosteroids and of neurocholesterol (currently believed to be of astrocyte origin (Pfrieger, 2003)) reopens interest for the physico-chemical state of *membrane fluidity* as a factor for neural regulation.

17.3 CLINICAL POTENTIALS OF NEUROSTEROIDS

Obviously, it is particularly difficult to assess the actual role of steroids synthesized in the nervous system in human beings, due to the simultaneous presence of the same steroids possibly originating from blood circulation. However, the same steroids and identical receptor systems as in animals have been demonstrated, justifying in part some extrapolation from animals to human beings.

17.3.1 Neuroprotection, Neurotrophicity, and (Re)myelination

Several studies have been published in these fields, and among them many relate to the neuroactivity of steroids such as estradiol, glucocorticosteroids that are here not considered as neurosteroids.

A number of reports indicate that after trauma, vascular defect (e.g., cerebral ischemia), neurotoxic aggression (e.g., glutamate, kainate, ethydium bromide), or oxidative stress, neurosteroids such as PROG (Stein, 2001), PREG, PREGS, DHEA, and DHEAS (Garcia-Estrada et al., 1999; Kimonides et al., 1998) display neuroprotective properties, the mechanisms of which are essentially unknown. Very intriguing is the effect of RU486 (mifepristone), an antiprogesterone and antiglucocorticosteroid compound, which protects rat Purkinje cells in culture from apoptosis (Ghoumari et al., 2003a): this result is even observed in the absence of PR and glucocorticosteroid receptor, and its significance in terms of neurosteroid function remains to be elucidated.

It is now well established that PROG displays *(re)myelinating activity*. This action has been demonstrated in the PNS with the sciatic nerve, as above-described. In this case, the antiprogesterone RU486 confirms the implication of a nuclear PR mechanism. This PR-dependent mechanism is also likely to be involved in the myelinating effect of PROG in the CNS, as observed in rat cerebellum explants in culture (Ghoumari et al., 2003b). A small but significant remyelinating effect of PROG has also been observed in the cerebellar peduncles *in vivo* in aged rats (Ibanez and Franklin, unpublished data).

The case of DHEA is not yet clear in terms of mechanisms. It may be argued that an antioxidative effect is involved in its neuroprotective activity, for example, during oxidative stress (Bastianetto et al., 1999). Whether DHEA is directly involved in an antiglucorticosteroid activity has yet to be demonstrated (Karishma and Herbert, 2002).

17.3.2 Depression and Anxiety

A number of pathophysiological states labeled as depression have been correlated to defects in steroid regulation, as revealed by epidemiological studies (Barrett-Connor et al., 1999; Berr et al., 1996; Fabian et al., 2001; Wolkowitz et al., 2001) and therapeutic interventions (Almeida and Barclay, 2001). For instance, treatment of major depression with DHEA has met with some success (Wolkowitz et al., 1999) and might result from its sigma 1 receptor-mediated enhancement of noradrenaline and serotonin neurotransmission (Van Broekhoven and Verkes, 2003) or the equilibration of the DHEAS/glucocorticosteroid ratio, especially during aging (Gallagher and Young, 2002; Van Broekhoven and Verkes, 2003; Young et al., 2002), as possibly also suggested by the therapeutic effect of RU486 (as anticortisol agent) in psychotic major depression (Belanoff et al., 2002). More generally, the role of low DHEAS levels in the occurrence of poor subjective health may also be evoked (Belanoff et al., 2002; Mazat et al., 2001). A possible role of allopregnanolone in antidepressant therapy has also been suggested in major depression (Romeo et al., 1998), postpartum "blues" (Nappi et al., 2001), and premenstrual syndrome (Freeman et al., 2002).

Allopregnanolone is so far the best-studied neurosteroid in relation to anxiety state. In animal models, the anxiolytic effect of allopregnanolone (Akwa et al., 1999; Bitran et al., 1991; Wieland et al., 1995) has been demonstrated and mainly attributed to its positive modulation of $GABA_A$ receptor activity (Bitran et al., 1995; Brot et al., 1997; Smith et al., 1998), but may also implicate the activation of PBR (Bitran et al., 2000). Clinically, there is still no steroidal derivative available for treatment of anxiety disorders. PROG, which is largely metabolized to allopregnanolone after systemic administration, currently remains the best candidate.

17.3.3 Memory Disturbances, toward Alzheimer's Disease ... the Role of PREGS and Analogs

A complete appreciation of both classification and understanding of memory disorders is difficult. In spite of progress in animal models to study different aspects of memory, it remains that extrapolation to the human problems in this field cannot be perfect (how to study the loss of words and concepts?). Therefore, studies deal mostly with spatial memory, visual discrimination, or avoidance paradigms (usually in two sessions — training and retention). On the other hand, it is currently almost, if not completely, impossible to distinguish early mild memory loss associated with aging (so-called mild cognitive impairment (MCI)), which may stay only a functional handicap, from the first stage of an evolution to Alzheimer's disease and its disastrous lesions: the mechanism of the memory impairment may be identical at the beginning.

Animal studies have demonstrated a spatial *memory-enhancing activity* by PREGS. This effect has been observed following steroid administration into the cerebral ventricles (Akwa et al., 2001; Darnaudéry et al., 2000; Ladurelle et al.,

2000) or directly into the basal magnocellularis nucleus (Mayo et al., 1993; Pallarès et al., 1998). In addition, PREGS injections into different regions of the limbic system also block pharmacologically induced memory deficits (Flood et al., 1995). PREGS is also effective in restoring age-related decrease of memory in rats (Reddy and Kulkarni, 1998; Vallée et al., 1997).

Among the possible mechanisms of action of PREGS on memory processes, the enhancement of glutamatergic neurotransmission through mainly NMDA receptors has to be considered. This has been shown in different behavioral tests (Akwa et al., 2001; Mathis et al., 1994, 1996) and has also been suggested by long-term potentiation (LTP) (a form a synaptic plasticity) experiments in hippocampal slices (Sliwinski and Morin-Surun, unpublished data). In addition, collateral studies have indicated that PREGS increases the acetylcholine release and counteracts scopolamine-induced amnesia (Meziane et al., 1996; Vallée et al., 2001), suggesting a role for central cholinergic systems. Alteration in the function of the sigma 1 receptor also affects the memory response to PREGS (Urani et al., 1998). The role of the GABAergic system, however, has not been directly examined. From most findings, it appears that the improvement of memory performance by PREGS (1) is dose dependent and follows an inverted U-shape dose–response, which is typically reported for memory-enhancing drugs, although the optimal range concentration depends on the steroid administration route, the type of memory, animal species, and gender; and (2) is time dependent since differences are observed after short- vs. long-delay posttraining or after acute vs. chronic pretraining administration of the steroid. Overall, it seems that PREGS acts on consolidation and retrieval processes.

Recent data showed that the synthetic total enantiomer analog of natural PREGS was even more potent in enhancing spatial memory performance in rodents (Akwa et al., 2001): this surprising result opens the way to a new pharmacological approach in the treatment of memory deficits. Steroid sulfates are indeed interesting molecules in this field: DHEAS and DHEA have also been found to counteract amnesia in animal experiments (Flood and Roberts, 1988; Flood et al., 1988; Urani et al., 1998); steroid sulfatase inhibitors improve memory abilities (Flood et al., 1999; Johnson et al., 2000; Li et al., 1995; Rhodes et al., 1997).

In recent studies with definitely characterized Alzheimer's disease brains vs. aged-matched nondemented controls, a decrease of PREGS concentrations in the striatum and cerebellum was found inversely correlated with the levels of cortical β-amyloid peptides (Weill-Engerer et al., 2002). A significant negative correlation was also found between the levels of DHEAS and phosphorylated tau proteins in the hypothalamus (Weill-Engerer et al., 2002). A neuroprotective role of these neurosteroids is therefore possible, as suggested by the reversal of neurotoxicity and memory deficits induced by β-amyloid peptides in the rat (Cardounel et al., 1999; Gursoy et al., 2001; Maurice et al., 1998). Their role in restoring memory impairment associated with Alzheimer's disease awaits further investigation.

17.4 CONCLUSIONS

After a long period where neuropeptides have been extensively studied, it seems reasonable to offer more space to research in the field of neurosteroids and neuroactive steroids. Much remains to be discovered, and the long beginning may be followed by important additional avenues to the treatment of functional and anatomical abnormalities of the nervous system with compounds of the steroid series.

REFERENCES

Akwa, Y., Morfin, R.F., Robel, P., and Baulieu, E.E., (1992), Neurosteroid metabolism: 7α-hydroxylation of dehydroepiandrosterone and pregnenolone by rat brain microsomes, *Biochem. J.*, 288, 959.

Akwa, Y., Purdy, R.H., Koob, G.F., and Britton, K.T., (1999), The amygdala mediates the anxiolytic-like effect of the neurosteroid allopregnanolone in rat, *Behav. Brain Res.*, 106, 119.

Akwa, Y., Ladurelle, N., Covey, D.F., and Baulieu, E.E., (2001), The synthetic enantiomer of pregnenolone sulfate is very active on memory in rats and mice, even more so than its physiological neurosteroid counterpart: distinct mechanisms? *Proc. Natl. Acad. Sci. U.S.A.*, 98, 14033.

Almeida, O.P. and Barclay, L., (2001), Sex hormones and their impact on dementia and depression: a clinical perspective, *Expert Opin. Pharmacother.*, 2, 527.

Barrett-Connor, E., von Muhlen, D., Laughlin, G.A., and Kripke, A., (1999), Endogenous levels of dehydroepiandrosterone sulfate, but not other sex hormones, are associated with depressed mood in older women: the Rancho Bernardo Study, *J. Am. Geriatr. Soc.*, 47, 685.

Bastianetto, S., Ramassamy, C., Poirier, J., and Quirion, R., (1999), Dehydroepiandrosterone (DHEA) protects hippocampal cells from oxidative stress-induced damage, *Brain Res. Mol. Brain Res.*, 66, 35.

Beaujean, D., Do-Rego, J.L., Galas, L., Mensah-Nyagan, A.G., Fredriksson, R., Larhammar, D., Fournier, A., Luu-The, V., Pelletier, G., and Vaudry, H., (2002), Neuropeptide Y inhibits the biosynthesis of sulfated neurosteroids in the hypothalamus through activation of Y1 receptors, *Endocrinology*, 143, 1950.

Belanoff, J.K., Rothschild, A.J., Cassidy, F., DeBattista, C., Baulieu, E.E., Schold, C., and Schatzberg, A.F., (2002), An open label trial of C-1073 (mifepristone) for psychotic major depression, *Biol. Psychiatry*, 52, 386.

Bergeron, R., de Montigny, C., and Debonnel, G., (1996), Potentiation of neuronal NMDA response induced by dehydroepiandrosterone and its suppression by progesterone: effects mediated via sigma receptors, *J. Neurosci.*, 16, 1193.

Berr, C., Lafont, S., Debuire, B., Dartigues, J.F., and Baulieu, E.E., (1996), Relationships of dehydroepiandrosterone sulfate in the elderly with functional, psychological, and mental status, and short-term mortality: a French community-based study, *Proc. Natl. Acad. Sci. U.S.A.*, 93, 13410.

Bitran, D., Hilvers, R.J., and Kellogg, C.K., (1991), Anxiolytic effects of 3α-hydroxy-5α[β]-pregnan-20-one: endogenous metabolites of progesterone that are active at the $GABA_A$ receptor, *Brain Res.*, 561, 157.

Bitran, D., Shiekh, M., and McLeod, M., (1995), Anxiolytic effect of progesterone is mediated by the neurosteroid allopregnanolone at brain $GABA_A$ receptors, *J. Neuroendocrinol.*, 7, 171.

Bitran, D., Foley, M., Audette, D., Leslie, N., and Frye, C.A., (2000), Activation of peripheral mitochondrial benzodiazepine receptors in the hippocampus stimulates allopregnanolone synthesis and produces anxiolytic-like effects in the rat, *Psychopharmacology* (Berl.) 151, 64.

Blackmore, P.F., Fisher, J.F., Spilman, C.H., and Bleasdale, J.E., (1996), Unusual steroid specificity of the cell surface progesterone receptor on human sperm, *Mol. Pharmacol.*, 49, 727.

Bowlby, M.R., (1993), Pregnenolone sulfate potentiation of N-methyl-D-aspartate receptor channels in hippocampal neurons, *Mol. Pharmacol.*, 43, 813.

Brot, M.D., Akwa, Y., Purdy, R.H., Koob, G.F., and Britton, K.T., (1997), The anxiolytic-like effects of the neurosteroid allopregnanolone: interactions with $GABA_A$ receptors, *Eur. J. Pharmacol.*, 325, 1.

Bullock, A.E., Clark, A.L., Grady, S.R., Robinson, S.F., Slobe, B.S., Marks, M.J., and Collins, A.C., (1997), Neurosteroids modulate nicotinic receptor function in mouse striatal and thalamic synaptosomes, *J. Neurochem.*, 68, 2412.

Cardounel, A., Regelson, W., and Kalimi, M., (1999), Dehydroepiandrosterone protects hippocampal neurons against neurotoxin-induced cell death: mechanism of action, *Proc. Soc. Exp. Biol. Med.*, 222, 145.

Coirini, H., Gouezou, M., Delespierre, B., Liere, P., Pianos, A., Eychenne, B., Schumacher, M., and Guennoun, R., (2003), Characterization and regulation of the 3β-hydroxysteroid dehydrogenase isomerase enzyme in the rat sciatic nerve, *J. Neurochem.*, 84, 119.

Darnaudéry, M., Koehl, M., Piazza, P.V., Le Moal, M., and Mayo, W., (2000), Pregnenolone sulfate increases hippocampal acetylcholine release and spatial recognition, *Brain Res.*, 852, 173.

Do-Rego, J.L., Mensah Nyagan, G.A., Beaujean, D., Vaudry, D., Sieghart, W., Luu The, V., Pelletier, G., and Vaudry, H., (2000), γ-Aminobutyric acid, acting through γ-aminobutyric acid type A receptors, inhibits the biosynthesis of neurosteroids in the frog hypothalamus, *Proc. Natl. Acad. Sci. U.S.A.*, 97, 13925.

El-Etr, M., Akwa, Y., Robel, P., and Baulieu, E.E., (1998), Opposing effects of different steroid sulfates on $GABA_A$ receptor-mediated chloride uptake, *Brain Res.*, 790, 334.

Fabian, T.J., Dew, M.A., Pollock, B.G., Reynolds, C.F., III, Mulsant, B.H., Butters, M.A., Zmuda, M.D., Linares, A.M., Trottini, M., and Kroboth, P.D., (2001), Endogenous concentrations of DHEA and DHEA-S decrease with remission of depression in older adults, *Biol. Psychiatry*, 50, 767.

Ferrara, G., Serra, M., Zammaretti, F., Pisu, M.G., Panzica, G.C., Biggio, G., and Eva, C., (2001), Increased expression of the neuropeptide Y receptor Y(1) gene in the medial amygdala of transgenic mice induced by long-term treatment with progesterone or allopregnanolone, *J. Neurochem.*, 79, 417.

Finidori-Lepicard, J., Schorderet-Slatkine, S., Hanoune, J., and Baulieu, E.E., (1981), Progesterone inhibits membrane-bound adenylate cyclase in Xenopus laevis oocytes, *Nature*, 292, 255.

Flood, J.F. and Roberts, E., (1988), Dehydroepiandrosterone sulfate improves memory in aging mice, *Brain Res.*, 448, 178.

Flood, J.F., Smith, G.E., and Roberts, E., (1988), Dehydroepiandrosterone and its sulfate enhance memory retention in mice, *Brain Res.*, 447, 269.

Flood, J.F., Scherrer, J.F., and Morley, J.E., (1995), Localized injections of various compounds effecting neurotransmitter activity in the mammillary complex enhance (T-maze) avoidance retention, *Eur. J. Pharmacol.*, 275, 223.

Flood, J.F., Farr, S.A., Johnson, D.A., Li, P.K., and Morley, J.E., (1999), Peripheral steroid sulfatase inhibition potentiates improvement of memory retention for hippocampally administered dehydroepiandrosterone sulfate but not pregnenolone sulfate, *Psychoneuroendocrinology*, 24, 799.

Freeman, E.W., Frye, C.A., Rickels, K., Martin, P.A., and Smith, S.S., (2002), Allopregnanolone levels and symptom improvement in severe premenstrual syndrome, *J. Clin. Psychopharmacol.*, 22, 516.

Furukawa, A., Miyatake, A., Ohnishi, T., and Ichikawa, Y., (1998), Steroidogenic acute regulatory protein (StAR) transcripts constitutively expressed in the adult rat central nervous system: colocalization of StAR, cytochrome P-450SCC (CYP XIA1), and 3β-hydroxysteroid dehydrogenase in the rat brain, *J. Neurochem.*, 71, 2231.

Gago, N., Akwa, Y., Sananes, N., Guennoun, R., Baulieu, E.E., El-Etr, M., and Schumacher, M., (2001), Progesterone and the oligodendroglial lineage: stage-dependent biosynthesis and metabolism, *Glia*, 36, 295.

Gallagher, P. and Young, A., (2002), Cortisol/DHEA ratios in depression, *Neuropsychopharmacology*, 26, 410.

Garcia-Estrada, J., Luquin, S., Fernandez, A.M., and Garcia-Segura, L.M., (1999), Dehydroepiandrosterone, pregnenolone and sex steroids down-regulate reactive astroglia in the male rat brain after a penetrating brain injury, *Int. J. Dev. Neurosci.*, 17, 145.

Ghoumari, A.M., Dusart, I., El-Etr, M., Tronche, F., Sotelo, C., Schumacher, M., and Baulieu, E., (2003a), Mifepristone (RU486) protects Purkinje cells from cell death in organotypic slice cultures of postnatal rat and mouse cerebellum, *Proc. Natl. Acad. Sci. U.S.A.*, 100, 7953.

Ghoumari, A.M., Ibanez, C., El-Etr, M., Leclerc, P., Eychenne, B., Baulieu, E., and Schumacher, M., (2003b), Progesterone and its metabolites increase myelin basic protein (MBP) expression in organotypic slice cultures of rat cerebellum, *J. Neurochem.*, 86, 848.

Gursoy, E., Cardounel, A., and Kalimi, M., (2001), Pregnenolone protects mouse hippocampal (HT-22) cells against glutamate and amyloid beta protein toxicity, *Neurochem. Res.*, 26, 15.

Hanner, M., Moebius, F.F., Flandorfer, A., Knaus, H.G., Striessnig, J., Kempner, E., and Glossmann, H., (1996), Purification, molecular cloning, and expression of the mammalian sigma1-binding site, *Proc. Natl. Acad. Sci. U.S.A.*, 93, 8072.

Johnson, D.A., Wu, T., Li, P., and Maher, T.J., (2000), The effect of steroid sulfatase inhibition on learning and spatial memory, *Brain Res.*, 865, 286.

Karishma, K.K. and Herbert, J., (2002), Dehydroepiandrosterone (DHEA) stimulates neurogenesis in the hippocampus of the rat, promotes survival of newly formed neurons and prevents corticosterone-induced suppression, *Eur. J. Neurosci.*, 16, 445.

Kimonides, V.G., Khatibi, N.H., Svendsen, C.N., Sofroniew, M.V., and Herbert, J., (1998), Dehydroepiandrosterone (DHEA) and DHEA-sulfate (DHEAS) protect hippocampal neurons against excitatory amino acid-induced neurotoxicity, *Proc. Natl. Acad. Sci. U.S.A.*, 95, 1852.

Kimoto, T., Tsurugizawa, T., Ohta, Y., Makino, J., Tamura, H., Hojo, Y., Takata, N., and Kawato, S., (2001), Neurosteroid synthesis by cytochrome p450-containing systems localized in the rat brain hippocampal neurons: N-methyl-D-aspartate and calcium-dependent synthesis, *Endocrinology*, 142, 3578.

King, S.R., Manna, P.R., Ishii, T., Syapin, P.J., Ginsberg, S.D., Wilson, K., Walsh, L.P., Parker, K.L., Stocco, D.M., Smith, R.G., and Lamb, D.J., (2002), An essential component in steroid synthesis, the steroidogenic acute regulatory protein, is expressed in discrete regions of the brain, *J. Neurosci.*, 22, 10613.

Köenig, H.L., Schumacher, M., Ferzaz, B., Thi, A.N., Ressouches, A., Guennoun, R., Jung-Testas, I., Robel, P., Akwa, Y. and Baulieu, E.E., (1995), Progesterone synthesis and myelin formation by Schwann cells, *Science*, 268, 1500.

Lacor, P., Gandolfo, P., Tonon, M.C., Brault, E., Dalibert, I., Schumacher, M., Benavides, J., and Ferzaz, B., (1999), Regulation of the expression of peripheral benzodiazepine receptors and their endogenous ligands during rat sciatic nerve degeneration and regeneration: a role for PBR in neurosteroidogenesis, *Brain Res.*, 815, 70.

Ladurelle, N., Eychenne, B., Denton, D., Blair-West, J., Schumacher, M., Robel, P., and Baulieu, E., (2000), Prolonged intracerebroventricular infusion of neurosteroids affects cognitive performances in the mouse, *Brain Res.*, 858, 371.

Li, P.K., Rhodes, M.E., Jagannathan, S., and Johnson, D.A., (1995), Reversal of scopolamine induced amnesia in rats by the steroid sulfatase inhibitor estrone-3-O-sulfamate, *Brain Res. Cogn. Brain Res.*, 2, 251.

MacLusky, N.J. and McEwen, B.S., (1978), Oestrogen modulates progestin receptor concentrations in some rat brain regions but not in others, *Nature*, 274, 276.

Maksay, G., Laube, B., and Betz, H., (2001), Subunit-specific modulation of glycine receptors by neurosteroids, *Neuropharmacology*, 41, 369.

Mathis, C., Paul, S.M., and Crawley, J.N., (1994), The neurosteroid pregnenolone sulfate blocks NMDA antagonist-induced deficits in a passive avoidance memory task, *Psychopharmacology*, 116, 201.

Mathis, C., Vogel, E., Cagniard, B., Criscuolo, F., and Ungerer, A., (1996), The neurosteroid pregnenolone sulfate blocks deficits induced by a competitive NMDA antagonist in active avoidance and lever-press learning tasks in mice, *Neuropharmacology*, 35, 1057.

Maurice, T., Su, T.P., and Privat, A., (1998), Sigma1 (σ1) receptor agonists and neurosteroids attenuate β25-35-amyloid peptide-induced amnesia in mice through a common mechanism, *Neuroscience*, 83, 413.

Mayo, W., Dellu, F., Robel, P., Cherkaoui, J., Le Moal, M., Baulieu, E.E., and Simon, H., (1993), Infusion of neurosteroids into the nucleus basalis magnocellularis affects cognitive processes in the rat, *Brain Res.*, 607, 324.

Mazat, L., Lafont, S., Berr, C., Debuire, B., Tessier, J.F., Dartigues, J.F., and Baulieu, E.E., (2001), Prospective measurements of dehydroepiandrosterone sulfate in a cohort of elderly subjects: relationship to gender, subjective health, smoking habits, and 10-year mortality, *Proc. Natl. Acad. Sci. U.S.A.*, 98, 8145.

McEwen, B.S., Coirini, H., and Schumacher, M., (1990), Steroid effects on neuronal activity: when is the genome involved? *Ciba Found. Symp.*, 153, 3.

Melcangi, R.C., Magnaghi, V., Cavarretta, I., Martini, L., and Piva, F., (1998), Age-induced decrease of glycoprotein Po and myelin basic protein gene expression in the rat sciatic nerve: repair by steroid derivatives, *Neuroscience*, 85, 569.

Melcangi, R.C., Magnaghi, V., Galbiati, M., Ghelarducci, B., Sebastiani, L., and Martini, L., (2000), The action of steroid hormones on peripheral myelin proteins: a possible new tool for the rebuilding of myelin? *J. Neurocytol.*, 29, 327.

Mensah-Nyagan, A.G., Do-Rego, J.L., Beaujean, D., Luu-The, V., Pelletier, G., and Vaudry, H., (1999), Neurosteroids: expression of steroidogenic enzymes and regulation of steroid biosynthesis in the central nervous system, *Pharmacol. Rev.*, 51, 63.

Meziane, H., Mathis, C., Paul, S.M., and Ungerer, A., (1996), The neurosteroid pregnenolone sulfate reduces learning deficits induced by scopolamine and has promnestic effects in mice performing an appetitive learning task, *Psychopharmacology*, 126, 323.

Monnet, F.P., Mahe, V., Robel, P., and Baulieu, E.E., (1995), Neurosteroids, via sigma receptors, modulate the [^{3}H]norepinephrine release evoked by N-methyl-D-aspartate in the rat hippocampus, *Proc. Natl. Acad. Sci. U.S.A.*, 92, 3774.

Murakami, K., Fellous, A., Baulieu, E.E., and Robel, P., (2000), Pregnenolone binds to microtubule-associated protein 2 and stimulates microtubule assembly, *Proc. Natl. Acad. Sci. U.S.A.*, 97, 3579.

Nappi, R.E., Petraglia, F., Luisi, S., Polatti, F., Farina, C., and Genazzani, A.R., (2001), Serum allopregnanolone in women with postpartum "blues," *Obstet. Gynecol.*, 97, 77.

Orchinik, M., Murray, T.F., and Moore, F.L., (1991), A corticosteroid receptor in neuronal membranes, *Science*, 252, 1848.

Pallarès, M., Darnaudéry, M., Day, J., Le Moal, M., and Mayo, W., (1998), The neurosteroid pregnenolone sulfate infused into the nucleus basalis increases both acetylcholine release in the frontal cortex or amygdala and spatial memory, *Neuroscience*, 87, 551.

Park-Chung, M., Wu, F.S., Purdy, R.H., Malayev, A.A., Gibbs, T.T., and Farb, D.H., (1997), Distinct sites for inverse modulation of N-methyl-D-aspartate receptors by sulfated steroids, *Mol. Pharmacol.*, 52, 1113.

Pfrieger, F.W., (2003), Outsourcing in the brain: do neurons depend on cholesterol delivery by astrocytes? *Bioessays*, 25, 72.

Reddy, D.S. and Kulkarni, S.K., (1998), The effects of neurosteroids on acquisition and retention of a modified passive-avoidance learning task in mice, *Brain Res.*, 791, 108.

Rhodes, M.E., Li, P.K., Burke, A.M., and Johnson, D.A., (1997), Enhanced plasma DHEAS, brain acetylcholine and memory mediated by steroid sulfatase inhibition, *Brain Res.*, 773, 28.

Romeo, E., Strohle, A., Spalletta, G., di Michele, F., Hermann, B., Holsboer, F., Pasini, A., and Rupprecht, R., (1998), Effects of antidepressant treatment on neuroactive steroids in major depression, *Am. J. Psychiatry*, 155, 910.

Schumacher, M., Guennoun, R., Mercier, G., Desarnaud, F., Lacor, P., Benavides, J., Ferzaz, B., Robert, F., and Baulieu, E.E., (2001), Progesterone synthesis and myelin formation in peripheral nerves, *Brain Res. Brain Res. Rev.*, 37, 343.

Selye, H., (1942), Correlations between the chemical structure and the pharmacological actions of the steroids, *Endocrinology*, 30, 437.

Serra, M., Madau, P., Chessa, M.F., Caddeo, M., Sanna, E., Trapani, G., Franco, M., Liso, G., Purdy, R.H., Barbaccia, M.L., and Biggio, G., (1999), 2-Phenyl-imidazo[1,2-a]pyridine derivatives as ligands for peripheral benzodiazepine receptors: stimulation of neurosteroid synthesis and anticonflict action in rats, *Br. J. Pharmacol.*, 127, 177.

Smith, S.S., Gong, Q.H., Li, X., Moran, M.H., Bitran, D., Frye, C.A., and Hsu, F.C., (1998), Withdrawal from 3α-OH-5α-pregnan-20-one using a pseudopregnancy model alters the kinetics of hippocampal $GABA_A$-gated current and increases the $GABA_A$ receptor $\alpha 4$ subunit in association with increased anxiety, *J. Neurosci.*, 18, 5275.

Stein, D.G., (2001), Brain damage, sex hormones and recovery: a new role for progesterone and estrogen? *Trends Neurosci.*, 24, 386.

Sterling, R.J., Gasc, J.M., Sharp, P.J., Tuohimaa, P., and Baulieu, E.E., (1984), Absence of nuclear progesterone receptor in LH releasing hormone in laying hens, *J. Endocrinol.*, 102, R5.

Ueda, H., Yoshida, A., Tokuyama, S., Mizuno, K., Maruo, J., Matsuno, K., and Mita, S., (2001), Neurosteroids stimulate G protein-coupled sigma receptors in mouse brain synaptic membrane, *Neurosci. Res.*, 41, 33.

Urani, A., Privat, A., and Maurice, T., (1998), The modulation by neurosteroids of the scopolamine-induced learning impairment in mice involves an interaction with sigma1 (σ1) receptors, *Brain Res.*, 799, 64.

Valera, S., Ballivet, M., and Bertrand, D., (1992), Progesterone modulates a neuronal nicotinic acetylcholine receptor, *Proc. Natl. Acad. Sci. U.S.A.*, 89, 9949.

Vallée, M., Mayo, W., Darnaudéry, M., Corpéchot, C., Young, J., Koehl, M., Le Moal, M., Baulieu, E.E., Robel, P., and Simon, H., (1997), Neurosteroids: deficient cognitive performance in aged rats depends on low pregnenolone sulfate levels in the hippocampus, *Proc. Natl. Acad. Sci. U.S.A.*, 94, 14865.

Vallée, M., Shen, W., Heinrichs, S.C., Zorumski, C.F., Covey, D.F., Koob, G.F., and Purdy, R.H., (2001), Steroid structure and pharmacological properties determine the anti-amnesic effects of pregnenolone sulphate in the passive avoidance task in rats, *Eur. J. Neurosci.*, 14, 2003.

Van Broekhoven, F. and Verkes, R.J., (2003), Neurosteroids in depression: a review, *Psychopharmacology*, 165, 97.

Weill-Engerer, S., David, J.P., Sazdovitch, V., Liere, P., Eychenne, B., Pianos, A., Schumacher, M., Delacourte, A., Baulieu, E.E., and Akwa, Y., (2002), Neurosteroid quantification in human brain regions: comparison between Alzheimer's and nondemented patients, *J. Clin. Endocrinol. Metab.*, 87, 5138.

Weill-Engerer, S., David, J.-P., Sazdovitch, V., Liere, P., Schumacher, M., Delacourte, A., Baulieu, E., and Akwa, Y., (2003), *In vitro* metabolism of dehydroepiandrosterone (DHEA) to 7α-OH-DHEA and Δ5-androstene-3β,17β-diol in specific regions of the aging brain from Alzheimer's and non-demented patients, *Brain Res.*, 969, 117.

Wieland, S., Belluzzi, J.D., Stein, L., and Lan, N.C., (1995), Comparative behavioral characterization of the neuroactive steroids 3α-OH,5a-pregnan-20-one and 3α-OH,5β-pregnan-20-one in rodents, *Psychopharmacology*, 118, 65.

Wolkowitz, O.M., Reus, V.I., Keebler, A., Nelson, N., Friedland, M., Brizendine, L., and Roberts, E., (1999), Double-blind treatment of major depression with dehydroepiandrosterone, *Am. J. Psychiatry*, 156, 646.

Wolkowitz, O.M., Epel, E.S., and Reus, V.I., (2001), Stress hormone-related psychopathology: pathophysiological and treatment implications, *World J. Biol. Psychiatry*, 2, 115.

Wu, F.S., Gibbs, T.T., and Farb, D.H., (1991), Pregnenolone sulfate: a positive allosteric modulator at the N-methyl-D-aspartate receptor, *Mol. Pharmacol.*, 40, 333.

Young, A.H., Gallagher, P., and Porter, R.J., (2002), Elevation of the cortisol-dehydroepiandrosterone ratio in drug-free depressed patients, *Am. J. Psychiatry*, 159, 1237.

Zwain, I.H. and Yen, S.S., (1999), Neurosteroidogenesis in astrocytes, oligodendrocytes, and neurons of cerebral cortex of rat brain, *Endocrinology*, 140, 3843.

Index

A

B

C

D

E

F

G

H

I

K

R

S

T

V